D0225608

VECTOR CALCULUS

VECTOR CALCULUS

JERROLD E. MARSDEN
UNIVERSITY OF CALIFORNIA, BERKELEY

and

ANTHONY J. TROMBA
UNIVERSITY OF CALIFORNIA, SANTA CRUZ

with the assistance of
Joanne Seitz and Michael Hoffman

W. H. FREEMAN AND COMPANY
SAN FRANCISCO

These clouds reveal turbulence and waves in westerly winds as they pass over the Sierra Nevada to meet the calm air of the Owens Valley in California. The notions of vector calculus can give insight into this and other complex phenomena. Photograph by Nancy Williams.

Library of Congress Cataloging in Publication Data

Marsden, Jerrold E
Vector calculus.

Includes index.
1. Calculus. 2. Vector analysis. I. Tromba, Anthony, joint author. II. Title.
QA303.M338 515′.63 75–17864
ISBN 0-7167-0462-5

Printed in the United States of America

AMS 1970 subject classifications: 26A60, 35-01

9 8 7 6 5 4 3 2 1

to Ralph Abraham,
teacher and friend

CONTENTS

PREFACE

This text is intended for a one-semester course in the calculus of functions of several variables and vector analysis, at the sophomore level. Normally the course is preceded by a beginning course in linear algebra,* but this is not an essential prerequisite. We require only the bare rudiments of matrix algebra, and the necessary concepts are summarized in the text. On the other hand, we do assume a knowledge of the fundamentals of one-variable calculus—differentiation and integration of the standard functions.

The text includes most of the basic theory as well as many concrete examples and problems. Teaching experience at this level shows that it is desirable to omit many of the technical proofs; they are hard for beginning students and are included in the text mainly for reference or supplementary reading. Section 2.2, on limits and continuity, is designed to be treated lightly and is deliberately brief. More sophisticated theoretical topics, such as compactness, delicate proofs in integration theory, and the implicit function theorem, have been left out, since they usually belong to a more advanced course.† A few optional or supplementary proofs and digressions, and some whole sections, are

* Such as is found, for example, in M. O'Nan, *Linear Algebra*, Harcourt Brace Jovanovich, New York, 1971.

† See, e.g., J. Marsden, *Elementary Classical Analysis*, W. H. Freeman and Company, San Francisco, 1974.

printed in smaller type. Optional sections are marked by an asterisk in the Contents, and especially difficult exercises by an asterisk in the text.

Computational skills and intuitive understanding are important at this level, and we have endeavored to meet this need. Thus the text is quite concrete, concentrating on basic skills and deliberately de-emphasizing proofs. Also, we include quite a large number of physical illustrations. Specifically, we have not hesitated to include examples and motivation from the physical sciences, such as fluid mechanics, gravitation, and electromagnetic theory, although prior knowledge of these subjects is not assumed.

Another pedagogical feature of the text is the early introduction of vector fields, divergence, and curl in Chapter 3, before integration. Vector analysis usually suffers in a course of this type, and the present arrangement is designed to offset this tendency. Even further in this direction, one might consider teaching Chapter 4 (Taylor's Theorem, maxima and minima, Lagrange multipliers) after Chapter 7 (vector analysis).

We have striven to make the book as concrete and student-oriented as possible. For example, although we formulate the definition of the derivative "correctly," it is done by making use of matrices of partial derivatives, rather than linear transformations. This device alone can save one or even two weeks of teaching time and can save endless headaches for those students whose linear algebra is not in top form.

There are many colleagues and students who have made valuable contributions and suggestions since this book was begun. An earlier draft of the book was written in collaboration with Ralph Abraham. We thank him for allowing us to use his work. The authors are also grateful to Harcourt Brace Jovanovich, Inc. as well as to Eagle Mathematics, Inc. for permission to draw upon portions of *Calculus*, by Kenneth McAloon and Anthony Tromba, which is in the Eagle Mathematics Series. The other members of Eagle Mathematics, Inc. (an association of mathematics authors) are thanked for their help and encouragement. It is impossible to cite all those who assisted with the book, but we wish especially to thank Joanne Seitz, Michael Hoffman, Keith Phillips, Anne Perleman, Peter Renz, Diane Sauvageot, Steve Wan, and John Wilker.

A final word of thanks goes to those who helped in the preparation of the manuscript and the production of the book. Especially we thank Nora Lee, Rosemarie Stampfel, and Ikuko Workman for their excellent typing of the manuscript and Herb Holden of Stanford Research Institute for suggesting and preparing the computer-generated figures.

May 1975

Jerrold E. Marsden
Anthony J. Tromba

NOTATION

The student is assumed to have studied the calculus of functions of a real variable, including analytic geometry in the plane. Some students may have had some exposure to matrices as well, although what we shall need is given in Sections 1.3 and 1.4.

The reader is also assumed to be familiar with functions of elementary calculus such as $\sin x$, $\cos x$, e^x and $\log x$ (we write $\log x$ for the natural logarithm, which is sometimes denoted $\ln x$ or $\log_e x$). The reader is expected to know, or to review as the course proceeds, the basic rules of differentiation and integration for functions of one variable, such as the Chain Rule, the quotient rule, integration by parts, and so forth.

We shall review here the notations to be used later on, often without explicit mention. This can be read through quickly now, then referred to later if the need arises.

The collection of all real numbers is denoted $\mathbb{R}$. Thus $\mathbb{R}$ includes the *integers*, $\dots, -3, -2, -1, 0, 1, 2, 3, \dots$; the *rational numbers*, p/q, where p and q are integers ($q \neq 0$); and the *irrational numbers*, like $\sqrt{2}$, π, e, etc. Members of $\mathbb{R}$ may be visualized as points on the real-number line as shown in Figure 0.1.

When we write $a \in \mathbb{R}$ we mean that a is a member of the set $\mathbb{R}$; in

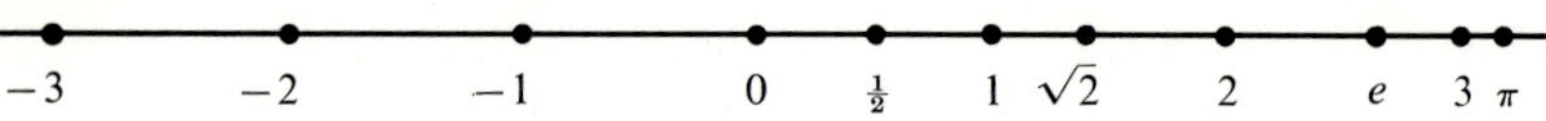

FIGURE 0.1
The geometric representation of points on the real number line.

other words, that a is a real number. Given two real numbers a and b with $a < b$ (that is, a is less than b), we can form the *closed interval*

$$[a, b], \text{ consisting of all } x \text{ such that } a \leq x \leq b$$

and the *open interval*

$$]a, b[, \text{ consisting of all } x \text{ such that } a < x < b$$

(In other books the open interval is sometimes denoted (a, b).) Similarly, we may form half-open intervals $]a, b]$ and $[a, b[$ (Figure 0.2).

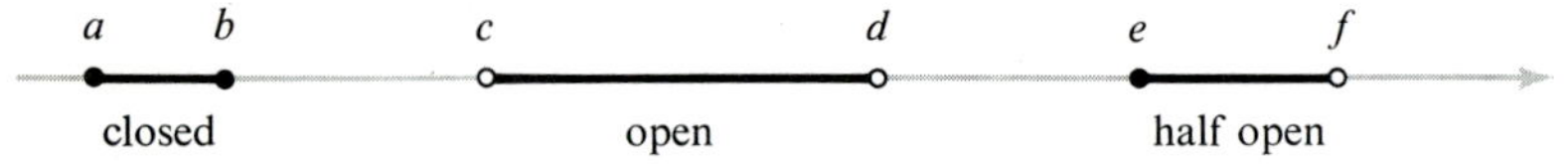

FIGURE 0.2
The geometric representation of the intervals $[a, b]$, $]c, d[$ *and* $[e, f[$.

The *absolute value* of a number $a \in \mathbb{R}$ is written $|a|$ and is defined

$$|a| = \begin{cases} a & \text{if } a \geq 0 \\ -a & \text{if } a < 0 \end{cases}$$

For example, $|3| = 3$, $|-3| = 3$, $|0| = 0$, and $|-6| = 6$. The inequality $|a + b| \leq |a| + |b|$ always holds. The *distance from a to b* is given by

$$|a - b|$$

Thus, the distance from 6 to 10 is 4 and from -6 to 3 is 9.

If we write $A \subset \mathbb{R}$, we mean A is a *subset* of $\mathbb{R}$. For example, A could equal the set of integers

$$A = \{\ldots, -3, -2, -1, 0, 1, 2, 3, \ldots\}.$$

Another example of a subset of $\mathbb{R}$ is the set $\mathbb{Q}$ of rational numbers. Generally, for two collections of objects (i.e., sets) A and B, $A \subset B$ means A is a subset of B; that is, every member of A is also a member of B.

The symbol $A \cup B$ means the *union* of A and B, the collection whose members are members of either A or B. Thus

$$\{\ldots, -3, -2, -1, 0\} \cup \{-1, 0, 1, 2, \ldots\}$$
$$= \{\ldots, -3, -2, -1, 0, 1, 2, \ldots\}$$

Similarly $A \cap B$ means the *intersection* of A and B; that is, this set consists of those members of A and B that are in both A and B. Thus the intersection of the two sets above is $\{-1, 0\}$.

We shall write $A \backslash B$ for those members of A that are not in B. Thus

$$\{\ldots, -3, -2, -1, 0\} \backslash \{-1, 0, 1, 2, \ldots\} = \{\ldots, -3, -2\}$$

We can also specify sets as in the following examples:

$$\{a \in \mathbb{R} \mid a \text{ is an integer}\} = \{\ldots, -3, -2, -1, 0, 1, 2, \ldots\}$$

$$\{a \in \mathbb{R} \mid a \text{ is an even integer}\} = \{\ldots, -2, 0, 2, 4, \ldots\}$$

$$\{x \in \mathbb{R} \mid a \leq x \leq b\} = [a, b]$$

A *function* $f\colon A \to B$ is a rule that assigns to each $a \in A$ one specific member $f(a)$ of B. Thus $f(x) = x^3/(1 - x)$ assigns a number to each $x \neq 1$ in $\mathbb{R}$. We can specify a function f by giving the rule for $f(x)$. Thus, the above function f can be defined by the rule $x \mapsto x^3/(1 - x)$.

If $A \subset \mathbb{R}$, $f\colon A \subset \mathbb{R} \to \mathbb{R}$ means that f assigns a value in $\mathbb{R}$, $f(x)$, to each $x \in A$. The set A is called the *domain* of f. The *graph* of f consists of the points $(x, f(x))$ in the plane (Figure 0.3). Generally, a *mapping* (= function = transformation = map) $f\colon A \to B$, where A and B are sets, is a rule that assigns to each $x \in A$ a specific point $f(x) \in B$.

The notation $\sum_{i=1}^{n} a_i$ means $a_1 + \cdots + a_n$, where $a_1, \ldots, a_n$ are given numbers. The sum of the first n integers is

$$1 + 2 + \cdots + n = \sum_{i=1}^{n} i = \frac{n(n+1)}{2}$$

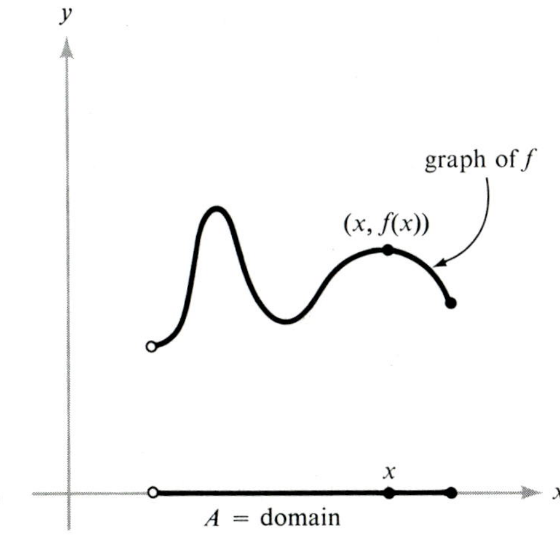

FIGURE 0.3
The graph of a function with the half-open interval A as domain.

NOTATION

The *derivative* of a function $f(x)$ is denoted $f'(x)$ or

$$\frac{df}{dx}$$

and the *definite integral* is written

$$\int_a^b f(x)\,dx \quad \text{or} \quad \int_a^b f$$

If we set $y = f(x)$, the derivative is also denoted by

$$\frac{dy}{dx}$$

There is a short table of derivatives and integrals at the back of the book, which is adequate for the needs of this text.

VECTOR CALCULUS

CHAPTER 1

THE GEOMETRY OF EUCLIDEAN SPACE

In this chapter we shall consider the basic operations on vectors in three-dimensional space: vector addition, scalar multiplication, and the dot and cross products. In Section 1.4 we shall generalize some of these notions to n-space, and briefly review some properties of matrices that will be needed in chapters 2 and 3.

1.1 VECTORS IN THREE-DIMENSIONAL SPACE

We recall that points P in the plane may be represented by ordered pairs of real numbers (a, b) called Cartesian coordinates. If we draw two perpendicular lines and label them the x- and y-axes, we may drop perpendiculars from P to these axes as in Figure 1.1.1; a is called the x-component of P and b is called the y-component.

Points in space may be similarly represented as ordered triples of real numbers. To construct such a representation, we choose three mutually perpendicular lines that meet at a point in space. These lines are called the x-axis, y-axis, and z-axis, and the point at which they meet is called the *origin* (this is our reference point). The set of axes is

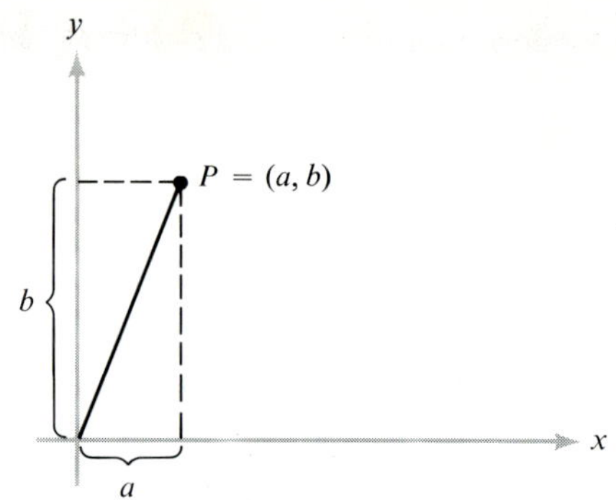

FIGURE 1.1.1
Cartesian coordinates in the plane.

often referred to as a *coordinate system*, and it is drawn as shown in Figure 1.1.2.

If we identify a real number with each point on these axis lines (as we did with the real number line), then we may assign to each point P in space a unique ordered triple of real numbers (a, b, c) and, conversely, to each triple we may assign a unique point in space.

Let the triple (0, 0, 0) correspond to the origin of the coordinate system and let the arrows on the axes indicate the positive directions.

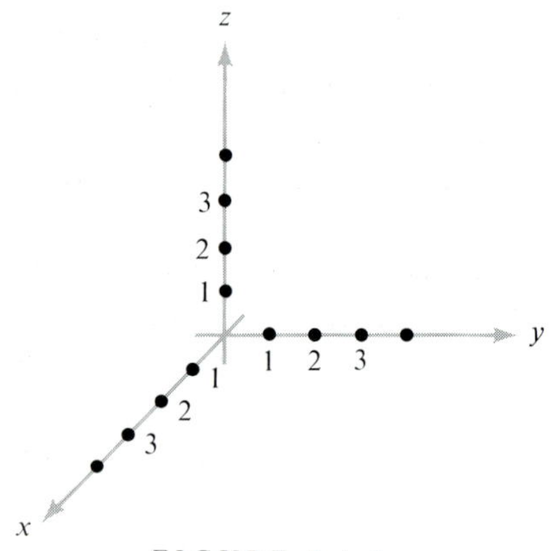

FIGURE 1.1.2
Cartesian coordinates in space.

Then, for example, the triple (2, 4, 4) represents a point 2 units from the origin in the positive direction along the x-axis, 4 units in the positive direction along the y-axis, and 4 units in the positive direction along the z-axis. This can be done in any order (Figure 1.1.3).

Because we can associate points in space and ordered triples in this way, we often use the expression "the point (a, b, c)" instead of the longer phrase "the point P that corresponds to the triple (a, b, c)." If the triple (a, b, c) corresponds to P, we say that a is the x-coordinate

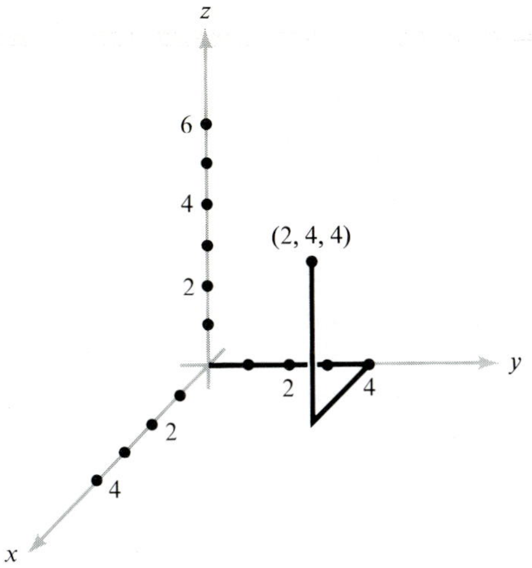

FIGURE 1.1.3
Geometric representation of the point (2, 4, 4) in Cartesian coordinates.

(or first coordinate), b is the y-coordinate (or second coordinate), and c is the z-coordinate (or third coordinate) of P. With this method of representing points in mind, we see that the x-axis consists of the points of the form $(a, 0, 0)$, where a is any real number, the y-axis consists of the points $(0, a, 0)$, and the z-axis consists of the points $(0, 0, a)$. It is also common to denote points in space with the letters x, y, and z in place of a, b, and c. Thus the triple (x, y, z) represents a point whose first coordinate is x, second coordinate is y, and third coordinate is z.

What we have just constructed is a model of three-dimensional space. In the next few paragraphs we shall examine the mathematical properties of this model, and then consider its geometric interpretation.

We employ the following notation for the line, the plane, and three-dimensional space:

(*i*) The real line is denoted $\mathbb{R}^1$ (thus, $\mathbb{R}$ and $\mathbb{R}^1$ are identical).

(*ii*) The set of all ordered pairs (x, y) of real numbers is denoted $\mathbb{R}^2$.

(*iii*) The set of all ordered triples (x, y, z) of real numbers is denoted $\mathbb{R}^3$.

When speaking of $\mathbb{R}^1$, $\mathbb{R}^2$, and $\mathbb{R}^3$ collectively, we write $\mathbb{R}^n$, $n = 1, 2$, or 3; or $\mathbb{R}^m$, $m = 1, 2, 3$.

Since elements of $\mathbb{R}^3$ are ordered triples of real numbers, the operation of addition can be extended to $\mathbb{R}^3$. Given the two triples (x, y, z) and (x', y', z'), we define their *sum* by

$$(x, y, z) + (x', y', z') = (x + x', y + y', z + z')$$

EXAMPLE 1.

$$(1, 1, 1) + (2, -3, 4) = (3, -2, 5)$$
$$(x, y, z) + (0, 0, 0) = (x, y, z)$$
$$(1, 7, 3) + (2, 0, 6) = (3, 7, 9)$$

The element $(0, 0, 0)$ is called the *zero element* of $\mathbb{R}^3$. The element $(-x, -y, -z)$ is called the *additive inverse* of (x, y, z), and we write $(x, y, z) - (x', y', z')$ for $(x, y, z) + (-x', -y', -z')$.

There are two very important product operations in $\mathbb{R}^3$. One of these, called the inner product, assigns a real number to each pair of elements of $\mathbb{R}^3$. We shall discuss the inner product in detail in Section 1.2. The other important product operation for $\mathbb{R}^3$ is called scalar multiplication (the word "scalar" is here a synonym for "real number"). This product combines scalars (real numbers) and elements of $\mathbb{R}^3$ (ordered triples) to yield elements of $\mathbb{R}^3$ as follows: given a scalar α and a triple (x, y, z), we define the scalar multiple or *scalar product* by

$$\alpha(x, y, z) = (\alpha x, \alpha y, \alpha z)$$

EXAMPLE 2.

$$2(4, e, 1) = (2 \cdot 4, 2 \cdot e, 2 \cdot 1) = (8, 2e, 2)$$
$$6(1, 1, 1) = (6, 6, 6)$$
$$1(x, y, z) = (x, y, z)$$
$$0(x, y, z) = (0, 0, 0)$$
$$\begin{aligned}(\alpha + \beta)(x, y, z) &= ((\alpha + \beta)x, (\alpha + \beta)y, (\alpha + \beta)z)\\ &= (\alpha x + \beta x, \alpha y + \beta y, \alpha z + \beta z)\\ &= \alpha(x, y, z) + \beta(x, y, z)\end{aligned}$$

It is an immediate result of the definitions that scalar multiplication and addition for $\mathbb{R}^3$ satisfy the following identities:

(*i*) $(\alpha\beta)(x, y, z) = \alpha(\beta(x, y, z))$ (associativity)

(*ii*) $(\alpha + \beta)(x, y, z) = \alpha(x, y, z) + \beta(x, y, z)$
(*iii*) $\alpha((x, y, z) + (x', y', z')) = \alpha(x, y, z) + \alpha(x', y', z')$ } (distributivity)

(*iv*) $\alpha(0, 0, 0) = (0, 0, 0)$
(*v*) $0(x, y, z) = (0, 0, 0)$ } (properties of zero elements)

(*vi*) $1(x, y, z) = (x, y, z)$ (property of identity element)

For $\mathbb{R}^2$, addition is defined just as in $\mathbb{R}^3$, by

$$(x, y) + (x', y') = (x + x', y + y')$$

and scalar multiplication is defined by

$$\alpha(x, y) = (\alpha x, \alpha y)$$

We often identify $\mathbb{R}^2$ with the set of triples $(x, y, 0)$ and speak of $\mathbb{R}^2$ geometrically as the plane (or more precisely, the xy-plane).

Now let us turn to the geometry of our model. One of the more fruitful tools of mathematics and its applications has been the notion of a *vector*. We define a (geometric) vector to be a directed line segment beginning at the origin, that is, a line segment with specified magnitude and direction, with initial point at the origin.* Figure 1.1.4 shows several vectors. Vectors may thus be thought of as arrows beginning at the origin. They are generally printed thus: **v**.

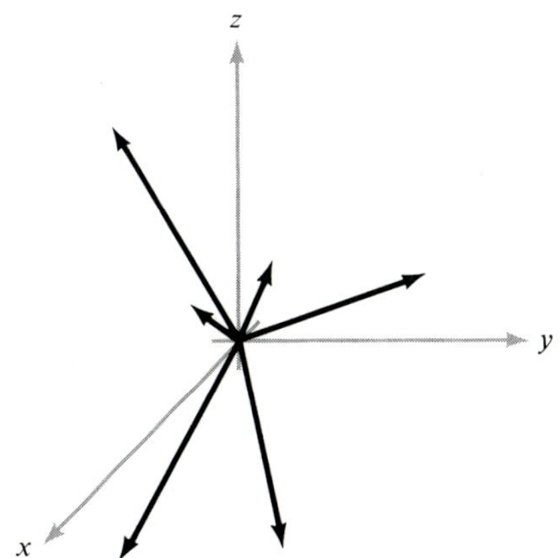

FIGURE 1.1.4
Geometrically, vectors are thought of as arrows emanating from the origin.

Using this definition of a vector, we may associate with each vector **v** the point in space (x, y, z) where **v** terminates, and conversely with each point (x, y, z) in space we can associate a vector **v**. Thus we shall identify **v** with (x, y, z) and write $\mathbf{v} = (x, y, z)$. For this reason, the elements of $\mathbb{R}^3$ not only are ordered triples of real numbers, but are also called vectors. The triple $(0, 0, 0)$ is denoted $\mathbf{0}$.

We say that two vectors are *equal* if and only if they have the same direction and the same magnitude. This condition may be expressed

* Have you heard a pilot say: "We are now vectoring in on the landing strip"? He is referring to the vector giving the direction and distance of the airplane from the landing strip. Needless to say, both direction and distance are important here.

algebraically by saying that if $\mathbf{v}_1 = (x, y, z)$ and $\mathbf{v}_2 = (x', y', z')$, then

$$\mathbf{v}_1 = \mathbf{v}_2 \quad \text{if and only if} \quad x = x',\ y = y',\ z = z'$$

Geometrically, we define vector addition as follows. In the plane containing the vectors $\mathbf{v}_1$ and $\mathbf{v}_2$ (see Figure 1.1.5), let us form the parallelogram having $\mathbf{v}_1$ as one side and $\mathbf{v}_2$ as its adjacent side. Then the sum $\mathbf{v}_1 + \mathbf{v}_2$ is the directed line segment along the diagonal of the parallelogram.* To show that our geometric definition of addition is consistent with our algebraic definition, we must demonstrate that $\mathbf{v}_1 + \mathbf{v}_2 = (x + x', y + y', z + z')$.

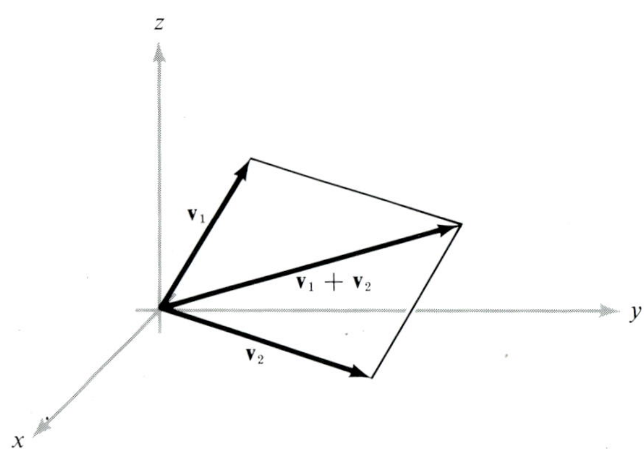

FIGURE 1.1.5
The geometry of vector addition.

We shall prove this result in the plane, and leave the reader to formulate the proposition for three-dimensional space. Thus we wish to show that if $\mathbf{v}_1 = (x, y)$ and $\mathbf{v}_2 = (x', y')$, then $\mathbf{v}_1 + \mathbf{v}_2 = (x + x', y + y')$.

In Figure 1.1.6 let $\mathbf{v}_1 = (x, y)$ be the vector ending at the point A, and let $\mathbf{v}_2 = (x', y')$ be the vector ending at point B. By definition, the vector $\mathbf{v}_1 + \mathbf{v}_2$ ends at the vertex C of parallelogram OBCA. Hence, to verify that $\mathbf{v}_1 + \mathbf{v}_2 = (x + x', y + y')$, it is sufficient to show that the coordinates of C are $(x + x', y + y')$.

From Figure 1.1.6, it is clear that triangle OAD is congruent to triangle BCG. By the congruence relation, $BG = OD$; and since BGFE is a rectangle, we have $EF = BG$. Furthermore, $OD = x$ and $OE = x'$. Hence $EF = BG = OD = x$. Since $OF = EF + OE$, it follows that $OF = x + x'$. This shows that the x-coordinate of C is $x + x'$. The proof for the y-coordinate is analogous. With a similar argument

* Vector addition arises in many physical situations, as we shall see in the text. For an easily visualized example, consider a bird or airplane flying through the air with velocity $\mathbf{v}_1$, in a wind with velocity $\mathbf{v}_2$. The resultant velocity, $\mathbf{v}_1 + \mathbf{v}_2$, is what you see.

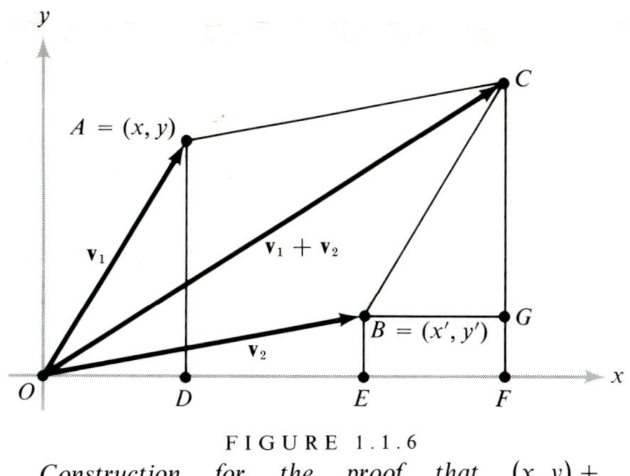

FIGURE 1.1.6
Construction for the proof that $(x, y) + (x', y') = (x + x', y + y')$.

for the other quadrants, we see that the geometric definition of vector addition is equivalent to the algebraic definition in terms of coordinates.

Figure 1.1.7(a) illustrates another way of looking at vector addition; that is, we translate (without rotation) the directed line segment representing the vector $\mathbf{v}_2$ so that it begins at the end of the vector $\mathbf{v}_1$. The endpoint of the resulting directed segment is the endpoint of the vector

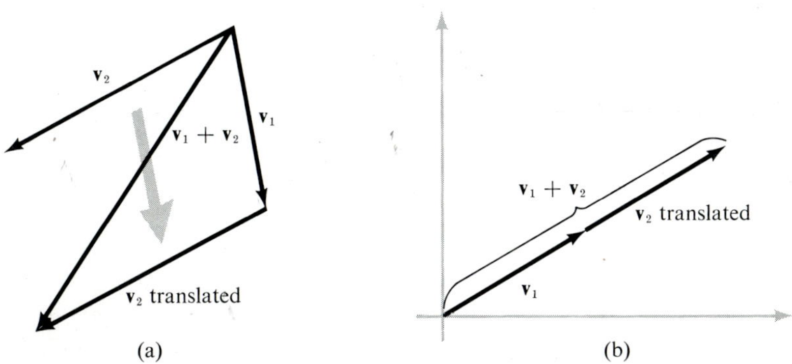

FIGURE 1.1.7
Vector addition may be visualized in terms of triangles as well as parallelograms.

$\mathbf{v}_1 + \mathbf{v}_2$. We note that when $\mathbf{v}_1$ and $\mathbf{v}_2$ are collinear, the triangle collapses. This situation is illustrated in Figure 1.1.7(b).

Scalar multiples of vectors have similar geometric interpretations. If α is a scalar and $\mathbf{v}$ a vector, we define $\alpha\mathbf{v}$ to be the vector which is α times as long as $\mathbf{v}$ with the same direction as $\mathbf{v}$ if $\alpha > 0$, but with the opposite direction if $\alpha < 0$. Figure 1.1.8 illustrates several examples.

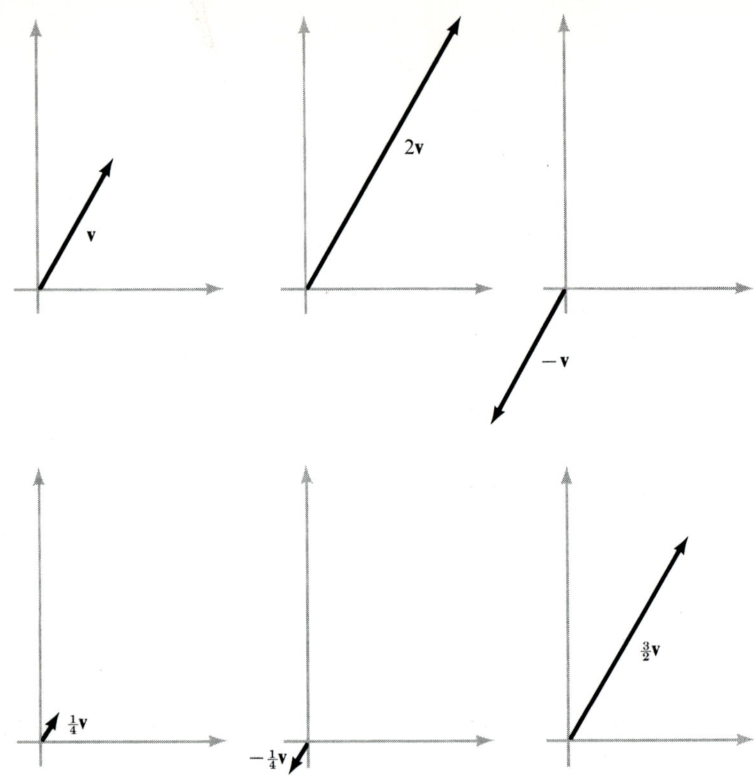

FIGURE 1.1.8
Some scalar multiples of a vector **v**.

Using an argument that depends on similar triangles we can prove that if $\mathbf{v} = (x, y, z)$, then

$$\alpha\mathbf{v} = (\alpha x, \alpha y, \alpha z) \tag{1}$$

that is, the geometric definition coincides with the algebraic one.

How do we represent the vector $\mathbf{b} - \mathbf{a}$ geometrically? Since $\mathbf{a} + (\mathbf{b} - \mathbf{a}) = \mathbf{b}$, $\mathbf{b} - \mathbf{a}$ is that vector which when added to $\mathbf{a}$ gives $\mathbf{b}$. In view of this, we may conclude that $\mathbf{b} - \mathbf{a}$ is the vector parallel to, and with the same magnitude as, the directed line segment beginning at the endpoint of $\mathbf{a}$ and terminating at the endpoint of $\mathbf{b}$ (see Figure 1.1.9).

Let us denote by $\mathbf{i}$ the vector that ends at (1, 0, 0), $\mathbf{j}$ the vector that ends at (0, 1, 0), and $\mathbf{k}$ the vector that ends at (0, 0, 1). Then using (1) above we find that if $\mathbf{v} = (x, y, z)$,

$$\mathbf{v} = x(1, 0, 0) + y(0, 1, 0) + z(0, 0, 1) = x\mathbf{i} + y\mathbf{j} + z\mathbf{k}$$

Hence we can represent any vector in three-dimensional space in terms of the vectors, $\mathbf{i}$, $\mathbf{j}$, and $\mathbf{k}$. For this reason the vectors $\mathbf{i}$, $\mathbf{j}$, and $\mathbf{k}$ are called the *standard basis vectors* for $\mathbb{R}^3$.

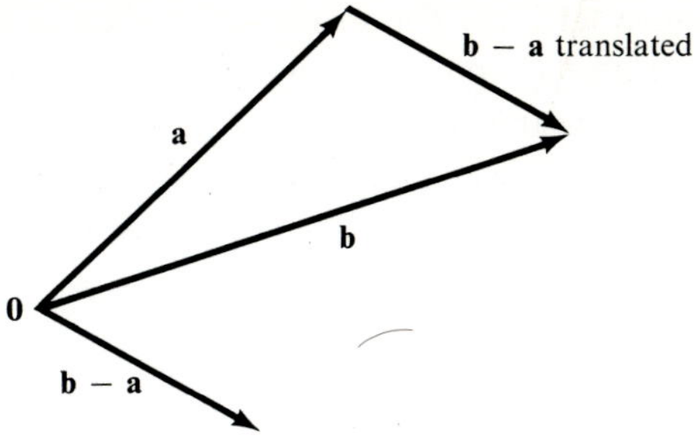

b − a translated
a
b
0
b − a

FIGURE 1.1.9
The geometry of vector subtraction.

EXAMPLE 3. The vector ending at (2, 3, 2) is $2\mathbf{i} + 3\mathbf{j} + 2\mathbf{k}$, and the vector ending at (0, −1, 4) is $-\mathbf{j} + 4\mathbf{k}$ (see Figure 1.1.10).

We also have the relationships

$$(x\mathbf{i} + y\mathbf{j} + z\mathbf{k}) + (x'\mathbf{i} + y'\mathbf{j} + z'\mathbf{k}) = (x + x')\mathbf{i} + (y + y')\mathbf{j} + (z + z')\mathbf{k}$$

and

$$\alpha(x\mathbf{i} + y\mathbf{j} + z\mathbf{k}) = (\alpha x)\mathbf{i} + (\alpha y)\mathbf{j} + (\alpha z)\mathbf{k}$$

Because of the correspondence between vectors and points, we may sometimes refer to a *point* **a** under circumstances in which **a** has been defined to be a vector. The reader should realize that by this statement we mean the *endpoint* of the vector **a**.

z
2k
(2, 3, 2)
k
i
j
y
3j
2i
x

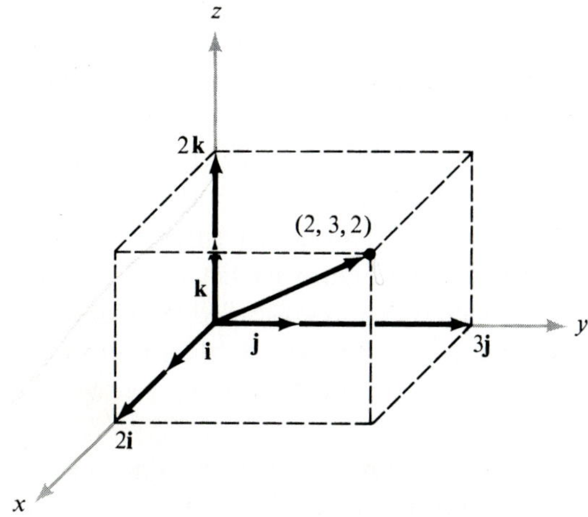

FIGURE 1.1.10
Representation of (2, 3, 2) *in terms of the standard basis* **i**, **j**, *and* **k**.

EXAMPLE 4. As an example of the use of vectors, let us describe the points that lie in the parallelogram whose adjacent sides are the vectors **a** and **b**.

Consider Figure 1.1.11. If P is any point in the given parallelogram and we construct lines l_1 and l_2 through P parallel to the vectors **a** and **b**, respectively, we see that l_1 intersects the side of the parallelogram determined by the vector **b** at some point $t\mathbf{b}$, where $0 \leq t \leq 1$. Likewise, l_2 intersects the side determined by the vector **a** at some point $s\mathbf{a}$, where $0 \leq s \leq 1$.

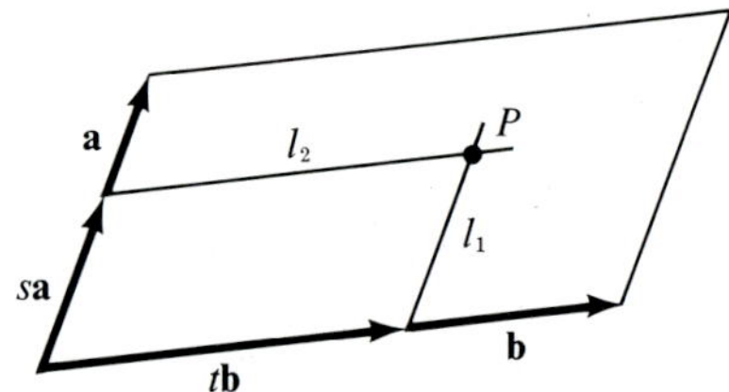

FIGURE 1.1.11
Describing points within the parallelogram formed by vectors **a** *and* **b**.

Since P is now the endpoint of the diagonal of a parallelogram having adjacent sides $s\mathbf{a}$ and $t\mathbf{b}$, if **v** denotes the vector ending at P, we see that $\mathbf{v} = s\mathbf{a} + t\mathbf{b}$. Thus, all the points in the given parallelogram are endpoints of vectors of the form $s\mathbf{a} + t\mathbf{b}$ for $0 \leq s \leq 1$ and $0 \leq t \leq 1$. By reversing our steps it is easy to see that all vectors of this form end within the parallelogram.

Since two lines through the origin determine a plane, two nonparallel vectors also determine a plane. If we apply the same reasoning as in the example above, it is not hard to see that the plane formed by two nonparallel vectors **v** and **w** consists of all points of the form $\alpha\mathbf{v} + \beta\mathbf{w}$ where α and β vary over the real numbers. This follows from the fact that any point P in the plane formed by the two vectors will be the opposite vertex of the parallelogram determined by $\alpha\mathbf{v}$ and $\beta\mathbf{w}$, where α and β are some scalars (see Figure 1.1.12).

The plane determined by **v** and **w** is called the plane *spanned by* **v** and **w**. When $\mathbf{v} = \gamma\mathbf{w}$, a multiple of **w** ($\mathbf{w} \neq \mathbf{0}$), then **v** and **w** are parallel and the plane degenerates to a straight line. When $\mathbf{v} = \mathbf{w} = \mathbf{0}$ (zero vectors) we obtain a single point.

There are three particular planes that arise naturally in a coordinate system and which will be of use to us later. We call the plane spanned by vectors **i** and **j** the *xy*-plane, the plane spanned by **j** and **k** the *yz*-plane, and the plane spanned by **i** and **k** the *xz*-plane. These planes are illustrated in Figure 1.1.13.

Planes and lines are geometric objects that can be represented by equations. We shall defer until Section 1.3 a study of equations representing planes. However, using the geometric interpretation of

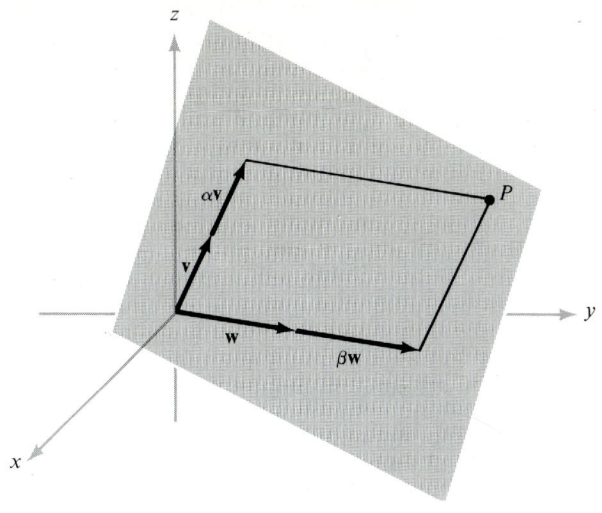

FIGURE 1.1.12
Describing points P in the plane formed from vectors **v** *and* **w**.

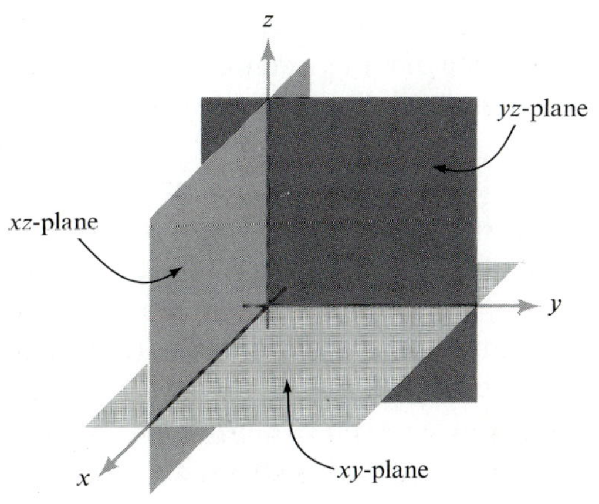

FIGURE 1.1.13
The three coordinate planes.

vector addition and scalar multiplication, we may find the equation of a line l that passes through the endpoint of the vector $\mathbf{a}$, with the direction of a vector $\mathbf{v}$ (see Figure 1.1.14). As t varies through all real values, the points of the form $t\mathbf{v}$ are all scalar multiples of the vector $\mathbf{v}$, and therefore exhaust the points of the line passing through the origin in the direction of $\mathbf{v}$. Since every point on l is the endpoint of the diagonal of a parallelogram with sides $\mathbf{a}$ and $t\mathbf{v}$, for some suitable value of t, we see that all the points on l are of the form $\mathbf{a} + t\mathbf{v}$. Thus, the line l may be expressed by the equation $\mathbf{l}(t) = \mathbf{a} + t\mathbf{v}$. We say that l is expressed *parametrically*, with t the parameter. At $t = 0$, $\mathbf{l}(t) = \mathbf{a}$. As t

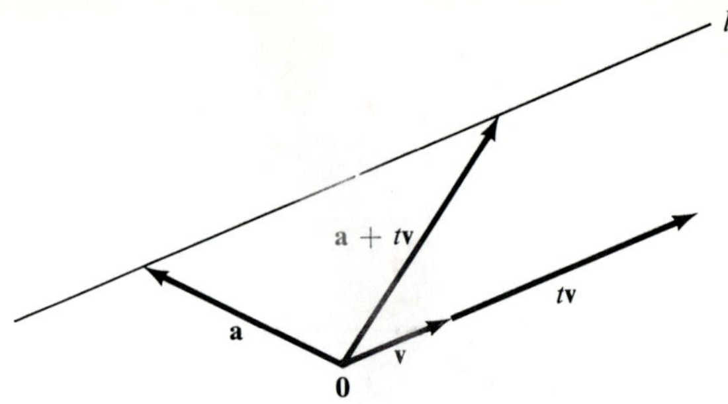

FIGURE 1.1.14
The line l, parametrically $\mathbf{l}(t) = \mathbf{a} + t\mathbf{v}$, *lies in the direction* $\mathbf{v}$ *and passes through the tip of* $\mathbf{a}$.

increases, the point $\mathbf{l}(t)$ moves away from $\mathbf{a}$ in the direction of $\mathbf{v}$. As t decreases from $t = 0$ through negative values, $\mathbf{l}(t)$ moves away from $\mathbf{a}$ in the direction of $-\mathbf{v}$.

Of course there are other parametrizations of the same line. These may be obtained by choosing instead of $\mathbf{a}$ a different point on the given line, and forming the parametric equation of the line beginning at that point and in the direction of $\mathbf{v}$. For example, the endpoint of $\mathbf{a} + \mathbf{v}$ is on the line $\mathbf{l}(t) = \mathbf{a} + t\mathbf{v}$, and thus $\mathbf{l}_1(t) = (\mathbf{a} + \mathbf{v}) + t\mathbf{v}$ represents the same line.

Still other parametrizations may be obtained by observing that if $\alpha \neq 0$, the vector $\alpha\mathbf{v}$ has the same (or opposite) direction as $\mathbf{v}$. Thus $\mathbf{l}_2(t) = \mathbf{a} + \alpha t\mathbf{v}$ is another parametrization of $\mathbf{l}(t) = \mathbf{a} + t\mathbf{v}$.

EXAMPLE 5. Determine the equation of the line passing through (1, 0, 0) in the direction of $\mathbf{j}$.

The desired line can be expressed parametrically as $\mathbf{l}(t) = \mathbf{i} + t\mathbf{j}$ (Figure 1.1.15). In terms of coordinates we have

$$\mathbf{l}(t) = (1, 0, 0) + t(0, 1, 0) = (1, t, 0)$$

We may also derive the equation of a line passing through the endpoints of two given vectors $\mathbf{a}$ and $\mathbf{b}$.

Since the vector $\mathbf{b} - \mathbf{a}$ is parallel to the directed line segment from $\mathbf{a}$ to $\mathbf{b}$, what we really wish to do here is calculate the parametric equations of the line passing through $\mathbf{a}$ in the direction of $\mathbf{b} - \mathbf{a}$ (Figure 1.1.16). Thus $\mathbf{l}(t) = \mathbf{a} + t(\mathbf{b} - \mathbf{a})$, that is, $\mathbf{l}(t) = (1 - t)\mathbf{a} + t\mathbf{b}$.

As t increases from 0 to 1, $t(\mathbf{b} - \mathbf{a})$ starts as the zero vector and increases in length (remaining in the direction of $\mathbf{b} - \mathbf{a}$) until at $t = 1$ it *is* the vector $\mathbf{b} - \mathbf{a}$. Thus for $\mathbf{l}(t) = \mathbf{a} + t(\mathbf{b} - \mathbf{a})$, as t increases from 0 to 1, the vector $\mathbf{l}(t)$ moves from the endpoint of $\mathbf{a}$ to the endpoint of $\mathbf{b}$ along the directed line segment from $\mathbf{a}$ to $\mathbf{b}$.

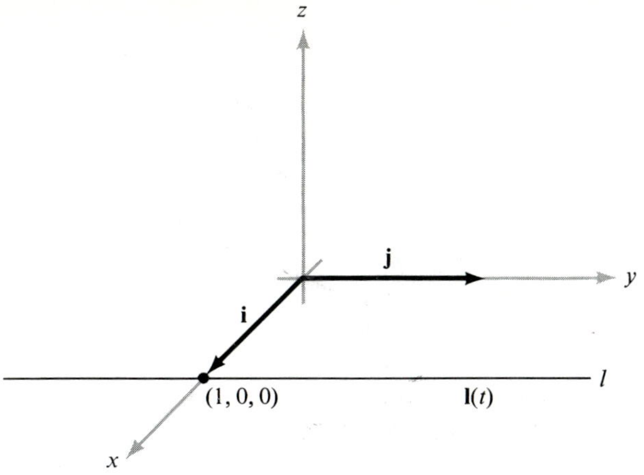

FIGURE 1.1.15
The line l, parametrically $\mathbf{l}(t) = \mathbf{i} + t\mathbf{j}$, *passes through the tip of* $\mathbf{i}$ *in the direction* $\mathbf{j}$.

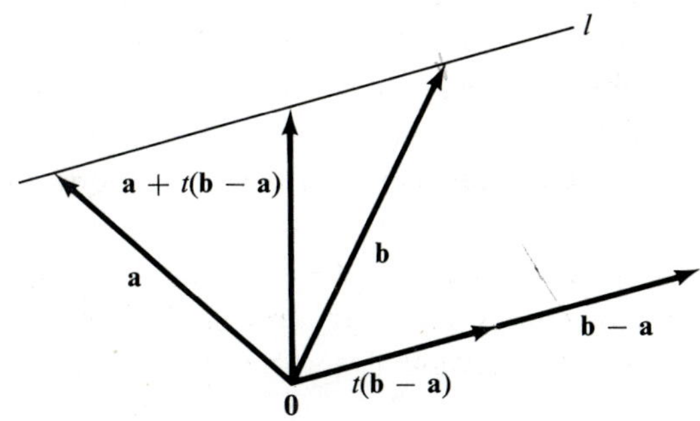

FIGURE 1.1.16
The line l, parametrically $\mathbf{l}(t) = \mathbf{a} + t(\mathbf{b} - \mathbf{a})$, *passing through the tips of* $\mathbf{a}$ *and* $\mathbf{b}$.

EXAMPLE 6. Find the equation of the line passing through $(-1, 1, 0)$ and $(0, 0, 1)$ (see Figure 1.1.17).

Letting $\mathbf{a} = -\mathbf{i} + \mathbf{j}$, $\mathbf{b} = \mathbf{k}$, we have

$$\begin{aligned} \mathbf{l}(t) &= (1 - t)(-\mathbf{i} + \mathbf{j}) + t\mathbf{k} \\ &= -(1 - t)\mathbf{i} + (1 - t)\mathbf{j} + t\mathbf{k} \end{aligned}$$

We note that any vector of the form $\mathbf{c} = \lambda\mathbf{a} + \mu\mathbf{b}$, where $\lambda + \mu = 1$, is on the line passing through the endpoints of $\mathbf{a}$ and $\mathbf{b}$. To see this, observe that $\mathbf{c} = (1 - \mu)\mathbf{a} + \mu\mathbf{b} = \mathbf{a} + \mu(\mathbf{b} - \mathbf{a})$.

EXAMPLE 7. As an example of the power of the vector concept, let us give a simple proof that the diagonals of a parallelogram bisect each other.

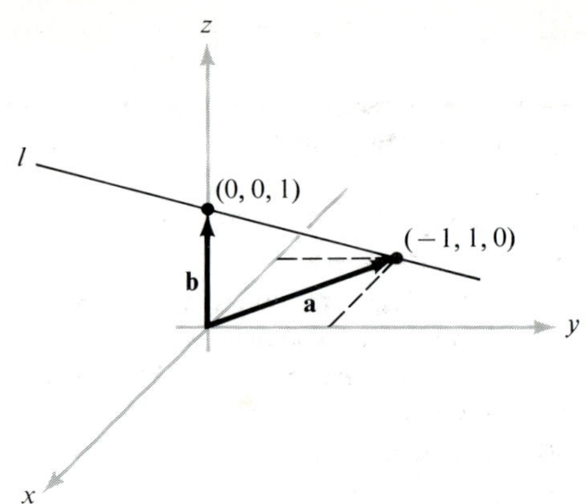

FIGURE 1.1.17
Special case of the preceding figure, in which $\mathbf{a} = (-1, 1, 0)$ *and* $\mathbf{b} = (0, 0, 1)$.

Let the adjacent sides of the parallelogram be represented by the vectors **a** and **b**, as shown in Figure 1.1.18. We first calculate the vector to the midpoint of PQ. Since $\mathbf{b} - \mathbf{a}$ is parallel and equal in length to the directed segment from P to Q, $(\mathbf{b} - \mathbf{a})/2$ is parallel and equal in length to the directed line segment from P to the midpoint of PQ. Thus, the vector $\mathbf{a} + (\mathbf{b} - \mathbf{a})/2 = (\mathbf{a} + \mathbf{b})/2$ ends at the midpoint of PQ.

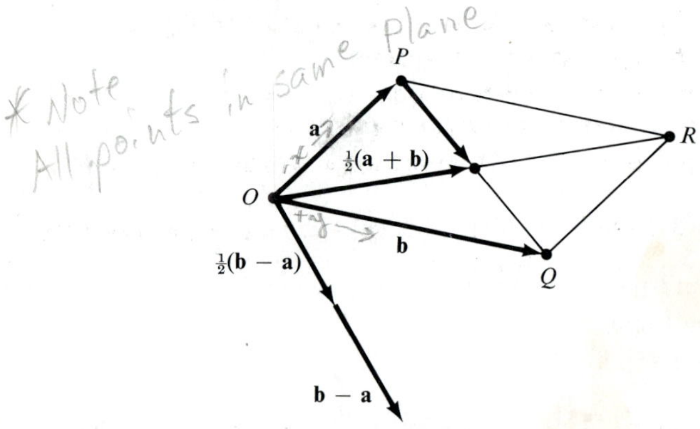

FIGURE 1.1.18
Constructions used in the proof that the diagonals of a parallelogram bisect each other.

Next, we calculate the vector to the midpoint of OR. We know $\mathbf{a} + \mathbf{b}$ ends at R; thus $(\mathbf{a} + \mathbf{b})/2$ ends at the midpoint of OR.

Since we have shown that the vector $(\mathbf{a} + \mathbf{b})/2$ ends at both the midpoint of OR and the midpoint of PQ, it follows that OR and PQ bisect each other.

EXERCISES

1. What restrictions must be made on x, y, and z, so that the triple (x, y, z) will represent a point on the y-axis? On the z-axis? In the xz-plane? In the yz-plane?
2. Sketch the vectors $\mathbf{v} = (2, 3, -6)$ and $\mathbf{w} = (-1, 1, 1)$. On your sketch, draw in $-\mathbf{v}$, $\mathbf{v} + \mathbf{w}$, $2\mathbf{v}$, and $\mathbf{v} - \mathbf{w}$.
3. (a) Show by geometrical construction that if $\mathbf{v}_1 = (x, y, z)$ and $\mathbf{v}_2 = (x', y', z')$ then $\mathbf{v}_1 + \mathbf{v}_2 = (x + x', y + y', z + z')$.
   (b) Supply the argument based on similar triangles proving that $\alpha\mathbf{v} = (\alpha x, \alpha y, \alpha z)$ when $\mathbf{v} = (x, y, z)$.

Complete the following computations:

4. $(3, 4, 5) + (6, 2, -6) = (?, ?, ?)$
5. $(-21, 23) - (?, 6) = (-25, ?)$
6. $3(133, -0.33, 0) + (-399, 0.99, 0) = (?, ?, ?)$
7. $(8\text{a}, -2\text{b}, 13\text{c}) = (52, 12, 11) + \frac{1}{2}(?, ?, ?)$
8. $(2, 3, 5) - 4\mathbf{i} + 3\mathbf{j} = (?, ?, ?)$
9. $800(0.03, 0, 0) = ?\mathbf{i} + ?\mathbf{j} + ?\mathbf{k}$

In exercises 10 to 16 use vector notation to describe the points that lie in the given configurations, as we did in Examples 4, 5, and 6.

10. The parallelogram whose adjacent sides are the vectors $\mathbf{i} + 3\mathbf{k}$ and $-2\mathbf{j}$.
11. The plane spanned by $\mathbf{v}_1 = (2, 7, 0)$ and $\mathbf{v}_2 = (0, 2, 7)$.
12. The line passing through $(0, 2, 1)$ in the direction of $2\mathbf{i} - \mathbf{k}$.
13. The line passing through $(-1, -1, -1)$ and $(1, -1, 2)$.
14. The parallelepiped with sides the vectors $\mathbf{a}$, $\mathbf{b}$, and $\mathbf{c}$. (See Figure 1.3.5 for a picture of the region we have in mind.)
15. The parallelogram with one corner at (x_0, y_0, z_0) whose sides meeting at that corner are parallel to the vectors $\mathbf{a}$ and $\mathbf{b}$.
16. The plane determined by the three points (x_0, y_0, z_0), (x_1, y_1, z_1), and (x_2, y_2, z_2).
17. Show that the medians of a triangle intersect at a point, and that this point divides each median in a ratio of 2 : 1.

1.2 THE INNER PRODUCT

In this section and the next we shall discuss two products of vectors, the inner product and the cross product, that are often very useful in physical applications and that have interesting geometric interpretations. The first product we shall consider is called the *inner product*. The name *dot product* is often used instead.

Suppose we have two vectors $\mathbf{a}$ and $\mathbf{b}$ (Figure 1.2.1) and we wish to determine the angle between them, that is, the smaller angle subtended by $\mathbf{a}$ and $\mathbf{b}$ in the plane that they span. The inner product enables us to do this. Let us first develop the concept formally and then prove that this product does what we claim.

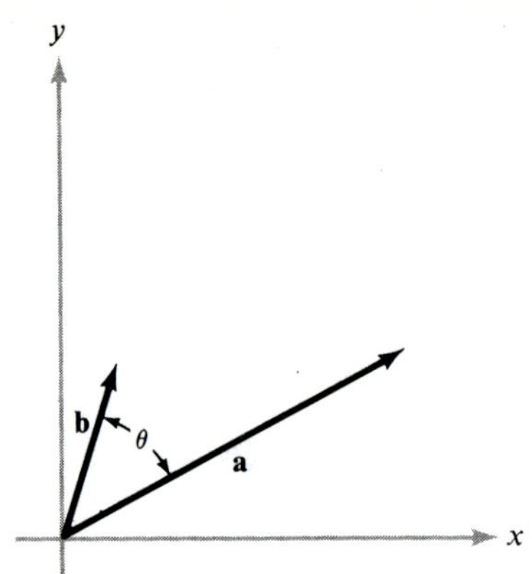

FIGURE 1.2.1
θ is the angle between the vectors **a** *and* **b**.

Let $\mathbf{a} = a_1\,\mathbf{i} + a_2\,\mathbf{j} + a_3\,\mathbf{k}$ and $\mathbf{b} = b_1\,\mathbf{i} + b_2\,\mathbf{j} + b_3\,\mathbf{k}$ be two vectors in $\mathbb{R}^3$. We define the *inner product* of **a** and **b**, written $\mathbf{a} \cdot \mathbf{b}$, to be the real number

$$\mathbf{a} \cdot \mathbf{b} = a_1 b_1 + a_2 b_2 + a_3 b_3$$

Note that the inner product of two vectors is a scalar quantity.

Certain properties of the inner product follow immediately from the definition. If **a**, **b**, and **c** are vectors in $\mathbb{R}^3$, and α and β are real numbers, then

(*i*) $\mathbf{a} \cdot \mathbf{a} \geq 0$
$\mathbf{a} \cdot \mathbf{a} = 0$ if and only if $\mathbf{a} = \mathbf{0}$

(*ii*) $\alpha\mathbf{a} \cdot \mathbf{b} = \alpha(\mathbf{a} \cdot \mathbf{b})$ and $\mathbf{a} \cdot \beta\mathbf{b} = \beta(\mathbf{a} \cdot \mathbf{b})$

(*iii*) $\mathbf{a} \cdot (\mathbf{b} + \mathbf{c}) = \mathbf{a} \cdot \mathbf{b} + \mathbf{a} \cdot \mathbf{c}$ and $(\mathbf{a} + \mathbf{b}) \cdot \mathbf{c} = \mathbf{a} \cdot \mathbf{c} + \mathbf{b} \cdot \mathbf{c}$

(*iv*) $\mathbf{a} \cdot \mathbf{b} = \mathbf{b} \cdot \mathbf{a}$

To prove (*i*), observe that if $\mathbf{a} = a_1\mathbf{i} + a_2\,\mathbf{j} + a_3\,\mathbf{k}$, then $\mathbf{a} \cdot \mathbf{a} = a_1^2 + a_2^2 + a_3^2$. Since a_1, a_2, and a_3 are real numbers we know $a_1^2 \geq 0$, $a_2^2 \geq 0$, $a_3^2 \geq 0$. Thus $\mathbf{a} \cdot \mathbf{a} \geq 0$. Moreover, if $a_1^2 + a_2^2 + a_3^2 = 0$, then $a_1 = a_2 = a_3 = 0$; therefore $\mathbf{a} = \mathbf{0}$ (zero vector). The proofs of the other properties of the inner product are also easily obtained.

It follows from the Pythagorean Theorem that the length of the vector $\mathbf{a} = a_1\mathbf{i} + a_2\,\mathbf{j} + a_3\,\mathbf{k}$ is $\sqrt{a_1^2 + a_2^2 + a_3^2}$ (see Figure 1.2.2). The length of the vector **a** is denoted by $\|\mathbf{a}\|$. This quantity is often called the *norm* of **a**. Since $\mathbf{a} \cdot \mathbf{a} = a_1^2 + a_2^2 + a_3^2$, it follows that

$$\|\mathbf{a}\| = (\mathbf{a} \cdot \mathbf{a})^{1/2}$$

Vectors with norm 1 are called *unit vectors*. For example, the vectors **i**,

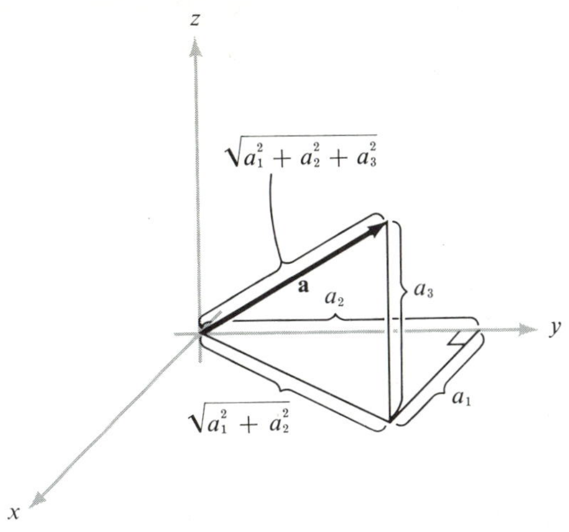

FIGURE 1.2.2
The length of the vector $\mathbf{a} = (a_1, a_2, a_3)$ *is given by the Pythagorean formula:* $\sqrt{a_1^2 + a_2^2 + a_3^2}$.

$\mathbf{j}$, $\mathbf{k}$ are unit vectors. Observe that for any nonzero vector $\mathbf{a}$, $\mathbf{a}/\|\mathbf{a}\|$ is a unit vector and we say that we have *normalized* $\mathbf{a}$.

In the plane, the vector $\mathbf{i}_\theta = (\cos\theta)\mathbf{i} + (\sin\theta)\mathbf{j}$ is the unit vector making an angle of θ degrees with the x-axis (see Figure 1.2.3). Clearly, $\|\mathbf{i}_\theta\| = (\sin^2\theta + \cos^2\theta)^{1/2} = 1$.

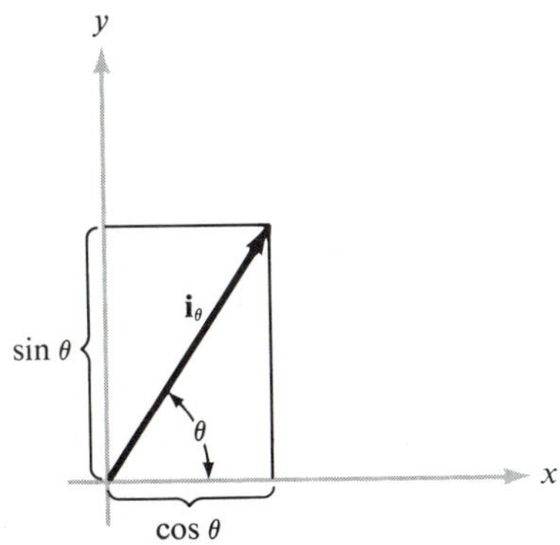

FIGURE 1.2.3
The coordinates of $\mathbf{i}_\theta$ *are* $\cos\theta$ *and* $\sin\theta$.

If $\mathbf{a}$ and $\mathbf{b}$ are vectors, we have seen that the vector $\mathbf{b} - \mathbf{a}$ is parallel to and has the same magnitude as the directed line segment from the endpoint of $\mathbf{a}$ to the endpoint of $\mathbf{b}$. It follows that the distance from the endpoint of $\mathbf{a}$ to the endpoint of $\mathbf{b}$ is $\|\mathbf{b} - \mathbf{a}\|$ (see Figure 1.2.4). For example, the distance from the endpoint of the vector $\mathbf{i}$, that is, the point $(1, 0, 0)$, to the endpoint of the vector $\mathbf{j}$, $(0, 1, 0)$, is $\sqrt{(0-1)^2 + (1-0)^2 + (0-0)^2} = \sqrt{2}$.

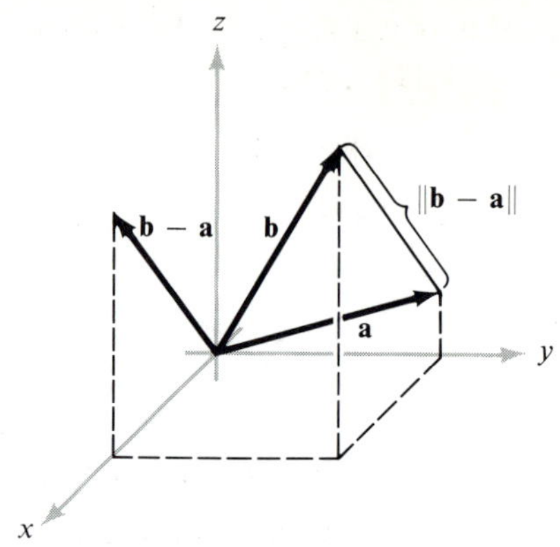

FIGURE 1.2.4
The distance between the tips of **a** *and* **b** *is* $\|\mathbf{b} - \mathbf{a}\|$.

Let us now show that the inner product does indeed measure the angle between two vectors.

Theorem 1. *Let* **a** *and* **b** *be two vectors in* $\mathbb{R}^3$ *and let* θ, $0 \leq \theta \leq \pi$, *be the angle between them (Figure 1.2.5). Then*

$$\mathbf{a} \cdot \mathbf{b} = \|\mathbf{a}\| \, \|\mathbf{b}\| \cos \theta$$

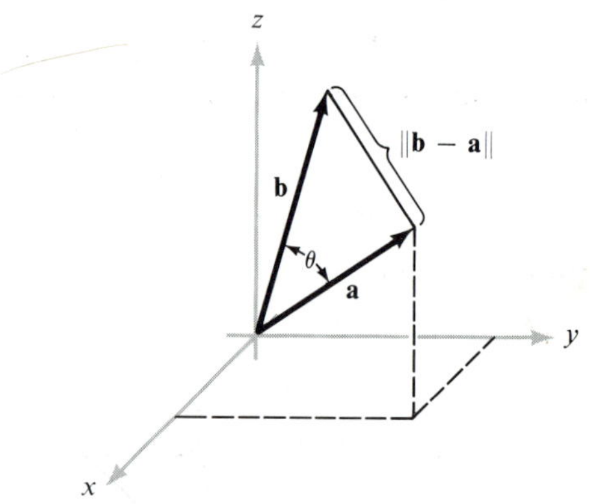

FIGURE 1.2.5
The vectors **a**, **b** *and the angle* θ *between them; the geometry for Theorem 1 and its proof.*

Thus we may express the angle between **a** *and* **b** *as*

$$\theta = \cos^{-1} \frac{\mathbf{a} \cdot \mathbf{b}}{\|\mathbf{a}\| \, \|\mathbf{b}\|}$$

Proof. If we apply the law of cosines from trigonometry to the triangle with one vertex at the origin and adjacent sides determined by the vectors **a** and **b**, it follows that

$$\|\mathbf{b}-\mathbf{a}\|^2 = \|\mathbf{a}\|^2 + \|\mathbf{b}\|^2 - 2\|\mathbf{a}\|\,\|\mathbf{b}\|\cos\theta$$

Since $\|\mathbf{b}-\mathbf{a}\|^2 = (\mathbf{b}-\mathbf{a})\cdot(\mathbf{b}-\mathbf{a})$, $\|\mathbf{a}\|^2 = \mathbf{a}\cdot\mathbf{a}$, and $\|\mathbf{b}\|^2 = \mathbf{b}\cdot\mathbf{b}$, we can rewrite the above equation as

$$(\mathbf{b}-\mathbf{a})\cdot(\mathbf{b}-\mathbf{a}) = \mathbf{a}\cdot\mathbf{a} + \mathbf{b}\cdot\mathbf{b} - 2\|\mathbf{a}\|\,\|\mathbf{b}\|\cos\theta$$

Now

$$\begin{aligned}(\mathbf{b}-\mathbf{a})\cdot(\mathbf{b}-\mathbf{a}) &= \mathbf{b}\cdot(\mathbf{b}-\mathbf{a}) - \mathbf{a}\cdot(\mathbf{b}-\mathbf{a})\\ &= \mathbf{b}\cdot\mathbf{b} - \mathbf{b}\cdot\mathbf{a} - \mathbf{a}\cdot\mathbf{b} + \mathbf{a}\cdot\mathbf{a}\\ &= \mathbf{a}\cdot\mathbf{a} + \mathbf{b}\cdot\mathbf{b} - 2\mathbf{a}\cdot\mathbf{b}\end{aligned}$$

Thus,

$$\mathbf{a}\cdot\mathbf{a} + \mathbf{b}\cdot\mathbf{b} - 2\mathbf{a}\cdot\mathbf{b} = \mathbf{a}\cdot\mathbf{a} + \mathbf{b}\cdot\mathbf{b} - 2\|\mathbf{a}\|\,\|\mathbf{b}\|\cos\theta$$

that is

$$\mathbf{a}\cdot\mathbf{b} = \|\mathbf{a}\|\,\|\mathbf{b}\|\cos\theta \quad \blacksquare$$

This result shows that the inner product of two vectors is the product of their lengths times the cosine of the angle between them. This relationship is often of value in problems of a geometric nature.

***Corollary* (*Cauchy-Schwarz inequality*).** *For any two vectors* **a** *and* **b**, *we have*

$$|\mathbf{a}\cdot\mathbf{b}| \le \|\mathbf{a}\|\,\|\mathbf{b}\|$$

with equality if and only if **a** *is a scalar multiple of* **b**.

Proof. If **a** is not a scalar multiple of **b**, then $|\cos\theta| < 1$ and the inequality holds. When **a** is a scalar multiple of **b** then $\theta = 0$ or π and $|\cos\theta| = 1$. $\blacksquare$

EXAMPLE 1. Find the angle between the vectors $\mathbf{i}+\mathbf{j}+\mathbf{k}$ and $\mathbf{i}+\mathbf{j}-\mathbf{k}$ (Figure 1.2.6).

Using Theorem 1 we have

$$(\mathbf{i}+\mathbf{j}+\mathbf{k})\cdot(\mathbf{i}+\mathbf{j}-\mathbf{k}) = \|\mathbf{i}+\mathbf{j}+\mathbf{k}\|\,\|\mathbf{i}+\mathbf{j}-\mathbf{k}\|\cos\theta$$

So

$$1+1-1 = (\sqrt{3})(\sqrt{3})\cos\theta$$

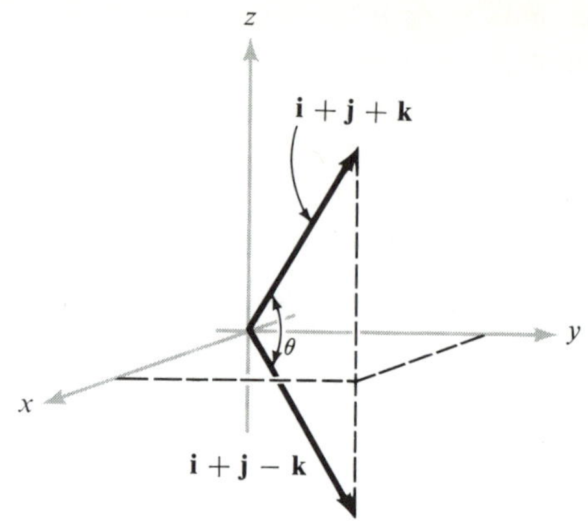

FIGURE 1.2.6
Finding the angle between $\mathbf{a} = \mathbf{i} + \mathbf{j} + \mathbf{k}$ *and* $\mathbf{b} = \mathbf{i} + \mathbf{j} - \mathbf{k}$.

Hence,

$$\cos\theta = \tfrac{1}{3}$$

that is

$$\theta = \cos^{-1}(\tfrac{1}{3}) \approx 1.23 \text{ radians } (71^\circ)$$

If $\mathbf{a}$ and $\mathbf{b}$ are nonzero vectors in $\mathbb{R}^3$ and θ is the angle between them, we see that $\mathbf{a} \cdot \mathbf{b} = 0$ if and only if $\cos\theta = 0$. From this it follows that the inner product of two nonzero vectors is zero if and only if the vectors are perpendicular. Often we say that perpendicular vectors are *orthogonal.* We shall adopt the convention that the zero vector is orthogonal to all vectors. Hence, the inner product provides us with a convenient method for determining if two vectors are orthogonal.

For example, the vectors $\mathbf{i}_\theta = (\cos\theta)\mathbf{i} + (\sin\theta)\mathbf{j}$ and $\mathbf{j}_\theta = -(\sin\theta)\mathbf{i} + (\cos\theta)\mathbf{j}$ are orthogonal, since

$$\mathbf{i}_\theta \cdot \mathbf{j}_\theta = -\cos\theta\sin\theta + \sin\theta\cos\theta = 0$$

(see Figure 1.2.7).

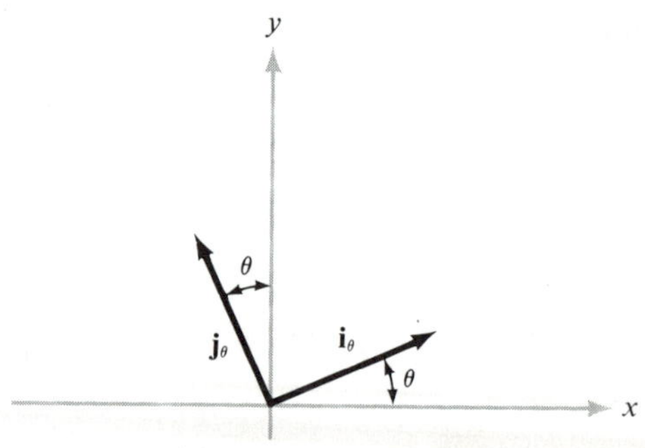

FIGURE 1.2.7
The vectors $\mathbf{i}_\theta$ *and* $\mathbf{j}_\theta$ *are orthogonal.*

EXAMPLE 2. Let **a** and **b** be two nonzero orthogonal vectors. Let **c** be another vector in the plane spanned by **a** and **b**. As we have seen, there are scalars α and β, such that $\mathbf{c} = \alpha\mathbf{a} + \beta\mathbf{b}$. The vector $\alpha\mathbf{a}$ is called the *component* of **c** along **a** (or tangent to **a**), and $\beta\mathbf{b}$ the component of **c** along **b**. Thus $\mathbf{c} = \alpha\mathbf{a} + \beta\mathbf{b}$ is a resolution of **c** into two orthogonal vectors. We use the inner product to determine α and β (see Figure 1.2.8).

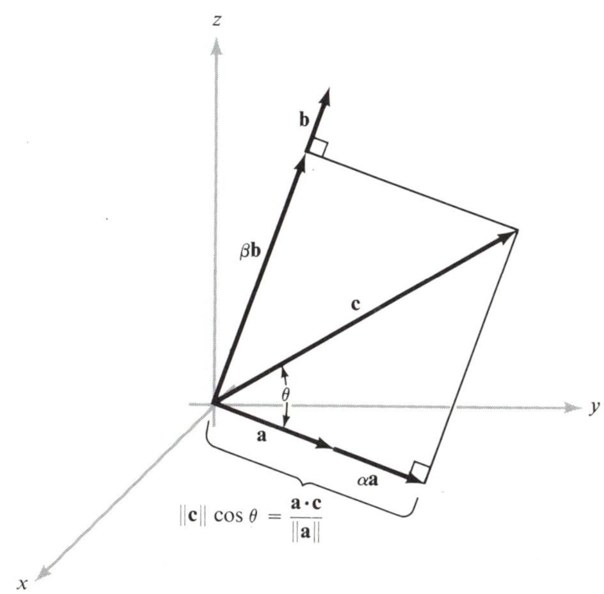

FIGURE 1.2.8
The geometry for finding α *and* β *where* $\mathbf{c} = \alpha\mathbf{a} + \beta\mathbf{b}$, *as in Example 2.*

Taking the inner product of **a** and **c**, we have

$$\mathbf{a} \cdot \mathbf{c} = \mathbf{a} \cdot (\alpha\mathbf{a} + \beta\mathbf{b}) = \alpha\mathbf{a} \cdot \mathbf{a} + \beta\mathbf{a} \cdot \mathbf{b}$$

Since **a** and **b** are orthogonal, $\mathbf{a} \cdot \mathbf{b} = 0$, and so,

$$\alpha = \frac{\mathbf{a} \cdot \mathbf{c}}{\mathbf{a} \cdot \mathbf{a}} = \frac{\mathbf{a} \cdot \mathbf{c}}{\|\mathbf{a}\|^2}$$

Similarly,

$$\beta = \frac{\mathbf{b} \cdot \mathbf{c}}{\mathbf{b} \cdot \mathbf{b}} = \frac{\mathbf{b} \cdot \mathbf{c}}{\|\mathbf{b}\|^2}$$

The result of Example 2 may be obtained using the geometric interpretation of the inner product. Let l be the distance, measured along the line determined by extending **a**, from the origin to the point where the perpendicular from **c** intersects the extension of **a**. It follows that

$$l = \|\mathbf{c}\| \cos\theta$$

where θ is the angle between **a** and **c**. Moreover, $l = \alpha\|\mathbf{a}\|$. Taken together, these results yield

$$\alpha\|\mathbf{a}\| = \|\mathbf{c}\|\cos\theta, \text{ or } \alpha = \frac{\|\mathbf{c}\|\cos\theta}{\|\mathbf{a}\|} = \frac{\|\mathbf{c}\|}{\|\mathbf{a}\|}\left(\frac{\mathbf{a}\cdot\mathbf{c}}{\|\mathbf{c}\|\,\|\mathbf{a}\|}\right) = \frac{\mathbf{a}\cdot\mathbf{c}}{\mathbf{a}\cdot\mathbf{a}}$$

In Example 2, the vector $\alpha\mathbf{a}$ is called the *projection* of **c** onto **a**. Similarly, the vector $\beta\mathbf{b}$ is the projection of **c** onto **b**. In general, the length of the projection of a vector **b** onto a vector **a**, where θ is the angle between **a** and **b**, is given by (Figure 1.2.9)

$$\|\mathbf{b}\|\,|\cos\theta| = \frac{|\mathbf{a}\cdot\mathbf{b}|}{\|\mathbf{a}\|}$$

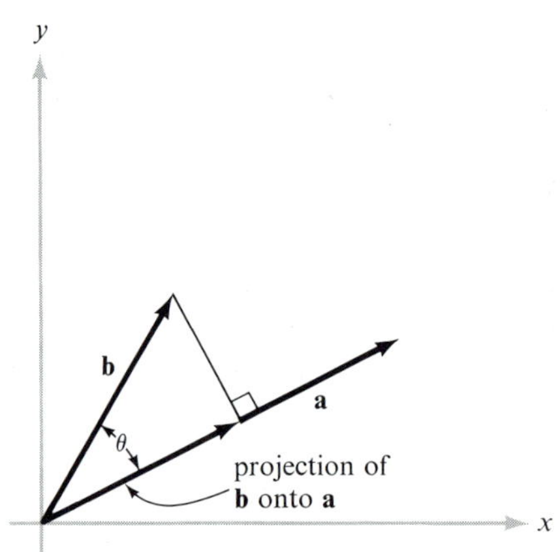

FIGURE 1.2.9

The projection of **b** *on* **a** *is* $(\mathbf{a}\cdot\mathbf{b}/\|\mathbf{a}\|^2)\mathbf{a}$.

EXERCISES

1. (a) Prove properties (*ii*) and (*iii*) of the inner product.
   (b) Prove that $\mathbf{a}\cdot\mathbf{b} = \mathbf{b}\cdot\mathbf{a}$.
2. Calculate $\mathbf{a}\cdot\mathbf{b}$ where $\mathbf{a} = 2\mathbf{i} + 10\mathbf{j} - 12\mathbf{k}$ and $\mathbf{b} = -3\mathbf{i} + 4\mathbf{k}$.
3. Find the angle between $7\mathbf{j} + 19\mathbf{k}$ and $-2\mathbf{i} - \mathbf{j}$ (to the nearest degree).
4. Compute $\mathbf{u}\cdot\mathbf{v}$, where $\mathbf{u} = \sqrt{3}\,\mathbf{i} - 315\mathbf{j} + 22\mathbf{k}$ and $\mathbf{v} = \mathbf{u}/\|\mathbf{u}\|$.
5. What is $\|8\mathbf{i} - 12\mathbf{k}\| \cdot \|6\mathbf{j} + \mathbf{k}\| - |(8\mathbf{i} - 12\mathbf{k})\cdot(6\mathbf{j} + \mathbf{k})|$ (to the nearest tenth)?

In exercises 6 to 11, compute $\|\mathbf{u}\|$, $\|\mathbf{v}\|$, and $\mathbf{u} \cdot \mathbf{v}$ for the given vectors.

6. $\mathbf{u} = 15\mathbf{i} - 2\mathbf{j} + 4\mathbf{k}$, $\mathbf{v} = \pi\mathbf{i} + 3\mathbf{j} - \mathbf{k}$
7. $\mathbf{u} = 2\mathbf{j} - \mathbf{i}$, $\mathbf{v} = -\mathbf{j} + \mathbf{i}$
8. $\mathbf{u} = 5\mathbf{i} - \mathbf{j} + 2\mathbf{k}$, $\mathbf{v} = \mathbf{i} + \mathbf{j} - \mathbf{k}$
9. $\mathbf{u} = -\mathbf{i} + 3\mathbf{j} + \mathbf{k}$, $\mathbf{v} = -2\mathbf{i} - 3\mathbf{j} - 7\mathbf{k}$
10. $\mathbf{u} = -\mathbf{i} + 3\mathbf{k}$, $\mathbf{v} = 4\mathbf{j}$
11. $\mathbf{u} = -\mathbf{i} + 2\mathbf{j} - 3\mathbf{k}$, $\mathbf{v} = -\mathbf{i} - 3\mathbf{j} + 4\mathbf{k}$
12. Normalize the vectors in exercises 6 to 8.
13. Find two nonparallel vectors both orthogonal to (1, 1, 1).

1.3 THE CROSS PRODUCT

In Section 1.2 we defined a product of vectors that was a scalar. In this section we shall define a product of vectors that is a vector; that is, we shall show how, given two vectors **a** and **b**, we can produce a third vector $\mathbf{a} \times \mathbf{b}$, called the *cross product* of **a** and **b**. This new vector will have the pleasing geometric property that it is perpendicular to the plane spanned (determined) by **a** and **b**.

We shall first develop the somewhat lengthy mathematical formalism necessary to state a useful definition of this product. Once this has been accomplished, we can study the geometric implications of the mathematical structure we have built.

We begin by defining a 2×2 *matrix* to be an array

$$\begin{bmatrix} a_{11} & a_{12} \\ a_{21} & a_{22} \end{bmatrix}$$

where a_{11}, a_{12}, a_{21}, a_{22} are four scalars. For example,

$$\begin{bmatrix} 2 & 1 \\ 0 & 4 \end{bmatrix}, \qquad \begin{bmatrix} -1 & 0 \\ 1 & 1 \end{bmatrix}, \qquad \text{and} \qquad \begin{bmatrix} 13 & 7 \\ 6 & 11 \end{bmatrix}$$

are 2×2 matrices. The *determinant*

$$\begin{vmatrix} a_{11} & a_{12} \\ a_{21} & a_{22} \end{vmatrix}$$

of such a matrix is defined by the equation

$$\begin{vmatrix} a_{11} & a_{12} \\ a_{21} & a_{22} \end{vmatrix} = a_{11}a_{22} - a_{12}a_{21} \tag{1}$$

EXAMPLE 1.

$$\begin{vmatrix} 1 & 1 \\ 1 & 1 \end{vmatrix} = 1 - 1 = 0$$

$$\begin{vmatrix} 1 & 2 \\ 3 & 4 \end{vmatrix} = 4 - 6 = -2$$

$$\begin{vmatrix} 5 & 6 \\ 7 & 8 \end{vmatrix} = 40 - 42 = -2$$

Next we pass to some properties of 3×3 matrices and determinants. A 3×3 matrix is an array

$$\begin{bmatrix} a_{11} & a_{12} & a_{13} \\ a_{21} & a_{22} & a_{23} \\ a_{31} & a_{32} & a_{33} \end{bmatrix}$$

again where each a_{ij} is a scalar; a_{ij} denotes the entry in the array that is in the ith row and the jth column. We define the determinant of a 3×3 matrix by the rule

$$\begin{vmatrix} a_{11} & a_{12} & a_{13} \\ a_{21} & a_{22} & a_{23} \\ a_{31} & a_{32} & a_{33} \end{vmatrix}$$

$$= a_{11} \begin{vmatrix} a_{22} & a_{23} \\ a_{32} & a_{33} \end{vmatrix} - a_{12} \begin{vmatrix} a_{21} & a_{23} \\ a_{31} & a_{33} \end{vmatrix} + a_{13} \begin{vmatrix} a_{21} & a_{22} \\ a_{31} & a_{32} \end{vmatrix} \quad (2)$$

Without some mnemonic device, formula (2) would be difficult to memorize. The rule to learn here is that you move along the first row, multiplying a_{1j} by the determinant of the 2×2 matrix obtained by cancelling out the first row and the jth column, and then you add these up, remembering to put a minus in front of the a_{12} term. For example, the matrix involved in the middle term of (2), namely

$$\begin{bmatrix} a_{21} & a_{23} \\ a_{31} & a_{33} \end{bmatrix}$$

is obtained by crossing out the first row and the second column of the given 3×3 matrix, thus:

$$\begin{bmatrix} \cancel{a_{11}} & \cancel{a_{12}} & \cancel{a_{13}} \\ a_{21} & \cancel{a_{22}} & a_{23} \\ a_{31} & \cancel{a_{32}} & a_{33} \end{bmatrix}$$

EXAMPLE 2.

$$\begin{vmatrix} 1 & 0 & 0 \\ 0 & 1 & 0 \\ 0 & 0 & 1 \end{vmatrix} = 1\begin{vmatrix} 1 & 0 \\ 0 & 1 \end{vmatrix} - 0\begin{vmatrix} 0 & 0 \\ 0 & 1 \end{vmatrix} + 0\begin{vmatrix} 0 & 1 \\ 0 & 0 \end{vmatrix} = 1$$

$$\begin{vmatrix} 1 & 2 & 3 \\ 4 & 5 & 6 \\ 7 & 8 & 9 \end{vmatrix} = 1\begin{vmatrix} 5 & 6 \\ 8 & 9 \end{vmatrix} - 2\begin{vmatrix} 4 & 6 \\ 7 & 9 \end{vmatrix} + 3\begin{vmatrix} 4 & 5 \\ 7 & 8 \end{vmatrix} = -3 + 12 - 9 = 0$$

An important property of determinants is that interchanging two rows or two columns results in a change of sign. This is an immediate consequence of the definition for the 2×2 case. For rows we have

$$\begin{vmatrix} a_{11} & a_{12} \\ a_{21} & a_{22} \end{vmatrix} = a_{11}a_{22} - a_{21}a_{12}$$

$$= -(a_{21}a_{12} - a_{11}a_{22}) = -\begin{vmatrix} a_{21} & a_{22} \\ a_{11} & a_{12} \end{vmatrix}$$

and for columns

$$\begin{vmatrix} a_{11} & a_{12} \\ a_{21} & a_{22} \end{vmatrix} = -(a_{12}a_{21} - a_{11}a_{22}) = -\begin{vmatrix} a_{12} & a_{11} \\ a_{22} & a_{21} \end{vmatrix}$$

We leave it to the reader to verify this property for the 3×3 case.

A second fundamental property of determinants is that we can factor scalars out of any row or column. For 2×2 determinants, this means

$$\begin{vmatrix} \alpha a_{11} & a_{12} \\ \alpha a_{21} & a_{22} \end{vmatrix} = \begin{vmatrix} a_{11} & \alpha a_{12} \\ a_{21} & \alpha a_{22} \end{vmatrix}$$

$$= \alpha\begin{vmatrix} a_{11} & a_{12} \\ a_{21} & a_{22} \end{vmatrix} = \begin{vmatrix} \alpha a_{11} & \alpha a_{12} \\ a_{21} & a_{22} \end{vmatrix} = \begin{vmatrix} a_{11} & a_{12} \\ \alpha a_{21} & \alpha a_{22} \end{vmatrix}$$

Similarly, for 3×3 determinants we have

$$\begin{vmatrix} \alpha a_{11} & \alpha a_{12} & \alpha a_{13} \\ a_{21} & a_{22} & a_{23} \\ a_{31} & a_{32} & a_{33} \end{vmatrix} = \alpha\begin{vmatrix} a_{11} & a_{12} & a_{13} \\ a_{21} & a_{22} & a_{23} \\ a_{31} & a_{32} & a_{33} \end{vmatrix} = \begin{vmatrix} a_{11} & \alpha a_{12} & a_{13} \\ a_{21} & \alpha a_{22} & a_{23} \\ a_{31} & \alpha a_{32} & a_{33} \end{vmatrix}$$

and so on. These results follow easily from the definitions. In particular, if any row or column consists of zeros, then the value of the determinant is zero.

A third fundamental fact about determinants is the following: If we change a row (respectively, column) by adding another row (respectively, column) to it, the value of the determinant remains the same.

For the 2 × 2 case this means that

$$\begin{vmatrix} a_1 & a_2 \\ b_1 & b_2 \end{vmatrix} = \begin{vmatrix} a_1 + b_1 & a_2 + b_2 \\ b_1 & b_2 \end{vmatrix} = \begin{vmatrix} a_1 & a_2 \\ b_1 + a_1 & b_2 + a_2 \end{vmatrix}$$

$$= \begin{vmatrix} a_1 + a_2 & a_2 \\ b_1 + b_2 & b_2 \end{vmatrix} = \begin{vmatrix} a_1 & a_1 + a_2 \\ b_1 & b_1 + b_2 \end{vmatrix}$$

For the 3 × 3 case, this means

$$\begin{vmatrix} a_1 & a_2 & a_3 \\ b_1 & b_2 & b_3 \\ c_1 & c_2 & c_3 \end{vmatrix} = \begin{vmatrix} a_1 + b_1 & a_2 + b_2 & a_3 + b_3 \\ b_1 & b_2 & b_3 \\ c_1 & c_2 & c_3 \end{vmatrix}$$

$$= \begin{vmatrix} a_1 + a_2 & a_2 & a_3 \\ b_1 + b_2 & b_2 & b_3 \\ c_1 + c_2 & c_2 & c_3 \end{vmatrix}$$

and so on. Again, this property can be proved using the definition of determinant.

EXAMPLE 3. Suppose

$$\mathbf{a} = \alpha\mathbf{b} + \beta\mathbf{c},\ \mathbf{a} = (a_1, a_2, a_3) = \alpha(b_1, b_2, b_3) + \beta(c_1, c_2, c_3)$$

Let us show that

$$\begin{vmatrix} a_1 & a_2 & a_3 \\ b_1 & b_2 & b_3 \\ c_1 & c_2 & c_3 \end{vmatrix} = 0$$

We shall prove the case $\alpha \neq 0, \beta \neq 0$. The case $\alpha = 0 = \beta$ is trivial, and the case where exactly one of α, β is zero is a simple modification of the one we prove. Using the fundamental properties of determinants, we have

$$\begin{vmatrix} \alpha b_1 + \beta c_1 & \alpha b_2 + \beta c_2 & \alpha b_3 + \beta c_3 \\ b_1 & b_2 & b_3 \\ c_1 & c_2 & c_3 \end{vmatrix}$$

$$= -\frac{1}{\alpha}\begin{vmatrix} \alpha b_1 + \beta c_1 & \alpha b_2 + \beta c_2 & \alpha b_3 + \beta c_3 \\ -\alpha b_1 & -\alpha b_2 & -\alpha b_3 \\ c_1 & c_2 & c_3 \end{vmatrix}$$

(factoring $-1/\alpha$ out of the second row)

$$= \left(-\frac{1}{\alpha}\right)\left(-\frac{1}{\beta}\right)\begin{vmatrix} \alpha b_1 + \beta c_1 & \alpha b_2 + \beta c_2 & \alpha b_3 + \beta c_3 \\ -\alpha b_1 & -\alpha b_2 & -\alpha b_3 \\ -\beta c_1 & -\beta c_2 & -\beta c_3 \end{vmatrix}$$

(factoring $-1/\beta$ out of the third row)

$$= \frac{1}{\alpha\beta} \begin{vmatrix} \beta c_1 & \beta c_2 & \beta c_3 \\ -\alpha b_1 & -\alpha b_2 & -\alpha b_3 \\ -\beta c_1 & -\beta c_2 & -\beta c_3 \end{vmatrix}$$

(adding the second row to the first row)

$$= \frac{1}{\alpha\beta} \begin{vmatrix} 0 & 0 & 0 \\ -\alpha b_1 & -\alpha b_2 & -\alpha b_3 \\ -\beta c_1 & -\beta c_2 & -\beta c_3 \end{vmatrix}$$

(adding the third row to the first row)

$$= 0$$

Now that we have established the necessary results about determinants, we return to products of vectors. Let $\mathbf{a} = a_1\mathbf{i} + a_2\mathbf{j} + a_3\mathbf{k}$ and $\mathbf{b} = b_1\mathbf{i} + b_2\mathbf{j} + b_3\mathbf{k}$ be vectors in $\mathbb{R}^3$. The *cross product* of **a** and **b**, denoted $\mathbf{a} \times \mathbf{b}$, is defined to be the vector

$$\mathbf{a} \times \mathbf{b} = \begin{vmatrix} a_2 & a_3 \\ b_2 & b_3 \end{vmatrix} \mathbf{i} - \begin{vmatrix} a_1 & a_3 \\ b_1 & b_3 \end{vmatrix} \mathbf{j} + \begin{vmatrix} a_1 & a_2 \\ b_1 & b_2 \end{vmatrix} \mathbf{k}$$

or, symbolically,

$$\mathbf{a} \times \mathbf{b} = \begin{vmatrix} \mathbf{i} & \mathbf{j} & \mathbf{k} \\ a_1 & a_2 & a_3 \\ b_1 & b_2 & b_3 \end{vmatrix}$$

Note that the cross product of two vectors is another vector; it is sometimes called the *vector product*.

Certain algebraic properties of the cross product follow immediately from the definition. If **a**, **b**, and **c** are vectors and α, β, and γ are scalars, then

(*i*) $\mathbf{a} \times \mathbf{b} = -(\mathbf{b} \times \mathbf{a})$

(*ii*) $\mathbf{a} \times (\beta\mathbf{b} + \gamma\mathbf{c}) = \beta(\mathbf{a} \times \mathbf{b}) + \gamma(\mathbf{a} \times \mathbf{c})$
$(\alpha\mathbf{a} + \beta\mathbf{b}) \times \mathbf{c} = \alpha(\mathbf{a} \times \mathbf{c}) + \beta(\mathbf{b} \times \mathbf{c})$

Note that $\mathbf{a} \times \mathbf{a} = -(\mathbf{a} \times \mathbf{a})$ by (*i*). Thus, $\mathbf{a} \times \mathbf{a} = \mathbf{0}$. Also,

$$\mathbf{i} \times \mathbf{j} = \mathbf{k}, \qquad \mathbf{j} \times \mathbf{k} = \mathbf{i}, \qquad \mathbf{k} \times \mathbf{i} = \mathbf{j}$$

which can be remembered by cyclicly permuting **i**, **j**, **k** like this:

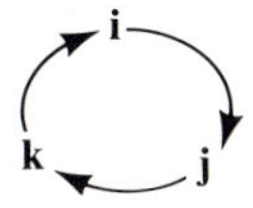

For example,

$$(3\mathbf{i} - \mathbf{j} + \mathbf{k}) \times (\mathbf{i} + 2\mathbf{j} - \mathbf{k}) = \begin{vmatrix} \mathbf{i} & \mathbf{j} & \mathbf{k} \\ 3 & -1 & 1 \\ 1 & 2 & -1 \end{vmatrix} = -\mathbf{i} + 4\mathbf{j} + 7\mathbf{k}$$

Our next goal is to provide a geometric interpretation of the cross product. To do this, we first introduce the triple product. Given three vectors $\mathbf{a}$, $\mathbf{b}$, and $\mathbf{c}$, the real number

$$\mathbf{a} \cdot (\mathbf{b} \times \mathbf{c})$$

is called the *triple product* of $\mathbf{a}$, $\mathbf{b}$, and $\mathbf{c}$ (in that order). Let us obtain a formula for the triple product $\mathbf{a} \cdot (\mathbf{b} \times \mathbf{c})$. If $\mathbf{a} = a_1\mathbf{i} + a_2\,\mathbf{j} + a_3\,\mathbf{k}$, $\mathbf{b} = b_1\mathbf{i} + b_2\,\mathbf{j} + b_3\,\mathbf{k}$, and $\mathbf{c} = c_1\mathbf{i} + c_2\,\mathbf{j} + c_3\,\mathbf{k}$, then

$$\begin{aligned} \mathbf{a} \cdot (\mathbf{b} \times \mathbf{c}) &= (a_1\mathbf{i} + a_2\,\mathbf{j} + a_3\,\mathbf{k}) \\ &\quad \cdot \left(\begin{vmatrix} b_2 & b_3 \\ c_2 & c_3 \end{vmatrix} \mathbf{i} - \begin{vmatrix} b_1 & b_3 \\ c_1 & c_3 \end{vmatrix} \mathbf{j} + \begin{vmatrix} b_1 & b_2 \\ c_1 & c_2 \end{vmatrix} \mathbf{k} \right) \\ &= a_1 \begin{vmatrix} b_2 & b_3 \\ c_2 & c_3 \end{vmatrix} - a_2 \begin{vmatrix} b_1 & b_3 \\ c_1 & c_3 \end{vmatrix} + a_3 \begin{vmatrix} b_1 & b_2 \\ c_1 & c_2 \end{vmatrix} \end{aligned}$$

This may be written more concisely as

$$\mathbf{a} \cdot (\mathbf{b} \times \mathbf{c}) = \begin{vmatrix} a_1 & a_2 & a_3 \\ b_1 & b_2 & b_3 \\ c_1 & c_2 & c_3 \end{vmatrix}$$

Now suppose that $\mathbf{a}$ is a vector in the plane spanned by the vectors $\mathbf{b}$ and $\mathbf{c}$. This means that the first row in the determinant expression for $\mathbf{a} \cdot (\mathbf{b} \times \mathbf{c})$ is of the form $\mathbf{a} = \alpha\mathbf{b} + \beta\mathbf{c}$, and therefore $\mathbf{a} \cdot (\mathbf{b} \times \mathbf{c}) = 0$ by Example 3. In other words, the vector $\mathbf{b} \times \mathbf{c}$ is orthogonal to any vector in the plane spanned by $\mathbf{b}$ and $\mathbf{c}$, in particular to both $\mathbf{b}$ and $\mathbf{c}$.

Next, we calculate the magnitude of $\mathbf{b} \times \mathbf{c}$. Note that

$$\begin{aligned} \|\mathbf{b} \times \mathbf{c}\|^2 &= \begin{vmatrix} b_2 & b_3 \\ c_2 & c_3 \end{vmatrix}^2 + \begin{vmatrix} b_1 & b_3 \\ c_1 & c_3 \end{vmatrix}^2 + \begin{vmatrix} b_1 & b_2 \\ c_1 & c_2 \end{vmatrix}^2 \\ &= (b_2 c_3 - b_3 c_2)^2 + (b_1 c_3 - c_1 b_3)^2 + (b_1 c_2 - c_1 b_2)^2 \end{aligned}$$

Writing out this last expression, we see that it is equal to

$$\begin{aligned} &(b_1^2 + b_2^2 + b_3^2)(c_1^2 + c_2^2 + c_3^2) - (b_1 c_1 + b_2 c_2 + b_3 c_3)^2 \\ &= \|\mathbf{b}\|^2\|\mathbf{c}\|^2 - (\mathbf{b} \cdot \mathbf{c})^2 = \|\mathbf{b}\|^2\|\mathbf{c}\|^2 - \|\mathbf{b}\|^2\|\mathbf{c}\|^2 \cos^2 \theta \\ &= \|\mathbf{b}\|^2\|\mathbf{c}\|^2 \sin^2 \theta \end{aligned}$$

where θ is the angle between $\mathbf{b}$ and $\mathbf{c}$, $0 \leq \theta \leq \pi$.

Combining our results, we conclude that $\mathbf{b} \times \mathbf{c}$ *is a vector perpendicular to the plane spanned by* $\mathbf{b}$ *and* $\mathbf{c}$ *with length* $\|\mathbf{b}\|\,\|\mathbf{c}\|\,|\sin \theta|$.

However, there are two possible vectors that satisfy these conditions, because there are two choices of direction that are perpendicular (or normal) to the plane P spanned by **b** and **c**. This is clear from Figure 1.3.1 which shows the two choices $\mathbf{n}_1$ and $-\mathbf{n}_1$ perpendicular to P, with $\|\mathbf{n}_1\| = \|-\mathbf{n}_1\| = \|\mathbf{b}\|\ \|\mathbf{c}\|\ |\sin\theta|$.

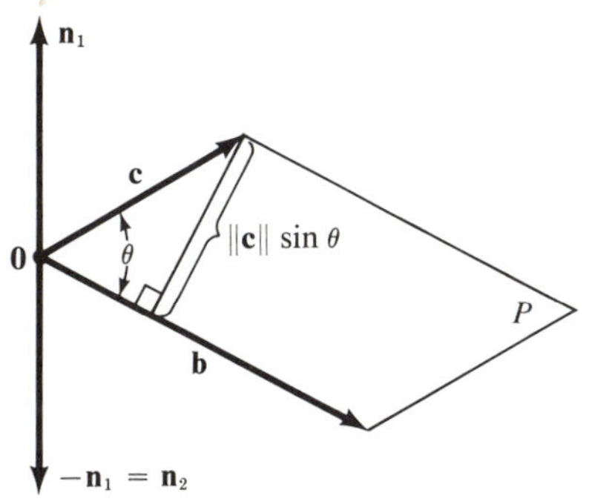

FIGURE 1.3.1
$\mathbf{n}_1$ and $\mathbf{n}_2$ are the two possible vectors orthogonal to both **b** *and* **c**, *and with norm $\|b\|\ \|c\|\ |\sin\theta|$.*

Which vector represents $\mathbf{b} \times \mathbf{c}$, $\mathbf{n}_1$ or $-\mathbf{n}_1$? The answer is $\mathbf{n}_1 = \mathbf{b} \times \mathbf{c}$. Try a few cases to see, such as $\mathbf{k} = \mathbf{i} \times \mathbf{j}$. The following "right-hand rule" determines the direction of $\mathbf{b} \times \mathbf{c}$.

Take the palm of your right hand and place it in such a way that your fingers curl from **b** in the direction of **c** through the angle θ. Then your thumb points in the direction of $\mathbf{b} \times \mathbf{c}$ (Figure 1.3.2).

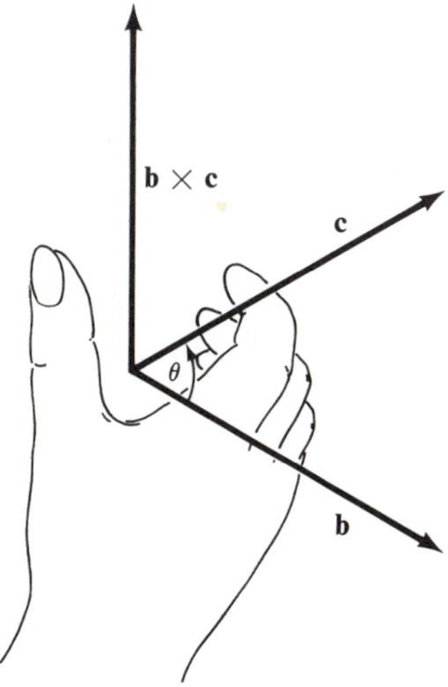

FIGURE 1.3.2
The right-hand rule for determining in which of the two possible directions $\mathbf{b} \times \mathbf{c}$ *points.*

If **b** and **c** are collinear, $\sin\theta = 0$, and so $\mathbf{b}\times\mathbf{c} = \mathbf{0}$; if **b** and **c** are not collinear, they span a plane and $\mathbf{b}\times\mathbf{c}$ is a vector perpendicular to this plane. *The length of* $\mathbf{b}\times\mathbf{c}$, $\|\mathbf{b}\|\,\|\mathbf{c}\|\,|\sin\theta|$, *is just the area of the parallelogram with adjacent sides the vectors* **b** *and* **c** (Figure 1.3.3).

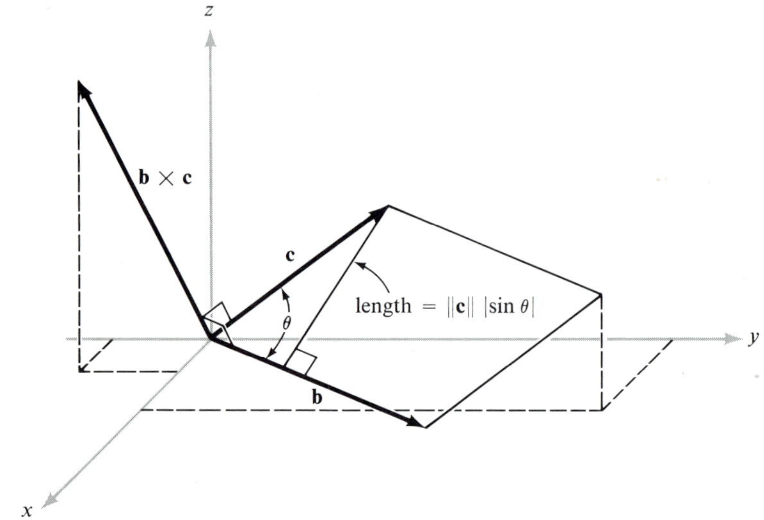

FIGURE 1.3.3
The length of $\mathbf{b}\times\mathbf{c}$ *equals the area of the parallelogram formed by* **b** *and* **c**.

EXAMPLE 4. Find a unit vector orthogonal to the vectors $\mathbf{i}+\mathbf{j}$ and $\mathbf{j}+\mathbf{k}$.

A vector perpendicular to both $\mathbf{i}+\mathbf{j}$ and $\mathbf{j}+\mathbf{k}$ is the vector

$$(\mathbf{i}+\mathbf{j})\times(\mathbf{j}+\mathbf{k}) = \begin{vmatrix} \mathbf{i} & \mathbf{j} & \mathbf{k} \\ 1 & 1 & 0 \\ 0 & 1 & 1 \end{vmatrix} = \mathbf{i}-\mathbf{j}+\mathbf{k}$$

Since $\|\mathbf{i}-\mathbf{j}+\mathbf{k}\| = \sqrt{3}$, the vector

$$\frac{1}{\sqrt{3}}(\mathbf{i}-\mathbf{j}+\mathbf{k})$$

is a unit vector perpendicular to $\mathbf{i}+\mathbf{j}$ and $\mathbf{j}+\mathbf{k}$.

Using the cross product, we may obtain the basic geometric interpretation of determinants. Let $\mathbf{b} = b_1\mathbf{i} + b_2\mathbf{j}$ and $\mathbf{c} = c_1\mathbf{i} + c_2\mathbf{j}$ be two vectors in the plane. If θ denotes the angle between **b** and **c**, we have seen that $\|\mathbf{b}\times\mathbf{c}\| = \|\mathbf{b}\|\,\|\mathbf{c}\|\,|\sin\theta|$. As noted above, $\|\mathbf{b}\|\,\|\mathbf{c}\|\,|\sin\theta|$ is the area of the parallelogram with adjacent sides **b** and **c** (see Figure 1.3.3). Using the definition of the cross product,

$$\mathbf{b} \times \mathbf{c} = \begin{vmatrix} \mathbf{i} & \mathbf{j} & \mathbf{k} \\ b_1 & b_2 & 0 \\ c_1 & c_2 & 0 \end{vmatrix} = \begin{vmatrix} b_1 & b_2 \\ c_1 & c_2 \end{vmatrix} \mathbf{k}$$

Thus $\|\mathbf{b} \times \mathbf{c}\|$ is the absolute value of the determinant

$$\begin{vmatrix} b_1 & b_2 \\ c_1 & c_2 \end{vmatrix} = b_1 c_2 - b_2 c_1$$

From this it follows that *the absolute value of the above determinant is the area of the parallelogram with adjacent sides the vectors* $\mathbf{b} = b_1\mathbf{i} + b_2\mathbf{j}$ *and* $\mathbf{c} = c_1\mathbf{i} + c_2\mathbf{j}$.

EXAMPLE 5. Find the area of the triangle with vertices at the points (1, 1), (0, 2), and (3, 2) (Figure 1.3.4).

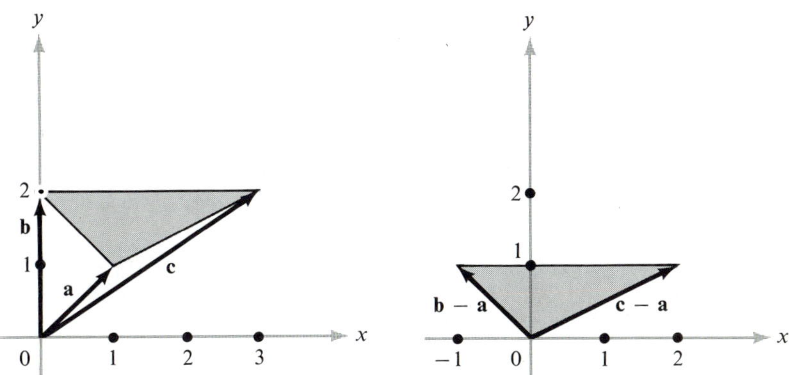

FIGURE 1.3.4
Problem: find the area A of the shaded triangle.
Solution: $A = \frac{1}{2}\|(\mathbf{b} - \mathbf{a}) \times (\mathbf{c} - \mathbf{a})\|$.

Let $\mathbf{a} = \mathbf{i} + \mathbf{j}$, $\mathbf{b} = 2\mathbf{j}$, and $\mathbf{c} = 3\mathbf{i} + 2\mathbf{j}$. It is clear that the triangle whose vertices are the endpoints of the vectors **a**, **b**, and **c** has the same area as the triangle with vertices at **0**, $\mathbf{b} - \mathbf{a}$, and $\mathbf{c} - \mathbf{a}$ (Figure 1.3.4). Indeed, the latter is merely a translation of the former triangle. Since the area of this translated triangle is one-half the area of the parallelogram with adjacent sides $\mathbf{b} - \mathbf{a}$ and $\mathbf{c} - \mathbf{a}$, we find that the area of the triangle with vertices (1, 1), (0, 2), and (3, 2) is the absolute value of

$$\frac{1}{2}\begin{vmatrix} -1 & 1 \\ 2 & 1 \end{vmatrix} = -\frac{3}{2}$$

that is, 3/2.

There is an interpretation of determinants of 3×3 matrices as volumes which is analogous to the interpretation of determinants of 2×2 matrices as areas. Let $\mathbf{a} = a_1\mathbf{i} + a_2\mathbf{j} + a_3\mathbf{k}$, $\mathbf{b} = b_1\mathbf{i} + b_2\mathbf{j} + b_3\mathbf{k}$, and $\mathbf{c} = c_1\mathbf{i} + c_2\mathbf{j} + c_3\mathbf{k}$ be vectors in $\mathbb{R}^3$. We will show that the volume of the parallelepiped with adjacent sides **a**, **b**, and **c** (Figure 1.3.5) is the absolute value of the determinant

$$D = \begin{vmatrix} a_1 & a_2 & a_3 \\ b_1 & b_2 & b_3 \\ c_1 & c_2 & c_3 \end{vmatrix}$$

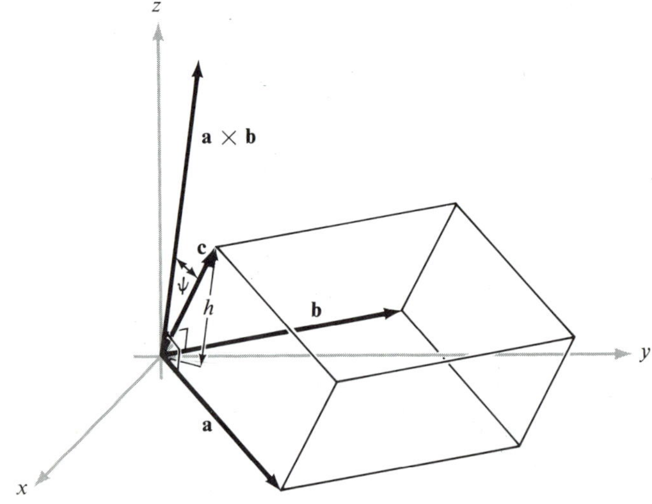

FIGURE 1.3.5
The volume of the parallelepiped formed by **a**, **b**, **c** *is the absolute value of the determinant of the* 3×3 *matrix with* **a**, **b**, **c** *as its rows.*

We know that $\|\mathbf{a} \times \mathbf{b}\|$ is the area of the parallelogram with adjacent sides **a** and **b**. Moreover, $(\mathbf{a} \times \mathbf{b}) \cdot \mathbf{c} = \|\mathbf{c}\| \, \|\mathbf{a} \times \mathbf{b}\| \cos \psi$, where ψ is the acute angle that **c** makes with the normal to the plane spanned by **a** and **b**. Since the volume of the parallelepiped with adjacent sides **a**, **b**, and **c** is the product of the area of the base $\|\mathbf{a} \times \mathbf{b}\|$ times the altitude $\|\mathbf{c}\| \cos \psi$, it follows that the volume is merely $|(\mathbf{a} \times \mathbf{b}) \cdot \mathbf{c}|$. We saw on p. 28 that $D = \mathbf{a} \cdot (\mathbf{b} \times \mathbf{c})$. By interchanging rows, $D = -\mathbf{c} \cdot (\mathbf{b} \times \mathbf{a}) = \mathbf{c} \cdot (\mathbf{a} \times \mathbf{b}) = (\mathbf{a} \times \mathbf{b}) \cdot \mathbf{c}$; therefore, the absolute value of D is the volume of the parallelepiped with adjacent sides **a**, **b**, and **c**.

To conclude this section, we shall use vector methods to determine the equation of a plane in space. Let P be a plane in space, **a** a vector ending on the plane, and **n** a vector normal to the plane (see Figure 1.3.6).

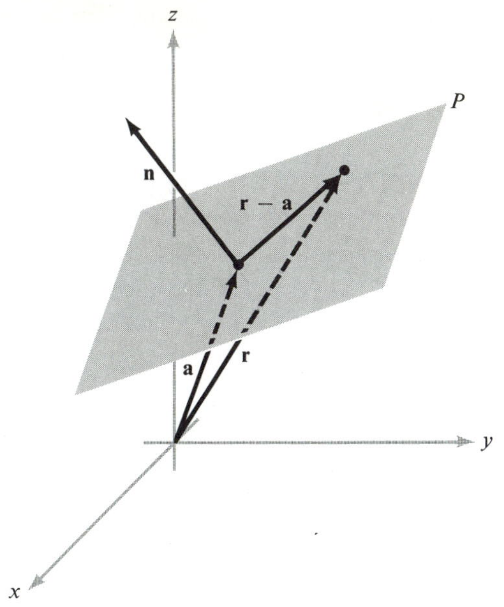

FIGURE 1.3.6
The points **r** *of the plane through* **a** *and perpendicular to* **n** *satisfy the equation* $(\mathbf{r} - \mathbf{a}) \cdot \mathbf{n} = 0$.

Now if **r** is a vector in $\mathbb{R}^3$, then the endpoint of **r** is on the plane P if and only if $\mathbf{r} - \mathbf{a}$ is parallel to P and, hence, if and only if $(\mathbf{r} - \mathbf{a}) \cdot \mathbf{n} = 0$ (**n** is perpendicular to any vector parallel to P—see Figure 1.3.6). Since the inner product is distributive, this last condition is equivalent to $\mathbf{r} \cdot \mathbf{n} = \mathbf{a} \cdot \mathbf{n}$. Therefore, if we let $\mathbf{a} = a_1\mathbf{i} + a_2\mathbf{j} + a_3\mathbf{k}$, $\mathbf{n} = A\mathbf{i} + B\mathbf{j} + C\mathbf{k}$, and $\mathbf{r} = x\mathbf{i} + y\mathbf{j} + z\mathbf{k}$, it follows that the endpoint of **r** lies on P if and only if

$$Ax + By + Cz = \mathbf{r} \cdot \mathbf{n} = \mathbf{a} \cdot \mathbf{n} = Aa_1 + Ba_2 + Ca_3 \tag{3}$$

Since **n** and **a** are fixed, the right-hand side of equation (3) is a constant, say $-D$. Thus an equation that determines the plane P is

$$Ax + By + Cz + D = 0 \tag{4}$$

where $A\mathbf{i} + B\mathbf{j} + C\mathbf{k}$ is normal to P; conversely, if A, B, and C are not all zero, the set of points (x, y, z) satisfying equation (4) is a plane with normal $A\mathbf{i} + B\mathbf{j} + C\mathbf{k}$.

The four numbers A, B, C, D are not determined uniquely by P. To see this, note that (x, y, z) satisfies equation (4) if and only if it also satisfies the relation

$$(\lambda A)x + (\lambda B)y + (\lambda C)z + (\lambda D) = 0$$

for $\lambda \neq 0$. If A, B, C, D and A', B', C', D', determine the same plane P then $A = \lambda A'$, $B = \lambda B'$, $C = \lambda C'$, $D = \lambda D'$ for a scalar λ. We say that A,

B, C, D are determined by P up to a scalar multiple. Conversely, given A, B, C, D and A', B', C', D', they determine the same plane if $A = \lambda A'$, $B = \lambda B'$, $C = \lambda C'$, $D = \lambda D'$ for some scalar λ. This fact will become more apparent in Example 7 below.

EXAMPLE 6. Determine the equation of the plane perpendicular to the vector $\mathbf{i} + \mathbf{j} + \mathbf{k}$ and containing the point (1, 0, 0).

From the above discussion it follows that the equation of the plane is of the form $x + y + z + D = 0$. Since (1, 0, 0) is on the plane, $1 + 0 + 0 + D = 0$, or $D = -1$. Thus, $x + y + z = 1$ is the equation of the plane.

EXAMPLE 7. Find an equation of the plane containing the points (1, 1, 1), (2, 0, 0), and (1, 1, 0).

Method 1. Any equation of the plane is of the form $Ax + By + Cz + D = 0$. Since the points (1, 1, 1), (2, 0, 0), and (1, 1, 0) lie on the plane, we have that

$$\begin{aligned} A + B + C + D &= 0 \\ 2A \qquad\quad + D &= 0 \\ A + B \qquad + D &= 0 \end{aligned}$$

Proceeding by elimination, we reduce the above system to the form

$$\begin{aligned} 2A + D &= 0 \\ 2B + D &= 0 \\ C &= 0 \end{aligned}$$

Since the numbers A, B, C, and D are determined only up to a scalar multiple, we can fix the value of one of them and then the others will be determined uniquely. If we let $D = -2$, then $A = +1$, $B = +1$, $C = 0$. Thus $x + y - 2 = 0$ is an equation of the plane that contains the given points.

Method 2. Let $\mathbf{a} = \mathbf{i} + \mathbf{j} + \mathbf{k}$, $\mathbf{b} = 2\mathbf{i}$, $\mathbf{c} = \mathbf{i} + \mathbf{j}$. Any vector normal to the plane must be orthogonal to the vectors $\mathbf{a} - \mathbf{b}$ and $\mathbf{c} - \mathbf{b}$, which are parallel to the plane since their endpoints lie on the plane. Thus, $\mathbf{n} = (\mathbf{a} - \mathbf{b}) \times (\mathbf{c} - \mathbf{b})$ is normal to the plane and we have

$$\mathbf{n} = \begin{vmatrix} \mathbf{i} & \mathbf{j} & \mathbf{k} \\ -1 & 1 & 1 \\ -1 & 1 & 0 \end{vmatrix} = -\mathbf{i} - \mathbf{j}$$

Thus, any equation of the plane is of the form $-x - y + D = 0$ (up to a scalar multiple). Since (2, 0, 0) lies on the plane, $D = +2$. After substituting, we obtain $x + y - 2 = 0$.

EXAMPLE 8. Let $Ax + By + Cz + D = 0$ be the equation of a plane P in $\mathbb{R}^3$. Determine the distance from the point $E = (x_1, y_1, z_1)$ to the plane (see Figure 1.3.7).

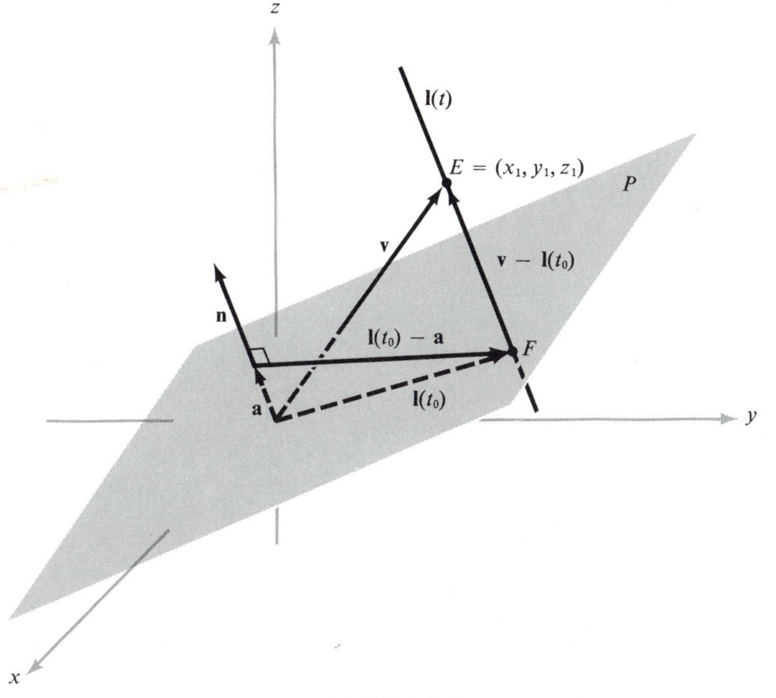

FIGURE 1.3.7
The geometry for determining the distance from point E to the plane P.

The vector

$$\mathbf{n} = \frac{A\mathbf{i} + B\mathbf{j} + C\mathbf{k}}{\sqrt{A^2 + B^2 + C^2}}$$

is a unit vector normal to the plane. If we write

$$\mathbf{a} = \frac{-D\mathbf{n}}{\sqrt{A^2 + B^2 + C^2}} = \frac{-DA\mathbf{i} - DB\mathbf{j} - DC\mathbf{k}}{A^2 + B^2 + C^2}$$

then we assert that **a** ends on the given plane. To see this note that in coordinates **a** is represented by

$$\left(\frac{-DA}{A^2 + B^2 + C^2}, \frac{-DB}{A^2 + B^2 + C^2}, \frac{-DC}{A^2 + B^2 + C^2}\right) = (x, y, z)$$

Now (x, y, z) lies on P if and only if $Ax + By + Cz + D = 0$. But

$$A\left(\frac{-DA}{A^2 + B^2 + C^2}\right) + B\left(\frac{-DB}{A^2 + B^2 + C^2}\right) + C\left(\frac{-DC}{A^2 + B^2 + C^2}\right) + D$$

$$= -D\frac{(A^2 + B^2 + C^2)}{A^2 + B^2 + C^2} + D = 0$$

and so $\mathbf{a}$ does indeed end on P.

The distance from (x_1, y_1, z_1) to P will be the length of the line segment EF (Figure 1.3.7) through (x_1, y_1, z_1) perpendicular to P and ending on P. So if we can find where the line perpendicular to P passing through (x_1, y_1, z_1) intersects P, we can then compute the length of the line segment EF and this will complete the problem.

If the endpoint of the vector $\mathbf{r} = x\mathbf{i} + y\mathbf{j} + z\mathbf{k}$ lies on the plane, we know that $(\mathbf{r} - \mathbf{a}) \cdot \mathbf{n} = 0$. Let $\mathbf{v} = x_1\mathbf{i} + y_1\mathbf{j} + z_1\mathbf{k}$ be the vector ending at (x_1, y_1, z_1). Then the line through (x_1, y_1, z_1) perpendicular to the plane is, in parametric form, $\mathbf{l}(t) = \mathbf{v} + t\mathbf{n}$. This line intersects the plane for some $t = t_0$ when

$$(\mathbf{l}(t_0) - \mathbf{a}) \cdot \mathbf{n} = 0$$

Substituting from $\mathbf{l}(t) = \mathbf{v} + t\mathbf{n}$ gives

$$(\mathbf{v} + t_0\,\mathbf{n} - \mathbf{a}) \cdot \mathbf{n} = 0 = \mathbf{v} \cdot \mathbf{n} + t_0(\mathbf{n} \cdot \mathbf{n}) - \mathbf{a} \cdot \mathbf{n} = t_0 + \mathbf{v} \cdot \mathbf{n} - \mathbf{a} \cdot \mathbf{n}$$

$$= t_0 - (\mathbf{a} - \mathbf{v}) \cdot \mathbf{n}$$

so

$$t_0 = (\mathbf{a} - \mathbf{v}) \cdot \mathbf{n}$$

The distance from the point (x_1, y_1, z_1) to the point at which $\mathbf{l}(t)$ intersects the plane, i.e., the distance from (x_1, y_1, z_1) to the plane is given by

$$\|\mathbf{v} - \mathbf{l}(t_0)\| = \|\mathbf{v} - (\mathbf{v} + t_0\,\mathbf{n})\| = |(\mathbf{a} - \mathbf{v}) \cdot \mathbf{n}| = |(\mathbf{a} \cdot \mathbf{n}) - (\mathbf{v} \cdot \mathbf{n})|$$

so we get our answer:

$$\text{distance} = \frac{|Ax_1 + By_1 + Cz_1 + D|}{\sqrt{A^2 + B^2 + C^2}}$$

EXERCISES

1. Verify that interchanging two rows or two columns of the 3×3 determinant

$$\begin{vmatrix} 1 & 2 & 1 \\ 3 & 0 & 1 \\ 2 & 0 & 2 \end{vmatrix}$$

changes the sign of the determinant (pick any two rows and any two columns).

2. Evaluate

(a) $\begin{vmatrix} 2 & -1 & 0 \\ 4 & 3 & 2 \\ 3 & 0 & 1 \end{vmatrix}$ (b) $\begin{vmatrix} 36 & 18 & 17 \\ 45 & 24 & 20 \\ 3 & 5 & -2 \end{vmatrix}$

(c) $\begin{vmatrix} 1 & 4 & 9 \\ 4 & 9 & 16 \\ 9 & 16 & 25 \end{vmatrix}$ (d) $\begin{vmatrix} 2 & 3 & 5 \\ 7 & 11 & 13 \\ 17 & 19 & 23 \end{vmatrix}$

3. Compute $\mathbf{a} \times \mathbf{b}$, where $\mathbf{a} = \mathbf{i} - 2\mathbf{j} + \mathbf{k}$, $\mathbf{b} = 2\mathbf{i} + \mathbf{j} + \mathbf{k}$.
4. Compute $\mathbf{a} \cdot (\mathbf{b} \times \mathbf{c})$, where $\mathbf{a}$ and $\mathbf{b}$ are as in Exercise 3 and $\mathbf{c} = 3\mathbf{i} - \mathbf{j} + 2\mathbf{k}$.
5. Find the area of the parallelogram with sides the vectors $\mathbf{a}$ and $\mathbf{b}$ given in Exercise 3.
6. What is the volume of the parallelepiped with sides $2\mathbf{i} + \mathbf{j} - \mathbf{k}$, $5\mathbf{i} - 3\mathbf{k}$, and $\mathbf{i} - 2\mathbf{j} - \mathbf{k}$?

In exercises 7 to 10, describe all unit vectors orthogonal to the given vectors.

7. $\mathbf{i} + \mathbf{j}$
8. $-5\mathbf{i} + 9\mathbf{j} - 4\mathbf{k}$, $7\mathbf{i} + 8\mathbf{j} + 9\mathbf{k}$
9. $-5\mathbf{i} + 9\mathbf{j} - 4\mathbf{k}$, $7\mathbf{i} + 8\mathbf{j} + 9\mathbf{k}$, $\mathbf{0}$
10. $2\mathbf{i} - 4\mathbf{j} + 3\mathbf{k}$, $-4\mathbf{i} + 8\mathbf{j} - 6\mathbf{k}$
11. Determine the distance from the plane $12x + 13y + 5z + 2 = 0$ to the point $(1, 1, -5)$.
12. Find the distance from the line through the origin and $(1, 1, 1)$, to the line through $(1, 2, -2)$ parallel to $2\mathbf{i} - \mathbf{j} + 2\mathbf{k}$.
13. Compute $\mathbf{u} + \mathbf{v}$, $\mathbf{u} \cdot \mathbf{v}$, $\|\mathbf{u}\|$, $\|\mathbf{v}\|$, and $\mathbf{u} \times \mathbf{v}$ where $\mathbf{u} = \mathbf{i} - 2\mathbf{j} + \mathbf{k}$, $\mathbf{v} = 2\mathbf{i} - \mathbf{j} + 2\mathbf{k}$.
14. Repeat Exercise 13 for $\mathbf{u} = 3\mathbf{i} + \mathbf{j} - \mathbf{k}$, $\mathbf{v} = -6\mathbf{i} - 2\mathbf{j} - 2\mathbf{k}$.
15. Find the equation of the plane that
    (a) is perpendicular to $\mathbf{v} = (1, 1, 1)$ and passes through $(1, 0, 0)$.
    (b) is perpendicular to $\mathbf{v} = (1, 2, 3)$ and passes through $(1, 1, 1)$.
16. Find the equation of the plane that passes through $(0, 0, 0)$, $(2, 0, -1)$, and $(0, 4, -3)$.
17. (a) Prove $(\mathbf{u} \times \mathbf{v}) \times \mathbf{w} = \mathbf{u} \times (\mathbf{v} \times \mathbf{w})$ if and only if $(\mathbf{u} \times \mathbf{w}) \times \mathbf{v} = \mathbf{0}$.
    (b) Prove $(\mathbf{u} \times \mathbf{v}) \times \mathbf{w} + (\mathbf{v} \times \mathbf{w}) \times \mathbf{u} + (\mathbf{w} \times \mathbf{u}) \times \mathbf{v} = \mathbf{0}$ (the *Jacobi identity*).
18. (a) Prove

$$(\mathbf{u} \times \mathbf{v}) \cdot (\mathbf{u}' \times \mathbf{v}') = (\mathbf{u} \cdot \mathbf{u}')(\mathbf{v} \cdot \mathbf{v}') - (\mathbf{u} \cdot \mathbf{v}')(\mathbf{u}' \cdot \mathbf{v}) = \begin{vmatrix} \mathbf{u} \cdot \mathbf{u}' & \mathbf{u} \cdot \mathbf{v}' \\ \mathbf{u}' \cdot \mathbf{v} & \mathbf{u} \cdot \mathbf{v}' \end{vmatrix}$$

    (b) Prove, without recourse to geometry, that

$$\mathbf{u} \cdot (\mathbf{v} \times \mathbf{w}) = \mathbf{v} \cdot (\mathbf{w} \times \mathbf{u}) = \mathbf{w} \cdot (\mathbf{u} \times \mathbf{v}) = -\mathbf{u} \cdot (\mathbf{w} \times \mathbf{v}) = -\mathbf{w} \cdot (\mathbf{v} \times \mathbf{u}) = -\mathbf{v} \cdot (\mathbf{u} \times \mathbf{w})$$

1.4 n-DIMENSIONAL EUCLIDEAN SPACE

In Sections 1.1 and 1.2 we studied the spaces $\mathbb{R}^1(\mathbb{R})$, $\mathbb{R}^2$, and $\mathbb{R}^3$, and gave geometric interpretations to them. For example, a point (x, y, z) in $\mathbb{R}^3$ can be thought of as a geometric object, namely, the directed line segment or vector emanating from the origin and ending at the point (x, y, z). We can therefore think of $\mathbb{R}^3$ in two ways:

(*i*) algebraically, as a set of triples (x, y, z) where x, y, and z are real numbers; or

(*ii*) geometrically, as a set of directed line segments.

These two ways of looking at $\mathbb{R}^3$ are equivalent. For generalization it is easier to use definition (*i*). Specifically, we can define $\mathbb{R}^n$, where n is a positive integer (possibly greater than 3), as the set of all ordered n-tuples $(x_1, x_2, \ldots, x_n)$, where the x_i are real numbers. For instance, $(1, \sqrt{5}, 2, \sqrt{3}) \in \mathbb{R}^4$. It is possible, but more difficult, to formulate a definition of $\mathbb{R}^n$ as a set of directed geometric objects.

The set $\mathbb{R}^n$ defined above is known as *Euclidean n-space* and its elements $\mathbf{x} = (x_1, x_2, \ldots, x_n)$ are known as *vectors* or *n-vectors*. By setting $n = 1$, 2, or 3, we recover the line, the plane, and three-dimensional space, $\mathbb{R}^3$, respectively.

We shall launch our study of Euclidean n-space by introducing several algebraic operations. These are completely analogous to those introduced in Section 1.1 for $\mathbb{R}^2$ and $\mathbb{R}^3$. The first two, addition and scalar multiplication, are defined by:

(*i*) $(x_1, x_2, \ldots, x_n) + (y_1, y_2, \ldots, y_n)$
$= (x_1 + y_1, x_2 + y_2, \ldots, x_n + y_n)$; and

(*ii*) for any real number α,

$$\alpha(x_1, x_2, \ldots, x_n) = (\alpha x_1, \alpha x_2, \ldots, \alpha x_n)$$

The geometric significance of these operations for $\mathbb{R}^2$ and $\mathbb{R}^3$ was discussed in Section 1.1.

The n vectors $\mathbf{e}_1 = (1, 0, 0, \ldots, 0)$, $\mathbf{e}_2 = (0, 1, 0, \ldots, 0)$, $\ldots$, $\mathbf{e}_n = (0, 0, \ldots, 0, 1)$ are called the *standard basis vectors* of $\mathbb{R}^n$. The vector $\mathbf{x} = (x_1, x_2, \ldots, x_n)$ can be written $\mathbf{x} = x_1\mathbf{e}_1 + x_2\mathbf{e}_2 + \cdots + x_n\mathbf{e}_n$.

For two vectors $\mathbf{x} = (x_1, x_2, x_3)$ and $\mathbf{y} = (y_1, y_2, y_3)$ in $\mathbb{R}^3$, we defined the *dot* or *inner product* $\mathbf{x} \cdot \mathbf{y}$ to be the real number

$$\mathbf{x} \cdot \mathbf{y} = x_1 y_1 + x_2 y_2 + x_3 y_3$$

This definition easily extends to $\mathbb{R}^n$; namely, for $\mathbf{x} = (x_1, x_2, \ldots, x_n)$, $\mathbf{y} = (y_1, y_2, \ldots, y_n)$ define $\mathbf{x} \cdot \mathbf{y} = x_1 y_1 + x_2 y_2 + \cdots + x_n y_n$. In $\mathbb{R}^n$, the notation $\langle \mathbf{x}, \mathbf{y} \rangle$ will often be used in place of $\mathbf{x} \cdot \mathbf{y}$ for the inner product. Continuing the analogy with $\mathbb{R}^3$, we are led to define the abstract notion of the *length* or *norm* of a vector $\mathbf{x}$ by the formula

$$\text{length of } \mathbf{x} = \|\mathbf{x}\| = \sqrt{\mathbf{x} \cdot \mathbf{x}} = \sqrt{x_1^2 + x_2^2 + \cdots + x_n^2}.$$

If $\mathbf{x}$ and $\mathbf{y}$ are two vectors in the plane $\mathbb{R}^2$, then we know that the angle θ between them is given by the formula

$$\cos \theta = \frac{\mathbf{x} \cdot \mathbf{y}}{\|\mathbf{x}\| \, \|\mathbf{y}\|}$$

The right side of this equation can be defined in $\mathbb{R}^n$ as well as in $\mathbb{R}^2$. It still represents the cosine of the angle between $\mathbf{x}$ and $\mathbf{y}$; this angle is well defined since $\mathbf{x}$ and $\mathbf{y}$ lie in a two-dimensional subspace of $\mathbb{R}^n$ (the plane determined by $\mathbf{x}$ and $\mathbf{y}$). The dot product is a powerful mathematical tool, which reflects the geometric notion of the angle between two vectors in $\mathbb{R}^n$.

It will be useful to have available some algebraic properties of the inner product. These are summarized in the next theorem (cf. formulas (*i*), (*ii*), (*iii*), and (*iv*) of Section 1.2).

Theorem 2. *For* $\mathbf{x}, \mathbf{y}, \mathbf{z} \in \mathbb{R}^n$ *and* α, β *real numbers, we have*

(*i*) $(\alpha\mathbf{x} + \beta\mathbf{y}) \cdot \mathbf{z} = \alpha(\mathbf{x} \cdot \mathbf{z}) + \beta(\mathbf{y} \cdot \mathbf{z})$

(*ii*) $\mathbf{x} \cdot \mathbf{y} = \mathbf{y} \cdot \mathbf{x}$

(*iii*) $\mathbf{x} \cdot \mathbf{x} \geq 0$

(*iv*) $\mathbf{x} \cdot \mathbf{x} = 0$ *if and only if* $\mathbf{x} = \mathbf{0}$.

Proof. Each assertion above can be proved by a simple computation. For example, to prove (*i*) we write

$$\begin{aligned}(\alpha\mathbf{x} + \beta\mathbf{y}) \cdot \mathbf{z} &= (\alpha x_1 + \beta y_1, \alpha x_2 + \beta y_2, \ldots, \alpha x_n + \beta y_n) \\ &\quad \cdot (z_1, z_2, \ldots, z_n) \\ &= (\alpha x_1 + \beta y_1)z_1 + (\alpha x_2 + \beta y_2)z_2 \\ &\quad + \cdots + (\alpha x_n + \beta y_n)z_n \\ &= \alpha x_1 z_1 + \beta y_1 z_1 + \alpha x_2 z_2 + \beta y_2 z_2 \\ &\quad + \cdots + \alpha x_n z_n + \beta y_n z_n \\ &= \alpha(\mathbf{x} \cdot \mathbf{z}) + \beta(\mathbf{y} \cdot \mathbf{z}).\end{aligned}$$

The other proofs are similar. ∎

In Section 1.2 we proved a far more interesting property of dot products, called the Cauchy-Schwarz inequality (sometimes called the Cauchy-Bunyakovskii-Schwarz inequality, or simply CBS inequality, because of apparently justifiable Russian claims that it was independently discovered by the Russian mathematician Bunyakovskii). For

$\mathbb{R}^2$ our proof required the use of the law of cosines. For $\mathbb{R}^n$ we could also use this method, by confining our attention to a plane in $\mathbb{R}^n$. However, we can also give a direct, completely algebraic proof.

Theorem 3. *Let* $\mathbf{x}, \mathbf{y}$ *be vectors in* $\mathbb{R}^n$. *Then*

$$|\mathbf{x} \cdot \mathbf{y}| \le \|\mathbf{x}\| \, \|\mathbf{y}\|$$

Proof. Let $a = \mathbf{y} \cdot \mathbf{y}$ and $b = -\mathbf{x} \cdot \mathbf{y}$. If $a = 0$ the theorem is clearly valid, since then $\mathbf{y} = \mathbf{0}$ and both sides of the inequality reduce to 0. Thus we may suppose $a \ne 0$. By Theorem 2 we have

$$\begin{aligned} 0 \le (a\mathbf{x} + b\mathbf{y}) \cdot (a\mathbf{x} + b\mathbf{y}) &= a^2\mathbf{x} \cdot \mathbf{x} + 2ab\mathbf{x} \cdot \mathbf{y} + b^2\mathbf{y} \cdot \mathbf{y} \\ &= (\mathbf{y} \cdot \mathbf{y})^2\mathbf{x} \cdot \mathbf{x} - (\mathbf{y} \cdot \mathbf{y})(\mathbf{x} \cdot \mathbf{y})^2 \end{aligned}$$

Dividing by $\mathbf{y} \cdot \mathbf{y}$ gives

$$0 \le (\mathbf{y} \cdot \mathbf{y})(\mathbf{x} \cdot \mathbf{x}) - (\mathbf{x} \cdot \mathbf{y})^2$$

or

$$(\mathbf{x} \cdot \mathbf{y})^2 \le (\mathbf{x} \cdot \mathbf{x})(\mathbf{y} \cdot \mathbf{y}) = \|\mathbf{x}\|^2 \, \|\mathbf{y}\|^2$$

Taking square roots on both sides of this equation yields the desired result. ▮

There is a very useful consequence of the Cauchy-Schwarz inequality in terms of lengths. The *triangle inequality* is geometrically clear in $\mathbb{R}^3$. In Figure 1.4.1, $\|\mathrm{OQ}\| = \|\mathbf{x} + \mathbf{y}\|$, $\|\mathrm{OP}\| = \|\mathbf{x}\| = \|\mathrm{RQ}\|$, and

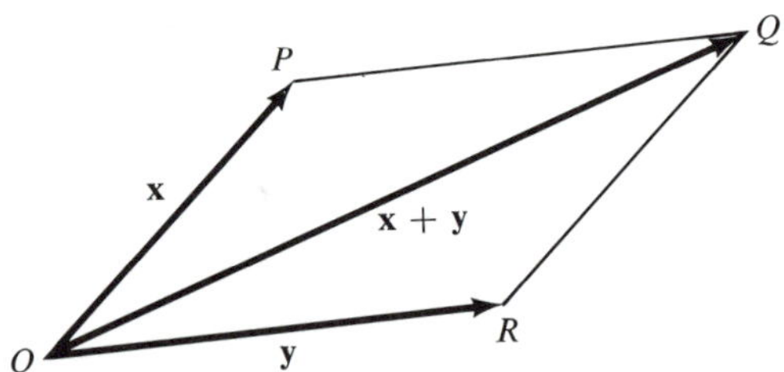

FIGURE 1.4.1
This geometry shows that $\|OQ\| \le \|OR\| + \|RQ\|$, *in vector notation* $\|\mathbf{x} + \mathbf{y}\| \le \|\mathbf{x}\| + \|\mathbf{y}\|$, *which is the triangle inequality.*

$\|\mathrm{OR}\| = \|\mathbf{y}\|$. Since the sum of the lengths of two sides of a triangle is greater than or equal to the length of the third, we have $\|\mathrm{OQ}\| \le \|\mathrm{OR}\| + \|\mathrm{RQ}\|$; that is, $\|\mathbf{x} + \mathbf{y}\| \le \|\mathbf{x}\| + \|\mathbf{y}\|$. The case for $\mathbb{R}^n$ is not as obvious, so we shall provide the analytic proof.

Corollary. *Let* **x**, **y** *be vectors in* $\mathbb{R}^n$. *Then*

$$\|\mathbf{x} + \mathbf{y}\| \le \|\mathbf{x}\| + \|\mathbf{y}\|$$

Proof. By Theorem 3, $\mathbf{x} \cdot \mathbf{y} \le |\mathbf{x} \cdot \mathbf{y}| \le \|\mathbf{x}\| \, \|\mathbf{y}\|$, so that

$$\|\mathbf{x} + \mathbf{y}\|^2 = \|\mathbf{x}\|^2 + 2\mathbf{x} \cdot \mathbf{y} + \|\mathbf{y}\|^2 \le \|\mathbf{x}\|^2 + 2\|\mathbf{x}\| \, \|\mathbf{y}\| + \|\mathbf{y}\|^2$$

Hence, we get $\|\mathbf{x} + \mathbf{y}\|^2 \le (\|\mathbf{x}\| + \|\mathbf{y}\|)^2$; taking square roots gives the result. ∎

If Theorem 3 and its corollary are written out algebraically they become the following rather unobvious inequalities:

$$\left|\sum_{i=1}^{n} x_i y_i\right| \le \left(\sum_{i=1}^{n} x_i^2\right)^{1/2} \left(\sum_{i=1}^{n} y_i^2\right)^{1/2}$$

$$\left(\sum_{i=1}^{n} (x_i + y_i)^2\right)^{1/2} \le \left(\sum_{i=1}^{n} x_i^2\right)^{1/2} + \left(\sum_{i=1}^{n} y_i^2\right)^{1/2}$$

These are very useful inequalities; we shall use them often in what follows, but primarily in vector form.

EXAMPLE 1. Let $\mathbf{x} = (1, 2, 0, -1)$, and $\mathbf{y} = (-1, 1, 1, 0)$. Then

$$\|\mathbf{x}\| = \sqrt{1^2 + 2^2 + 0^2 + (-1)^2} = \sqrt{6}$$

$$\|\mathbf{y}\| = \sqrt{(-1)^2 + 1^2 + 1^2 + 0^2} = \sqrt{3}$$

$$\langle \mathbf{x}, \mathbf{y} \rangle = 1(-1) + 2 \cdot 1 + 0 \cdot 1 + (-1)0 = 1$$

$$\mathbf{x} + \mathbf{y} = (0, 3, 1, -1)$$

$$\|\mathbf{x} + \mathbf{y}\| = \sqrt{0^2 + 3^2 + 1^2 + (-1)^2} = \sqrt{11}$$

We can directly verify Theorem 3 in this case:

$$|\langle \mathbf{x}, \mathbf{y} \rangle| = 1 \le 4.24 \approx \sqrt{6}\sqrt{3} = \|\mathbf{x}\| \, \|\mathbf{y}\|$$

Similarly, we can check its corollary:

$$\begin{aligned} \|\mathbf{x} + \mathbf{y}\| &= \sqrt{11} \approx 3.32 \le 4.18 \\ &= 2.45 + 1.73 \approx \sqrt{6} + \sqrt{3} = \|\mathbf{x}\| + \|\mathbf{y}\| \end{aligned}$$

By analogy with $\mathbb{R}^3$, we can define the notion of distance in $\mathbb{R}^n$; namely, if **x** and **y** are points in $\mathbb{R}^n$, the distance between **x** and **y** is defined to be $\|\mathbf{x} - \mathbf{y}\|$, or the length of the vector $\mathbf{x} - \mathbf{y}$. We remark that there is no cross product defined on $\mathbb{R}^n$ except for $n = 3$. It is only the dot product that generalizes.

Generalizing 2×2 and 3×3 matrices (see Section 1.3), we can consider $n \times n$ matrices, arrays of n^2 numbers:

$$A = \begin{bmatrix} a_{11} & a_{12} & \cdots & a_{1n} \\ a_{21} & a_{22} & \cdots & a_{2n} \\ \vdots & \vdots & & \vdots \\ a_{n1} & a_{n2} & \cdots & a_{nn} \end{bmatrix}$$

We shall also write A as (a_{ij}). Let us now define how $n \times n$ matrices are multiplied. If $A = (a_{ij})$, $B = (b_{ij})$ then $AB = C$ has entries given by

$$c_{ij} = \sum_{k=1}^{n} a_{ik} b_{kj}$$

which is the dot product of the ith row of A and the jth column of B:

$$\begin{bmatrix} a_{11} & \cdots & a_{1n} \\ a_{i1} & \cdots & a_{in} \\ \multicolumn{3}{c}{\longrightarrow} \\ a_{n1} & \cdots & a_{nn} \end{bmatrix} \begin{bmatrix} b_{11} & \cdots & b_{1j} & \downarrow & \cdots & b_{1n} \\ \vdots & & \vdots & & & \vdots \\ b_{n1} & \cdots & b_{nj} & & \cdots & b_{nn} \end{bmatrix}$$

EXAMPLE 2. Let

$$A = \begin{bmatrix} 1 & 0 & 3 \\ 2 & 1 & 0 \\ 1 & 0 & 0 \end{bmatrix} \text{ and } B = \begin{bmatrix} 0 & 1 & 0 \\ 1 & 0 & 0 \\ 0 & 1 & 1 \end{bmatrix}$$

Then

$$AB = \begin{bmatrix} 0 & 4 & 3 \\ 1 & 2 & 0 \\ 0 & 1 & 0 \end{bmatrix} \text{ and } BA = \begin{bmatrix} 2 & 1 & 0 \\ 1 & 0 & 3 \\ 3 & 1 & 0 \end{bmatrix}$$

Similarly, we can multiply an $n \times m$ matrix (n rows, m columns) by an $m \times p$ matrix (m rows, p columns) to obtain an $n \times p$ matrix (n rows, p columns) by the same rule. Note that for AB to be defined, the number of columns of A must equal the number of rows of B.

EXAMPLE 3. Let

$$A = \begin{bmatrix} 2 & 0 & 1 \\ 1 & 1 & 2 \end{bmatrix} \text{ and } B = \begin{bmatrix} 1 & 0 & 2 \\ 0 & 2 & 1 \\ 1 & 1 & 1 \end{bmatrix}$$

Then

$$AB = \begin{bmatrix} 3 & 1 & 5 \\ 3 & 4 & 5 \end{bmatrix}$$

and BA is not defined according to our method of matrix multiplication.

EXAMPLE 4. Let

$$A = \begin{bmatrix} 1 \\ 2 \\ 1 \\ 3 \end{bmatrix} \quad \text{and} \quad B = [2 \quad 2 \quad 1 \quad 2]$$

Then

$$AB = \begin{bmatrix} 2 & 2 & 1 & 2 \\ 4 & 4 & 2 & 4 \\ 2 & 2 & 1 & 2 \\ 6 & 6 & 3 & 6 \end{bmatrix} \quad \text{and} \quad BA = [13]$$

Any $m \times n$ matrix A determines a mapping of $\mathbb{R}^n$ to $\mathbb{R}^m$ as follows: let $\mathbf{x} = (x_1, \ldots, x_n) \in \mathbb{R}^n$; write $\mathbf{x}$ as a *column vector*

$$\mathbf{x} = \begin{pmatrix} x_1 \\ \vdots \\ x_n \end{pmatrix}$$

and multiply A by $\mathbf{x}$ (considered as an $n \times 1$ matrix) to get a new vector (in column form)

$$A\mathbf{x} = \begin{bmatrix} a_{11} & \cdots & a_{1n} \\ \vdots & & \vdots \\ a_{m1} & \cdots & a_{mn} \end{bmatrix} \begin{pmatrix} x_1 \\ \vdots \\ x_n \end{pmatrix} = \begin{pmatrix} y_1 \\ \vdots \\ y_m \end{pmatrix}$$

Thus we have a mapping $\mathbf{x} \mapsto A\mathbf{x}$ of $\mathbb{R}^n$ to $\mathbb{R}^m$. This mapping is *linear*; that is, it satisfies

$$A(\mathbf{x} + \mathbf{y}) = A\mathbf{x} + A\mathbf{y}$$

$$A(\alpha\mathbf{x}) = \alpha(A\mathbf{x})$$

as may be easily verified. One learns in a linear-algebra course that, conversely, any linear transformation of $\mathbb{R}^n$ to $\mathbb{R}^m$ is representable in this way by an $m \times n$ matrix.*

EXAMPLE 5. If

$$A = \begin{bmatrix} 1 & 0 & 3 \\ -1 & 0 & 1 \\ 2 & 1 & 2 \\ -1 & 2 & 1 \end{bmatrix}$$

* See, for instance, M. O'Nan, *Linear Algebra*, §5.2 (Theorem 2) and §5.8. (Harcourt Brace Janovich, New York, 1971.)

then $\mathbf{x} \mapsto A\mathbf{x}$ of $\mathbb{R}^3$ to $\mathbb{R}^4$ is the mapping defined by

$$\begin{pmatrix} x_1 \\ x_2 \\ x_3 \end{pmatrix} \mapsto \begin{pmatrix} x_1 + 3x_3 \\ -x_1 + x_3 \\ 2x_1 + x_2 + 2x_3 \\ -x_1 + 2x_2 + x_3 \end{pmatrix}$$

If $A = (a_{ij})$ is an $m \times n$ matrix and $\mathbf{e}_j$ is the jth standard basis vector of $\mathbb{R}^n$, then $A\mathbf{e}_j$ is a vector in $\mathbb{R}^m$ with components the same as the jth column of A. That is, the ith component of $A\mathbf{e}_j$ is a_{ij}. In symbols, $(A\mathbf{e}_j)_i = a_{ij}$.

EXAMPLE 6.

$$\begin{bmatrix} 4 & 2 & 9 \\ 3 & 5 & 4 \\ 1 & 2 & 3 \\ 0 & 1 & 2 \end{bmatrix} \begin{pmatrix} 0 \\ 1 \\ 0 \end{pmatrix} = \begin{pmatrix} 2 \\ 5 \\ 2 \\ 1 \end{pmatrix}$$

Matrix multiplication is not, in general, *commutative*: if A and B are $n \times n$ matrices, then generally

$$AB \neq BA$$

(see Examples 2, 3, and 4).

An $n \times n$ matrix is called *invertible* if there is an $n \times n$ matrix B such that

$$AB = BA = I_n\,.$$

where

$$I_n = \begin{bmatrix} 1 & 0 & 0 & \cdots & 0 \\ 0 & 1 & 0 & \cdots & 0 \\ 0 & 0 & 1 & \cdots & 0 \\ \vdots & \vdots & \vdots & & \vdots \\ 0 & 0 & 0 & \cdots & 1 \end{bmatrix}$$

is the $n \times n$ identity matrix: I_n has the property that $I_n C = CI_n = C$ for any $n \times n$ matrix C. We denote B by A^{-1} and call A^{-1} the *inverse* of A.

EXAMPLE 7. If

$$A = \begin{bmatrix} 2 & 4 & 0 \\ 0 & 2 & 1 \\ 3 & 0 & 2 \end{bmatrix} \quad \text{then} \quad A^{-1} = \frac{1}{20}\begin{bmatrix} 4 & -8 & 4 \\ 3 & 4 & -2 \\ -6 & 12 & 4 \end{bmatrix}$$

since $AA^{-1} = I_n = A^{-1}A$, as may be checked by matrix multiplication.

Methods of computing inverses are learned in linear algebra; we won't require these methods in this book. If A is invertible, the equation $A\mathbf{x} = \mathbf{y}$ can be solved for the vector $\mathbf{x}$ by multiplying both sides by A^{-1} to obtain $\mathbf{x} = A^{-1}\mathbf{y}$.

In Section 1.3 we defined the determinant of a 3×3 matrix. This can be generalized by induction to $n \times n$ determinants. We illustrate here how to write the determinant of a 4×4 matrix in terms of the determinants of 3×3 matrices:

$$\begin{vmatrix} a_{11} & a_{12} & a_{13} & a_{14} \\ a_{21} & a_{22} & a_{23} & a_{24} \\ a_{31} & a_{32} & a_{33} & a_{34} \\ a_{41} & a_{42} & a_{43} & a_{44} \end{vmatrix}$$

$$= a_{11} \begin{vmatrix} a_{22} & a_{23} & a_{24} \\ a_{32} & a_{33} & a_{34} \\ a_{42} & a_{43} & a_{44} \end{vmatrix} - a_{12} \begin{vmatrix} a_{21} & a_{23} & a_{24} \\ a_{31} & a_{33} & a_{34} \\ a_{41} & a_{43} & a_{44} \end{vmatrix}$$

$$+ a_{13} \begin{vmatrix} a_{21} & a_{22} & a_{24} \\ a_{31} & a_{32} & a_{34} \\ a_{41} & a_{42} & a_{44} \end{vmatrix} - a_{14} \begin{vmatrix} a_{21} & a_{22} & a_{23} \\ a_{31} & a_{32} & a_{33} \\ a_{41} & a_{42} & a_{43} \end{vmatrix}$$

(cf. formula (2) of Section 1.3; the signs alternate $+, -, +, -, \ldots$).

The basic properties of 3×3 determinants reviewed in Section 1.3 remain valid for $n \times n$ determinants. In particular, we note the fact that if A is an $n \times n$ matrix and B is the matrix formed by adding a scalar multiple of the kth row (column) of A to the lth row (respectively, column) of A, then the determinant of A (det A) is equal to the determinant of B (det B) (see Example 8 below).

A basic theorem of linear algebra states that an $n \times n$ matrix A is invertible if and only if the determinant of A is not zero. Another basic property is that $\det(AB) = (\det A)(\det B)$. In this text we shall not make use of many details of linear algebra, so we shall leave these assertions unproved.

EXAMPLE 8. Let

$$A = \begin{bmatrix} 1 & 0 & 1 & 0 \\ 1 & 1 & 1 & 1 \\ 2 & 1 & 0 & 1 \\ 1 & 1 & 0 & 2 \end{bmatrix}$$

Then adding $(-1) \times$ first column to the third column, we get

$$\det A = \begin{vmatrix} 1 & 0 & 0 & 0 \\ 1 & 1 & 0 & 1 \\ 2 & 1 & -2 & 1 \\ 1 & 1 & -1 & 2 \end{vmatrix} = 1 \begin{vmatrix} 1 & 0 & 1 \\ 1 & -2 & 1 \\ 1 & -1 & 2 \end{vmatrix}$$

Adding $(-1) \times$ first column to the third column of this 3×3 determinant gives

$$\det A = \begin{vmatrix} 1 & 0 & 0 \\ 1 & -2 & 0 \\ 1 & -1 & 1 \end{vmatrix} = \begin{vmatrix} -2 & 0 \\ -1 & 1 \end{vmatrix} = -2$$

Thus, $\det A = -2 \neq 0$, and A has an inverse.

If we have three matrices A, B, and C such that the products AB and BC are defined, then the products $(AB)C$ and $A(BC)$ will be defined and equal (that is, matrix multiplication is associative). We call this the *triple product* of matrices and denote it by ABC.

EXAMPLE 9. Let

$$A = \begin{bmatrix} 3 \\ 5 \end{bmatrix}, \qquad B = [1 \quad 1], \quad \text{and} \quad C = \begin{bmatrix} 1 \\ 2 \end{bmatrix}$$

Then

$$ABC = A(BC) = \begin{bmatrix} 3 \\ 5 \end{bmatrix}[3] = \begin{bmatrix} 9 \\ 15 \end{bmatrix}$$

EXAMPLE 10.

$$\begin{bmatrix} 2 & 0 \\ 0 & 1 \end{bmatrix}\begin{bmatrix} 1 & 1 \\ 1 & 1 \end{bmatrix}\begin{bmatrix} 0 & -1 \\ 1 & 0 \end{bmatrix} = \begin{bmatrix} 2 & 0 \\ 0 & 1 \end{bmatrix}\begin{bmatrix} 1 & -1 \\ 1 & -1 \end{bmatrix} = \begin{bmatrix} 2 & -2 \\ 1 & -1 \end{bmatrix}$$

Matrix multiplication is the most important nonelementary algebraic operation on matrices; there are also some elementary operations that deserve mention. Given two $m \times n$ matrices A and B, we can add (subtract) them to obtain a new $m \times n$ matrix $C = A + B$ $(C = A - B)$, whose ijth entry c_{ij} is the sum (difference) of a_{ij} and b_{ij}. It is clear that $A + B = B + A$.

EXAMPLE 11.

$$\begin{bmatrix} 2 & 1 & 0 \\ 3 & 4 & 1 \end{bmatrix} + \begin{bmatrix} -1 & 1 & 3 \\ 0 & 0 & 7 \end{bmatrix} = \begin{bmatrix} 1 & 2 & 3 \\ 3 & 4 & 8 \end{bmatrix}$$

EXAMPLE 12.

$$[1 \quad 2] + [0 \quad -1] = [1 \quad 1]$$

EXAMPLE 13.

$$\begin{bmatrix} 2 & 1 \\ 1 & 2 \end{bmatrix} - \begin{bmatrix} 1 & 0 \\ 0 & 1 \end{bmatrix} = \begin{bmatrix} 1 & 1 \\ 1 & 1 \end{bmatrix}$$

Given a scalar λ and an $m \times n$ matrix A, we can multiply A by λ to obtain a new $m \times n$ matrix $\lambda A = C$, whose ijth entry c_{ij} is the product λa_{ij}.

EXAMPLE 14.

$$3 \begin{bmatrix} 1 & -1 & 2 \\ 0 & 1 & 5 \\ 1 & 0 & 3 \end{bmatrix} = \begin{bmatrix} 3 & -3 & 6 \\ 0 & 3 & 15 \\ 3 & 0 & 9 \end{bmatrix}$$

EXERCISES

1. Prove (*ii*) to (*iv*) of Theorem 2.
2. In $\mathbb{R}^n$ show that
   (a) $2\|\mathbf{x}\|^2 + 2\|\mathbf{y}\|^2 = \|\mathbf{x} + \mathbf{y}\|^2 + \|\mathbf{x} - \mathbf{y}\|^2$
   (This is known as the parallelogram law.)
   (b) $\|\mathbf{x} + \mathbf{y}\| \, \|\mathbf{x} - \mathbf{y}\| \leq \|\mathbf{x}\|^2 + \|\mathbf{y}\|^2$
   (c) $4\langle \mathbf{x}, \mathbf{y} \rangle = \|\mathbf{x} + \mathbf{y}\|^2 - \|\mathbf{x} - \mathbf{y}\|^2$
   (This is called the polarization identity.)

   Interpret these results geometrically in terms of the parallelogram formed by $\mathbf{x}$ and $\mathbf{y}$.
3. Verify the CBS inequality and the triangle inequality for:
   (a) $\mathbf{x} = (2, 0, -1)$, $\mathbf{y} = (4, 0, -2)$
   (b) $\mathbf{x} = (1, 0, 2, 6)$, $\mathbf{y} = (3, 8, 4, 1)$
   (c) $\mathbf{x} = (1, -1, 1, -1, 1)$, $\mathbf{y} = (3, 0, 0, 0, 2)$
4. Verify that if A is an $n \times n$ matrix, the map $\mathbf{x} \mapsto A\mathbf{x}$ of $\mathbb{R}^n$ to $\mathbb{R}^n$ is linear.
5. Assuming the law $\det(AB) = (\det A)(\det B)$, verify that $(\det A) \times (\det A^{-1}) = 1$ and conclude that if A has an inverse, $\det A \neq 0$.
6. Compute AB, $\det A$, $\det B$, $\det(AB)$, and $\det(A + B)$ for

$$A = \begin{bmatrix} 3 & 0 & 1 \\ 1 & 2 & -1 \\ 1 & 0 & 1 \end{bmatrix} \qquad B = \begin{bmatrix} 1 & 0 & -1 \\ 2 & 0 & 1 \\ 0 & 1 & 0 \end{bmatrix}$$

7. Verify that the inverse of

$$\begin{bmatrix} a & b \\ c & d \end{bmatrix} \quad \text{is} \quad \frac{1}{ad - bc} \begin{bmatrix} d & -b \\ -c & a \end{bmatrix}$$

8. Use your answer in Exercise 7 to show that the solution of

$$ax + by = e$$

$$cx + dy = f$$

is

$$\begin{pmatrix} x \\ y \end{pmatrix} = \frac{1}{ad - bc} \begin{bmatrix} d & -b \\ -c & a \end{bmatrix} \begin{pmatrix} e \\ f \end{pmatrix}$$

9. Use induction on k to prove that if $\mathbf{x}_1, \ldots, \mathbf{x}_k \in \mathbb{R}^n$, then

$$\|\mathbf{x}_1 + \cdots + \mathbf{x}_k\| \le \|\mathbf{x}_1\| + \cdots + \|\mathbf{x}_k\|$$

10. Prove, using algebra, the *identity of Lagrange*: for real numbers $x_1, \ldots, x_n$ and $y_1, \ldots, y_n$

$$\left(\sum_{i=1}^{n} x_i y_i\right)^2 = \left(\sum_{i=1}^{n} x_i^2\right)\left(\sum_{i=1}^{n} y_i^2\right) - \sum_{i<j} (x_i y_j - x_j y_i)^2$$

Use this to give another proof of the Cauchy-Schwarz inequality in $\mathbb{R}^n$.

11. Prove by induction that if A is an $n \times n$ matrix
    (a) $\det(\lambda A) = \lambda^n \det A$
    (b) and B is a matrix obtained from A by multiplying any row or column by a scalar λ, then $\det B = \lambda \det A$.
12. Is $\det(A + B) = \det A + \det B$? Give a proof or counterexample.
13. Does $(A + B)(A - B) = A^2 - B^2 = (A)(A) - (B)(B)$?
14. Prove that $\det(ABC) = (\det A)(\det B)(\det C)$.

REVIEW EXERCISES FOR CHAPTER 1

1. Let $\mathbf{v} = 3\mathbf{i} + 4\mathbf{j} + 5\mathbf{k}$ and $\mathbf{w} = \mathbf{i} - \mathbf{j} + \mathbf{k}$. Compute $\mathbf{v} + \mathbf{w}$, $3\mathbf{v}$, $6\mathbf{v} + 8\mathbf{w}$, $-2\mathbf{v}$, $\mathbf{v} \cdot \mathbf{w}$, $\mathbf{v} \times \mathbf{w}$. Interpret each operation geometrically by graphing the vectors.
2. (a) Find the equation of the line through $(0, 1, 0)$ in the direction of $3\mathbf{i} + \mathbf{k}$.
   (b) Find the equation of the line passing through $(0, 1, 1)$ and $(0, 1, 0)$.
   (c) Find the equation of the plane perpendicular to $(-1, 1, -1)$ and passing through $(1, 1, 1)$.
3. Use vector notation to describe the triangle in space whose vertices are the origin and the endpoints of vectors $\mathbf{a}$ and $\mathbf{b}$.
4. Show that three vectors $\mathbf{a}, \mathbf{b}, \mathbf{c}$ lie in a plane if and only if there are three scalars α, β, γ, not all zero, such that $\alpha\mathbf{a} + \beta\mathbf{b} + \gamma\mathbf{c} = \mathbf{0}$.
5. For real numbers $a_1, a_2, a_3, b_1, b_2, b_3$ show that

$$(a_1 b_1 + a_2 b_2 + a_3 b_3)^2 \le (a_1^2 + a_2^2 + a_3^2)(b_1^2 + b_2^2 + b_3^2)$$

6. Let $\mathbf{u}, \mathbf{v}, \mathbf{w}$ be unit vectors that are orthogonal to each other. If $\mathbf{a} = \alpha\mathbf{u} + \beta\mathbf{v} + \gamma\mathbf{w}$, show that

$$\alpha = \mathbf{a} \cdot \mathbf{u}, \qquad \beta = \mathbf{a} \cdot \mathbf{v}, \qquad \gamma = \mathbf{a} \cdot \mathbf{w}$$

Interpret the results geometrically.

7. Let $\mathbf{a}, \mathbf{b}$ be two vectors in the plane, $\mathbf{a} = (a_1, a_2)$, $\mathbf{b} = (b_1, b_2)$, and let λ be a real number. Show that the area of the parallelogram determined by $\mathbf{a}$

and $\mathbf{b} + \lambda\mathbf{a}$ is the same as that determined by $\mathbf{a}$ and $\mathbf{b}$. Sketch. Relate this result to a known property of determinants.

8. Find the volume of the parallelepiped determined by the vertices $(0, 1, 0)$, $(1, 1, 1)$, $(0, 2, 0)$, $(3, 1, 2)$.

9. Given vectors $\mathbf{a}$ and $\mathbf{b}$ in $\mathbb{R}^3$, show that the vector

$$\mathbf{v} = \|\mathbf{a}\|\mathbf{b} + \|\mathbf{b}\|\mathbf{a}$$

bisects the angle between $\mathbf{a}$ and $\mathbf{b}$.

10. Use vector methods to prove that the distance from the point (x_1, y_1) to the line $ax + by = c$ is

$$\frac{|ax_1 + by_1 - c|}{\sqrt{a^2 + b^2}}$$

11. Verify that the direction of $\mathbf{b} \times \mathbf{c}$ is given by the right-hand rule, by considering $\mathbf{b}$, $\mathbf{c}$ to be the vectors $\mathbf{i}$, $\mathbf{j}$ or $\mathbf{k}$.

12. (a) Suppose $\mathbf{a} \cdot \mathbf{b} = \mathbf{a}' \cdot \mathbf{b}$ for all $\mathbf{b}$. Show that $\mathbf{a} = \mathbf{a}'$.
    (b) Suppose $\mathbf{a} \times \mathbf{b} = \mathbf{a}' \times \mathbf{b}$ for all $\mathbf{b}$. Is it true that $\mathbf{a} = \mathbf{a}'$?

13. (a) Using vector methods, show that the distance between two lines l_1 and l_2 is given by

$$d = \frac{|(\mathbf{v}_2 - \mathbf{v}_1) \cdot (\mathbf{a}_1 \times \mathbf{a}_2)|}{\|\mathbf{a}_1 \times \mathbf{a}_2\|}$$

where $\mathbf{v}_1$, $\mathbf{v}_2$ are any points on l_1 and l_2, and $\mathbf{a}_1$ and $\mathbf{a}_2$ are the directions of l_1 and l_2.
(HINT: Consider the plane through l_2 which is parallel to l_1. Show that $(\mathbf{a}_1 \times \mathbf{a}_2)/\|\mathbf{a}_1 \times \mathbf{a}_2\|$ is a unit normal for this plane; now project $\mathbf{v}_2 - \mathbf{v}_1$ onto this normal direction.)
    (b) Find the distance between the line l_1 determined by the points $(-1, -1, 1)$ and $(0, 0, 0)$ and the line l_2 determined by the points $(0, -2, 0)$ and $(2, 0, 5)$.

14. Show that two planes given by the equations $Ax + By + Cz + D_1 = 0$ and $Ax + By + Cz + D_2 = 0$ are parallel, and that the distance between two such planes is

$$\frac{|D_1 - D_2|}{\sqrt{A^2 + B^2 + C^2}}$$

15. (a) Prove that the area of the triangle in the plane with vertices (x_1, y_1), (x_2, y_2), (x_3, y_3) is the absolute value of

$$\frac{1}{2}\begin{vmatrix} 1 & 1 & 1 \\ x_1 & x_2 & x_3 \\ y_1 & y_2 & y_3 \end{vmatrix}$$

    (b) Find the area of the triangle with vertices $(1, 2)$, $(0, 1)$, $(-1, 1)$.

16. Verify the Cauchy-Schwarz and triangle inequalities for

$$\mathbf{x} = (3, 2, 1, 0) \quad \text{and} \quad \mathbf{y} = (1, 1, 1, 2)$$

17. Multiply the matrices

$$A = \begin{bmatrix} 3 & 0 & 1 \\ 2 & 0 & 1 \\ 1 & 0 & 1 \end{bmatrix} \quad \text{and} \quad B = \begin{bmatrix} 1 & 0 & 1 \\ 1 & 1 & 1 \\ 0 & 0 & 1 \end{bmatrix}$$

Does $AB = BA$?

18. (a) Show that for two $n \times n$ matrices A and B, and $\mathbf{x} \in \mathbb{R}^n$,

$$(AB)\mathbf{x} = A(B\mathbf{x})$$

(b) What does (a) imply about the mappings $\mathbf{x} \mapsto B\mathbf{x}$, $\mathbf{y} \mapsto A\mathbf{y}$?

19. Is

$$\begin{bmatrix} 1 & 0 & 1 \\ 1 & 1 & 1 \\ 0 & 0 & 1 \end{bmatrix} \quad \text{invertible?}$$

*20. Verify that any linear mapping of $\mathbb{R}^n$ to $\mathbb{R}^n$ is determined by an $n \times n$ matrix; that is, it comes from an $n \times n$ matrix in the manner explained on p. 43.

CHAPTER 2

DIFFERENTIATION

In this chapter we shall extend the differential calculus from functions of one variable to functions of several variables. We shall begin with a section on the geometry of real-valued functions. The study of their graphs is useful for visualizing these functions.

Section 2.2 gives some basic definitions relating to limits and continuity. This subject is treated briefly for the reason that it requires time and mathematical maturity to develop fully, and is therefore best left to a more advanced course.* Fortunately, a complete understanding of all the subtleties of the limit concept is not necessary; the student who has difficulty with this section should bear that in mind. However, we hasten to add that the notion of limit is central to the definition of the derivative (but not to the computation of derivatives in specific problems), as we already know from one-variable calculus.

Sections 2.3 and 2.4 deal with the definition of the derivative, and establish some basic rules of calculus; namely, how to differentiate a sum, product, quotient, or composition. (Some more technical facts related to the basic theory of limits and differential calculus are presented in the optional Section 2.7.) In Section 2.5 we study directional

* See, for example, J. Marsden, *Elementary Classical Analysis*, W. H. Freeman and Company, San Francisco, 1974.

derivatives and tangent planes, relating these ideas to those in Section 2.1. Finally, in Section 2.6 we consider some properties of higher-order derivatives.

In generalizing calculus from one to several dimensions, it is often convenient, though not absolutely essential, to use the language of linear algebra. What we need in this text has already been summarized in Section 1.4.

2.1 THE GEOMETRY OF REAL-VALUED FUNCTIONS

We launch our investigation of real-valued functions by developing methods of visualizing them. In particular, we shall introduce the notions of graph, level curve, and level surface of such functions.

Let f be a function with domain a subset A of $\mathbb{R}^n$ (i.e., $A \subset \mathbb{R}^n$) and with range contained in $\mathbb{R}^m$. By this we mean that to each $\mathbf{x} = (x_1, \ldots, x_n) \in A$, f assigns a value $f(\mathbf{x})$, an m-tuple in $\mathbb{R}^m$. For example, $f(x, y, z) = (x^2 + y^2 + z^2)^{-3/2}$ maps the set A of $(x, y, z) \neq (0, 0, 0)$ in $\mathbb{R}^3$ ($n = 3$ in this case) to $\mathbb{R}$ ($m = 1$). To denote f we sometimes write

$$f\colon (x, y, z) \mapsto (x^2 + y^2 + z^2)^{-3/2}$$

Note that in $\mathbb{R}^3$ we often use the notation (x, y, z) instead of (x_1, x_2, x_3). In general, the notation $\mathbf{x} \mapsto f(\mathbf{x})$ is useful for indicating the value to which a point $\mathbf{x} \in \mathbb{R}^n$ is sent. We write $f\colon A \subset \mathbb{R}^n \to \mathbb{R}^m$ to signify that A is the domain of f (in $\mathbb{R}^n$) and the range is contained in $\mathbb{R}^m$. We also use the expression f *maps* A into $\mathbb{R}^m$.*

Such functions f are called *functions of several variables* if $A \subset \mathbb{R}^n$, $n > 1$. As another example we can take the function $g\colon \mathbb{R}^6 \to \mathbb{R}^2$ defined by the rule

$$g(\mathbf{x}) = g(x_1, x_2, x_3, x_4, x_5, x_6) = \left(x_1 x_2 x_3 x_4 x_5 x_6, \sqrt{x_1^2 + x_6^2}\right)$$

The first coordinate of the value of g at $\mathbf{x}$ is the product of the coordinates of $\mathbf{x}$.

It is important to be aware at this point that functions from $\mathbb{R}^n$ to $\mathbb{R}^m$ are not just mathematical abstractions. They do arise naturally in problems studied in all the sciences. For example, to specify the temperature T in a region A of space requires a function $T\colon A \subset \mathbb{R}^3 \to \mathbb{R}$ ($n = 3$, $m = 1$). Thus $T(x, y, z)$ is the temperature at the point (x, y, z). To specify the velocity of a fluid moving in space requires a map $V\colon \mathbb{R}^4 \to \mathbb{R}^3$ where $V(x, y, z, t)$ is the velocity vector of the fluid at the

* Some mathematicians would write such an f in boldface: $\mathbf{f}(\mathbf{x})$ since it is vector-valued. We did not do so, as a matter of personal taste. We use boldface primarily for vector fields, introduced later.

point (x, y, z) in space at time t (see Figure 2.1.1.) To specify the reaction rate of a solution consisting of six reacting chemicals A, B, C, D, E, F in proportions x, y, z, w, u, v requires a map $\sigma: B \subset \mathbb{R}^6 \to \mathbb{R}$, where $\sigma(x, y, z, w, u, v)$ gives the rate when the chemicals are in the indicated proportions. To specify the cardiac vector (the vector giving the direction of electric current flow in the heart) at time t requires a map $\mathbf{c}: \mathbb{R} \to \mathbb{R}^3$, $t \mapsto \mathbf{c}(t)$.

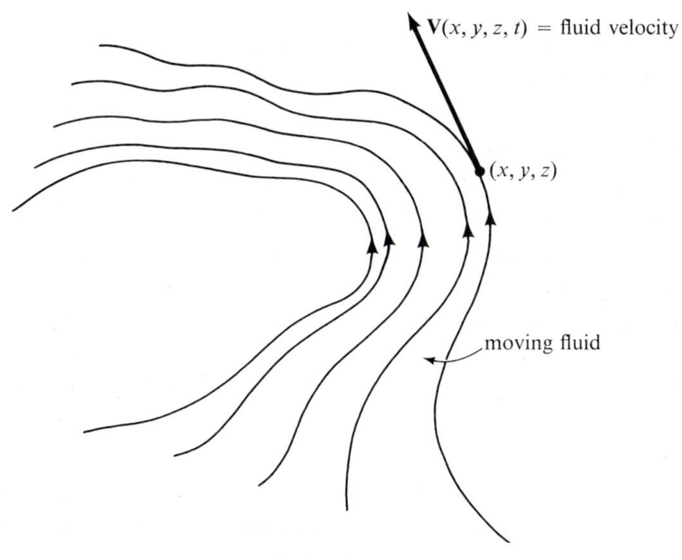

FIGURE 2.1.1
A fluid in motion defines a vector field $\mathbf{V}$ *by specifying the velocity of the fluid particles at each point in space and time.*

When $f: U \subset \mathbb{R}^n \to \mathbb{R}$, we say that f is a *real-valued function of n variables with domain U*. The reason for this is simply that we regard the coordinates of a point $\mathbf{x} = (x_1, \ldots, x_n) \in U$ as n variables, and $f(\mathbf{x}) = f(x_1, \ldots, x_n) \in \mathbb{R}$ depends on these variables. A good deal of our work will be with real-valued functions, so we shall give them special attention.

For $f: U \subset \mathbb{R} \to \mathbb{R}$ $(n = 1)$, the graph of f, as we know, is the subset of $\mathbb{R}^2$ consisting of all points $(x, f(x))$ for x in U. This subset can be thought of as a curve in $\mathbb{R}^2$. In symbols, we write this

$$\text{graph } f = \{(x, f(x)) \in \mathbb{R}^2 \mid x \in U\}$$

where the braces { } mean "the set of all" and the vertical bar is read as "such that." Similarly, if f is a function of two variables, its graph can be thought of as a surface in $\mathbb{R}^3$ (Figure 2.1.2). Let us formalize these ideas.

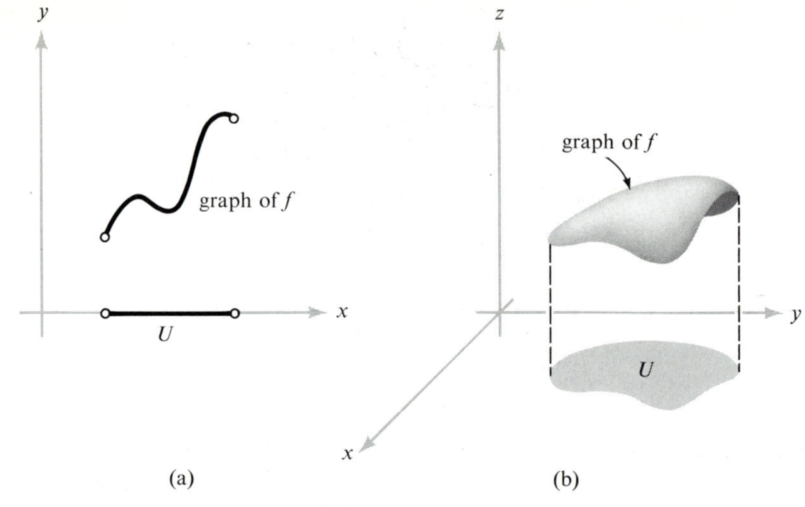

FIGURE 2.1.2
The graphs of (a) a function of one variable and (b) a function of two variables.

Definition. *Let $f: U \subset \mathbb{R}^n \to \mathbb{R}$. Define the* **graph** *of f to be the subset of $\mathbb{R}^{n+1}$ consisting of the points $(x_1, \ldots, x_n, f(x_1, \ldots, x_n))$ for $(x_1, \ldots, x_n)$ in U. In symbols,*

$$\text{graph } f = \{(x_1, \ldots, x_n, f(x_1, \ldots, x_n)) \in \mathbb{R}^{n+1} \mid (x_1, \ldots, x_n) \in U\}$$

For the case $n = 1$ the graph is, intuitively speaking, a curve, while for $n = 2$ it is a surface. For $n = 3$ it is difficult to visualize the graph because, being humans living in a three-dimensional world, it is hard for us to envisage sets in $\mathbb{R}^4$. To help overcome this handicap, we introduce the idea of a level set.

Suppose $f(x, y, z) = x^2 + y^2 + z^2$. Then a *level set* is a subset of $\mathbb{R}^3$ on which f is constant; for instance, the set where $x^2 + y^2 + z^2 = 1$. This we can visualize: it is just a sphere of radius 1 in $\mathbb{R}^3$. The behavior or structure of a function is partially determined by the shape of its level sets; consequently, understanding these sets aids us in understanding the function in question. The concept of level sets is also useful for understanding functions of two variables $f(x, y)$, in which case we speak of level *curves*.

The idea is very similar to that used to prepare contour maps, where one draws lines to represent lines of constant altitude; walking along such a line would mean walking on a level path. In the case of a hill rising from the xy-plane, a graph of all these level curves gives us a good idea of the function $h(x, y)$, which represents the height of the hill at point (x, y) (see Figure 2.1.3).

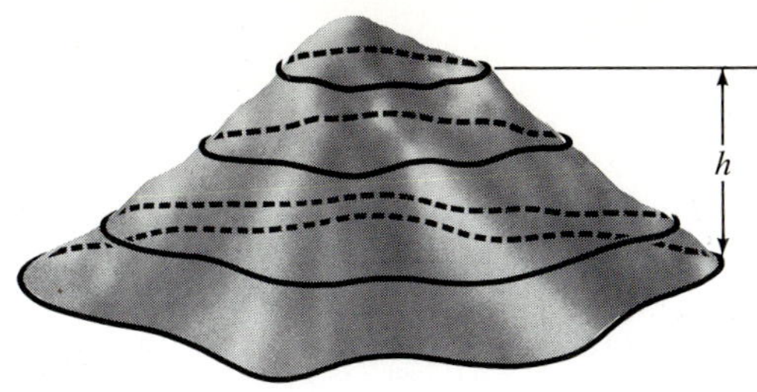

FIGURE 2.1.3
Level contours of a function are defined in the same manner as one obtains contour lines for a topographical map.

Definition. *Let $f\colon U \subset \mathbb{R}^n \to \mathbb{R}$ and let $c \in \mathbb{R}$. Then the* ***level set of value*** *c is defined to be those points $\mathbf{x} \in U$ at which $f(\mathbf{x}) = c$. If $n = 2$ we speak of a* ***level curve*** *(of value c), and if $n = 3$ we speak of a* ***level surface****. In symbols, the level set of value c is written*

$$\{\mathbf{x} \in U \mid f(\mathbf{x}) = c\} \subset \mathbb{R}^n$$

Note that the level set is always in the domain space.

EXAMPLE 1. The constant function $f\colon \mathbb{R}^2 \to \mathbb{R}$, $(x, y) \mapsto 2$, i.e. $f(x, y) = 2$, has as its graph the horizontal plane $z = 2$ in $\mathbb{R}^3$. The level curve of value c is empty if $c \neq 2$, and is the whole xy-plane if $c = 2$.

EXAMPLE 2. The function $f\colon \mathbb{R}^2 \to \mathbb{R}$, $(x, y) \mapsto x + y + 2$, has as its graph the inclined plane $z = x + y + 2$. This plane intersects the xy-plane $(z = 0)$ in the line $y = -x - 2$ and the z-axis in the point $(0, 0, 2)$. For any value $c \in \mathbb{R}$, the level curve of value c is the straight line $y = -x + (c - 2)$, or in symbols, the set

$$L_c = \{(x, y) \mid y = -x + (c - 2)\} \subset \mathbb{R}^2.$$

We indicate a few of these level curves in Figure 2.1.4. This is actually a contour map of the function f.

From the level curves, labelled with the value or "height" of the function, the shape of the graph may be inferred by mentally elevating each level curve to the appropriate height, without stretching, tilting, or sliding it. If this procedure is visualized for all level curves L_c, that is, for all values $c \in \mathbb{R}$, they will assemble to give the entire graph of f, as indicated in Figure 2.1.5. If it is visualized for only a finite number of level curves, as is usually the case, a sort of contour model of the graph is produced, as in Figure 2.1.4. However, if f is a smooth function its

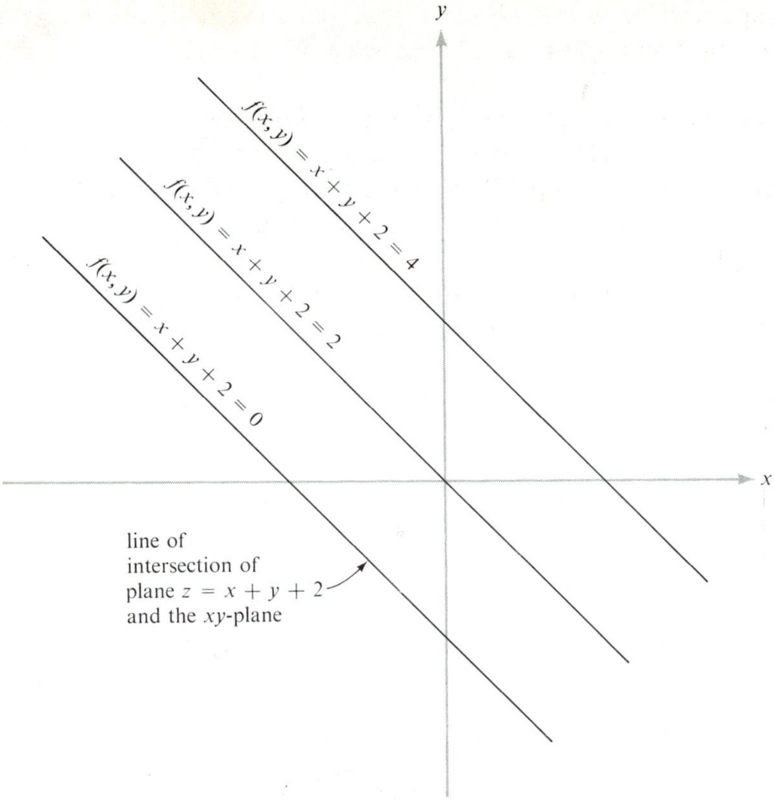

y
$f(x, y) = x + y + 2 = 4$
$f(x, y) = x + y + 2 = 2$
$f(x, y) = x + y + 2 = 0$
x
line of intersection of plane $z = x + y + 2$ and the xy-plane

FIGURE 2.1.4
The level curves of $f(x, y) = x + y + 2$ show the behavior of this function.

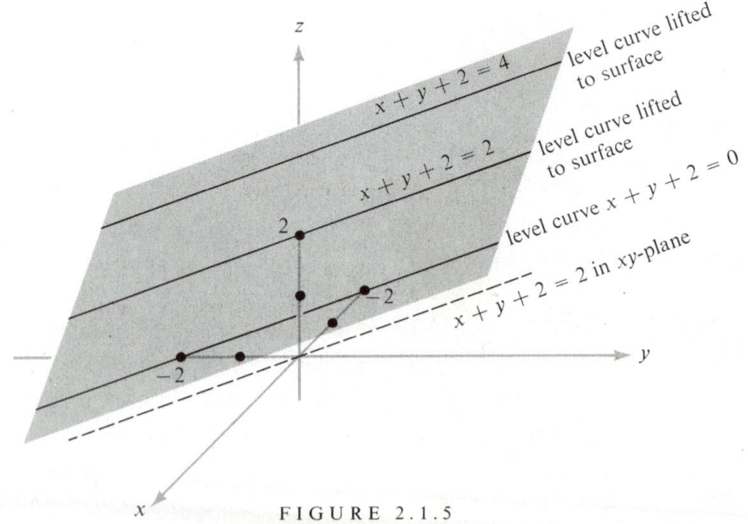

FIGURE 2.1.5
Relationship of level curves of Figure 2.1.4 to the graph of the function.

graph will be a smooth surface, so the contour model, mentally smoothed over, gives a good impression of the graph.

EXAMPLE 3. The quadratic function $f\colon \mathbb{R}^2 \to \mathbb{R}$, $(x, y) \mapsto x^2 + y^2$, has as its graph a paraboloid of revolution, oriented upwards from the origin, around the z-axis. The level curve of value c is empty for $c < 0$; for $c > 0$, if we write $c = a^2$, the level curve of value c is the set $\{(x, y) \mid x^2 + y^2 = a^2\}$, a circle of radius a centered at the origin. Thus at level c above the xy-plane the level set is a circle of radius $\sqrt{c}$, indicating a parabolic shape (see figures 2.1.6 and 2.1.7).

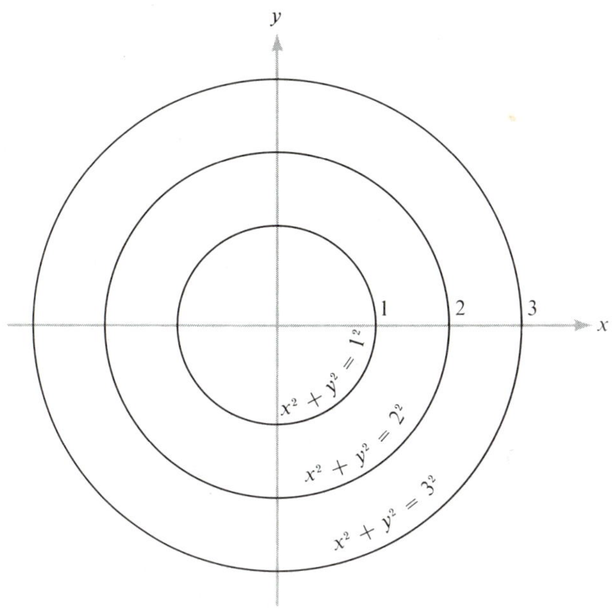

FIGURE 2.1.6
Some level curves for the function $f(x, y) = x^2 + y^2$.

The shape of a graph may also be determined by the *method of sections*. By a *section* of the graph of f we mean the intersection of the graph and a (vertical) plane. For example, if P_1 is the xz-plane in $\mathbb{R}^3$, defined by $y = 0$, then the section of the graph of f in Example 3 is the set

$$P_1 \cap \operatorname{graph} f = \{(x, y, z) \mid y = 0, z = x^2\}$$

which is a parabola in the xz-plane. Similarly, if P_2 denotes the yz-plane, defined by $x = 0$, then the section

$$P_2 \cap \operatorname{graph} f = \{(x, y, z) \mid x = 0, z = y^2\}$$

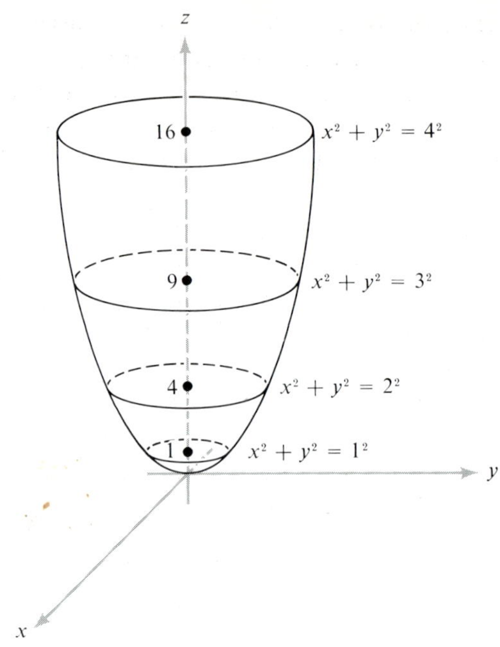

FIGURE 2.1.7
Level curves in Figure 2.1.6 raised to the graph.

is a parabola in the *yz*-plane (see Figure 2.1.8). It is usually helpful to compute at least one section to complement the information given by the level sets.

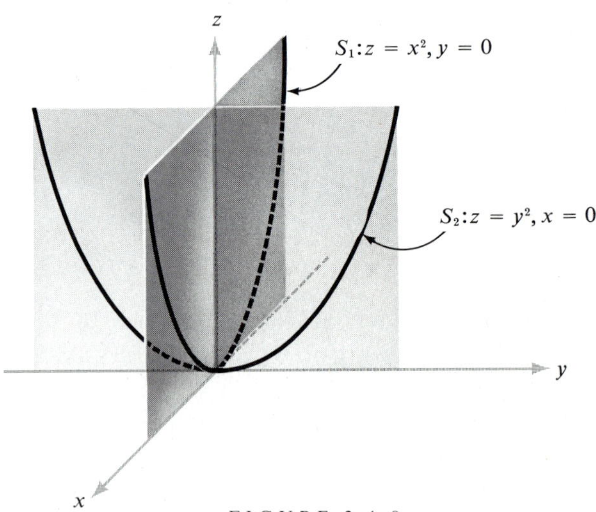

FIGURE 2.1.8
Two sections of the graph of $f(x, y) = x^2 + y^2$.

EXAMPLE 4. The quadratic function $f\colon \mathbb{R}^2 \to \mathbb{R}$, $(x, y) \mapsto x^2 - y^2$, has as its graph a surface called a *hyperbolic paraboloid*, or *saddle*, centered at the origin. In order to visualize this surface, we first draw the level curves. To determine the level curves, we must solve the equation $x^2 - y^2 = c$. Consider values of c near zero, say $c = 0, \pm 1, \pm 4$. For $c = 0$, we have $y^2 = x^2$, or $y = \pm x$, so this level set consists of two straight lines through the origin. For $c = 1$, the level curve is $x^2 - y^2 = 1$, or $y = \pm\sqrt{x^2 - 1}$, which is a hyperbola that passes vertically through the x-axis at the points $(\pm 1, 0)$ (see Figure 2.1.9). Similarly, for $c = 4$, the level curve is defined by $y = \pm\sqrt{x^2 - 4}$, the hyperbola passing vertically through the x-axis at $(\pm 2, 0)$. For $c = -1$, we obtain the curve $x^2 - y^2 = -1$, that is, $x = \pm\sqrt{y^2 - 1}$, the hyperbola passing horizontally through the y-axis at $(0, \pm 1)$. And for $c = -4$, the hyperbola through $(0, \pm 2)$ is obtained. These level curves are shown in Figure 2.1.9. Since the visualization of the graph of f is not easy from these data alone, we shall compute two sections, as in the previous example. For the section in the xz-plane, we have

$$P_1 \cap \text{graph } f = \{(x, y, z) \mid y = 0, z = x^2\}$$

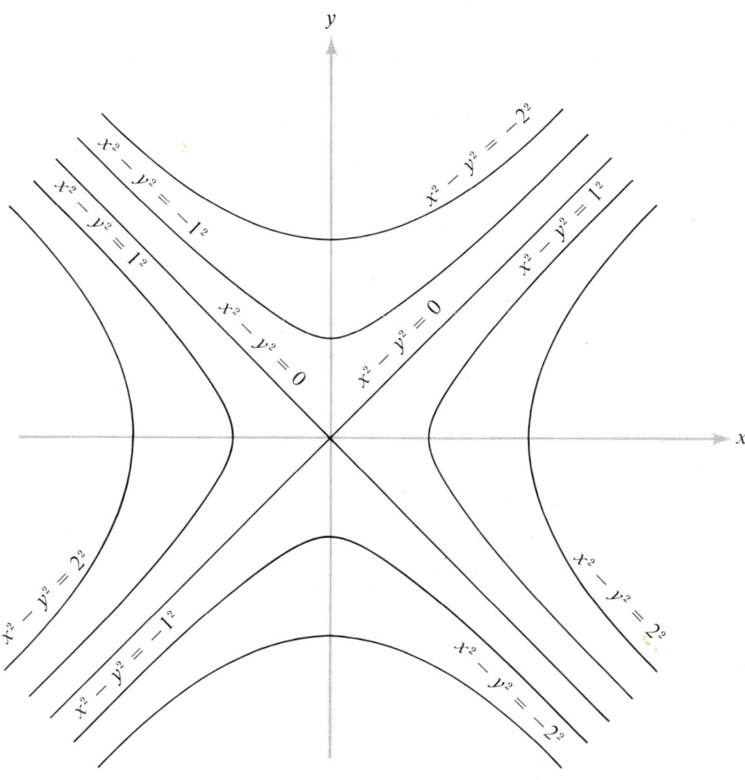

FIGURE 2.1.9
Level curves for the function $f(x, y) = x^2 - y^2$.

which is a parabola opening upward, and for the yz-plane

$$P_2 \cap \text{graph } f = \{(x, y, z) \mid x = 0, z = -y^2\}$$

which is a parabola opening downwards. The graph may now be visualized by lifting the level curves to the appropriate heights, and smoothing out the resulting surface. Their placement is aided by computing the parabolic sections. This procedure generates the hyperbolic saddle indicated in Figure 2.1.10.

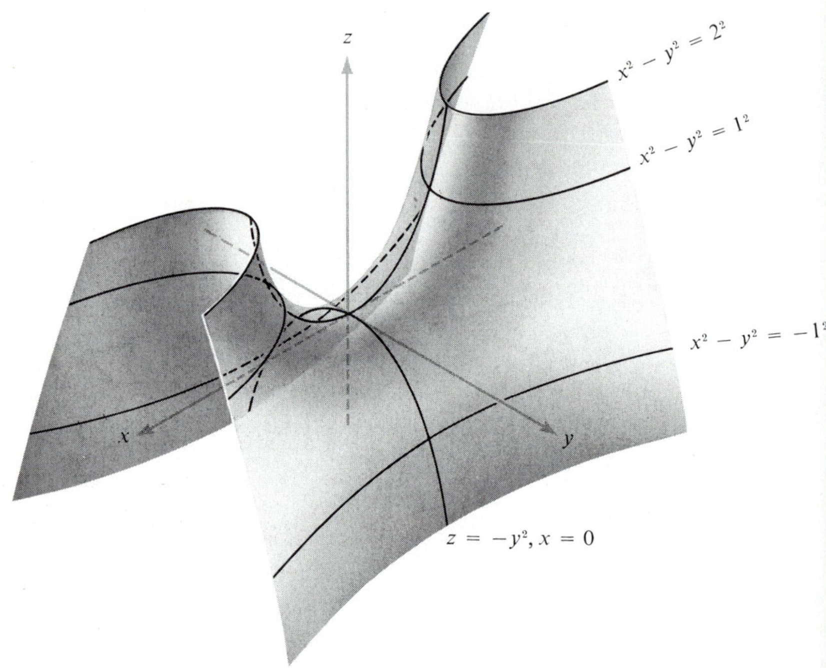

FIGURE 2.1.10
Some level curves on the graph of $f(x, y) = x^2 - y^2$.

EXAMPLE 5. Consider the function $f\colon \mathbb{R}^3 \to \mathbb{R}$, $(x, y, z) \mapsto x^2 + y^2 + z^2$. This is the three-dimensional analogue of Example 3. In this context, level sets are surfaces in the three-dimensional domain $\mathbb{R}^3$. The graph, in $\mathbb{R}^4$, cannot be visualized directly, but sections can nevertheless be computed analytically. The level set with value c is the set

$$L_c = \{(x, y, z) \mid x^2 + y^2 + z^2 = c\}$$

which is clearly the sphere of radius $\sqrt{c}$ for $c > 0$, a single point at the origin for $c = 0$, and is empty for $c < 0$. The level sets for $c = 0, 1, 4$, and 9 are indicated in Figure 2.1.11. Some additional information

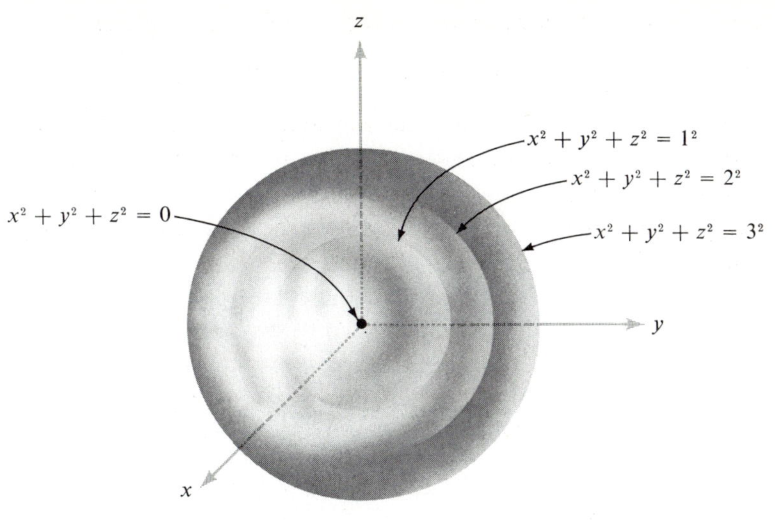

FIGURE 2.1.11
Some level surfaces for $f(x, y, z) = x^2 + y^2 + z^2$.

about the graph is given by computing a section. For example, if we write $S_{z=0} = \{(x, y, z, t) \mid z = 0\}$, then we can look at the section

$$S_{z=0} \cap \text{graph}\, f = \{(x, y, z, t) \mid t = x^2 + y^2, z = 0\}$$

Since z is held fixed at $z = 0$ here, we can visualize this section of the graph as a surface in $\mathbb{R}^3$ in the variables x, y, t (see Figure 2.1.12). The surface is a *paraboloid of revolution.*

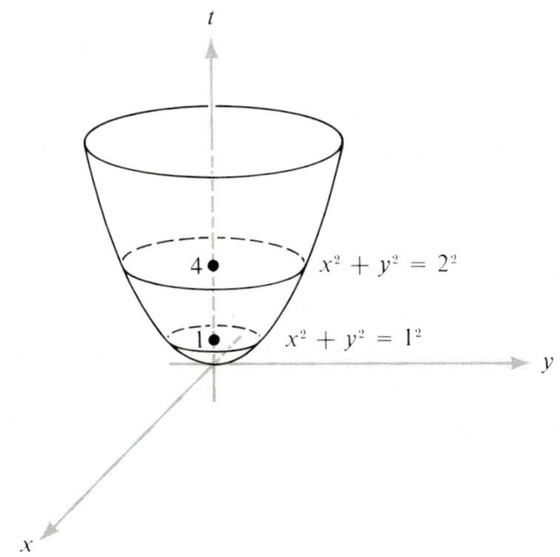

FIGURE 2.1.12
The $z = 0$ section of the graph of $f(x, y, z) = x^2 + y^2 + z^2$.

EXAMPLE 6. The function $f\colon \mathbb{R}^3 \to \mathbb{R}$, $(x, y, z) \mapsto x^2 + y^2 - z^2$, is the three-dimensional analogue of Example 4, and is also called a *saddle*. The level surfaces are defined by

$$L_c = \{(x, y, z) \mid x^2 + y^2 - z^2 = c\}$$

For $c = 0$, this is the cone $z = \pm\sqrt{x^2 + y^2}$ centered around the z-axis. For c negative, say $c = -a^2$, we obtain $z = \pm\sqrt{x^2 + y^2 + a^2}$, which is a hyperboloid of two sheets around the z-axis, passing through the z-axis at the points $(0, 0, \pm a)$. For c positive, say $c = b^2$, the level surface is the *single-sheeted hyperboloid of revolution* around the z-axis defined by $z = \pm\sqrt{x^2 + y^2 - b^2}$, which intersects the xy-plane in the circle of radius $|b|$. These level surfaces are sketched in Figure 2.1.13.

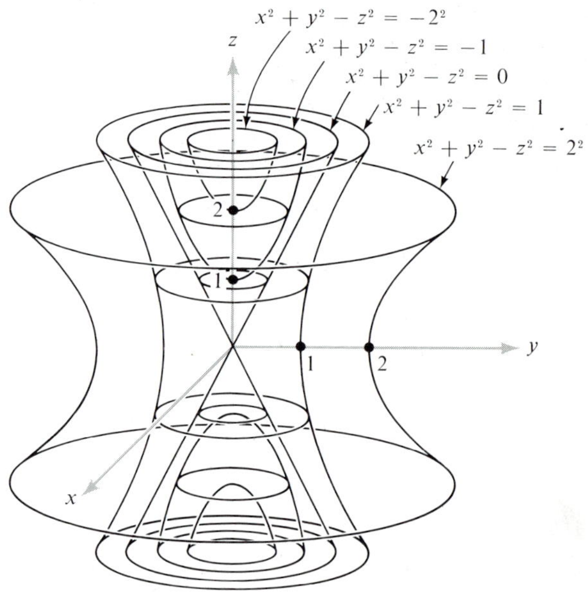

FIGURE 2.1.13
Some level surfaces of the function $f(x, y, z) = x^2 + y^2 - z^2$.

Another view of the graph may be obtained from a section. For example, the subspace $S_{y=0} = \{(x, y, z, t) \mid y = 0\}$ intersects the graph in the section

$$S_{y=0} \cap \text{graph } f = \{(x, y, z, t) \mid y = 0,\ t = x^2 - z^2\}$$

that is, the set of points of the form $(x, 0, z, x^2 - z^2)$, which may be considered, as in the previous example, as a surface in xzt-space (see Figure 2.1.14).

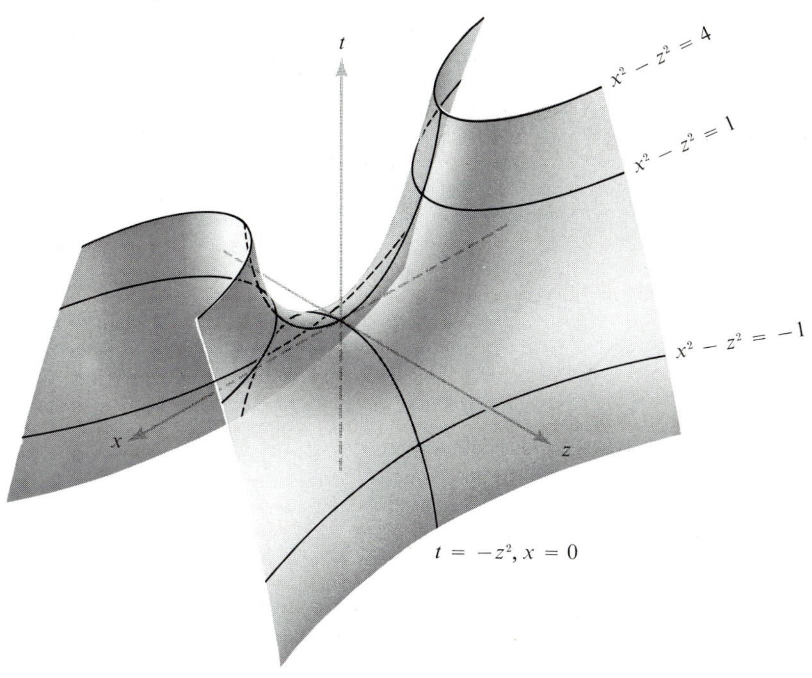

FIGURE 2.1.14
The $y = 0$ *section of the graph of* $f(x, y, z) = x^2 + y^2 - z^2$.

Many computer programs are available for plotting a given function. For functions of one variable this is just a matter of calculating selected values of the function and plotting points. For functions of two variables, the method of sections is used: to plot $f(x, y)$, the computer selects sections parallel to the axes by assigning values to, say, y and plotting the corresponding graph, then changing y and repeating. This way a whole piece of the graph can be swept out. Some examples are given in Figure 2.1.15. This particular program is also capable of giving the plot perspective so that we can see it at any desired elevation and deflection.

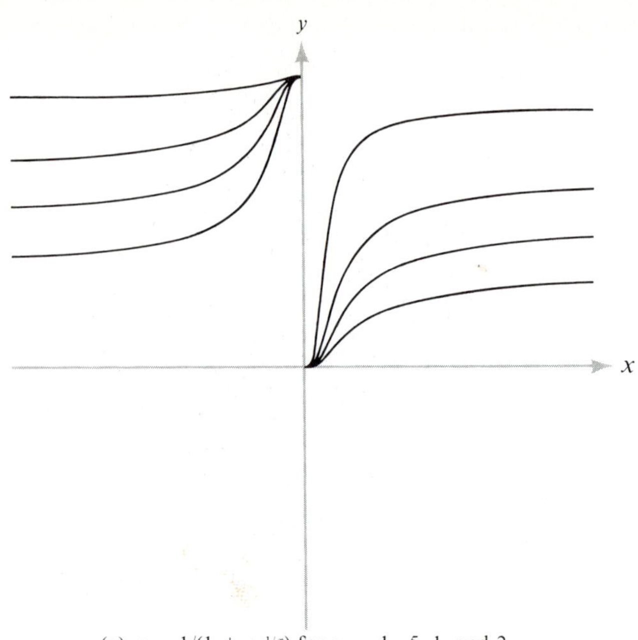

(a) $y = 1/(1 + ce^{1/x})$ for $c = .1, .5, 1$, and 2

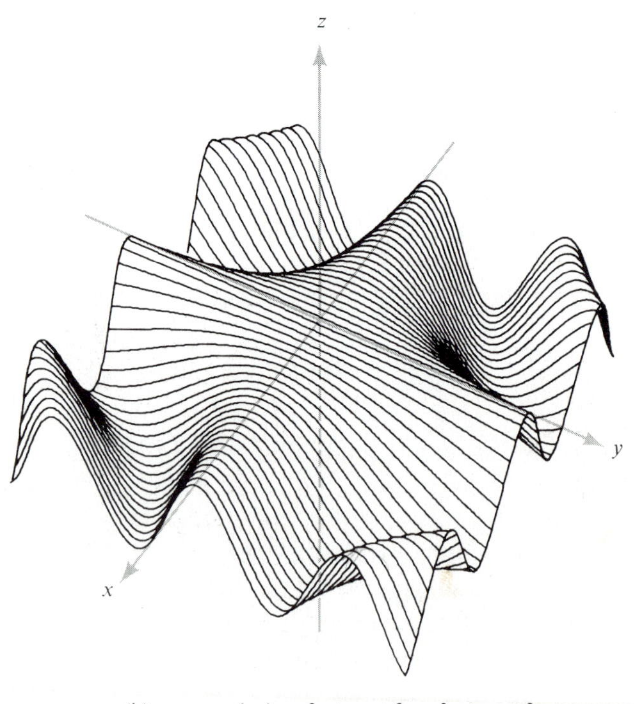

(b) $z = \cos(xy), -3 \leq x \leq 3, -3 \leq y \leq 3$

FIGURE 2.1.15

Some computer-generated graphs.

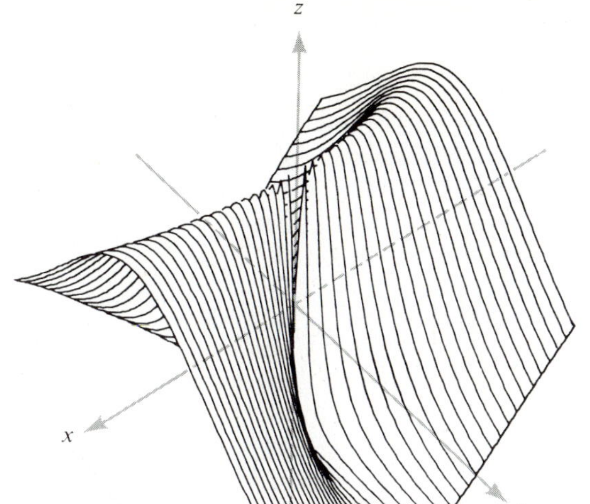

(c) $z = \dfrac{x^2 - y^2}{x^2 + y^2} \quad -1 \leq x \leq 1, -1 \leq y \leq 1$

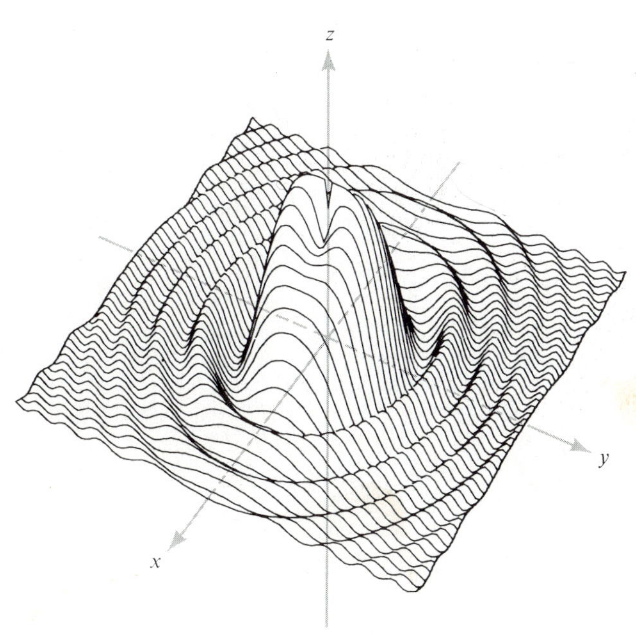

(d) $z = \dfrac{\sin(2x^2 + 3y^2)}{x^2 + y^2} \quad -3 \leq x \leq 3, -3 \leq y \leq 3$

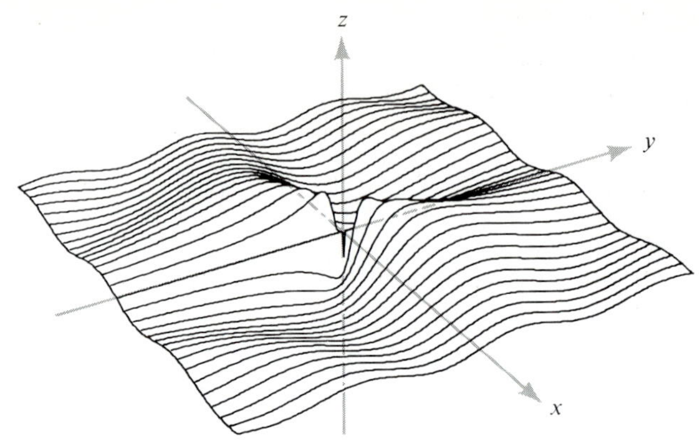

(e) $z = \dfrac{\sin xy}{x^2 + y^2} \qquad -3 \le x \le 3, -3 \le y \le 3$

EXERCISES

1. Determine the level curves and graphs of the following functions.
   (a) $f\colon \mathbb{R}^2 \to \mathbb{R}$, $(x, y) \mapsto x - y + 2$
   (b) $f\colon \mathbb{R}^2 \to \mathbb{R}$, $(x, y) \mapsto x^2 + 4y^2$
   (c) $f\colon \mathbb{R}^2 \to \mathbb{R}$, $(x, y) \mapsto -xy$
2. For Examples 2, 3, and 4, compute the section of the graph defined by the plane

$$S_\theta = \{(x, y, z) \mid y = x \tan \theta\}$$

   for a given constant θ. Determine which of these functions f have the property that the shape of $S_\theta \cap \operatorname{graph} f$ is independent of θ.

In exercises 3 to 8 draw the level curves (in the xy-plane) for the given function f and specified values of c. Sketch the graphs of $z = f(x, y)$.

3. $f(x, y) = (100 - x^2 - y^2)^{1/2}$, $c = 0, 2, 4, 6, 8, 10$
4. $f(x, y) = (x^2 + y^2)^{1/2}$, $c = 0, 1, 2, 3, 4, 5$
5. $f(x, y) = x^2 + y^2$, $c = 0, 1, 2, 3, 4, 5$
6. $f(x, y) = 3x - 7y$, $c = 0, 1, 2, 3, -1, -2, -3$
7. $f(x, y) = x^2 + xy$, $c = 0, 1, 2, 3, -1, -2, -3$
8. $f(x, y) = x/y$, $c = 0, 1, 2, 3, -1, -2, -3$

In exercises 9 to 11, determine the level surfaces and a section of the graph of each function.

9. $f\colon \mathbb{R}^3 \to \mathbb{R}$, $(x, y, z) \mapsto -x^2 - y^2 - z^2$
10. $f\colon \mathbb{R}^3 \to \mathbb{R}$, $(x, y, z) \mapsto 4x^2 + y^2 + 9z^2$
11. $f\colon \mathbb{R}^3 \to \mathbb{R}$, $(x, y, z) \mapsto x^2 + y^2$

In exercises 12 to 16, describe the graph of each function by computing some level surfaces and sections.

12. $f\colon \mathbb{R}^3 \to \mathbb{R}$, $(x, y, z) \mapsto xy$
13. $f\colon \mathbb{R}^3 \to \mathbb{R}$, $(x, y, z) \mapsto xy + yz$
14. $f\colon \mathbb{R}^3 \to \mathbb{R}$, $(x, y, z) \mapsto xy + z^2$

15. $f\colon \mathbb{R}^2 \to \mathbb{R}, (x, y) \mapsto |y|$
16. $f\colon \mathbb{R}^2 \to \mathbb{R}, (x, y) \mapsto \max(|x|, |y|)$
17. Describe the level curves of the function

$$f\colon \mathbb{R}^2 \to \mathbb{R}, (x, y) \mapsto \begin{cases} \dfrac{2xy}{x^2 + y^2} & \text{if } (x, y) \neq (0, 0) \\ 0 & \text{if } (x, y) = (0, 0) \end{cases}$$

18. Let $f\colon \mathbb{R}^2 \backslash \{\mathbf{0}\} \to \mathbb{R}$ be given in polar coordinates by $f(r, \theta) = (\cos 2\theta)/r^2$. Sketch a few level curves.

2.2 LIMITS AND CONTINUITY

In order to formulate the ideas of limit, continuity, and differentiability in a coherent manner, it will first be necessary to consider the notion of an open set. The principal characteristic of $\mathbb{R}^n$ upon which the idea of open set depends is the norm $\|\mathbf{x}\|$ of a vector $\mathbf{x}$, which was considered in Chapter 1.

We begin formulating the concept of open set by defining an open disc. Let $\mathbf{x}_0 \in \mathbb{R}^n$ and let r be a positive real number. The *open disc* of radius r and center $\mathbf{x}_0$ is defined to be the set of all points $\mathbf{x}$ such that $\|\mathbf{x} - \mathbf{x}_0\| < r$. This set is denoted $D_r(\mathbf{x}_0)$, and is the set of points $\mathbf{x}$ in $\mathbb{R}^n$ whose distance (see Section 1.4) from $\mathbf{x}_0$ is less than r. Notice that we include only those $\mathbf{x}$ for which *strict* inequality holds. The disc $D_r(\mathbf{x}_0)$ is illustrated in Figure 2.2.1 for $n = 1, 2, 3$. For the case $n = 1$, $x_0 \in \mathbb{R}$, the open disc $D_r(x_0)$ is the open interval $]x_0 - r, x_0 + r[$, which consists of all numbers $x \in \mathbb{R}$ *strictly* between $x_0 - r$ and $x_0 + r$.

We are now ready to define the concept of an open set.

Definition. *Let $U \subset \mathbb{R}^n$ (i.e., U is a subset of $\mathbb{R}^n$). We call U an* ***open set*** *when for every point $\mathbf{x}_0$ in U there exists some $r > 0$ such that $D_r(\mathbf{x}_0)$ is contained within U; in symbols, $D_r(\mathbf{x}_0) \subset U$ (see Figure 2.2.2).*

It is important to realize that the number $r > 0$ depends on the point $\mathbf{x}_0$, and generally r will shrink down as $\mathbf{x}_0$ gets closer to the "edge" of U. Intuitively speaking, a set is open when the "boundary" points of U do not lie in U. In Figure 2.2.2, the dashed line is *not* included in U.

We shall also establish the convention that the empty set $\varnothing$ (the set consisting of no elements) is open.

We have defined open disc and open set. From our choice of terms it would seem that an open disc should also be an open set. A little thought shows that this fact requires some proof.

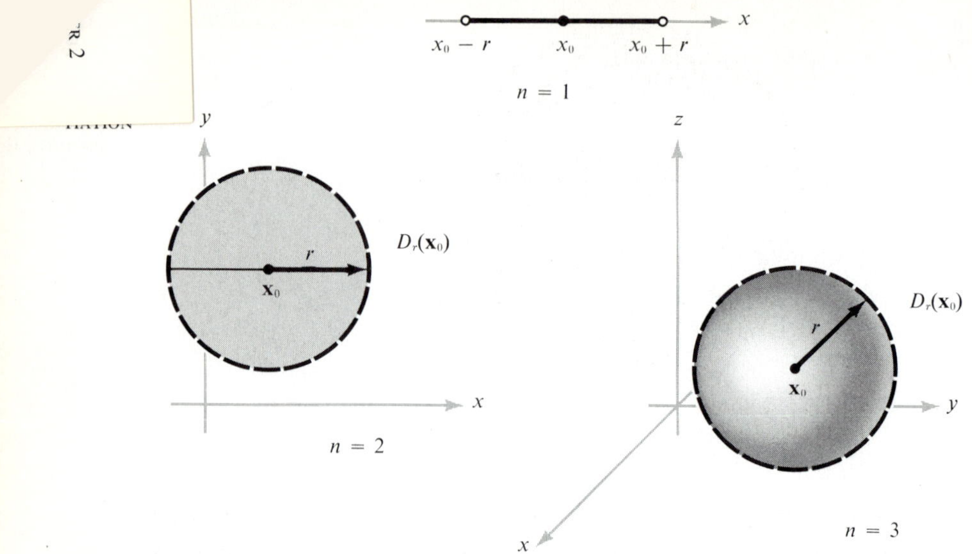

FIGURE 2.2.1
What discs $D_r(\mathbf{x}_0)$ look like in 1, 2, and 3 dimensions.

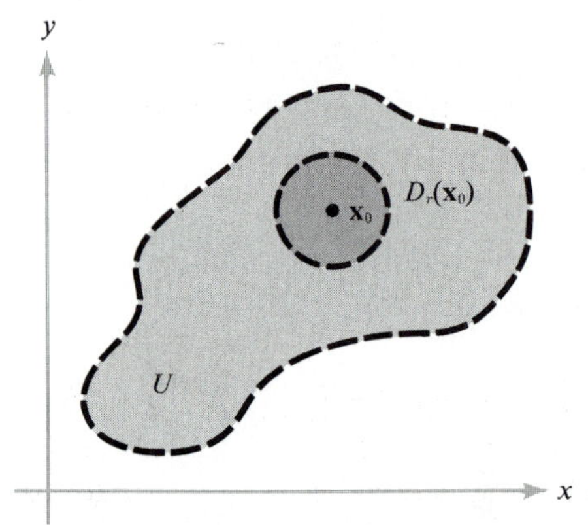

FIGURE 2.2.2
An open set U is one that completely encloses some disc $D_r(\mathbf{x}_0)$ about each of its points $\mathbf{x}_0$.

Theorem 1. *For each $\mathbf{x}_0 \in \mathbb{R}^n$ and $r > 0$, $D_r(\mathbf{x}_0)$ is an open set.*

Proof. Let $\mathbf{x} \in D_r(\mathbf{x}_0)$, that is, $\|\mathbf{x} - \mathbf{x}_0\| < r$. According to the definition of open set we must find an $s > 0$ such that $D_s(\mathbf{x}) \subset D_r(\mathbf{x}_0)$. Referring to Figure 2.2.3 we see that $s = r - \|\mathbf{x} - \mathbf{x}_0\|$ is a reasonable choice; note that $s > 0$ but that s becomes smaller if $\mathbf{x}$ is nearer the edge of $D_r(\mathbf{x}_0)$.

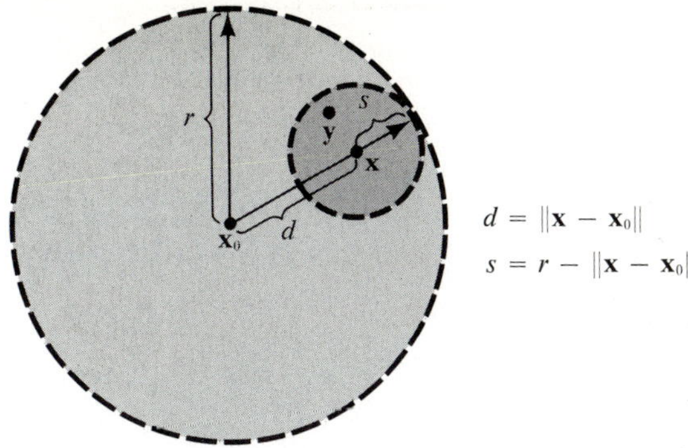

FIGURE 2.2.3
The geometry of the proof that an open disc is an open set.

To prove $D_s(\mathbf{x}) \subset D_r(\mathbf{x}_0)$, let $\mathbf{y} \in D_s(\mathbf{x})$; that is, $\|\mathbf{y} - \mathbf{x}\| < s$. We want to prove that $\mathbf{y} \in D_r(\mathbf{x}_0)$ as well. Proving this by definition of an r-disc entails showing that $\|\mathbf{y} - \mathbf{x}_0\| < r$. This is done by using the triangle inequality for vectors in $\mathbb{R}^n$ (see the corollary to Theorem 3, Section 1.4):

$$\begin{aligned}\|\mathbf{y} - \mathbf{x}_0\| &= \|(\mathbf{y} - \mathbf{x}) + (\mathbf{x} - \mathbf{x}_0)\| \\ &\leq \|\mathbf{y} - \mathbf{x}\| + \|\mathbf{x} - \mathbf{x}_0\| < s + \|\mathbf{x} - \mathbf{x}_0\| = r\end{aligned}$$

Hence $\|\mathbf{y} - \mathbf{x}_0\| < r$. ▮

The following example illustrates some techniques that are useful in establishing the openness of sets.

EXAMPLE 1. Prove that $A = \{(x, y) \in \mathbb{R}^2 \mid x > 0\}$ is an open set.

The set is pictured in Figure 2.2.4. Intuitively, this set is open since no points on the "boundary," $x = 0$, are contained in the set. Such an argument will often suffice after one has gotten used to the ideas. At first, however, we should give all the details. In order to prove A is open, we must show that for every point $(x, y) \in A$ there exists an $r > 0$ such that $D_r(x, y) \subset A$. If $(x, y) \in A$, then $x > 0$. Choose $r = x$. If $(x_1, y_1) \in D_r(x, y)$ we have

$$|x_1 - x| = \sqrt{(x_1 - x)^2} \leq \sqrt{(x_1 - x)^2 + (y_1 - y)^2} < r = x$$

and so $x_1 - x < x$ and $x - x_1 < x$. The latter inequality implies $x_1 > 0$. Hence $D_r(x, y) \subset A$, and A is open (see Figure 2.2.5).

It is useful to have a special name for an open set containing a given point $\mathbf{x}$, since this idea arises very often in the study of limits and

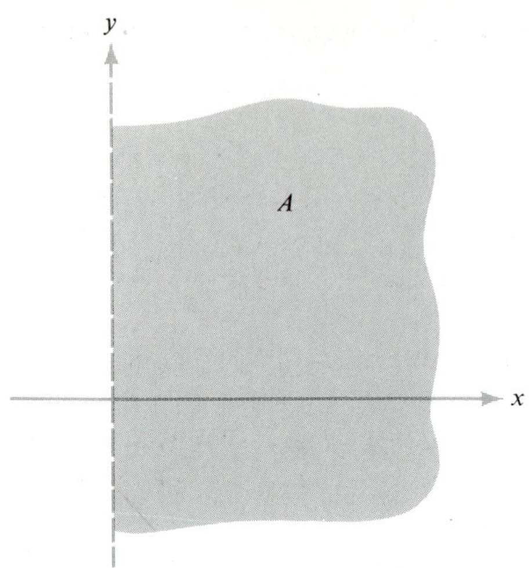

FIGURE 2.2.4
Problem: show that A is an open set.

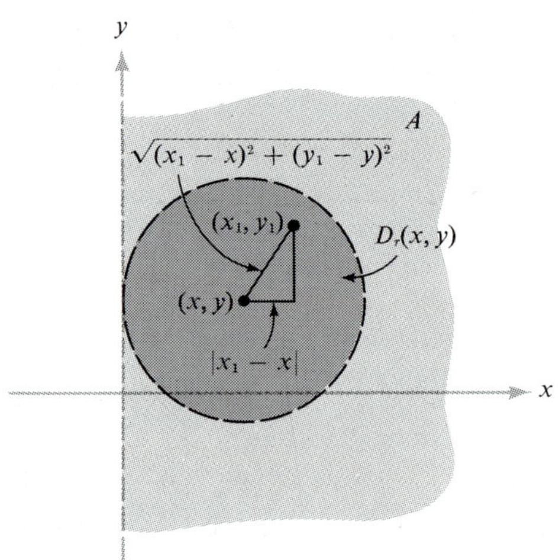

FIGURE 2.2.5
The construction of a disc about a point in A that is completely enclosed in A.

continuity. Thus, by a *neighborhood* of $\mathbf{x} \in \mathbb{R}^n$ we merely mean an open set U containing the point $\mathbf{x}$. For example $D_r(\mathbf{x}_0)$ is a neighborhood of $\mathbf{x}_0$ for any $r > 0$. The set A in Example 1 is a neighborhood of $(3, -10)$.

Now let us formally introduce the concept of a boundary point, which we have alluded to in Example 1.

Definition. *Let $A \subset \mathbb{R}^n$. A point $\mathbf{x} \in \mathbb{R}^n$ is called a **boundary point** of A* *if every neighborhood of* $\mathbf{x}$ *contains at least one point in A and at least one point not in A.*

In this definition $\mathbf{x}$ may or may not be in A; if $\mathbf{x} \in A$ then $\mathbf{x}$ is a boundary point if every neighborhood contains at least one point *not* in A (it already contains a point of A, namely $\mathbf{x}$). Similarly, if $\mathbf{x}$ is not in A it is a boundary point if every neighborhood of $\mathbf{x}$ contains at least one point of A.

We shall be particularly interested in boundary points of open sets. By the definition of open set, no point of an open set A can be a boundary point of A. Thus *a point* $\mathbf{x}$ *is a boundary point of an open set A if and only if* $\mathbf{x}$ *is not in A and every neighborhood of* $\mathbf{x}$ *has non-empty intersection with A.*

This expresses in precise terms the intuitive idea that a boundary point of A is a point just on the "edge" of A. In most examples it is perfectly clear what the boundary points are.

EXAMPLE 2(a). Let $A =]a, b[$ in $\mathbb{R}$. Then the boundary points of A consist of the points a and b. A consideration of Figure 2.2.6 and the

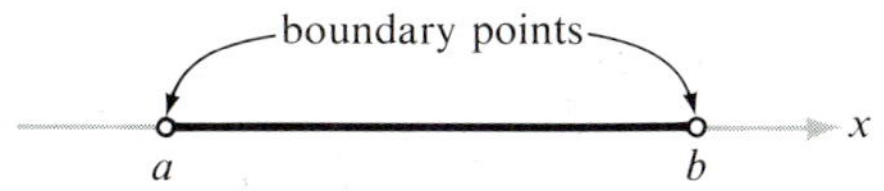

FIGURE 2.2.6
The boundary points of the interval $]a, b[$.

definition will make this clear. (The reader will be asked to prove this as Exercise 4.)

(b) Let $A = D_r(x_0, y_0)$ be an r-disc about (x_0, y_0) in the plane. The boundary consists of points (x, y) with $(x - x_0)^2 + (y - y_0)^2 = r^2$ (Figure 2.2.7).

(c) Let $A = \{(x, y) \in \mathbb{R}^2 \mid x > 0\}$. Then the boundary of A consists of all points on the y-axis (see Figure 2.2.8).

(d) Let A be $D_r(\mathbf{x}_0)$ minus the point $\mathbf{x}_0$ (a "punctured" disc about $\mathbf{x}_0$). Then $\mathbf{x}_0$ is a boundary point of A.

We shall now turn our attention to the concept of limit. Throughout the following discussions the domain of definition of the function f will be an open set A. We shall be interested in finding the limit of f as $\mathbf{x} \in A$ approaches either a point of A or a boundary point of A.

The reader should appreciate the fact that the limit concept is a basic and useful tool for the analysis of functions; it enables us to study derivatives, and hence maxima and minima, asymptotes, improper

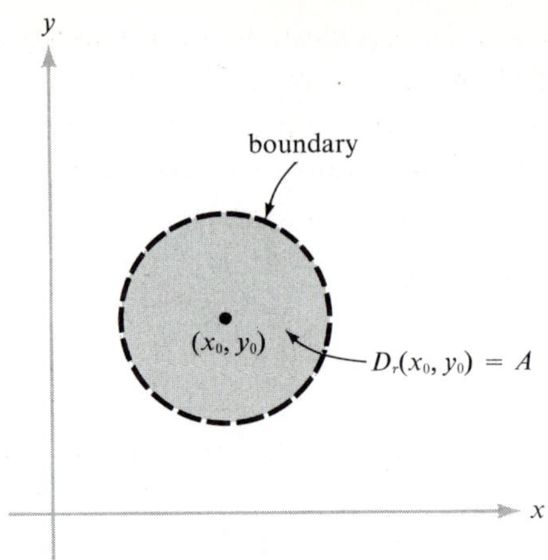

FIGURE 2.2.7
The boundary of A consists of points on the edge of A.

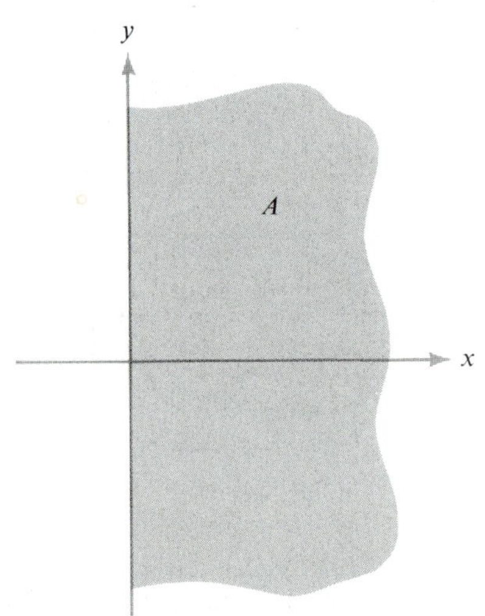

FIGURE 2.2.8
The boundary of A consists of all points on the y-axis.

integrals, and other important features of functions, as well as being useful for infinite series and sequences. We wish to present, briefly, a theory of limits for functions of several variables that includes the theory for functions of one variable as a special case.

In one-variable calculus the student has undoubtedly encountered the definition of $\lim_{x \to x_0} f(x) = l$, for $f: A \subset \mathbb{R} \to \mathbb{R}$ a function from a subset A of the real numbers to the real numbers. Intuitively, this means that as x gets closer and closer to x_0, the values $f(x)$ get closer and closer to l. To put this intuitive idea on a firm mathematical foundation, the notions of epsilon (ε) and delta (δ) or the concept of neighborhood are usually introduced—notions that are essential to a precise and workable definition of $\underset{x \to x_0}{\text{limit}}\, f(x)$. The same is true for functions of several variables.

Definition of Limit. *Let $f\colon A \subset \mathbb{R}^n \to \mathbb{R}^m$, where A is an open set. Let $\mathbf{x}_0$ be in A or be a boundary point of A, and let N be a neighborhood of $\mathbf{b} \in \mathbb{R}^m$. We say f is **eventually in** N **as** $\mathbf{x}$ **approaches** $\mathbf{x}_0$ if there exists a neighborhood U of $\mathbf{x}_0$ such that $\mathbf{x} \neq \mathbf{x}_0$, $\mathbf{x} \in U$, and $\mathbf{x} \in A$ imply $f(\mathbf{x}) \in N$. (The geometrical meaning of this assertion is illustrated in Figure 2.2.9 in terms of the graph* of a function of one variable where*

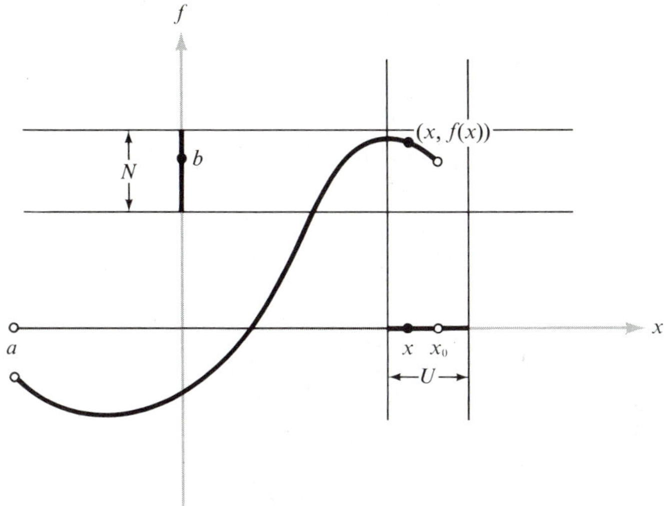

FIGURE 2.2.9
Limits in terms of neighborhoods; if x is in U, then $f(x)$ will be in N.

*$A =]a, x_0[$; note that $\mathbf{x}_0$ need not be in the set A, so that $f(\mathbf{x}_0)$ is not necessarily defined.) We say $f(\mathbf{x})$ **approaches** $\mathbf{b}$ as $\mathbf{x}$ approaches $\mathbf{x}_0$, or in symbols*

$$\underset{\mathbf{x} \to \mathbf{x}_0}{\text{limit}}\, f(\mathbf{x}) = \mathbf{b} \quad \text{or} \quad f(\mathbf{x}) \to \mathbf{b} \quad \textit{as} \quad \mathbf{x} \to \mathbf{x}_0$$

*when, given **any** neighborhood N of $\mathbf{b}$, f is eventually in N as $\mathbf{x}$ approaches $\mathbf{x}_0$ (that is, $f(\mathbf{x})$ is close to $\mathbf{b}$ if $\mathbf{x}$ is close to $\mathbf{x}_0$). It may be that as $\mathbf{x}$*

* The reader should attempt to draw similar pictures for functions of two variables; see Section 2.1 and below.

approaches $\mathbf{x}_0$ *the values* $f(\mathbf{x})$ *do not get close to any particular number. In this case we say that* $\operatorname{limit}_{\mathbf{x}\to\mathbf{x}_0} f(\mathbf{x})$ *does not exist.*

Henceforth, whenever we consider the notion $\operatorname{limit}_{\mathbf{x}\to\mathbf{x}_0} f(\mathbf{x})$ we shall always assume that $\mathbf{x}_0$ either belongs to some open set on which f is defined or is on the boundary of such a set. Let us review two simple limits from one-variable calculus.

EXAMPLE 3(a). Consider the function $f\colon \mathbb{R}\to\mathbb{R}$ defined by

$$f(x) = \begin{cases} 1 & \text{if } x > 0 \\ -1 & \text{if } x \le 0 \end{cases}$$

Then $\operatorname{limit}_{x\to 0} f(x)$ does not exist since there are points x_1 arbitrarily close to 0 with $f(x_1) = 1$ and also points x_2 arbitrarily close to 0 with $f(x_2) = -1$; that is, there is no single number that f is close to when x is close to 0 (see Figure 2.2.10). If f is restricted to the domain $]0, 1[$ or $]-1, 0[$ then the limit does exist. Can you see why?

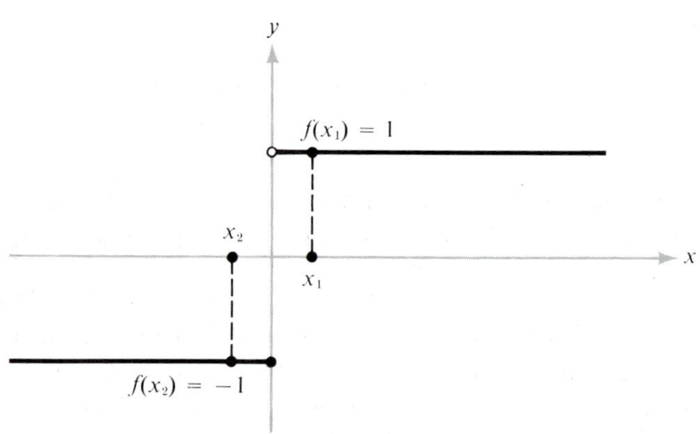

FIGURE 2.2.10
The limit of this function as $x \to 0$ *does not exist.*

One reason we insist on $\mathbf{x} \ne \mathbf{x}_0$ in the definition of limit will become clear if we remember from one-variable calculus that we want to be able to define

$$f'(x_0) = \operatorname{limit}_{x\to x_0} \frac{f(x) - f(x_0)}{x - x_0}$$

and this expression is not defined at $x = x_0$.

(b) Define $f\colon \mathbb{R} \to \mathbb{R}$ by

$$f(x) = \begin{cases} 0 & \text{if} \quad x \neq 0 \\ 1 & \text{if} \quad x = 0 \end{cases}$$

It is true that $\lim_{x \to 0} f(x) = 0$, since for any neighborhood U of 0, $x \in U$ and $x \neq 0$ implies that $f(x) = 0$. If we did not insist on $x \neq x_0$, then the limit (assuming we use the above definition of limit without the condition $\mathbf{x} \neq \mathbf{x}_0$) would not exist. Thus we are really interested in what value f approaches as $x \to 0$; as one sees from the graph in Figure 2.2.11, f approaches 0 as $x \to 0$, and we do not care that f happens to take on some other value at 0, or is not defined there.

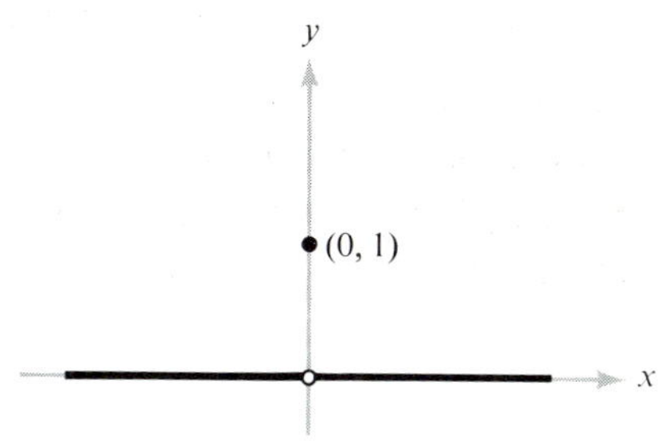

FIGURE 2.2.11
The limit of this function as $x \to 0$ is zero.

EXAMPLE 4. Let us use the definition to verify that the "obvious limit" $\lim_{\mathbf{x} \to \mathbf{x}_0} \mathbf{x} = \mathbf{x}_0$ holds, where $\mathbf{x}$ and $\mathbf{x}_0 \in \mathbb{R}^n$. Let f be the function $f\colon \mathbf{x} \mapsto \mathbf{x}$ and let N be any neighborhood of $\mathbf{x}_0$. We must show that $f(\mathbf{x})$ is eventually in N as $\mathbf{x} \to \mathbf{x}_0$. Thus, by the definition we must find a neighborhood U of $\mathbf{x}_0$ with the property that if $\mathbf{x} \neq \mathbf{x}_0$ and $\mathbf{x} \in U$ then $f(\mathbf{x}) \in N$. Pick $U = N$. If $\mathbf{x} \in U$, then $\mathbf{x} \in N$; and since $\mathbf{x} = f(\mathbf{x})$ it follows immediately that $f(\mathbf{x}) \in N$. Thus we have shown that $\lim_{\mathbf{x} \to \mathbf{x}_0} \mathbf{x} = \mathbf{x}_0$.

EXAMPLE 5. Find $\lim_{x \to 1} g(x)$ where

$$g\colon x \mapsto \frac{x - 1}{\sqrt{x} - 1}$$

This function is graphed in Figure 2.2.12. We see that $g(1)$ is not defined, since division by zero is not defined. However, if we multiply

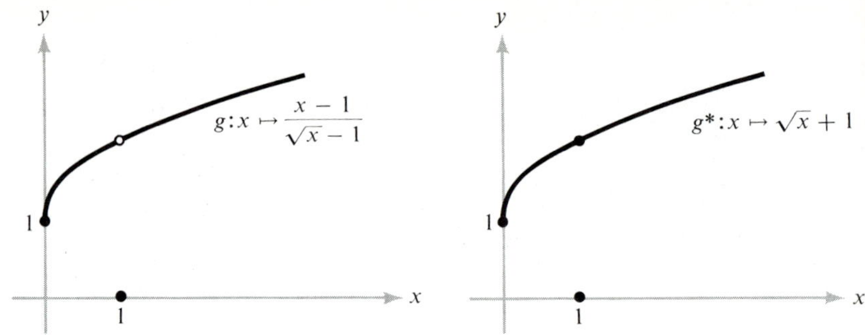

FIGURE 2.2.12
These graphs are the same except that g is undefined at x = 1.

the numerator and denominator of $g(x)$ by $\sqrt{x}+1$, we find that for all x in the domain of g we have

$$g(x) = \frac{x-1}{\sqrt{x}-1} = \sqrt{x}+1$$

The expression $g^*(x) = \sqrt{x}+1$ is defined and takes the value 2 at $x = 1$; we can easily see that $g^*(x) \to 2$ as $x \to 1$. But since $g^*(x) = g(x)$ for all $x \geq 0$, $x \neq 1$, we must have as well that $g(x) \to 2$ as $x \to 1$.

We shall now state a theorem giving a formulation of the notion of limit in terms of epsilons and deltas. This new formulation is quite useful; it is often taken as the definition of limit. It is another way of making precise the intuitive statement that "$f(\mathbf{x})$ is close to $\mathbf{b}$ when $\mathbf{x}$ is close to $\mathbf{x}_0$."

Theorem 2. *Let $f: A \subset \mathbb{R}^n \to \mathbb{R}^m$ and let $\mathbf{x}_0$ be in A or be a boundary point of A. Then* $\lim_{\mathbf{x}\to\mathbf{x}_0} f(\mathbf{x}) = \mathbf{b}$ *if and only if for every number $\varepsilon > 0$ there is a $\delta > 0$ such that for $\mathbf{x} \in A$ and $0 < \|\mathbf{x} - \mathbf{x}_0\| < \delta$ we have $\|f(\mathbf{x}) - \mathbf{b}\| < \varepsilon$ (see Figure 2.2.13).*

This and other limit theorems stated below will be proved in Section 2.7. Their proofs are deferred since it is not important at this stage to master their technical details.

In order that we can properly speak of *the* limit, we should establish that f can have at most one limit as $\mathbf{x} \to \mathbf{x}_0$. This is intuitively clear and we now state it formally.

Theorem 3 (Uniqueness of limits). *If* $\lim_{\mathbf{x}\to\mathbf{x}_0} f(\mathbf{x}) = \mathbf{b}_1$ *and* $\lim_{\mathbf{x}\to\mathbf{x}_0} f(\mathbf{x}) = \mathbf{b}_2$, *then* $\mathbf{b}_1 = \mathbf{b}_2$.

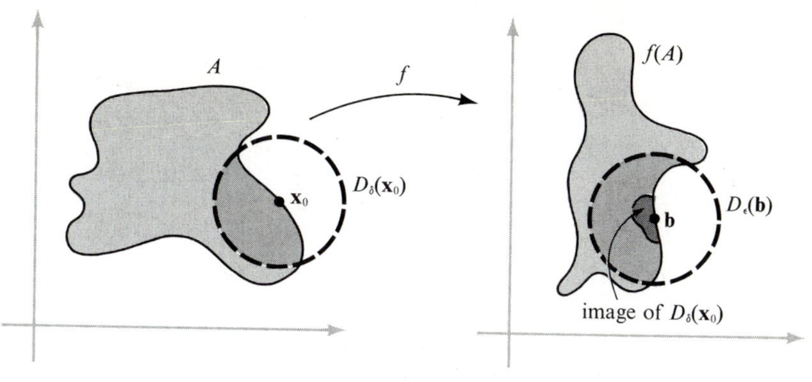

FIGURE 2.2.13
The geometry of the ε-δ definition of limit.

To illustrate the epsilon-delta technique, we consider the following examples.

EXAMPLE 6. Show that $\lim_{(x, y) \to (0, 0)} x = 0$.

Note that if $\delta > 0$, $\|(x, y) - (0, 0)\| = \sqrt{x^2 + y^2} < \delta$ implies $|x - 0| = |x| = \sqrt{x^2} \le \sqrt{x^2 + y^2} < \delta$. Thus if $\|(x, y) - (0, 0)\| < \delta$, then $|x - 0|$ is also less than δ. Given $\varepsilon > 0$ we must find a $\delta > 0$ (generally depending on ε) with the property that $0 < \|(x, y) - (0, 0)\| < \delta$ implies $|x - 0| < \varepsilon$. What are we to pick as our δ? From the above calculation, we see that if we choose $\delta = \varepsilon$, then $\|(x, y) - (0, 0)\| < \delta$ implies $|x - 0| < \varepsilon$. This shows that $\lim_{(x, y) \to (0, 0)} x = 0$. We could have also chosen $\delta = \varepsilon/2$ or $\varepsilon/3$, but it suffices to find just one δ satisfying the requirements of the definition of limit.

More generally, we can see in the same way that

$$\lim_{(x, y) \to (x_0, y_0)} x = x_0$$

EXAMPLE 7. Consider the function

$$f(x, y) = \frac{\sin(x^2 + y^2)}{x^2 + y^2}$$

Even though f is not defined at $(0, 0)$ we can ask whether $f(x, y)$ approaches some number as (x, y) approaches $(0, 0)$. From elementary calculus we know that

$$\lim_{\alpha \to 0} \frac{\sin \alpha}{\alpha} = 1$$

Thus it is reasonable to guess that

$$\lim_{\mathbf{v}\to(0,0)} f(\mathbf{v}) = \lim_{\mathbf{v}\to(0,0)} \frac{\sin\|\mathbf{v}\|^2}{\|\mathbf{v}\|^2} = 1$$

Indeed, since $\lim_{\alpha\to 0} (\sin \alpha)/\alpha = 1$, given $\varepsilon > 0$ we can find a $\delta > 0$, with $1 > \delta > 0$, such that $0 < |\alpha| < \delta$ implies that $|(\sin \alpha)/\alpha - 1| < \varepsilon$. If $\|\mathbf{v}\| < \delta$, then $\|\mathbf{v}\|^2 < \delta^2 < \delta$, and

$$|f(\mathbf{v}) - 1| = \left| \frac{\sin\|\mathbf{v}\|^2}{\|\mathbf{v}\|^2} - 1 \right| < \varepsilon$$

Thus $\lim_{\mathbf{v}\to(0,0)} f(\mathbf{v}) = 1$. Indeed, if we plot $(\sin(x^2 + y^2))/(x^2 + y^2)$ on a computer, we get a graph that is well-behaved near (0, 0) (Figure 2.2.14).

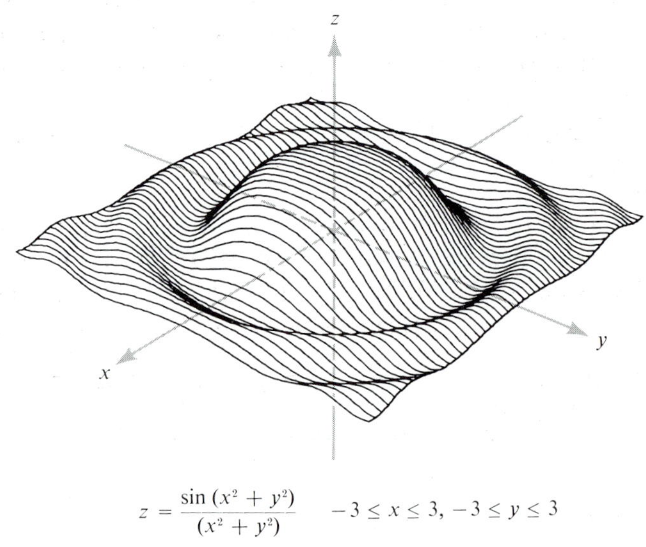

$$z = \frac{\sin(x^2 + y^2)}{(x^2 + y^2)} \qquad -3 \le x \le 3, -3 \le y \le 3$$

FIGURE 2.2.14
Computer-generated graph.

In order to carry out practical computations with limits we need to know some rules that limits obey. For example, we want to prove that the limit of a sum is the sum of the limits. These rules are all intuitively plausible and are summarized in the following theorem (see Section 2.7 for the proofs).

Theorem 4. Let $f: A \subset \mathbb{R}^n \to \mathbb{R}^m$, $g: A \subset \mathbb{R}^n \to \mathbb{R}^m$, $\mathbf{x}_0$ *be in* A *or be a boundary point of* A, $\mathbf{b} \in \mathbb{R}^m$, *and* $c \in \mathbb{R}$; *then*

(*i*) *if* $\lim_{\mathbf{x}\to\mathbf{x}_0} f(\mathbf{x}) = \mathbf{b}$, *then* $\lim_{\mathbf{x}\to\mathbf{x}_0} cf(\mathbf{x}) = c\mathbf{b}$, *where* $cf: A \to \mathbb{R}^m$ *is defined by* $\mathbf{x} \mapsto c(f(\mathbf{x}))$.

(*ii*) *if* $\lim_{\mathbf{x}\to\mathbf{x}_0} f(\mathbf{x}) = \mathbf{b}_1$ *and* $\lim_{\mathbf{x}\to\mathbf{x}_0} g(\mathbf{x}) = \mathbf{b}_2$, *then* $\lim_{\mathbf{x}\to\mathbf{x}_0} (f+g)(\mathbf{x}) = \mathbf{b}_1 + \mathbf{b}_2$, *where* $(f+g)\colon A \to \mathbb{R}^m$, $\mathbf{x} \mapsto f(\mathbf{x}) + g(\mathbf{x})$.

(*iii*) *if* $m = 1$, $\lim_{\mathbf{x}\to\mathbf{x}_0} f(\mathbf{x}) = b_1$, *and* $\lim_{\mathbf{x}\to\mathbf{x}_0} g(\mathbf{x}) = b_2$, *then* $\lim_{\mathbf{x}\to\mathbf{x}_0} (fg)(\mathbf{x}) = b_1 b_2$ *where* $(fg)\colon A \to \mathbb{R}$, $\mathbf{x} \mapsto f(\mathbf{x})g(\mathbf{x})$.

(*iv*) *if* $m = 1$, $\lim_{\mathbf{x}\to\mathbf{x}_0} f(\mathbf{x}) = b \neq 0$, *and* $f(\mathbf{x}) \neq 0$ *for all* $\mathbf{x} \in A$, *then* $\lim_{\mathbf{x}\to\mathbf{x}_0} 1/f = 1/b$ *where* $1/f\colon A \to \mathbb{R}$, $\mathbf{x} \mapsto 1/f(\mathbf{x})$.

(*v*) *if* $f(\mathbf{x}) = (f_1(\mathbf{x}), \ldots, f_m(\mathbf{x}))$ *where* $f_i\colon A \to \mathbb{R}$, $i = 1, \ldots, m$, *are the component functions of* f, *then* $\lim_{\mathbf{x}\to\mathbf{x}_0} f(\mathbf{x}) = \mathbf{b} = (b_1, \ldots, b_m)$ *if and only if* $\lim_{\mathbf{x}\to\mathbf{x}_0} f_i(\mathbf{x}) = b_i$ *for each* $i = 1, \ldots, m$.

These results ought to be intuitively clear. For instance (*ii*) says nothing more than if $f(\mathbf{x})$ is close to $\mathbf{b}_1$ and $g(\mathbf{x})$ is close to $\mathbf{b}_2$ when $\mathbf{x}$ is close to $\mathbf{x}_0$, then $f(\mathbf{x}) + g(\mathbf{x})$ is close to $\mathbf{b}_1 + \mathbf{b}_2$ when $\mathbf{x}$ is close to $\mathbf{x}_0$.

EXAMPLE 8. Let $f\colon \mathbb{R}^2 \to \mathbb{R}$, $(x, y) \mapsto x^2 + y^2 + 2$. Compute $\lim_{(x,y)\to(0,1)} f(x, y)$.

Here f is the sum of the three functions $(x, y) \mapsto x^2$, $(x, y) \mapsto y^2$, and $(x, y) \mapsto 2$. Now the limit of a sum is the sum of the limits, and the limit of a product is the product of the limits (Theorem 4). Hence, using the fact that $\lim_{(x,y)\to(x_0,y_0)} x = x_0$ (Example 6), we obtain: $\lim_{(x,y)\to(x_0,y_0)} x^2 = \left(\lim_{(x,y)\to(x_0,y_0)} x\right)\left(\lim_{(x,y)\to(x_0,y_0)} x\right) = x_0^2$ and, using the same reasoning $\lim_{(x,y)\to(x_0,y_0)} y^2 = y_0^2$. Consequently $\lim_{(x,y)\to(0,1)} f(x, y) = 0^2 + 1^2 + 2 = 3$.

EXAMPLE 9 (Optional). Evaluate

$$\lim_{(x,y)\to(0,0)} \frac{e^{xy} - 1}{x}$$

We cannot directly use the limit theorems to work this example because the limit of the denominator is zero. However, we may use the Mean Value Theorem from calculus. Let us recall its statement:

Mean Value Theorem. *Let $f(x)$ be continuous on $[a, b]$ and be differentiable in the open interval $]a, b[$. Then there is at least one number c between a and b such that*

$$f(b) - f(a) = f'(c)(b - a)$$

Thus, keeping x fixed and applying the Mean Value Theorem to e^{xy} as a function of y, we get

$$e^{xy} - 1 = xe^{x\alpha}y$$

for some α between 0 and y. Hence

$$\frac{e^{xy} - 1}{x} = ye^{x\alpha}$$

Given $\varepsilon > 0$, choose $\delta =$ minimum of 1 and ε/e. Then $0 < \|(x, y) - (0, 0)\| < \delta$ implies $\sqrt{x^2 + y^2} < 1$, so $|x| < 1$ and $|y| < 1$, and $e^{x\alpha} < e$. Also $\sqrt{x^2 + y^2} < \delta$ implies $|y| < \varepsilon/e$, so $|ye^{x\alpha}| < (\varepsilon/e)e = \varepsilon$. Hence, given $\varepsilon > 0$ we have found a $\delta > 0$ such that $0 < \|(x, y)\| < \delta$ implies $|ye^{x\alpha}| = |(e^{xy} - 1)/x| < \varepsilon$. Hence

$$\underset{(x, y) \to (0, 0)}{\text{limit}} \frac{e^{xy} - 1}{x} = 0$$

EXAMPLE 10. Does $\underset{(x, y) \to (0, 0)}{\text{limit}} x^2/(x^2 + y^2)$ exist?

If the limit exists, $x^2/(x^2 + y^2)$ should approach a definite value, say a, as (x, y) gets near $(0, 0)$. In particular, if (x, y) approaches zero along any given path then $x^2/(x^2 + y^2)$ should approach the limiting value a. If (x, y) approaches $(0, 0)$ along the line $y = 0$, the limiting value is clearly 1 (just set $y = 0$ in the above expression to get $x^2/x^2 = 1$). If (x, y) approaches $(0, 0)$ along the line $x = 0$, the limiting value is

$$\underset{(x, y) \to (0, 0)}{\text{limit}} \frac{0^2}{0^2 + y^2} = 0 \neq 1$$

Hence, $\underset{(x, y) \to (0, 0)}{\text{limit}} x^2/(x^2 + y^2)$ does not exist (see Figure 2.2.15).

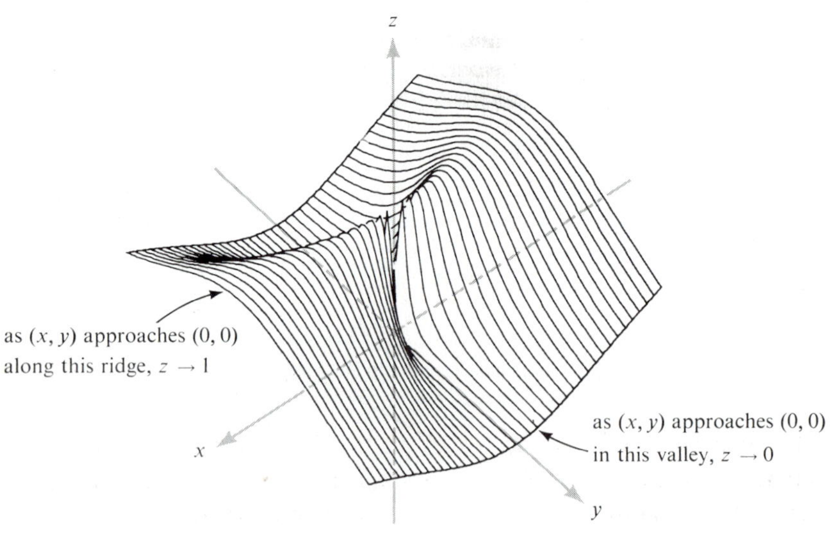

$$z = \frac{x^2}{x^2 + y^2} \qquad -1 \le x \le 1, -1 \le y \le 1$$

FIGURE 2.2.15
This function has no limit at (0, 0).

EXAMPLE 11. Prove (see Figure 2.2.16)

$$\lim_{(x, y)\to(0, 0)} \frac{x^2y}{x^2 + y^2} = 0$$

Indeed, note that

$$\left|\frac{x^2y}{x^2 + y^2}\right| \le \left|\frac{x^2y}{x^2}\right| = |y|$$

Thus, given $\varepsilon > 0$, choose $\delta = \varepsilon$; so $0 < \|(x, y) - (0, 0)\| = \sqrt{x^2 + y^2} < \delta$ implies $|y| < \delta$, and thus

$$\left|\frac{x^2y}{x^2 + y^2} - 0\right| < \varepsilon$$

For more examples of limits, see Section 2.7.

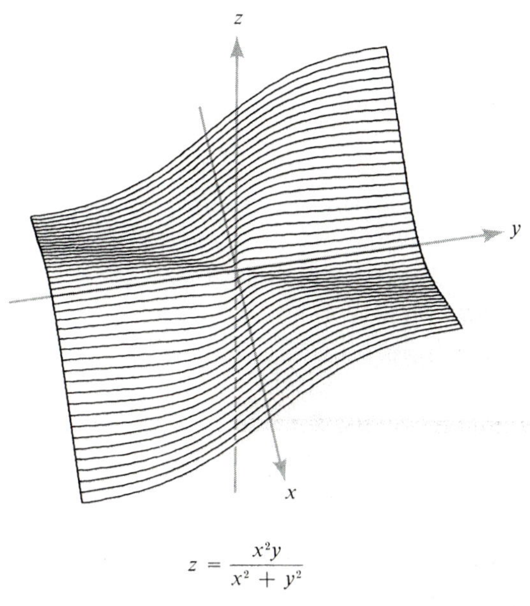

FIGURE 2.2.16
This function has limit 0 *at* (0, 0).

In elementary calculus we learned that the idea of a continuous function is based on the intuitive notion of an unbroken curve, that is, a curve whose graph has no *jumps*, such as would be traced out by a particle in motion or by a moving pencil point that is not lifted from the paper.

To perform a detailed analysis of functions we need concepts more precise than the rather vague notions mentioned above. An example may clarify these ideas. Consider the function $f: \mathbb{R} \to \mathbb{R}$ defined by

$f(x) = -1$ if $x \leq 0$ and $f(x) = 1$ if $x > 0$. The graph of f is shown in Figure 2.2.17. (The little open circle denotes the fact that (0, 1) does *not* lie on the graph of f.) Clearly the graph of f is broken at $x = 0$.

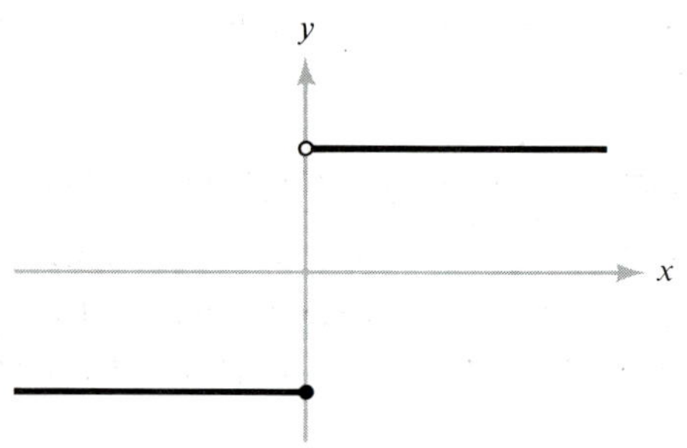

FIGURE 2.2.17
This function is not continuous because its value suddenly jumps as x crosses 0.

Consider also the function $g: x \mapsto x^2$. This function is pictured in Figure 2.2.18. The graph of g is not broken at any point. If one examines examples of functions like f above, whose graphs are broken at some point x_0, and functions like g above, whose graphs are not broken, one sees that the principal difference between them is that for a

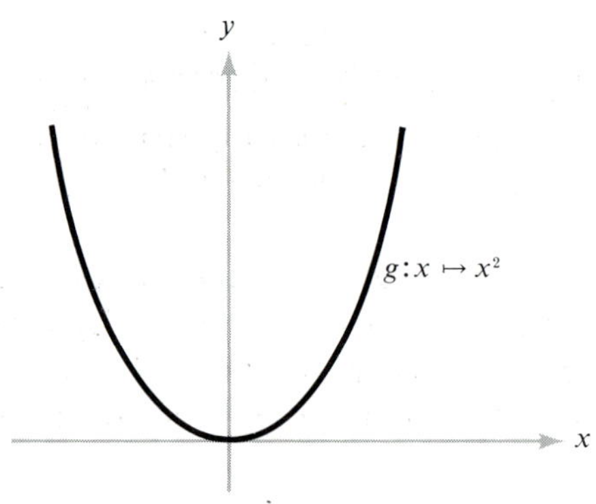

FIGURE 2.2.18
This function is continuous.

function g whose graph is *unbroken*, the values of $g(x)$ get closer and closer to $g(x_0)$ as x gets closer and closer to x_0. The same idea works for functions of several variables. But the notion of closer and closer does not suffice as a mathematical definition; thus we shall formulate these concepts precisely in terms of limits.

Definition. *Let $f\colon A \subset \mathbb{R}^n \to \mathbb{R}^m$ be a given function with domain A. Let $\mathbf{x}_0 \in A$. We say f is* ***continuous at*** $\mathbf{x}_0$ *if and only if*

$$\operatorname*{limit}_{\mathbf{x}\to\mathbf{x}_0} f(\mathbf{x}) = f(\mathbf{x}_0)$$

If we just say that f is ***continuous****, we shall mean that f is continuous at each point $\mathbf{x}_0$ of A.*

Since the condition $\operatorname*{limit}_{\mathbf{x}\to\mathbf{x}_0} f(\mathbf{x}) = f(\mathbf{x}_0)$ means that $f(\mathbf{x})$ is close to $f(\mathbf{x}_0)$ when $\mathbf{x}$ is close to $\mathbf{x}_0$, we see that our definition does indeed correspond to the requirement that the graph of f is unbroken (see Figure 2.2.19 where we illustrate the case $f\colon \mathbb{R} \to \mathbb{R}$). The case of several

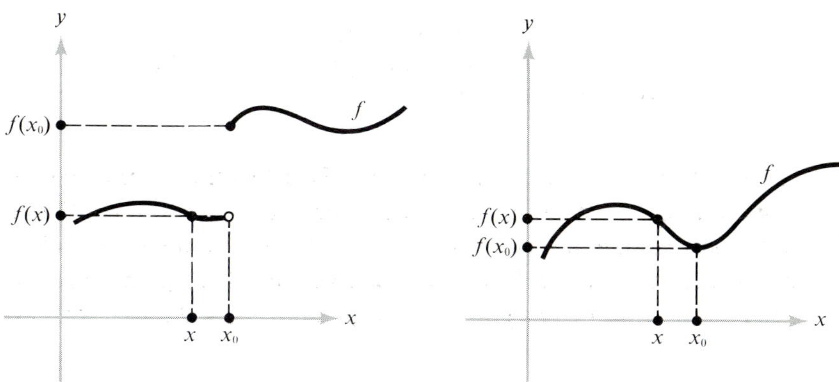

FIGURE 2.2.19
A discontinuous function (left), for which $\operatorname*{limit}_{x\to x_0} f(x)$ *does not exist, and a continuous function (right), for which this limit exists* and *equals* $f(x_0)$.

variables is easiest to visualize if we deal with real-valued functions; say $f\colon \mathbb{R}^2 \to \mathbb{R}$. In this case we can visualize f by drawing its graph, which consists of all points (x, y, z) with $z = f(x, y)$ in $\mathbb{R}^3$. The continuity of f thus means that its graph has no "breaks" in it (see Figure 2.2.20).

EXAMPLE 12. Any polynomial $p(x) = a_0 + a_1 x + \cdots + a_n x^n$ is continuous from $\mathbb{R}$ to $\mathbb{R}$. Indeed, from Theorem 4 and Example 6

$$\begin{aligned}
&\operatorname*{limit}_{x\to x_0} (a_0 + a_1 x + \cdots + a_n x^n) \\
&\qquad = \operatorname*{limit}_{x\to x_0} a_0 + \operatorname*{limit}_{x\to x_0} a_1 x + \cdots + \operatorname*{limit}_{x\to x_0} a_n x^n \\
&\qquad = a_0 + a_1 x_0 + \cdots + a_n x_0^n
\end{aligned}$$

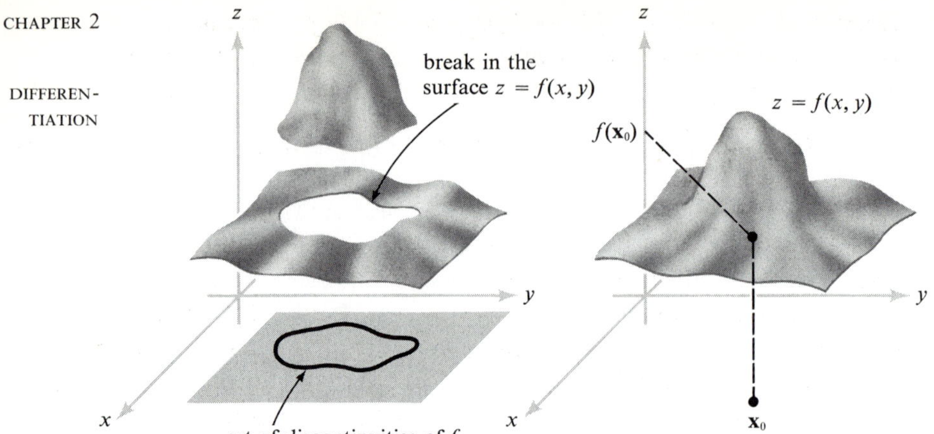

FIGURE 2.2.20
A discontinuous function of two variables (left) and a continuous function (right).

since

$$\lim_{x \to x_0} x^n = \left(\lim_{x \to x_0} x\right) \cdots \left(\lim_{x \to x_0} x\right) = x_0^n .$$

EXAMPLE 13. Let $f\colon \mathbb{R}^2 \to \mathbb{R}$, $f(x, y) = xy$. Then f is continuous since, by the limit theorems and Example 6,

$$\lim_{(x, y) \to (x_0, y_0)} xy = \left(\lim_{(x, y) \to (x_0, y_0)} x\right)\left(\lim_{(x, y) \to (x_0, y_0)} y\right) = x_0 y_0$$

One can see by the same method that any polynomial $p(x, y)$ in x and y is continuous.

EXAMPLE 14. The function $f\colon \mathbb{R}^2 \to \mathbb{R}$ defined by

$$f(x, y) = \begin{cases} 1 & \text{if } x \leq 0 \quad \text{or} \quad y \leq 0 \\ 0 & \text{otherwise} \end{cases}$$

is not continuous at $(0, 0)$ or at any point on the positive x-axis or positive y-axis. Indeed, if $(x_0, y_0) = \mathbf{u}$ is such a point and $\delta > 0$, there are points $(x, y) \in D_\delta(\mathbf{u})$ with $f(x, y) = 1$ and other points $(x, y) \in D_\delta(\mathbf{u})$ with $f(x, y) = 0$. Thus it is *not* true that $f(x, y) \to f(x_0, y_0) = 1$ as $(x, y) \to (x_0, y_0)$.

Remember that we can express limits in ε-δ notation. Using this, we are led to the following convenient reformulation of the definition of continuity.

Theorem 5. *Let $f\colon A \subset \mathbb{R}^n \to \mathbb{R}^m$ be given. Then f is continuous at $\mathbf{x}_0 \in A$ if and only if for every number $\varepsilon > 0$ there is a number $\delta > 0$ such that*

$$\mathbf{x} \in A \textit{ and } \|\mathbf{x} - \mathbf{x}_0\| < \delta \quad \textit{implies} \quad \|f(\mathbf{x}) - f(\mathbf{x}_0)\| < \varepsilon$$

The proof is almost immediate. However, notice that in Theorem 2 we insisted that $0 < \|\mathbf{x} - \mathbf{x}_0\|$, that is, $\mathbf{x} \neq \mathbf{x}_0$. That is *not* imposed here; indeed, the conclusion of Theorem 5 is certainly valid when $\mathbf{x} = \mathbf{x}_0$, so there is no need to exclude this case. Here we do care about the value of f at $\mathbf{x}_0$; we want f at nearby points to be close to this value.

In order to prove that specific functions are continuous we can avail ourselves of the limit theorems (see Theorem 4). If we transcribe those results in terms of continuity, we are led at once to the following:

Theorem 6. *Let $f\colon A \subset \mathbb{R}^n \to \mathbb{R}^m$, $g\colon A \subset \mathbb{R}^n \to \mathbb{R}^m$, and c a real number.*

(*i*) *If f is continuous at $\mathbf{x}_0$, so is cf, where $(cf)(\mathbf{x}) = c(f(\mathbf{x}))$.*
(*ii*) *If f and g are continuous at $\mathbf{x}_0$, so is $f + g$, where $(f + g)(\mathbf{x}) = f(\mathbf{x}) + g(\mathbf{x})$.*
(*iii*) *If f and g are continuous at $\mathbf{x}_0$ and $m = 1$, then the product function fg defined by $(fg)(\mathbf{x}) = f(\mathbf{x})g(\mathbf{x})$ is continuous at $\mathbf{x}_0$.*
(*iv*) *If $f\colon A \subset \mathbb{R}^n \to \mathbb{R}$ is continuous at $\mathbf{x}_0$ and nowhere zero on A, then the quotient $1/f$ is continuous at $\mathbf{x}_0$, where $(1/f)(\mathbf{x}) = 1/f(\mathbf{x})$.**
(*v*) *If $f\colon A \subset \mathbb{R}^n \to \mathbb{R}^m$ and $f(\mathbf{x}) = (f_1(\mathbf{x}), \ldots, f_m(\mathbf{x}))$ then f is continuous at $\mathbf{x}_0$ if and only if each of the real-valued functions $f_1, \ldots, f_m$ are continuous at $\mathbf{x}_0$.*

EXAMPLE 15. Let $f\colon \mathbb{R}^2 \to \mathbb{R}^2$, $(x, y) \mapsto (x^2 y, (y + x^3)/(1 + x^2))$. Show that f is continuous.

To see this, it is sufficient, by (*v*) above, to show that each component is continuous. So we first show that $(x, y) \mapsto x^2 y$ is continuous. Now $(x, y) \mapsto x$ is continuous (see Example 6), so by (*iii*) $(x, y) \mapsto x^2$ is continuous, and by (*iii*) again $(x, y) \mapsto x^2 y$ is continuous. Since $1 + x^2$ is continuous and nonzero, by (*iv*) $1/(1 + x^2)$ is continuous; hence $(y + x^3)/(1 + x^2)$ is a product of continuous functions, and by (*iii*) is continuous.

We have not yet discussed another basic operation that we can perform on functions. This operation is called *composition*. If f maps A to B and g maps B to C, the composition of f with g, or of g on f,

* Another way of stating (*iv*) is often used: If $f(\mathbf{x}_0) \neq 0$ and f is continuous, then $f(\mathbf{x}) \neq 0$ in a neighborhood of $\mathbf{x}_0$ so $1/f$ is defined in that neighborhood, and $1/f$ is continuous at $\mathbf{x}_0$.

denoted by $g \circ f$, maps A to C by sending $\mathbf{x} \mapsto g(f(\mathbf{x}))$ (see Figure 2.2.21). For example, $\sin x^2$ is the composition of $x \mapsto x^2$ with $y \mapsto \sin y$. The reader should be familiar with this idea from calculus.

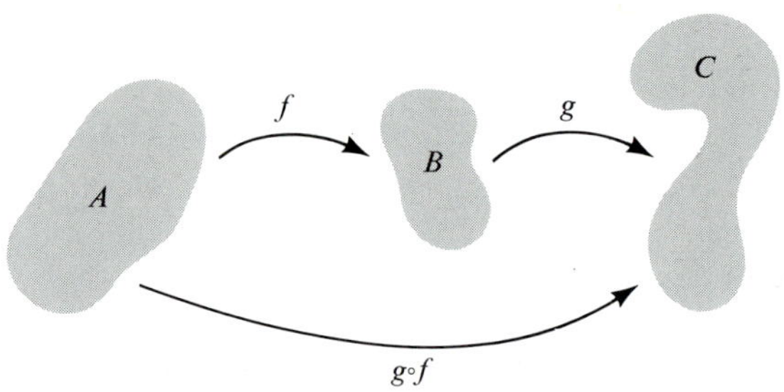

FIGURE 2.2.21
The composition of g on f.

Theorem 7. *Let $f\colon A \subset \mathbb{R}^n \to \mathbb{R}^m$ and let $g\colon B \subset \mathbb{R}^m \to \mathbb{R}^p$. Suppose $f(A) \subset B$ so that $g \circ f$ is defined on A. If f is continuous at $\mathbf{x}_0 \in A$ and g is continuous at $\mathbf{y}_0 = f(\mathbf{x}_0)$, then $g \circ f$ is continuous at $\mathbf{x}_0$.*

The intuitive proof of this is easy; the formal proof in Section 2.7 follows a similar pattern. Intuitively we must show that as $\mathbf{x}$ gets close to $\mathbf{x}_0$, $g(f(\mathbf{x}))$ gets close to $g(f(\mathbf{x}_0))$. But as $\mathbf{x}$ gets close to $\mathbf{x}_0$, $f(\mathbf{x})$ gets close to $f(\mathbf{x}_0)$ (by continuity of f at $\mathbf{x}_0$) and as $f(\mathbf{x})$ gets close to $f(\mathbf{x}_0)$, $g(f(\mathbf{x}))$ gets close to $g(f(\mathbf{x}_0))$ (by continuity of g at $f(\mathbf{x}_0)$).

EXAMPLE 16. Let $f(x, y, z) = (x^2 + y^2 + z^2)^{30} + \sin z^3$. Show that f is continuous.

Here we can write f as a sum of two functions $(x^2 + y^2 + z^2)^{30}$ and $\sin z^3$, so it suffices to show that each is continuous. The first is the composite $(x, y, z) \mapsto (x^2 + y^2 + z^2)$ with $u \mapsto u^{30}$ and the second is $(x, y, z) \mapsto z^3$ with $u \mapsto \sin u$, so we have continuity by Theorem 7.

EXERCISES

1. Show that the following subsets of the plane are open
   (a) $A = \{(x, y) \mid -1 < x < 1,\ -1 < y < 1\}$
   (b) $B = \{(x, y) \mid x > 0\}$
   (c) $C = \{(x, y) \mid 2 < x^2 + y^2 < 4\}$
   (d) $A \cup B \cup C$
   (e) $D = \{(x, y) \mid x \neq 0 \quad \text{and} \quad y \neq 0\}$
2. Prove that for $\mathbf{x} \in \mathbb{R}^n$ and $s < t$, $D_s(\mathbf{x}) \subset D_t(\mathbf{x})$.
3. Prove that if U and V are neighborhoods of $\mathbf{x} \in \mathbb{R}^n$, then so are $U \cap V$ and $U \cup V$.

4. Prove that the boundary points of $]a, b[\subset \mathbb{R}$ are the points a and b.
5. Use the ε-δ formulation of limits to prove that $x^2 \to 4$ as $x \to 2$. Give a shorter proof, using Theorem 4.
6. Compute the following limits
   (a) $\underset{(x,y)\to(0,1)}{\text{limit}}\ x^3 y$
   (b) $\underset{(x,y)\to(0,1)}{\text{limit}}\ e^x y$
   (c) $\underset{x\to 0}{\text{limit}}\ \dfrac{\sin^2 x}{x}$
   (d) $\underset{x\to 0}{\text{limit}}\ \dfrac{\sin^2 x}{x^2}$
7. Compute $\underset{\mathbf{x}\to\mathbf{x}_0}{\text{limit}}\ f(\mathbf{x})$, if it exists, for the following cases:
   (a) $f\colon \mathbb{R} \to \mathbb{R},\ x \mapsto |x|,\ x_0 = 1$
   (b) $f\colon \mathbb{R}^n \to \mathbb{R},\ \mathbf{x} \mapsto \|\mathbf{x}\|$, arbitrary $\mathbf{x}_0$
   (c) $f\colon \mathbb{R} \to \mathbb{R}^2,\ x \mapsto (x^2, e^x),\ x_0 = 1$
   (d) $f\colon\ \mathbb{R}^2\backslash\{(0,\ 0)\} \to \mathbb{R}^2,\ (x,\ y) \mapsto (\sin(x - y),\ e^{x(y+1)} - x - 1)/\|(x,\ y)\|$, $\mathbf{x}_0 = (0, 0)$
8. Let $A \subset \mathbb{R}^2$ be the unit disc $D_1(0, 0)$ with the point $\mathbf{x}_0 = (1, 0)$ added, and $f\colon\ A \to \mathbb{R},\ \mathbf{x} \mapsto f(\mathbf{x})$ be the constant function $f(\mathbf{x}) = 1$. Show that $\underset{\mathbf{x}\to\mathbf{x}_0}{\text{limit}}\ f(\mathbf{x}) = 1$.
9. Let $f\colon \mathbb{R}^3 \to \mathbb{R},\ f(x, y, z) = (x^2 + 3y^2)/(x + 1)$. Compute
$$\underset{(x,y,z)\to(0,0,0)}{\text{limit}}\ f(x, y, z).$$
10. Let $f\colon A \subset \mathbb{R}^n \to \mathbb{R}$ be given and let $\mathbf{x}_0$ be a boundary point of A. We say that $\underset{\mathbf{x}\to\mathbf{x}_0}{\text{limit}}\ f(\mathbf{x}) = \infty$ if for every $N > 0$ there is a $\delta > 0$ such that $0 < \|\mathbf{x} - \mathbf{x}_0\| < \delta$ implies $f(\mathbf{x}) > N$.
    (a) Prove that $\underset{x\to 1}{\text{limit}}\ (x - 1)^{-2} = \infty$.
    (b) Prove that $\underset{x\to 0}{\text{limit}}\ 1/|x| = \infty$. Is it true that $\underset{x\to 0}{\text{limit}}\ 1/x = \infty$?
    (c) Prove that $\underset{(x,y)\to(0,0)}{\text{limit}}\ 1/(x^2 + y^2) = \infty$.
11. Let $f\colon \mathbb{R} \to \mathbb{R}$ be a function. We write $\underset{x\to b-}{\text{limit}}\ f(x) = L$ and say that L is the *left-hand limit* of f at b, if for every $\varepsilon > 0$, there is a $\delta > 0$ such that $x < b$ and $0 < |x - b| < \delta$ implies $|f(x) - L| < \varepsilon$.
    (a) Formulate a definition of *right-hand limit*, or $\underset{x\to b+}{\text{limit}}\ f(x)$.
    (b) Find $\underset{x\to 0-}{\text{limit}}\ 1/(1 + e^{1/x})$ and $\underset{x\to 0+}{\text{limit}}\ 1/(1 + e^{1/x})$.
    (c) Sketch the graph of $1/(1 + e^{1/x})$.
12. (a) Prove that there is a number $\delta > 0$ such that if $|a| < \delta$, then $|a^3 + 3a^2 + a| < 1/100$.
    (b) Prove that there is a number $\delta > 0$ such that if $x^2 + y^2 < \delta^2$, then $|x^2 + y^2 + 3xy + 180xy^5| < 1/10{,}000$.
13. Compute the following limits (review properties of the relevant functions).
    (a) $\underset{x\to 3}{\text{limit}}\ (x^2 - 3x + 5)$

(b) $\lim_{x\to 0} \sin x$

(c) $\lim_{h\to 0} \frac{(x+h)^2 - x^2}{h}$

(d) $\lim_{h\to 0} \frac{e^h - 1}{h}$

(e) $\lim_{x\to 0} \frac{\cos x - 1}{x^2}$

14. Compute the following limits.
(a) $\lim_{(x,y)\to(0,0)} (x^2 + y^2 + 3)$

(b) $\lim_{(x,y)\to(0,0)} \frac{xy}{x^2 + y^2 + 2}$

(c) $\lim_{(x,y)\to(0,0)} \frac{e^{xy}}{x+1}$

15. Prove that $\lim_{(x,y)\to(0,0)} (\sin xy)/xy = 1$.
16. Show that the map $f\colon \mathbb{R} \to \mathbb{R}$, $x \mapsto x^2e^x/(2 - \sin x)$ is continuous.
17. Show that f is continuous at $\mathbf{x}_0$ if and only if

$$\lim_{\mathbf{x}\to\mathbf{x}_0} \|f(\mathbf{x}) - f(\mathbf{x}_0)\| = 0$$

18. Show that $f\colon \mathbb{R} \to \mathbb{R}$, $x \mapsto (1-x)^8 + \cos(1 + x^3)$ is continuous.
19. If $f\colon \mathbb{R}^n \to \mathbb{R}$ and $g\colon \mathbb{R}^n \to \mathbb{R}$ are continuous, show that the functions

$$f^2g\colon \mathbb{R}^n \to \mathbb{R},\ \mathbf{x} \mapsto (f(\mathbf{x}))^2 g(\mathbf{x})$$

and

$$f^2 + g\colon \mathbb{R}^n \to \mathbb{R},\ \mathbf{x} \mapsto (f(\mathbf{x}))^2 + g(\mathbf{x})$$

are continuous.
20. Prove that $f\colon \mathbb{R}^2 \to \mathbb{R}$, $(x, y) \mapsto ye^x + \sin x + (xy)^4$ is continuous.

2.3 DIFFERENTIATION

In Section 2.1 we considered a few methods for graphing simple functions. However, by these methods alone it is impossible in practice to compute enough information, in a reasonable time, to grasp even the general features of a complicated function. From elementary calculus we know that the idea of the derivative can greatly aid us in this task; for example, it enables us to locate maxima and minima and to compute rates of change. The derivative also has many applications beyond this, as the student surely has discovered in elementary calculus.

Intuitively, we know from our work in Section 2.2 that a continuous

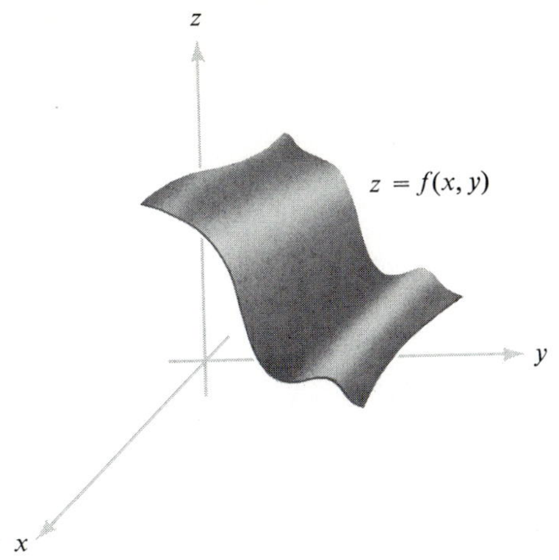

FIGURE 2.3.1
A smooth graph.

function is one that has no "breaks" in its graph. A differentiable function from $\mathbb{R}^2$ to $\mathbb{R}$ ought to be such that not only are there no "breaks" in its graph, but there is a well defined tangent plane to the graph at each point. Thus, there must not be any sharp folds, corners, or peaks in the graph (see figures 2.3.1, 2.3.2). In other words, the graph must be *smooth.*

In order to make these ideas precise, we need a sound definition of what we mean by "$f(x_1, \ldots, x_n)$ is differentiable at $\mathbf{x}_0 = (x_{01}, \ldots, x_{0n})$." Actually this is not quite as simple to do as one might think. Towards this end, however, let us introduce the notion of the *partial derivative.*

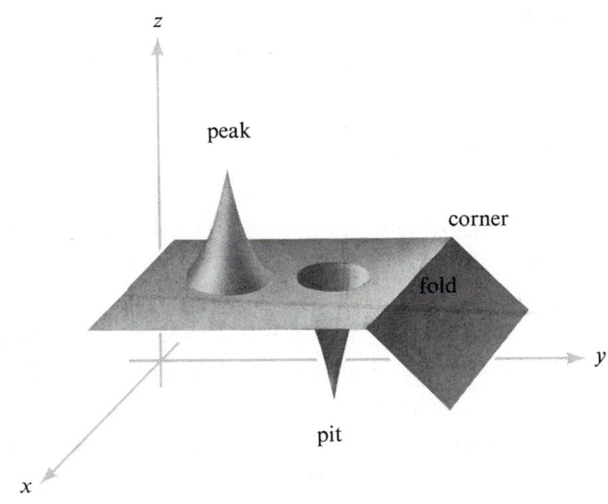

FIGURE 2.3.2
This graph is not smooth.

This notion relies only on our knowledge of one-variable calculus. (A quick review of the definition of the derivative in a one-variable calculus text might be advisable at this point.)

Definition. *Let $f\colon U \subset \mathbb{R}^n \to \mathbb{R}$ be a given real-valued function. Then $\partial f/\partial x_1, \ldots, \partial f/\partial x_n$, the* ***partial derivatives*** *of f with respect to the first, second, ..., nth variable are also real-valued functions of n variables, which, at the point $(x_1, \ldots, x_n) = \mathbf{x}$, are defined by*

$$\frac{\partial f}{\partial x_j}(x_1, \ldots, x_n)$$

$$= \operatorname*{limit}_{h\to 0} \frac{f(x_1, x_2, \ldots, x_j + h, \ldots, x_n) - f(x_1, \ldots, x_n)}{h}$$

$$= \operatorname*{limit}_{h\to 0} \frac{f(\mathbf{x} + h\mathbf{e}_j) - f(\mathbf{x})}{h}$$

if the limits exist, where $1 \le j \le n$ and $\mathbf{e}_j$ is the jth standard basis vector $\mathbf{e}_j = (0, \ldots, 1, \ldots, 0)$, with 1 in the jth slot (see Section 1.4).

In other words, $\partial f/\partial x_j$ is just the derivative of f with respect to the variable x_j, with the other variables held fixed. If $f\colon \mathbb{R}^3 \to \mathbb{R}$, we shall often use the notation $\partial f/\partial x$, $\partial f/\partial y$, $\partial f/\partial z$ in place of $\partial f/\partial x_1$, $\partial f/\partial x_2$, $\partial f/\partial x_3$. If $f\colon U \subset \mathbb{R}^n \to \mathbb{R}^m$, then we can write $f(x_1, \ldots, x_n) = (f_1(x_1, \ldots, x_n), \ldots, f_m(x_1, \ldots, x_n))$ so that we can speak of the partial derivatives of each component; for example, $\partial f_m/\partial x_n$ is the partial derivative of the mth component with respect to x_n, the nth variable.

The existence of the partial derivatives of f at a point $\mathbf{x}$ is the first definition one might think of to give precise meaning to the phrase: "f is differentiable at $\mathbf{x}$." Actually, as we shall see below, to obtain the most satisfactory theory one must strengthen this condition somewhat. Meanwhile, it is essential that the student become thoroughly accustomed to computing partial derivatives. We present some examples to aid in this effort.

EXAMPLE 1. If $f(x, y) = x^2y + y^3$, find $\partial f/\partial x$ and $\partial f/\partial y$.

To find $\partial f/\partial x$ we hold y constant and differentiate only with respect to x; this yields

$$\frac{\partial f}{\partial x} = \frac{d(x^2y + y^3)}{dx} = 2xy$$

Similarly, to find $\partial f/\partial y$ we hold x constant and differentiate only with respect to y.

$$\frac{\partial f}{\partial y} = \frac{d(x^2y + y^3)}{dy} = x^2 + 3y^2$$

If we wish to indicate that a partial derivative is to be evaluated at a particular point, for example at (x_0, y_0), we write

$$\frac{\partial f}{\partial x}(x_0, y_0)$$

EXAMPLE 2. If $z = \cos xy + x \cos y = f(x, y)$, find $(\partial z/\partial x)(x_0, y_0)$ and $(\partial z/\partial y)(x_0, y_0)$. (Note we are using the notation $\partial z/\partial x$, $\partial z/\partial y$ for $\partial f/\partial x$, $\partial f/\partial y$.)

First we fix y_0 and differentiate with respect to x. So

$$\begin{aligned}\frac{\partial z}{\partial x}(x_0, y_0) &= \left.\frac{d(\cos xy_0 + x \cos y_0)}{dx}\right|_{x=x_0} \\ &= -y_0 \sin xy_0 + \cos y_0 \Big|_{x=x_0} \\ &= -y_0 \sin x_0 y_0 + \cos y_0\end{aligned}$$

Similarly, we fix x_0 and differentiate with respect to y to obtain

$$\begin{aligned}\frac{\partial z}{\partial y}(x_0, y_0) &= \left.\frac{d(\cos x_0 y + x_0 \cos y)}{dy}\right|_{y=y_0} \\ &= -x_0 \sin x_0 y - x_0 \sin y \Big|_{y=y_0} \\ &= -x_0 \sin x_0 y_0 - x_0 \sin y_0\end{aligned}$$

EXAMPLE 3. Find $\partial f/\partial x$ if $f(x, y) = xy/\sqrt{x^2 + y^2}$.

We have by the quotient rule

$$\begin{aligned}\frac{\partial f}{\partial x} &= \frac{y\sqrt{x^2 + y^2} - xy(x/\sqrt{x^2 + y^2})}{x^2 + y^2} \\ &= \frac{y(x^2 + y^2) - xy(x)}{(x^2 + y^2)^{3/2}} = \frac{y^3}{(x^2 + y^2)^{3/2}}\end{aligned}$$

Let us now consider what problems arise by defining "differentiability at **x**" to mean just the existence of partial derivatives. Consider the following example.

EXAMPLE 4. Let $f(x, y) = x^{1/3}y^{1/3}$.

By definition

$$\frac{\partial f}{\partial x}(0, 0) = \operatorname*{limit}_{h \to 0} \frac{f(h, 0) - f(0, 0)}{h} = 0$$

and similarly, $\partial f/\partial y(0, 0) = 0$. It is necessary to use the original definition of partial derivative because the functions $x^{1/3}$ and $y^{1/3}$ are not themselves differentiable at 0. But suppose we restrict f to the line $y = x$ to get $f(x, x) = x^{2/3}$ (see Figure 2.3.3). We can view the substitution $y = x$ as the composite of the function $g\colon \mathbb{R} \to \mathbb{R}^2$, given by $g(x) = (x, x)$, with $f\colon \mathbb{R}^2 \to \mathbb{R}$, given by $f(x, y) = x^{1/3}y^{1/3}$.

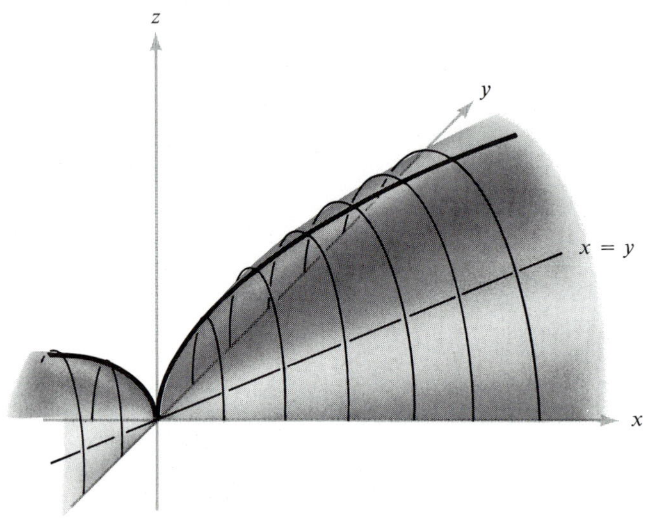

FIGURE 2.3.3
The "upper" part of the graph of $x^{1/3}y^{1/3}$.

Thus $f \circ g\colon \mathbb{R} \to \mathbb{R}$, with $(f \circ g)(x) = x^{2/3}$. Now g and f are both differentiable at 0, but $f \circ g$ is not differentiable at $x = 0$, since $d(x^{2/3})/dx = 2x^{-1/3}/3$ is undefined at $x = 0$. In other words, the composition of differentiable functions would not be differentiable if we used our provisional definition of differentiability, in contrast to what is true of functions of one variable. It would be much more useful to have a definition of differentiability by which the composition of differentiable functions was differentiable.

There is another reason for not wanting to call this function differentiable at (0, 0). Namely, there is no tangent plane, in any reasonable sense, to the graph at (0, 0). The xy-plane is tangent to the graph along the x- and y-axes because f has slope zero at (0, 0) along these axes; that is, $\partial f/\partial x = 0$ and $\partial f/\partial y = 0$ at (0, 0). Thus, if there is a tangent plane, it must be the xy-plane. However, as is evident from Figure 2.3.3, the xy-plane is not tangent to the graph in other directions, and so cannot be said to be tangent to the graph of f.

To motivate our final definition of differentiability, let us compute what the equation of the plane tangent to the graph of $f\colon \mathbb{R}^2 \to \mathbb{R}$,

$(x, y) \mapsto f(x, y)$ at (x_0, y_0) ought to be if f is smooth enough. In $\mathbb{R}^3$, a plane has an equation of the form

$$z = ax + by + c$$

If it is to be the plane tangent to the graph of f, the slopes along the x- and y-axes must be equal to $\partial f/\partial x$ and $\partial f/\partial y$, the rates of change of f with respect to x and y. Thus, $a = \partial f/\partial x$, $b = \partial f/\partial y$. Finally, we may determine c from the fact that $z = f(x_0, y_0)$ when $x = x_0$, $y = y_0$. Thus we finally get

$$z = f(x_0, y_0) + \frac{\partial f}{\partial x}(x_0, y_0)(x - x_0) + \frac{\partial f}{\partial y}(x_0, y_0)(y - y_0) \qquad (1)$$

which should be the equation of the plane tangent to the graph of f at (x_0, y_0), if f is "smooth enough" (see Figure 2.3.4).

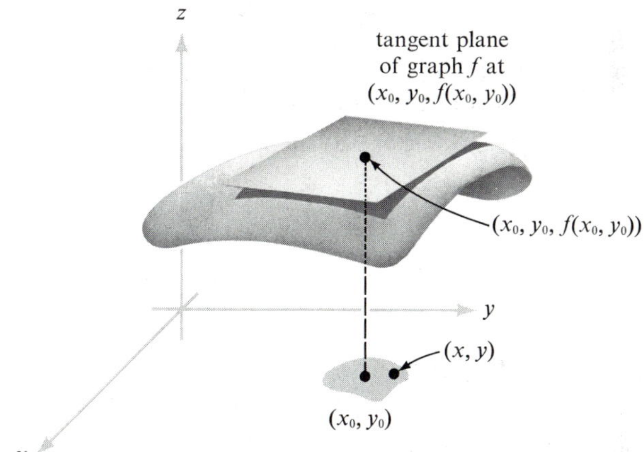

FIGURE 2.3.4
For points (x, y) near (x_0, y_0) the graph of the tangent plane is close to the graph of f.

Our definition of differentiability will amount to the plane (1) being a "good" approximation to f near (x_0, y_0). To get an idea of what one might mean by a good approximation, let us return for a moment to one-variable calculus. If f is differentiable at a point x_0 then we know that

$$\operatorname*{limit}_{\Delta x \to 0} \frac{f(x_0 + \Delta x) - f(x_0)}{\Delta x} = f'(x_0)$$

Let us rewrite this as

$$\operatorname*{limit}_{x \to x_0} \frac{f(x) - f(x_0)}{x - x_0} = f'(x_0)$$

where $x = x_0 + \Delta x$.

Since $\lim_{x \to x_0} f'(x_0) = f'(x_0)$ we can rewrite the above equation as

$$\lim_{x \to x_0} \frac{f(x) - f(x_0)}{x - x_0} = \lim_{x \to x_0} f'(x_0)$$

or

$$\lim_{x \to x_0} \left(\frac{f(x) - f(x_0)}{x - x_0} - f'(x_0) \right) = 0$$

or

$$\lim_{x \to x_0} \frac{f(x) - f(x_0) - f'(x_0)(x - x_0)}{x - x_0} = 0$$

Thus the tangent line through $(x_0, f(x_0))$ with slope $f'(x_0)$ is close to f in the sense that the difference between $f(x)$ and $f(x_0) + f'(x_0)(x - x_0)$ goes to zero *even* when divided by $x - x_0$. This is the notion of "good approximation" that we will try to adapt to functions of several variables, with the tangent line replaced by the tangent plane (see equation (1) above).

Definition. *Let $f\colon \mathbb{R}^2 \to \mathbb{R}$. We say f is **differentiable at** (x_0, y_0), if $\partial f/\partial x$ and $\partial f/\partial y$ exist at (x_0, y_0) **and if***

$$\frac{\left| f(x, y) - f(x_0, y_0) - \dfrac{\partial f}{\partial x}(x_0, y_0)(x - x_0) - \dfrac{\partial f}{\partial y}(x_0, y_0)(y - y_0) \right|}{\|(x, y) - (x_0, y_0)\|} \to 0$$

$$\text{as } (x, y) \to (x_0, y_0) \tag{2}$$

This equation expresses what we mean by saying that

$$f(x_0, y_0) + \frac{\partial f}{\partial x}(x_0, y_0)(x - x_0) + \frac{\partial f}{\partial y}(x_0, y_0)(y - y_0)$$

*is a **good approximation** to the function f.*

Actually, it is not always easy to use this definition to see if f is differentiable, but we shall be able to easily use another criterion given in Theorem 9 below.

We have used the informal notion of the plane tangent to the graph of a function in order to motivate our definition of differentiability. Now we are ready to adopt a formal definition of the tangent plane.

Definition. *Let f: $\mathbb{R}^2 \to \mathbb{R}$ be differentiable at $\mathbf{x}_0 = (x_0, y_0)$. Then the plane in $\mathbb{R}^3$ defined by equation (1)*

$$z = f(x_0, y_0) + \frac{\partial f}{\partial x}(x_0, y_0)(x - x_0) + \frac{\partial f}{\partial y}(x_0, y_0)(y - y_0)$$

is called the ***tangent plane*** *of the graph of f at the point (x_0, y_0).*

Let us write $Df(x_0, y_0)$ for the row matrix

$$\begin{bmatrix} \dfrac{\partial f}{\partial x}(x_0, y_0) & \dfrac{\partial f}{\partial y}(x_0, y_0) \end{bmatrix}$$

so that the definition of differentiability asserts that

$$f(x_0, y_0) + Df(x_0, y_0)\begin{pmatrix} x - x_0 \\ y - y_0 \end{pmatrix} = f(x_0, y_0) + \frac{\partial f}{\partial x}(x_0, y_0)(x - x_0) + \frac{\partial f}{\partial y}(x_0, y_0)(y - y_0) \qquad (3)$$

(matrix multiplication!) is our good approximation to f near (x_0, y_0). (As above, "good" in the sense that (3) differs from $f(x, y)$ by something small times $\sqrt{(x - x_0)^2 + (y - y_0)^2}$.) We say that (3) is the *best linear approximation* to f near (x_0, y_0).

Now we are ready to give a definition of differentiability for maps f of $\mathbb{R}^n$ to $\mathbb{R}^m$, using the above discussion as motivation. The derivative $Df(\mathbf{x}_0)$ of $f = (f_1, \ldots, f_m)$ at a point $\mathbf{x}_0$ is a matrix with elements $t_{ij} = \partial f_i / \partial x_j$ evaluated at $\mathbf{x}_0$.*

Definition. *Let U be an open set in $\mathbb{R}^n$ and let f: $U \subset \mathbb{R}^n \to \mathbb{R}^m$ be a given function. We say that f is* ***differentiable*** *at $\mathbf{x}_0 \in U$ if the partial derivatives of f exist at $\mathbf{x}_0$* ***and*** *if*

$$\operatorname*{limit}_{\mathbf{x} \to \mathbf{x}_0} \frac{\| f(\mathbf{x}) - f(\mathbf{x}_0) - T(\mathbf{x} - \mathbf{x}_0) \|}{\| \mathbf{x} - \mathbf{x}_0 \|} = 0 \qquad (4)$$

where T is the matrix with matrix elements $\partial f_i / \partial x_j$ evaluated at $\mathbf{x}_0$. We call T the ***derivative*** *of f at $\mathbf{x}_0$ and denote it by $Df(\mathbf{x}_0)$.*

We shall always denote the derivative T of f at $\mathbf{x}_0$ by $Df(\mathbf{x}_0)$, although in some books it is denoted $df(\mathbf{x}_0)$ and referred to as the *differential* of f.

* It miraculously turns out that we need only to postulate the existence of *some* matrix giving the best linear approximation near $\mathbf{x}_0 \in \mathbb{R}^n$, because in fact this matrix is *necessarily* the matrix whose ijth entry is $\partial f_i / \partial x_j$ (see Section 2.7 below).

In the case $m = 1$, the matrix T is just the row matrix

$$\left[\frac{\partial f}{\partial x_1}(\mathbf{x}_0) \cdots \frac{\partial f}{\partial x_n}(\mathbf{x}_0)\right]$$

Setting $n = 2$ and putting this fact back in equation (4) we see that conditions (2) and (4) do agree. Thus, letting $\mathbf{h} = \mathbf{x} - \mathbf{x}_0$, a real-valued function f of n variables is differentiable at a point $\mathbf{x}_0$ if

$$\underset{\mathbf{h}\to 0}{\text{limit}} \frac{1}{\|\mathbf{h}\|}\left| f(\mathbf{x}_0 + \mathbf{h}) - f(\mathbf{x}_0) - \sum_{j=1}^{n} \frac{\partial f}{\partial x_j}(\mathbf{x}_0)h_j \right| = 0$$

since

$$T(\mathbf{h}) = \sum_{j=1}^{n} h_j \frac{\partial f}{\partial x_j}(\mathbf{x}_0)$$

For the general case of f mapping a subset of $\mathbb{R}^n$ to $\mathbb{R}^m$, the derivative is the $m \times n$ matrix given by

$$Df(\mathbf{x}_0) = \begin{bmatrix} \dfrac{\partial f_1}{\partial x_1} & \cdots & \dfrac{\partial f_1}{\partial x_n} \\ \vdots & & \vdots \\ \dfrac{\partial f_m}{\partial x_1} & \cdots & \dfrac{\partial f_m}{\partial x_n} \end{bmatrix}$$

where $\partial f_i/\partial x_j$ is evaluated at $\mathbf{x}_0$.

EXAMPLE 5. Compute the plane tangent to the graph of $z = x^2 + y^4 + e^{xy}$ at the point (1, 0, 2).

Here we can use formula (1), so that

$$z = 2(x - 1) + 1(y - 0) + 2$$

that is

$$z = 2x + y$$

since

$$\frac{\partial z}{\partial x} = 2x + ye^{xy} \qquad \text{and} \qquad \frac{\partial z}{\partial y} = 4y^3 + xe^{xy}$$

Definition. *Consider again for a moment the special case $f\colon U \subset \mathbb{R}^n \to \mathbb{R}$. Here $Df(\mathbf{x})$ is a $1 \times n$ matrix*

$$Df(\mathbf{x}) = \left[\frac{\partial f}{\partial x_1} \quad \cdots \quad \frac{\partial f}{\partial x_n}\right]$$

*We can form the corresponding vector $(\partial f/\partial x_1, \ldots, \partial f/\partial x_n)$. This vector is called the **gradient** of f and is denoted* grad f *or* ∇f.

EXAMPLE 6. Let $f: \mathbb{R}^3 \to \mathbb{R}$, $f(x, y, z) = xe^y$. Then

$$\text{grad } f = \left(\frac{\partial f}{\partial x}, \frac{\partial f}{\partial y}, \frac{\partial f}{\partial z}\right) = (e^y, xe^y, 0)$$

From the definition we see that for $f: \mathbb{R}^3 \to \mathbb{R}$,

$$\nabla f = \frac{\partial f}{\partial x}\mathbf{i} + \frac{\partial f}{\partial y}\mathbf{j} + \frac{\partial f}{\partial z}\mathbf{k}$$

while for $f: \mathbb{R}^2 \to \mathbb{R}$

$$\nabla f = \frac{\partial f}{\partial x}\mathbf{i} + \frac{\partial f}{\partial y}\mathbf{j}$$

EXAMPLE 7(a). If $f: \mathbb{R}^2 \to \mathbb{R}$ is given by $(x, y) \mapsto e^{xy} + \sin xy$, then

$$\begin{aligned}\nabla f(x, y) &= (ye^{xy} + y \cos xy)\mathbf{i} + (xe^{xy} + x \cos xy)\mathbf{j} \\ &= (e^{xy} + \cos xy)(y\mathbf{i} + x\mathbf{j})\end{aligned}$$

(b) If $f: \mathbb{R}^2 \to \mathbb{R}^2$, $f(x, y) = (e^{x+y} + y, y^2x)$, then $f_1(x, y) = e^{x+y} + y$ and $f_2(x, y) = y^2x$. Hence

$$Df(x, y) = \begin{bmatrix} e^{x+y} & e^{x+y} + 1 \\ y^2 & 2xy \end{bmatrix}$$

For the rest of this section we shall discuss some general theorems that extend some of the basic results of one-variable calculus to several variables.

In one-variable calculus it is shown that if f is differentiable then f is continuous. We will state in Theorem 8 that this is also true for differentiable functions of several variables. As we know, there are plenty of functions that are continuous but not differentiable, such as $f(x) = |x|$.

Before stating the result, let us give an example of a function whose partial derivatives exist at a point but which is not continuous at that point.

EXAMPLE 8. Let $f: \mathbb{R}^2 \to \mathbb{R}$ be defined by

$$f(x, y) = \begin{cases} 1 & \text{if } \quad x = 0 \quad \text{or if} \quad y = 0 \\ 0 & \text{otherwise} \end{cases}$$

Since f is constant on the x- and y-axes, where it equals 1,

$$\frac{\partial f}{\partial x}(0, 0) = 0 \qquad \text{and} \qquad \frac{\partial f}{\partial y}(0, 0) = 0$$

But f is not continuous at (0, 0), because $\lim_{(x, y)\to(0, 0)} f(x, y)$ does not exist.

However, it is possible to conclude somewhat more about differentiable functions, as the following theorem states.

Theorem 8. *Let $f\colon U \subset \mathbb{R}^n \to \mathbb{R}^m$ be differentiable at $\mathbf{x}_0$. Then f is continuous at $\mathbf{x}_0$.*

Consult Section 2.7 for the proof.

As we have seen, it is usually easy to tell when the partial derivatives of a function exist—we just use what we know from one-variable calculus. However the definition of differentiability looks somewhat complicated and the required approximation condition (4) may seem difficult to verify. Fortunately there is a simple criterion, given in the following theorem, that tells us when a function is differentiable.

Theorem 9. *Let $f\colon U \subset \mathbb{R}^n \to \mathbb{R}^m$. Suppose the partial derivatives $\partial f_i/\partial x_j$ of f all exist and are continuous in a neighborhood of a point $\mathbf{x} \in U$. Then f is differentiable at $\mathbf{x}$ (continuity of the partials means that for each i, j the mapping $U \to \mathbb{R}^m$ given by*

$$\mathbf{x} \mapsto \frac{\partial f_i}{\partial x_j}(\mathbf{x})$$

is continuous).

We shall give the proof in Section 2.7. Notice the following hierarchy:

$$\begin{array}{ccccc} & & \text{Theorem 9} & & \text{Definition of derivative} \\ & & \downarrow & & \downarrow \\ \text{Continuous partials} & \Rightarrow & \text{Differentiable} & \Rightarrow & \text{Partials exist} \end{array}$$

Each converse statement, obtained by reversing an arrow, is invalid.*

A function whose partial derivatives exist and are continuous is said to be of *class C^1*. Thus Theorem 9 says that any *C^1 function is differentiable.*

EXAMPLE 9. Let

$$f(x, y) = \frac{\cos x + e^{xy}}{x^2 + y^2}.$$

* For a counter-example to the converse of the first implication, use $f(x) = x^2 \sin(1/x)$, $f(0) = 0$; for the second, see Example 2, Section 2.7.

Show that f is differentiable at all points $(x, y) \neq (0, 0)$.

This is so because, by what we know from Section 2.2, the partials

$$\frac{\partial f}{\partial x} = \frac{(x^2 + y^2)(ye^{xy} - \sin x) - 2x(\cos x + e^{xy})}{(x^2 + y^2)^2}$$

$$\frac{\partial f}{\partial y} = \frac{(x^2 + y^2)xe^{xy} - 2y(\cos x + e^{xy})}{(x^2 + y^2)^2}$$

are continuous except when $x = 0$ and $y = 0$.

EXERCISES

1. Find $\partial f/\partial x$, $\partial f/\partial y$ if
   (a) $f(x, y) = xy$
   (b) $f(x, y) = e^{xy}$
   (c) $f(x, y) = x \cos x \cos y$
   (d) $f(x, y) = (x^2 + y^2)\log(x^2 + y^2)$
2. Evaluate the partial derivatives $\partial z/\partial x$, $\partial z/\partial y$ for the given function at the indicated points.
   (a) $z = \sqrt{a^2 - x^2 - y^2}$; $(0, 0)$, $(a/2, a/2)$
   (b) $z = \log \sqrt{1 + xy}$; $(1, 2)$, $(0, 0)$
   (c) $z = e^{ax} \cos(bx + y)$; $(2\pi/b, 0)$
3. In each case following, find the partial derivatives $\partial w/\partial x$, $\partial w/\partial y$.
   (a) $w = xe^{x^2 + y^2}$ (b) $w = \dfrac{x^2 + y^2}{x^2 - y^2}$
   (c) $w = e^{xy} \log(x^2 + y^2)$ (d) $w = x/y$
   (e) $w = \cos(ye^{xy})\sin x$
4. Show that each of the following functions is differentiable at each point in its domain. Decide which of the functions are C^1.
   (a) $f(x, y) = \dfrac{2xy}{(x^2 + y^2)^2}$ (b) $f(x, y) = \dfrac{x}{y} + \dfrac{y}{x}$
   (c) $f(r, \theta) = \frac{1}{2}r \sin 2\theta$, $r > 0$ (d) $f(x, y) = \dfrac{xy}{\sqrt{x^2 + y^2}}$
   (e) $f(x, y) = \dfrac{x^2 y}{x^4 + y^2}$
5. Show that the equation of the plane tangent to the surface $z = x^2 + y^3$ at $(3, 1, 10)$ is $z = 6x + 3y - 11$.
6. Compute the plane tangent to the graphs of the functions in Exercise 1 at the indicated points.
   (a) $(0, 0)$ (b) $(0, 1)$ (c) $(0, \pi)$ (d) $(0, 1)$

7. Compute the derivatives of the following functions.
   (a) $f\colon \mathbb{R}^2 \to \mathbb{R}^2, f(x, y) = (x, y)$
   (b) $f\colon \mathbb{R}^2 \to \mathbb{R}^3, f(x, y) = (xe^y + \cos y, x, x + e^y)$
   (c) $f\colon \mathbb{R}^3 \to \mathbb{R}^2, f(x, y, z) = (x + e^z + y, yx^2)$
8. Argue that the graphs of $f(x, y) = x^2 + y^2$ and $g(x, y) = -x^2 - y^2 + xy^3$ should be tangent at $(0, 0)$.
9. Let $f(x, y) = e^{xy}$. Show that $x(\partial f/\partial x) = y(\partial f/\partial y)$.
10. Approximate $(.99\, e^{.02})^8$ using the expression (1), p. 93, to approximate a suitable function $f(x, y)$.
11. Compute the gradients of the following functions.
    (a) $f(x, y, z) = x \exp(-x^2 - y^2 - z^2) \qquad (\exp u = e^u)$
    (b) $f(x, y, z) = \dfrac{xyz}{x^2 + y^2 + z^2}$

2.4 PROPERTIES OF THE DERIVATIVE

In elementary calculus we learned how to differentiate sums, products, quotients, and composite functions. We now want to generalize these ideas to functions of several variables, and we want to pay particular attention to the differentiation of composite functions. The rule for differentiating composites, called the Chain Rule, takes on a more profound form for functions of several variables than for those of one variable. Thus, for example, if f is a real-valued function of one variable, written as $z = f(y)$, and y is a function of x, $y = y(x)$, z becomes a function of x through substitution and we have the familiar formula

$$\frac{dz}{dx} = \frac{df}{dy}\frac{dy}{dx}$$

If f is a real-valued function of three variables u, v, and w, written in the form $z = f(u, v, w)$, and the variables u, v, w are each functions of x, $u = u(x)$, $v = v(x)$, $w = w(x)$, then by substituting $u(x)$, $v(x)$, $w(x)$ for u, v, and w, z becomes a function of x, $z = f(u(x), v(x), w(x))$, and we get

$$\frac{dz}{dx} = \frac{\partial f}{\partial u}\frac{du}{dx} + \frac{\partial f}{\partial v}\frac{dv}{dx} + \frac{\partial f}{\partial w}\frac{dw}{dx}$$

One of the objects of this section is to explain such formulas in detail.

We shall begin with the differentiation rules for sums, products, and quotients.

Theorem 10

(*i*) *Let $f: U \subset \mathbb{R}^n \to \mathbb{R}^m$ be differentiable at $\mathbf{x}_0$ and let c be a real number. Then $h(\mathbf{x}) = cf(\mathbf{x})$ is differentiable at $\mathbf{x}_0$ and*

$$Dh(\mathbf{x}_0) = cDf(\mathbf{x}_0) \quad \textit{(equality of matrices)}$$

(*ii*) *Let $f: U \subset \mathbb{R}^n \to \mathbb{R}^m$ and $g: U \subset \mathbb{R}^n \to \mathbb{R}^m$ be differentiable at $\mathbf{x}_0$. Then $h(\mathbf{x}) = f(\mathbf{x}) + g(\mathbf{x})$ is differentiable at $\mathbf{x}_0$ and*

$$Dh(\mathbf{x}_0) = Df(\mathbf{x}_0) + Dg(\mathbf{x}_0) \quad \textit{(sum of matrices)}$$

(*iii*) *Let $f: U \subset \mathbb{R}^n \to \mathbb{R}$ and $g: U \subset \mathbb{R}^n \to \mathbb{R}$ be differentiable at $\mathbf{x}_0$ and let $h(\mathbf{x}) = g(\mathbf{x})f(\mathbf{x})$. Then $h: U \subset \mathbb{R}^n \to \mathbb{R}$ is differentiable at $\mathbf{x}_0$ and*

$$Dh(\mathbf{x}_0) = g(\mathbf{x}_0)Df(\mathbf{x}_0) + f(\mathbf{x}_0)Dg(\mathbf{x}_0)$$

Note that each side of this equation is an $1 \times n$ matrix.

(*iv*) *With the same hypotheses as (iii), let $h(\mathbf{x}) = f(\mathbf{x})/g(\mathbf{x})$ and suppose g is never zero on U. Then h is differentiable at $\mathbf{x}_0$ and*

$$Dh(\mathbf{x}_0) = \frac{g(\mathbf{x}_0)Df(\mathbf{x}_0) - f(\mathbf{x}_0)Dg(\mathbf{x}_0)}{(g(\mathbf{x}_0))^2}$$

Proof. The proofs of (*i*) through (*iv*) proceed almost exactly as in the one-variable case (although with a slight difference in notation). To show this, we shall prove (*i*) and (*ii*), leaving the proofs of (*iii*) and (*iv*) as an exercise (Exercise 16).

(*i*) To show that $Dh(\mathbf{x}_0) = cDf(\mathbf{x}_0)$, we must show

$$\operatorname*{limit}_{\mathbf{x}\to\mathbf{x}_0} \frac{\|h(\mathbf{x}) - h(\mathbf{x}_0) - cDf(\mathbf{x}_0)(\mathbf{x} - \mathbf{x}_0)\|}{\|\mathbf{x} - \mathbf{x}_0\|} = 0$$

that is

$$\operatorname*{limit}_{\mathbf{x}\to\mathbf{x}_0} \frac{\|cf(\mathbf{x}) - cf(\mathbf{x}_0) - cDf(\mathbf{x}_0)(\mathbf{x} - \mathbf{x}_0)\|}{\|\mathbf{x} - \mathbf{x}_0\|} = 0$$

(see equation (4) of Section 2.3). This is certainly true since f is differentiable (see Theorem 4(*i*), Section 2.2).

(*ii*) By the triangle inequality, we may write

$$\frac{\|h(\mathbf{x}) - h(\mathbf{x}_0) - (Df(\mathbf{x}_0) + Dg(\mathbf{x}_0))(\mathbf{x} - \mathbf{x}_0)\|}{\|\mathbf{x} - \mathbf{x}_0\|}$$

$$\leq \frac{\|f(\mathbf{x}) - f(\mathbf{x}_0) - Df(\mathbf{x}_0)(\mathbf{x} - \mathbf{x}_0)\|}{\|\mathbf{x} - \mathbf{x}_0\|} + \frac{\|g(\mathbf{x}) - g(\mathbf{x}_0) - Dg(\mathbf{x}_0)(\mathbf{x} - \mathbf{x}_0)\|}{\|\mathbf{x} - \mathbf{x}_0\|}$$

which approaches 0 as $\mathbf{x} \to \mathbf{x}_0$. Hence (*ii*) holds. ▮

EXAMPLE 1. Verify the formula for Dh in (*iv*) above with $f(x, y, z) = x^2 + y^2 + z^2$ and $g(x, y, z) = x^2 + 1$. Here

$$h(x, y, z) = \frac{x^2 + y^2 + z^2}{x^2 + 1}$$

so that directly

$$\begin{aligned} Dh(x, y, z) &= \begin{bmatrix} \dfrac{\partial h}{\partial x} & \dfrac{\partial h}{\partial y} & \dfrac{\partial h}{\partial z} \end{bmatrix} \\ &= \begin{bmatrix} \dfrac{(x^2+1)2x - (x^2+y^2+z^2)2x}{(x^2+1)^2} & \dfrac{2y}{x^2+1} & \dfrac{2z}{x^2+1} \end{bmatrix} \\ &= \begin{bmatrix} \dfrac{2x(1-y^2-z^2)}{(x^2+1)^2} & \dfrac{2y}{x^2+1} & \dfrac{2z}{x^2+1} \end{bmatrix} \end{aligned}$$

By the formula in (*iv*) we get

$$Dh = \frac{gDf - fDg}{g^2} = \frac{(x^2+1)(2x, 2y, 2z) - (x^2+y^2+z^2)(2x, 0, 0)}{(x^2+1)^2}$$

which is the same as what we obtained directly.

As we mentioned above, it is in the differentiation of composite functions that we meet apparently substantial alterations of the one-variable formula. However if we use the D notation, that is, matrix notation, the several-variable Chain Rule looks similar to the one-variable rule. Recall (see Example 4, Section 2.3) that it was the nonvalidity of the Chain Rule that forced us to change our provisional definition of differentiability.

***Theorem 11* (*Chain Rule*).** *Let $U \subset \mathbb{R}^n$ and $V \subset \mathbb{R}^m$ be open. Let $f\colon U \subset \mathbb{R}^n \to \mathbb{R}^m$ and $g\colon V \subset \mathbb{R}^m \to \mathbb{R}^p$ be given functions such that f maps U into V, so that $g \circ f$ is defined. Suppose f is differentiable at $\mathbf{x}_0$ and g is differentiable at $\mathbf{y}_0 = f(\mathbf{x}_0)$. Then $g \circ f$ is differentiable at $\mathbf{x}_0$ and*

$$D(g \circ f)(\mathbf{x}_0) = Dg(\mathbf{y}_0)Df(\mathbf{x}_0) \tag{1}$$

The right-hand side is a matrix product.

We shall defer the proof of Theorem 11 until Section 2.7, not because it is especially hard, but so as not to overburden the text with technical proofs at this stage. Now let us see how formula (1) works.

Corollary 1. *Let $(x, y, z) \mapsto (u(x, y, z), v(x, y, z), w(x, y, z))$ be a differentiable map of an open set U of $\mathbb{R}^3$ to $\mathbb{R}^3$ and let f be a differentiable* *function of* $\mathbb{R}^3$ *to* $\mathbb{R}$. *Let* $h(x, y, z) = f(u(x, y, z), v(x, y, z), w(x, y, z))$. *Then*

$$\frac{\partial h}{\partial x} = \frac{\partial f}{\partial u}\frac{\partial u}{\partial x} + \frac{\partial f}{\partial v}\frac{\partial v}{\partial x} + \frac{\partial f}{\partial w}\frac{\partial w}{\partial x} \tag{2}$$

with similar formulas for $\partial h/\partial y$, $\partial h/\partial z$ (*one sometimes writes, although it abuses the notation,* $\partial f/\partial x$ *for* $\partial h/\partial x$).

Proof. We can apply Theorem 11 to h, which is the composite of f on $k(x, y, z) = (u(x, y, z), v(x, y, z), w(x, y, z))$ to get

$$Dh(x, y, z) = Df(u, v, w)Dk(x, y, z)$$

that is

$$\begin{bmatrix} \dfrac{\partial h}{\partial x} & \dfrac{\partial h}{\partial y} & \dfrac{\partial h}{\partial z} \end{bmatrix} = \begin{bmatrix} \dfrac{\partial f}{\partial u} & \dfrac{\partial f}{\partial v} & \dfrac{\partial f}{\partial w} \end{bmatrix} \begin{bmatrix} \dfrac{\partial u}{\partial x} & \dfrac{\partial u}{\partial y} & \dfrac{\partial u}{\partial z} \\ \dfrac{\partial v}{\partial x} & \dfrac{\partial v}{\partial y} & \dfrac{\partial v}{\partial z} \\ \dfrac{\partial w}{\partial x} & \dfrac{\partial w}{\partial y} & \dfrac{\partial w}{\partial z} \end{bmatrix}$$

Multiplying out the matrices on the right-hand side and equating components gives us formula (2). ∎

The pattern of this rule will become clear once the student has worked some additional examples. For instance,

$$\frac{\partial}{\partial x} f(u(x, y), v(x, y), w(x, y), z(x, y))$$

$$= \frac{\partial f}{\partial u}\frac{\partial u}{\partial x} + \frac{\partial f}{\partial v}\frac{\partial v}{\partial x} + \frac{\partial f}{\partial w}\frac{\partial w}{\partial x} + \frac{\partial f}{\partial z}\frac{\partial z}{\partial x}$$

with a similar formula for $\partial f/\partial y$.

We state explicitly another case that will be of interest to us later.

Corollary 2. *Let* $\mathbf{c}\colon \mathbb{R} \to \mathbb{R}^3$ *and* $f\colon \mathbb{R}^3 \to \mathbb{R}$ *be differentiable. Then if* $h(t) = f(\mathbf{c}(t)) = f(x(t), y(t), z(t))$ *where* $\mathbf{c}(t) = (x(t), y(t), z(t))$, *we have the formula*

$$\frac{dh}{dt} = \frac{\partial f}{\partial x}\frac{dx}{dt} + \frac{\partial f}{\partial y}\frac{dy}{dt} + \frac{\partial f}{\partial z}\frac{dz}{dt} \tag{3}$$

that is, $dh/dt = \langle \nabla f, \mathbf{c}'(t) \rangle$ *where* $\mathbf{c}'(t) = (x'(t), y'(t), z'(t))$.

Proof. This follows at once from Theorem 11 since

$$Df = \begin{bmatrix} \frac{\partial f}{\partial x} & \frac{\partial f}{\partial y} & \frac{\partial f}{\partial z} \end{bmatrix} \quad \text{and} \quad D\mathbf{c} = \begin{bmatrix} \frac{dx}{dt} \\ \frac{dy}{dt} \\ \frac{dz}{dt} \end{bmatrix} \quad \blacksquare$$

The map **c** in Corollary 2 represents a curve (Figure 2.4.1) and $\mathbf{c}'(t)$ can be thought of as a tangent vector (or velocity vector) of the curve. Although this idea is studied in greater detail in Chapter 3, we can

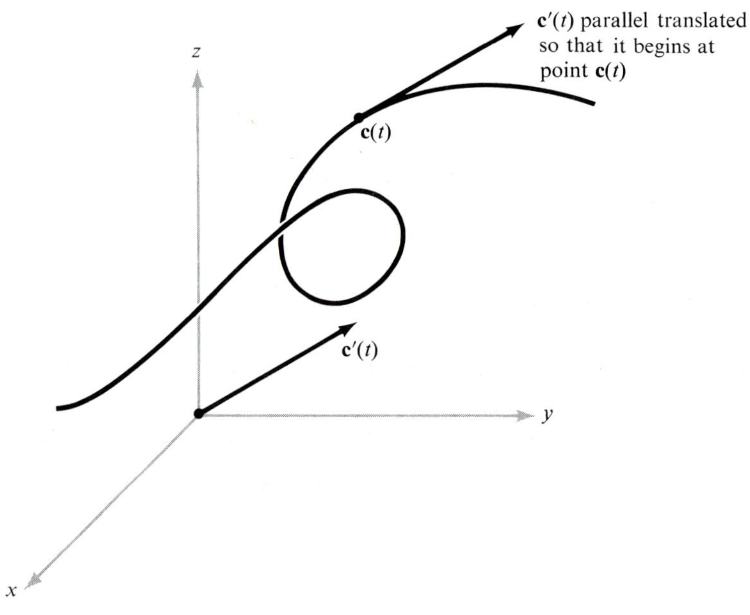

FIGURE 2.4.1
The vector $\mathbf{c}'(t)$ *represents the tangent vector (or velocity vector) of the curve* $\mathbf{c}(t)$.

briefly indicate here the reason for this interpretation. Using the definition of the derivative of a function of a single variable, we see that

$$\mathbf{c}'(t) = \underset{h \to 0}{\text{limit}} \frac{\mathbf{c}(t+h) - \mathbf{c}(t)}{h}$$

The quotient represents a secant that approximates a tangent vector as $h \to 0$ (see Figure 2.4.2).

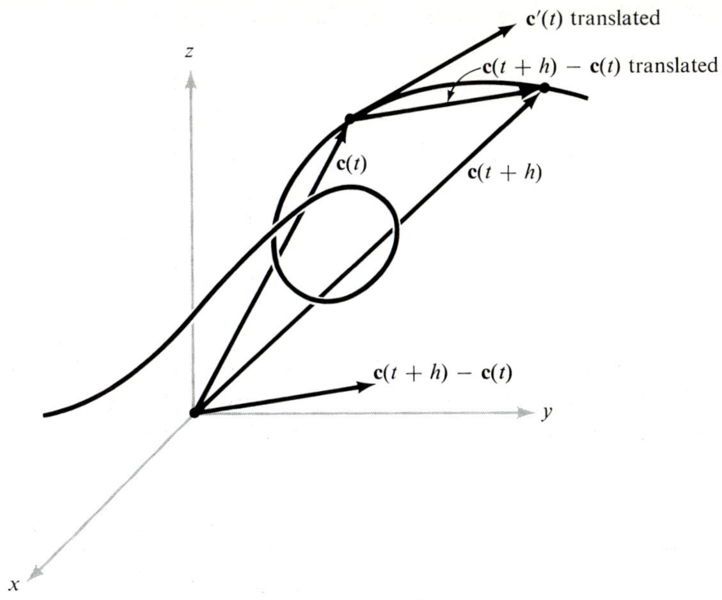

FIGURE 2.4.2
Geometry associated with the formula
$\lim_{h\to 0} (\mathbf{c}(t+h) - \mathbf{c}(t))/h = \mathbf{c}'(t)$.

EXAMPLE 2. Compute a tangent vector to the curve $\mathbf{c}(t) = (t, t^2, e^t)$ at $t = 0$.

Here $\mathbf{c}'(t) = (1, 2t, e^t)$, so at $t = 0$ a tangent vector is $(1, 0, 1)$.

EXAMPLE 3. Verify the Chain Rule in the form of formula (2) for

$$f(u, v, w) = u^2 + v^2 - w$$

$$u(x, y, z) = x^2y$$

$$v(x, y, z) = y^2$$

$$w(x, y, z) = e^{-xz}$$

Here

$$\begin{aligned} h(x, y, z) &= f(u(x, y, z), v(x, y, z), w(x, y, z)) \\ &= (x^2y)^2 + y^4 - e^{-xz} \\ &= x^4y^2 + y^4 - e^{-xz} \end{aligned}$$

Thus

$$\frac{\partial h}{\partial x} = 4x^3y^2 + ze^{-xz}$$

On the other hand

$$\frac{\partial f}{\partial u}\frac{\partial u}{\partial x}+\frac{\partial f}{\partial v}\frac{\partial v}{\partial x}+\frac{\partial f}{\partial w}\frac{\partial w}{\partial x}=2u(2xy)+2v\cdot 0+ze^{-xz}$$
$$=(2x^2y)(2xy)+ze^{-xz}$$

which is the same as the above.

EXAMPLE 4. Let $f\colon U\subset\mathbb{R}^n\to\mathbb{R}^m$ be differentiable, with $f=(f_1,\ldots,f_m)$, and let $g(\mathbf{x})=\sin\langle f(\mathbf{x}), f(\mathbf{x})\rangle$. Compute $Dg(\mathbf{x})$.

By the Chain Rule, $Dg(\mathbf{x})=\cos\langle f(\mathbf{x}), f(\mathbf{x})\rangle Dh(\mathbf{x})$ where $h(\mathbf{x})=\langle f(\mathbf{x}), f(\mathbf{x})\rangle=f_1^2(\mathbf{x})+\cdots+f_m^2(\mathbf{x})$. Then

$$Dh(\mathbf{x})=\left[\frac{\partial h}{\partial x_1}\cdots\frac{\partial h}{\partial x_n}\right]$$
$$=\left[2f_1\frac{\partial f_1}{\partial x_1}+\cdots+2f_m\frac{\partial f_m}{\partial x_1}\quad\cdots\quad 2f_1\frac{\partial f_1}{\partial x_n}+\cdots+2f_m\frac{\partial f_m}{\partial x_n}\right]$$

But $Dh(\mathbf{x})$ can be written as $2f^T(\mathbf{x})Df(\mathbf{x})$, where

$$f^T=[f_1\ \cdots\ f_m]\text{ and }Df=\begin{bmatrix}\frac{\partial f_1}{\partial x_1} & \cdots & \frac{\partial f_1}{\partial x_n}\\ \vdots & & \vdots\\ \frac{\partial f_m}{\partial x_1} & \cdots & \frac{\partial f_m}{\partial x_n}\end{bmatrix}$$

Thus $Dg(\mathbf{x})=2(\cos\langle f(\mathbf{x}), f(\mathbf{x})\rangle)f^T(\mathbf{x})Df(\mathbf{x})$.

EXAMPLE 5. Given $f(x, y)=(x^2+1, y^2)$ and $g(u, v)=(u+v, u, v^2)$, compute the derivative of $g\circ f$ at $(1, 1)$ using the Chain Rule.

Indeed,

$$Dg(u, v)=\begin{bmatrix}\frac{\partial g_1}{\partial u} & \frac{\partial g_1}{\partial v}\\ \frac{\partial g_2}{\partial u} & \frac{\partial g_2}{\partial v}\\ \frac{\partial g_3}{\partial u} & \frac{\partial g_3}{\partial v}\end{bmatrix}=\begin{bmatrix}1 & 1\\ 1 & 0\\ 0 & 2v\end{bmatrix}\quad\text{and}\quad Df(x, y)=\begin{bmatrix}2x & 0\\ 0 & 2y\end{bmatrix}$$

When $(x, y)=(1, 1)$, $f(x, y)=(u, v)=(2, 1)$. Hence

$$D(g\circ f)(1, 1)=Dg(2, 1)Df(1, 1)=\begin{bmatrix}1 & 1\\ 1 & 0\\ 0 & 2\end{bmatrix}\begin{bmatrix}2 & 0\\ 0 & 2\end{bmatrix}=\begin{bmatrix}2 & 2\\ 2 & 0\\ 0 & 4\end{bmatrix}$$

is the required derivative.

EXAMPLE 6. Let $f(x, y)$ be given and make the substitution $x = r\cos\theta$, $y = r\sin\theta$ (polar coordinates). Write a formula for $\partial f/\partial\theta$.

By the Chain Rule

$$\frac{\partial f}{\partial\theta} = \frac{\partial f}{\partial x}\frac{\partial x}{\partial\theta} + \frac{\partial f}{\partial y}\frac{\partial y}{\partial\theta}$$

that is

$$\frac{\partial f}{\partial\theta} = -r\sin\theta\frac{\partial f}{\partial x} + r\cos\theta\frac{\partial f}{\partial y}$$

which is the formula sought.

EXERCISES

1. If $f\colon U \subset \mathbb{R}^n \to \mathbb{R}$ is differentiable, prove that $\mathbf{x} \mapsto f^2(\mathbf{x}) + 2f(\mathbf{x})$ is differentiable as well, and compute its derivative in terms of $Df(\mathbf{x})$.
2. Prove that the following functions are differentiable, and find their derivatives at an arbitrary point.
   (a) $f\colon \mathbb{R}^2 \to \mathbb{R}$, $(x, y) \mapsto 2$
   (b) $f\colon \mathbb{R}^2 \to \mathbb{R}$, $(x, y) \mapsto x + y$
   (c) $f\colon \mathbb{R}^2 \to \mathbb{R}$, $(x, y) \mapsto 2 + x + y$
   (d) $f\colon \mathbb{R}^2 \to \mathbb{R}$, $(x, y) \mapsto x^2 + y^2$
   (e) $f\colon \mathbb{R}^2 \to \mathbb{R}$, $(x, y) \mapsto e^{xy}$
   (f) $f\colon U \to \mathbb{R}$, $(x, y) \mapsto \sqrt{1 - x^2 - y^2}$, $U = \{(x, y) \mid x^2 + y^2 < 1\}$
   (g) $f\colon \mathbb{R}^2 \to \mathbb{R}$, $(x, y) \mapsto x^4 - y^4$
3. Write out the Chain Rule for each of the following functions.
   (a) $\dfrac{\partial h}{\partial x}$ where $h(x, y) = f(x, u(x, y))$
   (b) $\dfrac{dh}{dx}$ where $h(x) = f(x, u(x), v(x))$
   (c) $\dfrac{\partial h}{\partial x}$ where $h(x, y, z) = f(u(x, y, z), v(x, y), w(x))$

   Now carefully justify your answer in each case by using Theorem 11.
4. Verify the Chain Rule for $\partial h/\partial x$ where $h(x, y) = f(u(x, y), v(x, y))$ and
   $$f(u, v) = \frac{u^2 + v^2}{u^2 - v^2}, \qquad u(x, y) = e^{-x-y}, \qquad v(x, y) = e^{xy}$$
5. Verify the validity of the Chain Rule as stated in Corollary 2 for the composition $f \circ \mathbf{c}$ in each of the cases below.
   (a) $f(x, y) = xy$, $\mathbf{c}(t) = (e^t, \cos t)$
   (b) $f(x, y) = e^{xy}$, $\mathbf{c}(t) = (3t^2, t^3)$
   (c) $f(x, y) = (x^2 + y^2)\log\sqrt{x^2 + y^2}$, $\mathbf{c}(t) = (e^t, e^{-t})$
   (d) $f(x, y) = x\exp(x^2 + y^2)$, $\mathbf{c}(t) = (t, -t)$
6. What is the velocity vector for each curve $\mathbf{c}(t)$ in Exercise 5?
7. Let $f\colon \mathbb{R}^3 \to \mathbb{R}$, $g\colon \mathbb{R}^3 \to \mathbb{R}$ be differentiable. Prove that
   $$\nabla(fg) = f\nabla g + g\nabla f$$

8. Let $f\colon \mathbb{R}^3 \to \mathbb{R}$ be differentiable. Making the substitution

$$x = r\cos\theta\sin\phi, \qquad y = r\sin\theta\sin\phi, \qquad z = r\cos\phi$$

(spherical coordinates) into $f(x, y, z)$, compute $\partial f/\partial r$, $\partial f/\partial\theta$, and $\partial f/\partial\phi$.

9. This exercise gives another example of the fact that the Chain Rule is not applicable if f is not differentiable. Consider the function

$$f(x, y) = \begin{cases} \dfrac{xy^2}{x^2 + y^2} & (x, y) \neq (0, 0) \\ 0 & (x, y) = (0, 0) \end{cases}$$

Show that

(a) $\dfrac{\partial f}{\partial x}(0, 0)$ and $\dfrac{\partial f}{\partial y}(0, 0)$ exist;

(b) if $\mathbf{g}(t) = (at, bt)$ then $f \circ \mathbf{g}$ is differentiable and $(f \circ \mathbf{g})'(0) = ab^2/(a^2 + b^2)$, but $\nabla f(0, 0) \cdot \mathbf{g}'(0) = 0$.

10. Find $(\partial/\partial s)(f \circ T)(0, 0)$, where $f(u, v) = \cos u \sin v$ and $T(s, t) = (\cos t^2 s, \log\sqrt{1 + s^2})$.

11. (a) Let $y(x)$ be defined implicitly by $G(x, y(x)) = 0$, where G is a given function of two variables. Prove that if $y(x)$ and G are differentiable, then

$$\frac{dy}{dx} = -\frac{\partial G/\partial x}{\partial G/\partial y} \quad \text{if} \quad \frac{\partial G}{\partial y} \neq 0$$

(b) Obtain a formula analogous to that in (a) if y_1, y_2 are defined implicitly by

$$G_1(x, y_1(x), y_2(x)) = 0$$

$$G_2(x, y_1(x), y_2(x)) = 0$$

(c) Let y be defined implicitly by

$$x^2 + y^3 + e^y = 0$$

Compute dy/dx in terms of x and y.

12. Let $f\colon \mathbb{R}^n \to \mathbb{R}^m$ be a linear mapping. Prove that $Df(\mathbf{x}) = f$.

13. Prove that if $f\colon U \subset \mathbb{R}^n \to \mathbb{R}$ is differentiable at $\mathbf{x}_0 \in U$, there is a neighborhood V of $\mathbf{0} \in \mathbb{R}^n$ and a function $R_1\colon V \to \mathbb{R}$ such that for all $\mathbf{h} \in V$, $\mathbf{x}_0 + \mathbf{h} \in U$,

$$f(\mathbf{x}_0 + \mathbf{h}) = f(\mathbf{x}_0) + Df(\mathbf{x}_0) \cdot \mathbf{h} + R_1(\mathbf{h})$$

and

$$R_1(\mathbf{h})/\|\mathbf{h}\| \to 0 \quad \text{as} \quad \mathbf{h} \to 0$$

*14. For what $p > 0$ is

$$f(x) = \begin{cases} x^p \sin(1/x) & x \neq 0 \\ 0 & x = 0 \end{cases}$$

differentiable? For what p is the derivative continuous?

15. Let $f(x, y) = 1/\sqrt{x^2 + y^2}$. Compute $\nabla f(x, y)$. In what direction is this vector pointing?
16. Prove (*iii*) and (*iv*) of Theorem 10. (HINT: Use the same addition and subtraction tricks as in the one-variable case.)

2.5 GRADIENTS AND DIRECTIONAL DERIVATIVES

In Section 2.1 we studied the graphs of real-valued functions. Now we take up this study again, using the methods of calculus. We shall use gradients to obtain a formula for the plane tangent to a level surface. Let us begin by recalling how the gradient is defined.

Definition. *If $f\colon U \subset \mathbb{R}^3 \to \mathbb{R}$ is differentiable, the* ***gradient*** *of f at (x, y, z) is the vector in the domain space $\mathbb{R}^3$ given by*

$$\operatorname{grad} f = \left(\frac{\partial f}{\partial x}, \frac{\partial f}{\partial y}, \frac{\partial f}{\partial z}\right)$$

This vector is also denoted ∇f or $\nabla f(x, y, z)$. Thus ∇f is just the matrix of the derivative Df, written as a vector.

EXAMPLE 1. Let $f(x, y, z) = \sqrt{x^2 + y^2 + z^2} = r$, the distance from $\mathbf{0}$ to (x, y, z). Then

$$\begin{aligned}\nabla f(x, y, z) &= \left(\frac{\partial f}{\partial x}, \frac{\partial f}{\partial y}, \frac{\partial f}{\partial z}\right)\\ &= \left(\frac{x}{\sqrt{x^2 + y^2 + z^2}}, \frac{y}{\sqrt{x^2 + y^2 + z^2}}, \frac{z}{\sqrt{x^2 + y^2 + z^2}}\right) = \frac{\mathbf{r}}{r}\end{aligned}$$

where $\mathbf{r}$ is the point (x, y, z). Thus ∇f is the unit vector in the direction of (x, y, z).

EXAMPLE 2. If $f(x, y, z) = xy + z$, then

$$\nabla f(x, y, z) = \left(\frac{\partial f}{\partial x}, \frac{\partial f}{\partial y}, \frac{\partial f}{\partial z}\right) = (y, x, 1)$$

Suppose $f\colon \mathbb{R}^3 \to \mathbb{R}$ is a real-valued function. Let $\mathbf{v}$ and $\mathbf{x} \in \mathbb{R}^3$ be fixed vectors and consider the function from $\mathbb{R}$ to $\mathbb{R}$ defined by $t \mapsto f(\mathbf{x} + t\mathbf{v})$. The set of points of the form $\mathbf{x} + t\mathbf{v}$, $t \in \mathbb{R}$, is just a line L through the point $\mathbf{x}$ parallel to the vector $\mathbf{v}$ (see Figure 2.5.1). Therefore, the function $t \mapsto f(\mathbf{x} + t\mathbf{v})$ is equivalent to the function f restricted to the line L. We may ask: How fast are the values of f changing

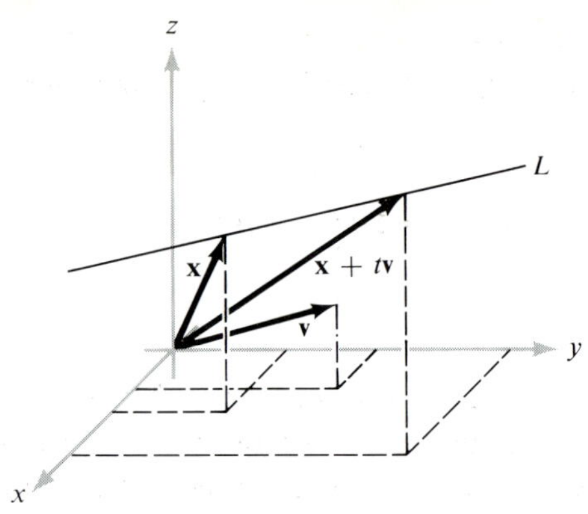

FIGURE 2.5.1
The equation of L is $\mathbf{l}(t) = \mathbf{x} + t\mathbf{v}$.

along the line L at the point $\mathbf{x}$? Since the rate of change of a function is given by a derivative we could say that the answer to this question is the value of the derivative of the given function of t at $t = 0$ (when $t = 0$, $\mathbf{x} + t\mathbf{v}$ reduces to $\mathbf{x}$). This would be the derivative of f at the point $\mathbf{x}$ in the direction of L, that is, of $\mathbf{v}$. We can formalize this concept as follows.

Definition. *If* $f\colon \mathbb{R}^3 \to \mathbb{R}$, *the* ***directional derivative*** *of* f *at* $\mathbf{x}$ *in the direction of a unit vector* $\mathbf{v}$ *is given by*

$$\left.\frac{d}{dt} f(\mathbf{x} + t\mathbf{v})\right|_{t=0}$$

if this exists.

From the definition we can see that the directional derivative can also be defined by the formula

$$\operatorname*{limit}_{h\to 0} \frac{f(\mathbf{x} + h\mathbf{v}) - f(\mathbf{x})}{h}$$

Theorem 12. *If* $f\colon \mathbb{R}^3 \to \mathbb{R}$ *is differentiable, then all directional derivatives exist. The directional derivative at* $\mathbf{x}$ *in direction* $\mathbf{v}$ *is given by*

$$\begin{aligned} Df(\mathbf{x})\mathbf{v} &= \langle \operatorname{grad} f(\mathbf{x}), \mathbf{v}\rangle = \langle \nabla f(\mathbf{x}), \mathbf{v}\rangle \\ &= \frac{\partial f}{\partial x}(\mathbf{x})v_1 + \frac{\partial f}{\partial y}(\mathbf{x})v_2 + \frac{\partial f}{\partial z}(\mathbf{x})v_3 \end{aligned}$$

where $\mathbf{v} = (v_1, v_2, v_3)$ *and* $\|\mathbf{v}\| = 1$.

Proof. Let $\mathbf{c}(t) = \mathbf{x} + t\mathbf{v}$, so that $f(\mathbf{x} + t\mathbf{v}) = f(\mathbf{c}(t))$. By Corollary 2, page 103, $(d/dt)f(\mathbf{c}(t)) = \langle \nabla f(\mathbf{c}(t)), \mathbf{c}'(t)\rangle$. However, $\mathbf{c}(0) = \mathbf{x}$ and $\mathbf{c}'(0) = \mathbf{v}$, so

$$\frac{d}{dt} f(\mathbf{x} + t\mathbf{v})\bigg|_{t=0} = \langle \nabla f(\mathbf{x}), v\rangle$$

as we were required to prove. ∎

We should explain why in the definition of directional derivative we have chosen $\mathbf{v}$ to be a unit vector. There are two reasons. First of all, if α is any positive real number, $\alpha\mathbf{v}$ is a vector that points in the same direction as $\mathbf{v}$, but may be either longer (if $\alpha > 1$) or shorter than $\mathbf{v}$ (if $\alpha < 1$) (see Figure 2.5.2). By Theorem 12 the directional derivative of f in the direction $\mathbf{v}$ is

$$\langle \nabla f(\mathbf{x}), \mathbf{v}\rangle = \frac{\partial f}{\partial x}(\mathbf{x})v_1 + \frac{\partial f}{\partial y}(\mathbf{x})v_2 + \frac{\partial f}{\partial z}(\mathbf{x})v_3$$

The derivative of f "in the direction" $\alpha\mathbf{v}$ is $\langle \nabla f(\mathbf{x}), \alpha\mathbf{v}\rangle = \alpha\langle \nabla f(\mathbf{x}), \mathbf{v}\rangle$, which is α times the directional derivative in direction $\mathbf{v}$, and

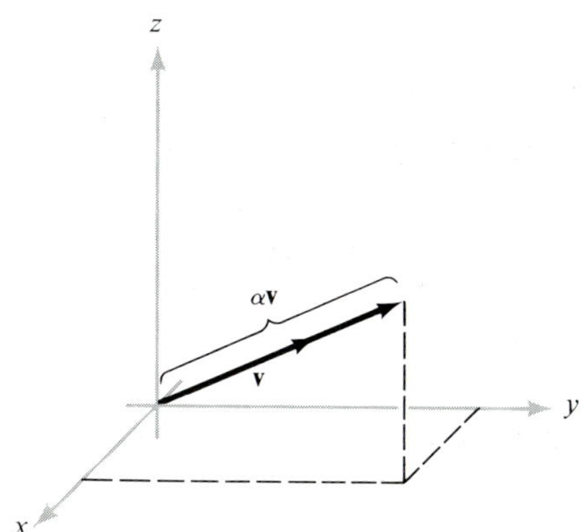

FIGURE 2.5.2
Multiplication of a vector $\mathbf{v}$ *by a scalar* α *rescales the length of* $\mathbf{v}$.

therefore not equal to it. So the directional derivative, if defined for *all* $\alpha\mathbf{v}$, would not depend only on a point $\mathbf{x}$ and a direction. To resolve this problem *we have required that the vector* $\mathbf{v}$ *be of standard length*, that is, of length 1. Then the vector $\mathbf{v}$ determines a direction, the same direction determined by $\alpha\mathbf{v}$ if $\alpha > 0$, but now the directional derivative is *uniquely* defined by $\langle \nabla f(\mathbf{x}), \mathbf{v}\rangle$.

Secondly, we may interpret $\langle \nabla f(\mathbf{x}), \mathbf{v}\rangle$ as the rate of change of f in the direction $\mathbf{v}$, for when $\|\mathbf{v}\| = 1$, the point $\mathbf{x} + t\mathbf{v}$ moves a distance s when t is increased by s; so we have, in effect, chosen a scale on L in Figure 2.5.1.

From Theorem 12 we can see that it is easy to compute the directional derivative in terms of the partial derivatives.

EXAMPLE 3. Let $f(x, y, z) = x^2e^{-yz}$. Compute the rate of change of f in the direction

$$\mathbf{v} = \left(\frac{1}{\sqrt{3}}, \frac{1}{\sqrt{3}}, \frac{1}{\sqrt{3}}\right) \qquad \text{at } (1, 0, 0)$$

(Note that $\|\mathbf{v}\| = 1$.)

This is just

$$\langle \operatorname{grad} f, \mathbf{v}\rangle = \langle (2xe^{-yz}, -x^2ze^{-yz}, -x^2ye^{-yz}), (1/\sqrt{3}, 1/\sqrt{3}, 1/\sqrt{3})\rangle$$

which at (1, 0, 0) becomes

$$\langle (2, 0, 0), (1/\sqrt{3}, 1/\sqrt{3}, 1/\sqrt{3})\rangle = 2/\sqrt{3}$$

From Theorem 12 we can also obtain the geometrical significance of the gradient:

Theorem 13. *Assume* grad $f(\mathbf{x}) \neq \mathbf{0}$. *Then* grad $f(\mathbf{x})$ *points in the direction along which f is increasing the fastest.*

Proof. If $\mathbf{n}$ is a unit vector, the rate of change of f in direction $\mathbf{n}$ is $\langle \operatorname{grad} f(\mathbf{x}), \mathbf{n}\rangle = \|\operatorname{grad} f(\mathbf{x})\| \cos\theta$ where θ is the angle between $\mathbf{n}$ and grad $f(\mathbf{x})$. This is maximum when $\theta = 0$; that is, when $\mathbf{n}$ and grad f are parallel. (If grad $f(\mathbf{x}) = \mathbf{0}$ this rate of change is 0 for any $\mathbf{n}$.) ▮

In other words, if one wishes to move in a direction in which f will increase most quickly, one should proceed in the direction $\nabla f(\mathbf{x})$. This idea will be discussed more fully following Example 4.

Consider a C^1 map $f\colon \mathbb{R}^3 \to \mathbb{R}$ and a level surface in $\mathbb{R}^3$ defined by

$$f(x, y, z) = k, \qquad k \text{ a constant}$$

Let S denote this surface. Then:

Theorem 14. *Let (x_0, y_0, z_0) lie on surface S. Then* grad $f(x_0, y_0, z_0)$ *is normal to the surface in the following sense: if $\mathbf{v}$ is the tangent vector at $t = 0$ of a path $\mathbf{c}(t)$ in S with $\mathbf{c}(0) = (x_0, y_0, z_0)$, then $\langle \operatorname{grad} f, \mathbf{v}\rangle = 0$ (see Figure 2.5.3).*

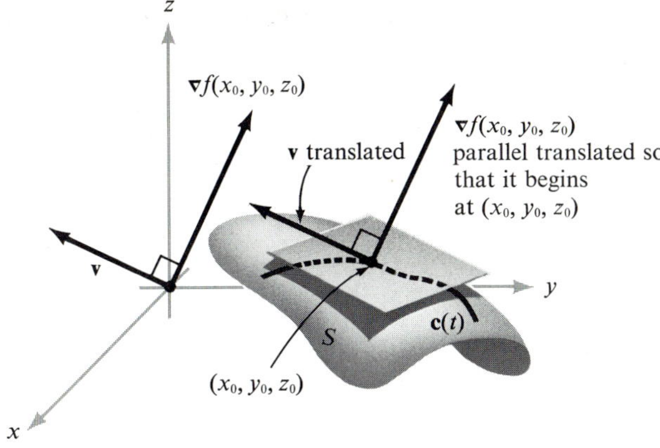

FIGURE 2.5.3
Geometric significance of the gradient: ∇f is orthogonal to the surface S on which f is constant.

Proof. Let $\mathbf{c}(t)$ lie in S. Thus $\mathbf{v} = \mathbf{c}'(0)$ and $f(\mathbf{c}(t)) = k$, since $\mathbf{c}(t)$ lies in S. Hence

$$\frac{d}{dt} f(\mathbf{c}(t))\bigg|_{t=0} = 0$$

By the Chain Rule, we get $Df(x_0, y_0, z_0)\mathbf{v} = 0$, that is, $\langle \nabla f, \mathbf{v} \rangle = 0$. ∎

If we study the conclusion of Theorem 14, we see that it is not unreasonable to *define* the plane tangent to S as follows:

Definition. *Let S be the surface consisting of those (x, y, z) such that $f(x, y, z) = k$, for k a constant. The* ***tangent plane*** *of S at a point (x_0, y_0, z_0) of S is defined by the equation*

$$\langle \nabla f(x_0, y_0, z_0), (x - x_0, y - y_0, z - z_0) \rangle = 0 \qquad (1)$$

if $\nabla f(x_0, y_0, z_0) \neq \mathbf{0}$. That is, the tangent plane is the set of points (x, y, z) that satisfy (1).

This extends the definition we gave earlier for the tangent plane of the graph of a function (see Exercise 7 below).

In Theorem 14 and the above definition we could have just as well worked in two dimensions as in three. Thus if we have $f: \mathbb{R}^2 \to \mathbb{R}$ and consider a level curve

$$C = \{(x, y) \mid f(x, y) = k\}$$

then $\nabla f(x_0, y_0)$ is perpendicular to C for any point (x_0, y_0) on C. Likewise, the tangent line to C at (x_0, y_0) has the equation

$$\langle \nabla f(x_0, y_0), (x - x_0, y - y_0) \rangle = 0 \qquad (2)$$

that is, the tangent line is the set of points (x, y) that satisfy equation (2) (see Figure 2.5.4).

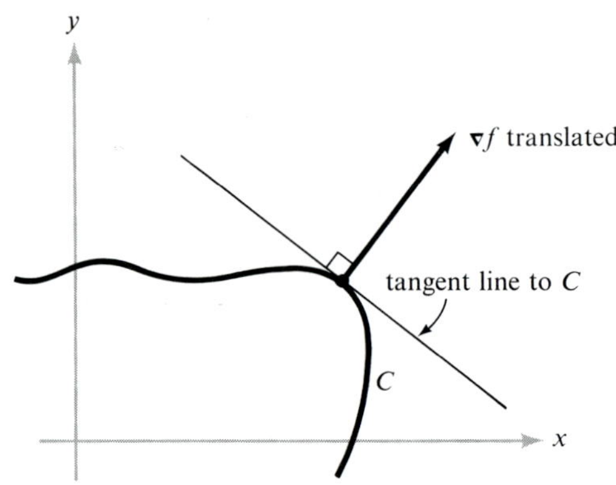

FIGURE 2.5.4
In the plane, the gradient ∇f is orthogonal to the curve $f =$ constant.

EXAMPLE 4. Compute the equation of the plane tangent to the surface defined by $3xy + z^2 = 4$ at $(1, 1, 1)$.

Here $f(x, y, z) = 3xy + z^2$ and $\nabla f = (3y, 3x, 2z)$ which at $(1, 1, 1)$ is the vector $(3, 3, 2)$. Thus the tangent plane is

$$\langle(x - 1, y - 1, z - 1), (3, 3, 2)\rangle = 0$$

$$3x + 3y + 2z = 8$$

We often speak of ∇f as a *gradient vector field*. Note that ∇f assigns a vector to each point (in the domain of f). See Figure 2.5.5. In this figure we describe the function ∇f not by drawing its graph, which would be a subset of $\mathbb{R}^6$, but by representing $\nabla f(P)$, for each point P, as a vector emanating from the point P rather than from the origin. Like a graph, this pictorial method of depicting ∇f contains the point P and the value $\nabla f(P)$ in the same picture.

The gradient vector field has important geometric significance. It tells us the direction in which f is increasing the fastest and, in addition, the direction that is orthogonal to the level surfaces of f. That it does both of these at once is intuitively quite plausible.

To see this, imagine a hill as shown in Figure 2.5.6(a). Let h be the height function, a function of two variables. If we draw level curves of h, these are just level contours of the hill. We could imagine them as level paths on the hill (see Figure 2.5.6(b)). Now one thing should be obvious to anyone who has gone for a hike: to get to the top of the hill

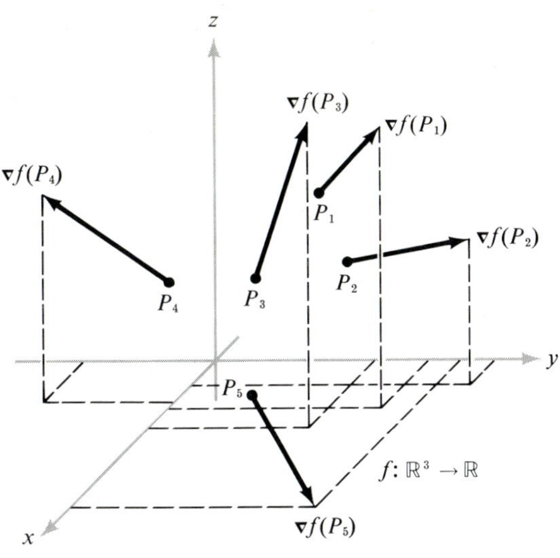

FIGURE 2.5.5
The gradient ∇f *of a function* $f: \mathbb{R}^3 \to \mathbb{R}$ *is a vector field on* $\mathbb{R}^3$; *at each point* P_i, $\nabla f(P_i)$ *is a vector emanating from* P_i.

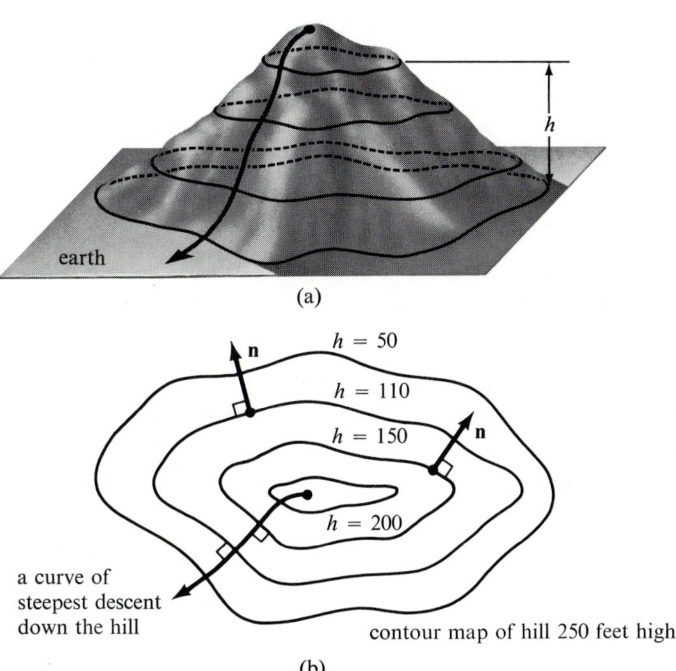

FIGURE 2.5.6
A physical illustration of the two facts (a) ∇f *is the direction of fastest increase of f and (b)* ∇f *is orthogonal to the level curves.*

the fastest, we should walk perpendicular to the level contours. This is consistent with the above theorems, which state that the direction of fastest increase (the gradient) is orthogonal to the level curves.

EXAMPLE 5. The gravitational force on a unit mass at $(x, y, z) = \mathbf{r}$ produced by a mass M at the origin in $\mathbb{R}^3$ is, according to Newton's law of gravitation, given by

$$\mathbf{F} = -\frac{GM}{r^2}\mathbf{n}$$

where G is a constant, $r = \|\mathbf{r}\| = \sqrt{x^2 + y^2 + z^2}$ is the distance of (x, y, z) from the origin, and $\mathbf{n} = \mathbf{r}/r$, the unit vector in the direction of $\mathbf{r}$.

Let us note that $\mathbf{F} = \nabla(GM/r) = -\nabla V$, that is, $\mathbf{F}$ is minus the gradient of the gravitational potential $V = -GM/r$. This can be verified as in Example 1. Notice that $\mathbf{F}$ is directed inwards towards the origin. Also, the level surfaces of V are spheres. $\mathbf{F}$ is normal to these spheres, which confirms the result of Theorem 14.

EXAMPLE 6. Find a unit vector normal to the surface S given by $z = x^2y^2 + y + 1$ at the point $(0, 0, 1)$.

Let $f(x, y, z) = x^2y^2 + y + 1 - z$, and consider the surface defined by $f(x, y, z) = 0$. Since this is the set of points (x, y, z) with $z = x^2y^2 + y + 1$ we see that this is the surface S. The gradient is given by

$$\begin{aligned}\nabla f(x, y, z) &= \frac{\partial f}{\partial x}\mathbf{i} + \frac{\partial f}{\partial y}\mathbf{j} + \frac{\partial f}{\partial z}\mathbf{k}\\ &= 2xy^2\mathbf{i} + (2x^2y + 1)\mathbf{j} - \mathbf{k}\end{aligned}$$

and so

$$\nabla f(0, 0, 1) = \mathbf{j} - \mathbf{k}$$

This vector is perpendicular to S at $(0, 0, 1)$, so to find a unit normal $\mathbf{n}$ we divide this vector by its length to obtain

$$\mathbf{n} = \frac{\nabla f(0, 0, 1)}{\|\nabla f(0, 0, 1)\|} = \frac{1}{\sqrt{2}}(\mathbf{j} - \mathbf{k})$$

EXAMPLE 7. Consider two conductors, one charged positively and the other negatively. Between them, there is an electric potential set up. This potential is a function $\varphi\colon \mathbb{R}^3 \to \mathbb{R}$. The electric field is given by $\mathbf{E} = -\nabla\varphi$. From Theorem 14 we know that $\mathbf{E}$ is perpendicular to level surfaces of φ. These level surfaces are called equipotential surfaces since the potential is constant on them (see Figure 2.5.7).

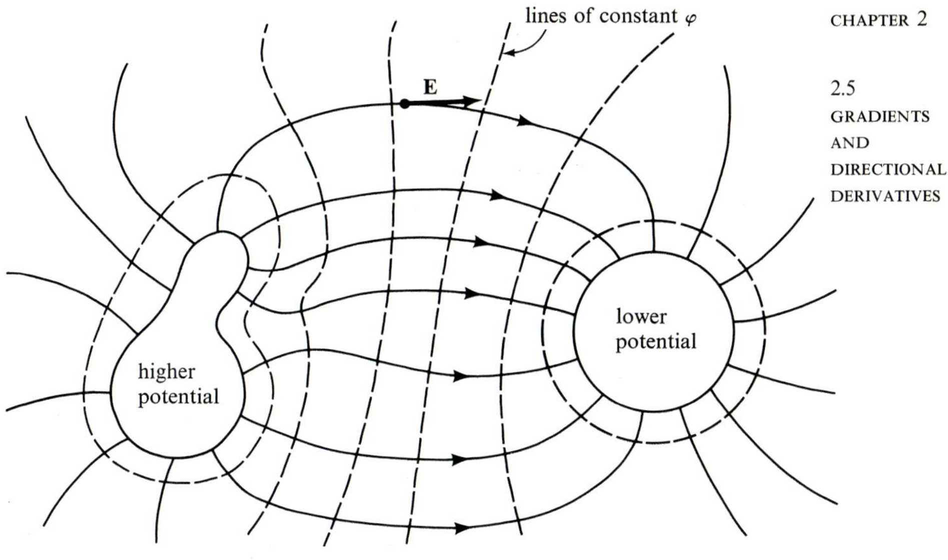

FIGURE 2.5.7
Equipotential surfaces are orthogonal to the electric force field **E**.

EXERCISES

1. Prove that the directional derivative of $f(x, y, z) = z^2x + y^3$ at $(1, 1, 2)$ in direction $(1/\sqrt{5})\mathbf{i} + (2/\sqrt{5})\mathbf{j}$ is $2\sqrt{5}$.
2. Compute the directional derivatives of the following functions at the indicated points in the given directions.
   (a) $f(x, y) = x + 2xy - 3y^2$, $(x_0, y_0) = (1, 2)$, $\mathbf{v} = \frac{3}{5}\mathbf{i} + \frac{4}{5}\mathbf{j}$
   (b) $f(x, y) = \log(\sqrt{x^2 + y^2})$, $(x_0, y_0) = (1, 0)$, $\mathbf{v} = (1/\sqrt{5})(2\mathbf{i} + \mathbf{j})$
   (c) $f(x, y) = e^x \cos(\pi y)$, $(x_0, y_0) = (0, -1)$, $\mathbf{v} = -(1/\sqrt{5})\mathbf{i} + (2/\sqrt{5})\mathbf{j}$
   (d) $f(x, y) = xy^2 + x^3y$, $(x_0, y_0) = (4, -2)$, $\mathbf{v} = (1/\sqrt{10})\mathbf{i} + (3/\sqrt{10})\mathbf{j}$
   (e) $f(x, y) = x^y$, $(x_0, y_0) = (e, e)$, $\mathbf{v} = \frac{5}{13}\mathbf{i} + \frac{12}{13}\mathbf{j}$
   (f) $f(x, y, z) = e^x + yz$, $(x_0, y_0, z_0) = (1, 1, 1)$, $\mathbf{v} = (1/\sqrt{3}, -1/\sqrt{3}, 1/\sqrt{3})$
   (g) $f(x, y, z) = xyz$, $(x_0, y_0, z_0) = (1, 0, 1)$, $\mathbf{v} = (1/\sqrt{2}, 0, -1/\sqrt{2})$
3. Find the planes tangent to the following surfaces at the indicated points.
   (a) $x^2 + 2y^2 + 3zx = 10$, $(1, 2, \frac{1}{3})$
   (b) $y^2 - x^2 = 3$, $(1, 2, 8)$
   (c) $xyz = 1$, $(1, 1, 1)$
   (d) $z = x^3 + y^3 - 6xy$, $(1, 2, -3)$
   (e) $z = (\cos x)(\cos y)$, $(0, \pi/2, 0)$
   (f) $z = (\cos x)(\sin y)$, $(0, \pi/2, 1)$
4. Let $f\colon \mathbb{R}^3 \to \mathbb{R}$ and regard $Df(x, y, z)$ as a linear map of $\mathbb{R}^3$ to $\mathbb{R}$. Show that its kernel (null space) is the linear subspace of $\mathbb{R}^3$ orthogonal to ∇f.
5. Verify Theorems 13 and 14 for $f(x, y, z) = x^2 + y^2 + z^2$.

6. Show that a unit normal to the surface $x^3y^3 + y - z + 2 = 0$ at $(0, 0, 2)$ is $\mathbf{n} = (1/\sqrt{2})(\mathbf{j} - \mathbf{k})$.
7. Show that Theorem 14 yields, as a special case, the formula for the plane tangent to the graph of $f(x, y)$ by regarding the graph as a level surface of $F(x, y, z) = f(x, y) - z$ (see Section 2.3).
8. Compute the gradient ∇f for each of the following functions.
   (a) $f(x, y, z) = \sqrt{x^2 + y^2 + z^2}$
   (b) $f(x, y, z) = xy + yz + xz$
   (c) $f(x, y, z) = \dfrac{1}{x^2 + y^2 + z^2}$
9. For the following functions $f\colon \mathbb{R}^3 \to \mathbb{R}$ and $\mathbf{g}\colon \mathbb{R} \to \mathbb{R}^3$ find ∇f and $\mathbf{g}'$ and evaluate $(f \circ \mathbf{g})'(1)$.
   (a) $f(x, y, z) = xz + yz + xy$, $\mathbf{g}(t) = (e^t, \cos t, \sin t)$
   (b) $f(x, y, z) = e^{xyz}$, $\mathbf{g}(t) = (6t, 3t^2, t^3)$
   (c) $f(x, y, z) = (x^2 + y^2 + z^2)\log\sqrt{x^2 + y^2 + z^2}$, $\mathbf{g}(t) = (e^t, e^{-t}, t)$
10. Compute the directional derivative of f in the given direction at the given point.
   (a) $f(x, y, z) = xy^2 + y^2z^3 + z^3x$, $P = (4, -2, -1)$,
   $\mathbf{v} = 1/\sqrt{14}(\mathbf{i} + 3\mathbf{j} + 2\mathbf{k})$
   (b) $f(x, y, z) = x^{yz}$, $P = (e, e, 0)$, $\mathbf{v} = \frac{12}{13}\mathbf{i} + \frac{3}{13}\mathbf{j} + \frac{4}{13}\mathbf{k}$
11. Prove

$$\nabla\left(\frac{1}{r}\right) = -\frac{\mathbf{r}}{r^3}, \quad \text{if} \quad r = \|\mathbf{r}\|.$$

12. Captain Ralph finds himself on the sunny side of Mercury and notices his space suit is melting. The temperature in a rectangular coordinate system in his vicinity is

$$T(x, y, z) = e^{-x} + e^{-2y} + e^{-3z}$$

   If he is at $(1, 1, 1)$, in what direction should he start to move in order to cool down the fastest?

2.6 ITERATED PARTIAL DERIVATIVES

In the preceding sections we have developed considerable information concerning the derivative of a map, and we have also investigated the geometry associated with the derivative of real-valued functions by making use of the gradient. We shall now proceed to study higher-order derivatives. Although this topic will be taken up in detail in Chapter 4, we shall consider it briefly here. Our main goal is to prove a theorem that asserts the equality of the "mixed partial derivatives." This result holds for functions of any number of variables, but we shall confine our attention here to real-valued functions of two or three variables.

Let $f\colon \mathbb{R}^3 \to \mathbb{R}$ be of class C^1. Remember this means that $\partial f/\partial x$, $\partial f/\partial y$, and $\partial f/\partial z$ exist and are continuous; and this implies f is differentiable

(Theorem 9). If these derivatives, in turn, have continuous partials, we say f is of *class* C^2, or is *twice continuously differentiable*. These higher derivatives are written as follows:

$$\frac{\partial^2 f}{\partial x^2} = \frac{\partial}{\partial x}\left(\frac{\partial f}{\partial x}\right)$$

$$\frac{\partial^2 f}{\partial x\, \partial y} = \frac{\partial}{\partial x}\left(\frac{\partial f}{\partial y}\right)$$

$$\frac{\partial^2 f}{\partial z\, \partial y} = \frac{\partial}{\partial z}\left(\frac{\partial f}{\partial y}\right)$$

etc. The process can of course be repeated for third-order derivatives and so on. If f: is a function of only x and y and $\partial f/\partial x$, $\partial f/\partial y$ are continuously differentiable, then by taking second partials we get the four functions

$$\frac{\partial^2 f}{\partial x^2}, \quad \frac{\partial^2 f}{\partial y^2}, \quad \frac{\partial^2 f}{\partial x\, \partial y}, \quad \text{and} \quad \frac{\partial^2 f}{\partial y\, \partial x}$$

EXAMPLE 1. Let $f(x, y) = xy + (x + 2y)^2$. Then

$$\frac{\partial f}{\partial x} = y + 2(x + 2y), \qquad \frac{\partial f}{\partial y} = x + 4(x + 2y)$$

$$\frac{\partial^2 f}{\partial x^2} = 2, \qquad \frac{\partial^2 f}{\partial y^2} = 8$$

$$\frac{\partial^2 f}{\partial x\, \partial y} = 5, \qquad \frac{\partial^2 f}{\partial y\, \partial x} = 5$$

EXAMPLE 2. If $f(x, y) = \sin x \sin^2 y$, then

$$\frac{\partial f}{\partial x} = \cos x \sin^2 y, \qquad \frac{\partial f}{\partial y} = 2 \sin x \sin y \cos y = \sin x \sin 2y$$

$$\frac{\partial^2 f}{\partial x^2} = -\sin x \sin^2 y, \qquad \frac{\partial^2 f}{\partial y^2} = 2 \sin x \cos 2y$$

$$\frac{\partial^2 f}{\partial x\, \partial y} = \cos x \sin 2y, \qquad \frac{\partial^2 f}{\partial y\, \partial x} = 2 \cos x \sin y \cos y = \cos x \sin 2y$$

EXAMPLE 3. Let $f(x, y, z) = e^{xy} + z \cos x$. Then, for example,

$$\frac{\partial f}{\partial x} = ye^{xy} - z \sin x, \qquad \frac{\partial f}{\partial y} = xe^{xy}, \qquad \frac{\partial f}{\partial z} = \cos x$$

$$\frac{\partial^2 f}{\partial z\, \partial x} = -\sin x, \qquad \frac{\partial^2 f}{\partial x\, \partial z} = -\sin x$$

In all these examples note that the mixed partials, like $\partial^2 f/(\partial x\, \partial y)$ and $\partial^2 f/(\partial y\, \partial x)$, or $\partial^2 f/(\partial z\, \partial x)$ and $\partial^2 f/(\partial x\, \partial z)$, are equal. It is a very basic and perhaps surprising fact that this is usually the case. We shall prove this in the next theorem for functions $f(x, y)$ of two variables, but the proof can easily be extended to functions of n variables.

Theorem 15. *If f is C^2 (twice continuously differentiable) then the mixed partials are equal; that is,*

$$\frac{\partial^2 f}{\partial x\, \partial y} = \frac{\partial^2 f}{\partial y\, \partial x}$$

Proof. Consider the expression

$$f(x_0 + \Delta x, y_0 + \Delta y) - f(x_0 + \Delta x, y_0) - f(x_0, y_0 + \Delta y) + f(x_0, y_0) \tag{1}$$

We fix y_0, Δy, and introduce the function

$$g(x) = f(x, y_0 + \Delta y) - f(x, y_0)$$

so that (1) equals $g(x_0 + \Delta x) - g(x_0)$. By the Mean Value Theorem for functions of one variable (p. 79) this equals $g'(\bar{x})\, \Delta x$ for $\bar{x}$ between x_0 and $x_0 + \Delta x$. Hence (1) equals

$$\left[\frac{\partial f}{\partial x}(\bar{x}, y_0 + \Delta y) - \frac{\partial f}{\partial x}(\bar{x}, y_0)\right] \Delta x$$

Applying the Mean Value Theorem again we get for (1)

$$\frac{\partial^2 f}{\partial y\, \partial x}(\bar{x}, \bar{y})\, \Delta x\, \Delta y$$

Since $\partial^2 f/\partial y\, \partial x$ is continuous, it follows that

$$\frac{\partial^2 f}{\partial y\, \partial x}(x_0, y_0) = \lim_{(\Delta x, \Delta y) \to (0, 0)} \frac{1}{\Delta x\, \Delta y}[f(x_0 + \Delta x, y_0 + \Delta y) - f(x_0 + \Delta x, y_0) - f(x_0, y_0 + \Delta y) + f(x_0, y_0)] \tag{2}$$

The right-hand side of (2) is symmetric in x and y, so that in this derivation we could have reversed the roles of x and y. In other words, in the same manner one proves that $\partial^2 f/\partial x\, \partial y$ is given by the same limit, and we obtain the desired result. ∎

From this theorem it follows, for example, that for a C^3 function

$$\frac{\partial^3 f}{\partial x\, \partial y\, \partial z} = \frac{\partial^3 f}{\partial z\, \partial y\, \partial x} = \frac{\partial^3 f}{\partial y\, \partial z\, \partial x} \text{ etc.}$$

In other words, we may compute iterated partial derivatives in any order we please.

EXAMPLE 4. Verify the equality of the mixed second partial derivatives for the function

$$f(x, y) = xe^y + yx^2$$

Here

$$\frac{\partial f}{\partial x} = e^y + 2xy, \qquad \frac{\partial f}{\partial y} = xe^y + x^2$$

$$\frac{\partial^2 f}{\partial y\, \partial x} = e^y + 2x, \qquad \frac{\partial^2 f}{\partial x\, \partial y} = e^y + 2x$$

and so we have

$$\frac{\partial^2 f}{\partial y\, \partial x} = \frac{\partial^2 f}{\partial x\, \partial y}$$

EXERCISES

1. Compute the second partial derivatives $\partial^2 f/\partial x^2$, $\partial^2 f/\partial x\, \partial y$, $\partial^2 f/\partial y\, \partial x$, $\partial^2 f/\partial y^2$ for each of the following functions. Verify Theorem 15 in each case:
   (a) $f(x, y) = 2xy/((x^2 + y^2)^2)$
   (b) $f(x, y, z) = e^z + 1/x + xe^{-y}$
   (c) $f(x, y) = \cos(xy^2)$
   (d) $f(x, y) = e^{-xy^2} + y^3x^4$
   (e) $f(x, y) = 1/(\cos^2 x + e^{-y})$
2. Let

$$f(x, y) = \begin{cases} xy(x^2 - y^2)/(x^2 + y^2), & (x, y) \neq (0, 0) \\ 0, & (x, y) = (0, 0) \end{cases}$$

   (a) If $(x, y) \neq (0, 0)$ calculate $\partial f/\partial x$ and $\partial f/\partial y$.
   (b) Show that $(\partial f/\partial x)(0, 0) = 0 = (\partial f/\partial y)(0, 0)$
   (c) Show that $(\partial^2 f/\partial x\, \partial y)(0, 0) = 1$, $(\partial^2 f/\partial y\, \partial x)(0, 0) = -1$
   What went wrong? Why are the mixed partials not equal?
3. Let $f(x, y)$ be a given function on $\mathbb{R}^2$ and let $\mathbf{c}(t)$ be a curve in $\mathbb{R}^2$. Write a formula for $(d^2/dt^2)\,(f \circ \mathbf{c}(t))$ using the Chain Rule twice.
4. Show that $u(x, y) = e^x \sin y$ satisfies Laplace's equation

$$\frac{\partial^2 u}{\partial x^2} + \frac{\partial^2 u}{\partial y^2} = 0$$

5. Find the equation of the plane tangent to $z = x^2 + 2y^3$ at $(1, 1, 3)$.

6. Verify that

$$\frac{\partial^3 f}{\partial x\,\partial y\,\partial z} = \frac{\partial^3 f}{\partial z\,\partial y\,\partial x}$$

for $f(x, y, z) = ze^{xy} + yz^3x^2$.

7. Evaluate all first and second partial derivatives of the following functions.
   (a) $f(x, y) = x \arctan(x/y)$
   (b) $f(x, y) = \cos\sqrt{x^2 + y^2}$
   (c) $f(x, y) = \exp(-x^2 - y^2)$
8. Find the directional derivatives of the functions in Exercise 7 at the point $(1, 1)$ in the direction $(1/\sqrt{2}, 1/\sqrt{2})$.
9. Use Theorem 15 to show that if $f(x, y, z)$ is of class C^3, then

$$\frac{\partial^3 f}{\partial x\,\partial y\,\partial z} = \frac{\partial^3 f}{\partial y\,\partial z\,\partial x}$$

10. Let g and f be differentiable functions from $\mathbb{R}$ to $\mathbb{R}$. Set $\phi(x, t) = f(x - t) + g(x + t)$. Prove that ϕ satisfies the equation

$$\frac{\partial^2 \phi}{\partial t^2} = \frac{\partial^2 \phi}{\partial x^2}$$

(This is called the *wave equation.*)

2.7 SOME TECHNICAL THEOREMS

In this section we shall examine the definition of the derivative in further detail and shall supply proofs omitted from sections 2.2, 2.3, and 2.4.

We shall begin by supplying the proofs of the limit theorems in section 2.2. Let us recall the definition of limit.

Definition of limit. *Let $f\colon A \subset \mathbb{R}^n \to \mathbb{R}^m$ where A is open. Let $\mathbf{x}_0$ be in A or be a boundary point of A, and let N be a neighborhood of $\mathbf{b} \in \mathbb{R}^m$. We say f is* ***eventually in N as $\mathbf{x}$ approaches $\mathbf{x}_0$*** *if there exists a neighborhood U of $\mathbf{x}_0$ such that $\mathbf{x} \neq \mathbf{x}_0$, $\mathbf{x} \in U$, and $\mathbf{x} \in A$ implies $f(\mathbf{x}) \in N$. We say $f(\mathbf{x})$* ***approaches*** $\mathbf{b}$ *as $\mathbf{x}$ approaches $\mathbf{x}_0$, or, in symbols,*

$$\lim_{\mathbf{x}\to\mathbf{x}_0} f(\mathbf{x}) = \mathbf{b} \qquad \text{or} \qquad f(\mathbf{x}) \to \mathbf{b} \quad \text{as} \quad \mathbf{x} \to \mathbf{x}_0$$

when, given ***any*** *neighborhood N of $\mathbf{b}$, f is eventually in N as $\mathbf{x}$ approaches $\mathbf{x}_0$. It may be that as $\mathbf{x}$ approaches $\mathbf{x}_0$ the values $f(\mathbf{x})$ do not get close to any particular number. In this case we say that the $\lim_{\mathbf{x}\to\mathbf{x}_0} f(\mathbf{x})$* ***does not exist.***

Let us now establish that this is equivalent to the ε-δ formulation of limits.

Theorem 2. *Let $f\colon A \subset \mathbb{R}^n \to \mathbb{R}^m$ and let $\mathbf{x}_0$ be in A or be a boundary point of A. Then $\lim_{\mathbf{x}\to\mathbf{x}_0} f(\mathbf{x}) = \mathbf{b}$ if and only if for every number $\varepsilon > 0$ there is a $\delta > 0$ such that for $\mathbf{x} \in A$ and $0 < \|\mathbf{x} - \mathbf{x}_0\| < \delta$ we have $\|f(\mathbf{x}) - \mathbf{b}\| < \varepsilon$.*

Proof. First let us assume that $\lim_{\mathbf{x}\to\mathbf{x}_0} f(\mathbf{x}) = \mathbf{b}$. Let $\varepsilon > 0$ be given, and consider the ε-neighborhood $N = D_\varepsilon(\mathbf{b})$, the ball or disc of radius ε with center $\mathbf{b}$. By the definition of limit, f is eventually in $D_\varepsilon(\mathbf{b})$ as $\mathbf{x}$ approaches $\mathbf{x}_0$, which means there is a neighborhood U of $\mathbf{x}_0$ such that $f(\mathbf{x}) \in D_\varepsilon(\mathbf{b})$ if $\mathbf{x} \in U$, $\mathbf{x} \in A$, and $\mathbf{x} \neq \mathbf{x}_0$. Now since U is open and $\mathbf{x}_0 \in U$, there is a $\delta > 0$ such that $D_\delta(\mathbf{x}_0) \subset U$. Consequently, $0 < \|\mathbf{x} - \mathbf{x}_0\| < \delta$ and $\mathbf{x} \in A$ implies $\mathbf{x} \in D_\delta(\mathbf{x}_0) \subset U$. Thus $f(\mathbf{x}) \in D_\varepsilon(\mathbf{b})$, which means that $\|f(\mathbf{x}) - \mathbf{b}\| < \varepsilon$. This is the ε-δ assertion we wanted to prove.

We still have to prove the converse. Assume that for every $\varepsilon > 0$ there is a $\delta > 0$ such that $0 < \|\mathbf{x} - \mathbf{x}_0\| < \delta$ and $\mathbf{x} \in A$ implies $\|f(\mathbf{x}) - \mathbf{b}\| < \varepsilon$. Let N be a neighborhood of $\mathbf{b}$. We have to show that f is eventually in N as $\mathbf{x} \to \mathbf{x}_0$, that is, we must find an open set $U \subset \mathbb{R}^n$ such that $\mathbf{x} \in U$, $\mathbf{x} \in A$, and $\mathbf{x} \neq \mathbf{x}_0$ implies $f(\mathbf{x}) \in N$. Now since N is open, there is an $\varepsilon > 0$ such that $D_\varepsilon(\mathbf{b}) \subset N$. If we choose $U = D_\delta(\mathbf{x})$ (according to our assumption), then $\mathbf{x} \in U$, $\mathbf{x} \in A$, and $\mathbf{x} \neq \mathbf{x}_0$ means $\|f(\mathbf{x}) - \mathbf{b}\| < \varepsilon$, that is, $f(\mathbf{x}) \in D_\varepsilon(\mathbf{b}) \subset N$. ▮

***Theorem 3* (*Uniqueness of limits*).** *If* $\lim_{\mathbf{x}\to\mathbf{x}_0} f(\mathbf{x}) = \mathbf{b}_1$ *and* $\lim_{\mathbf{x}\to\mathbf{x}_0} f(\mathbf{x}) = \mathbf{b}_2$, *then* $\mathbf{b}_1 = \mathbf{b}_2$.

Proof. It is convenient to use the ε-δ formulation of Theorem 2. Let us suppose $f(\mathbf{x}) \to \mathbf{b}_1$ and $f(\mathbf{x}) \to \mathbf{b}_2$ as $\mathbf{x} \to \mathbf{x}_0$. Given $\varepsilon > 0$ we can, by assumption, find $\delta_1 > 0$ such that if $\mathbf{x} \in A$ and $0 < \|\mathbf{x} - \mathbf{x}_0\| < \delta_1$, then $\|f(\mathbf{x}) - \mathbf{b}_1\| < \varepsilon$, and similarly, $\delta_2 > 0$ such that $0 < \|\mathbf{x} - \mathbf{x}_0\| < \delta_2$ implies $\|f(\mathbf{x}) - \mathbf{b}_2\| < \varepsilon$. Let δ be the smaller of δ_1 and δ_2. Choose $\mathbf{x}$ such that $0 < \|\mathbf{x} - \mathbf{x}_0\| < \delta$ and $\mathbf{x} \in A$. Such $\mathbf{x}$'s exist, because $\mathbf{x}_0$ is in A or is a boundary point of A. Thus, using the triangle inequality, we have

$$\begin{aligned}\|\mathbf{b}_1 - \mathbf{b}_2\| &= \|(\mathbf{b}_1 - f(\mathbf{x})) + (f(\mathbf{x}) - \mathbf{b}_2)\| \\ &\leq \|\mathbf{b}_1 - f(\mathbf{x})\| + \|f(\mathbf{x}) - \mathbf{b}_2\| \\ &< \varepsilon + \varepsilon = 2\varepsilon\end{aligned}$$

Thus for *every* $\varepsilon > 0$ we have $\|\mathbf{b}_1 - \mathbf{b}_2\| < 2\varepsilon$. Hence $\mathbf{b}_1 = \mathbf{b}_2$, for if $\mathbf{b}_1 \neq \mathbf{b}_2$ we could let $\varepsilon = \|\mathbf{b}_1 - \mathbf{b}_2\|/2 > 0$ and we would have $\|\mathbf{b}_1 - \mathbf{b}_2\| < \|\mathbf{b}_1 - \mathbf{b}_2\|$, an impossibility. ▮

Theorem 4. *If* $f\colon A \subset \mathbb{R}^n \to \mathbb{R}^m$, $g\colon A \subset \mathbb{R}^n \to \mathbb{R}^m$, $\mathbf{x}_0$ *is in* A *or is a boundary point of* A, $\mathbf{b} \in \mathbb{R}^m$, *and* $c \in \mathbb{R}$; *then*

(*i*) *if* $\lim_{\mathbf{x}\to\mathbf{x}_0} f(\mathbf{x}) = \mathbf{b}$, *then* $\lim_{\mathbf{x}\to\mathbf{x}_0} cf(\mathbf{x}) = c\mathbf{b}$ *where* $cf\colon A \to \mathbb{R}^m$ *is defined by* $\mathbf{x} \mapsto c(f(\mathbf{x}))$.

(*ii*) *if* $\lim_{\mathbf{x}\to\mathbf{x}_0} f(\mathbf{x}) = \mathbf{b}_1$ *and* $\lim_{\mathbf{x}\to\mathbf{x}_0} g(\mathbf{x}) = \mathbf{b}_2$, *then* $\lim_{\mathbf{x}\to\mathbf{x}_0} (f+g)(\mathbf{x}) = \mathbf{b}_1 + \mathbf{b}_2$, *where* $(f+g)\colon A \to \mathbb{R}^m$, $\mathbf{x} \mapsto f(\mathbf{x}) + g(\mathbf{x})$.

(*iii*) *if* $m = 1$, *and* $\lim_{\mathbf{x}\to\mathbf{x}_0} f(\mathbf{x}) = b_1$ *and* $\lim_{\mathbf{x}\to\mathbf{x}_0} g(\mathbf{x}) = b_2$ *then* $\lim_{\mathbf{x}\to\mathbf{x}_0} (fg)(\mathbf{x}) = b_1 b_2$ *where* $(fg)\colon A \to \mathbb{R}$, $\mathbf{x} \mapsto f(\mathbf{x})g(\mathbf{x})$.

(*iv*) *if* $m = 1$ *and* $\lim_{\mathbf{x}\to\mathbf{x}_0} f(\mathbf{x}) = b$, $b \neq 0$, *and* $f(\mathbf{x}) \neq 0$ *for* $\mathbf{x} \in A$, *then* $\lim_{\mathbf{x}\to\mathbf{x}_0} 1/f = 1/b$ *where* $1/f\colon A \to \mathbb{R}$, $\mathbf{x} \mapsto 1/f(\mathbf{x})$.

(*v*) *if* $f(\mathbf{x}) = (f_1(\mathbf{x}), \ldots, f_m(\mathbf{x}))$, *where* $f_i\colon A \to \mathbb{R}$, $i = 1, \ldots, m$, *are the component functions of f, then* $\lim_{\mathbf{x}\to\mathbf{x}_0} f(\mathbf{x}) = \mathbf{b} = (b_1, \ldots, b_m)$ *if and only if* $\lim_{\mathbf{x}\to\mathbf{x}_0} f_i(\mathbf{x}) = b_i$ *for each* $i = 1, \ldots, m$.

Proof. We shall illustrate the technique of proof by proving (*i*) and (*ii*). The proofs of the other assertions are just a bit more complicated and may be supplied by the reader. In each case, the ε-δ formulation of Theorem 2 is probably the most convenient approach.

To prove (*i*), let $\varepsilon > 0$ be given; we must produce a number $\delta > 0$ such that $\|cf(\mathbf{x}) - c\mathbf{b}\| < \varepsilon$ if $0 < \|\mathbf{x} - \mathbf{x}_0\| < \delta$. If $c = 0$, any δ will do, so we can suppose $c \neq 0$. Let $\varepsilon' = \varepsilon/|c|$; from the definition of limit, there is a δ with the property that $0 < \|\mathbf{x} - \mathbf{x}_0\| < \delta$ implies $\|f(\mathbf{x}) - \mathbf{b}\| < \varepsilon' = \varepsilon/|c|$. Thus $0 < \|\mathbf{x} - \mathbf{x}_0\| < \delta$ implies $\|cf(\mathbf{x}) - c\mathbf{b}\| = |c|\,\|f(\mathbf{x}) - \mathbf{b}\| < \varepsilon$. This proves (a).

To prove (*ii*), let $\varepsilon > 0$ be given again. Choose $\delta_1 > 0$ such that $0 < \|\mathbf{x} - \mathbf{x}_0\| < \delta_1$ implies $\|f(\mathbf{x}) - \mathbf{b}_1\| < \varepsilon/2$. Similarly, choose $\delta_2 > 0$ such that $0 < \|\mathbf{x} - \mathbf{x}_0\| < \delta_2$ implies $\|g(\mathbf{x}) - \mathbf{b}_2\| < \varepsilon/2$. Let δ be the lesser of δ_1 and δ_2. Then $0 < \|\mathbf{x} - \mathbf{x}_0\| < \delta$ implies

$$\begin{aligned}\|f(\mathbf{x}) + g(\mathbf{x}) - \mathbf{b}_1 - \mathbf{b}_2\| &\leq \|f(\mathbf{x}) - \mathbf{b}_1\| + \|g(\mathbf{x}) - \mathbf{b}_2\| \\ &< \varepsilon/2 + \varepsilon/2 = \varepsilon\end{aligned}$$

Thus we have proved that $(f + g)(\mathbf{x}) \to \mathbf{b}_1 + \mathbf{b}_2$ as $\mathbf{x} \to \mathbf{x}_0$. ∎

Let us recall the definition of a continuous function.

Definition. *Let* $f\colon A \subset \mathbb{R}^n \to \mathbb{R}^m$ *be a given function with domain A. Let* $\mathbf{x}_0 \in A$. *We say f is* ***continuous at*** $\mathbf{x}_0$ *if and only if*

$$\lim_{\mathbf{x}\to\mathbf{x}_0} f(\mathbf{x}) = f(\mathbf{x}_0)$$

If we just say that f is ***continuous****, we shall mean that f is continuous at each point* $\mathbf{x}_0$ *of A.*

From Theorem 2, we get Theorem 5: f is continuous at $\mathbf{x}_0 \in A$ if and only if for every number $\varepsilon > 0$ there is a number $\delta > 0$ such that

$$\mathbf{x} \in A \text{ and } \|\mathbf{x} - \mathbf{x}_0\| < \delta \quad \text{implies} \quad \|f(\mathbf{x}) - f(\mathbf{x}_0)\| < \varepsilon$$

One of the properties of continuous functions stated without proof in Section 2.2 was the following:

Theorem 7. *Let* $f\colon A \subset \mathbb{R}^n \to \mathbb{R}^m$ *and let* $g\colon B \subset \mathbb{R}^m \to \mathbb{R}^p$. *Suppose* $f(A) \subset B$ *so that* $g \circ f$ *is defined on A. If f is continuous at* $\mathbf{x}_0 \in A$ *and g is continuous at* $\mathbf{y}_0 = f(\mathbf{x}_0)$, *then* $g \circ f$ *is continuous at* $\mathbf{x}_0$.

Proof. We shall use the ε-δ criterion for continuity. Thus, given $\varepsilon > 0$, we must find $\delta > 0$ such that for $\mathbf{x} \in A$,

$$\|\mathbf{x} - \mathbf{x}_0\| < \delta \text{ implies } \|(g \circ f)(\mathbf{x}) - (g \circ f)(\mathbf{x}_0)\| < \varepsilon$$

Since g is continuous at $f(\mathbf{x}_0) = \mathbf{y}_0 \in B$, there is a $\gamma > 0$ such that for $\mathbf{y} \in B$,

$$\|\mathbf{y} - \mathbf{y}_0\| < \gamma \text{ implies } \|g(\mathbf{y}) - g(\mathbf{y}_0)\| < \varepsilon$$

As f is continuous at $\mathbf{x}_0 \in A$, there is, for this γ, a $\delta > 0$ such that for $\mathbf{x} \in A$,

$$\|\mathbf{x} - \mathbf{x}_0\| < \delta \text{ implies } \|f(\mathbf{x}) - f(\mathbf{x}_0)\| < \gamma$$

which in turn implies

$$\|g(f(\mathbf{x})) - g(f(\mathbf{x}_0))\| < \varepsilon$$

which is the desired conclusion. ∎

To simplify the exposition in Section 2.3 we assumed, as part of the definition of $Df(\mathbf{x}_0)$, that the partial derivatives of f existed. Our next objective is to show that this assumption can be omitted. Let us begin by redefining "differentiable." Theorem 16 will show that this definition is equivalent to the old one.

Definition. *Let U be an open set in $\mathbb{R}^n$ and let $f\colon U \subset \mathbb{R}^n \to \mathbb{R}^m$ be a given function. We say that f is **differentiable** at $\mathbf{x}_0 \in U$ if and only if there exists an $m \times n$ matrix T such that*

$$\operatorname*{limit}_{\mathbf{x} \to \mathbf{x}_0} \frac{\|f(\mathbf{x}) - f(\mathbf{x}_0) - T(\mathbf{x} - \mathbf{x}_0)\|}{\|\mathbf{x} - \mathbf{x}_0\|} = 0 \tag{1}$$

*We call T the **derivative** of f at $\mathbf{x}_0$ and denote it by $Df(\mathbf{x}_0)$. In matrix notation, $T(\mathbf{x} - \mathbf{x}_0)$ stands for (see Section 1.4)*

$$\begin{bmatrix} t_{11} & t_{12} & \cdots & t_{1n} \\ t_{21} & t_{22} & \cdots & t_{2n} \\ \vdots & \vdots & & \vdots \\ t_{m1} & t_{m2} & \cdots & t_{mn} \end{bmatrix} \begin{pmatrix} x_1 - x_{01} \\ \\ \vdots \\ x_n - x_{0n} \end{pmatrix}$$

where $\mathbf{x} = (x_1, \ldots, x_n)$ and $\mathbf{x}_0 = (x_{01}, \ldots, x_{0n})$. Sometimes we write $T(\mathbf{y})$ as $T \cdot \mathbf{y}$ or just $T\mathbf{y}$.

Condition (1) can be rewritten as

$$\operatorname*{limit}_{\mathbf{h} \to \mathbf{0}} \frac{\|f(\mathbf{x}_0 + \mathbf{h}) - f(\mathbf{x}_0) - T\mathbf{h}\|}{\|\mathbf{h}\|} = 0 \tag{2}$$

by letting $\mathbf{h} = \mathbf{x} - \mathbf{x}_0$. Written in terms of ε-δ notation, (2) says that for every $\varepsilon > 0$ there is a $\delta > 0$ such that $0 < \|\mathbf{h}\| < \delta$ implies

$$\frac{\|f(\mathbf{x}_0 + \mathbf{h}) - f(\mathbf{x}_0) - T\mathbf{h}\|}{\|\mathbf{h}\|} < \varepsilon$$

or in other words,

$$\|f(\mathbf{x}_0 + \mathbf{h}) - f(\mathbf{x}_0) - T\mathbf{h}\| < \varepsilon\|\mathbf{h}\|$$

Notice that because U is open, as long as δ is small enough, $\|\mathbf{h}\| < \delta$ implies $\mathbf{x}_0 + \mathbf{h} \in U$.

Our task is to show that the matrix T is necessarily the matrix of partial derivatives, and hence that this abstract definition agrees with the definition of differentiability given in Section 2.3.

Theorem 16. *Suppose $f\colon U \subset \mathbb{R}^n \to \mathbb{R}^m$ is differentiable at $\mathbf{x}_0 \in \mathbb{R}^n$. Then all the partial derivatives of f exist at the point $\mathbf{x}_0$ and the $m \times n$ matrix T given by*

$$(t_{ij}) = \left(\frac{\partial f_i}{\partial x_j}\right)$$

that is,

$$Df(\mathbf{x}_0) = \begin{bmatrix} \dfrac{\partial f_1}{\partial x_1} & \cdots & \dfrac{\partial f_1}{\partial x_n} \\ \vdots & & \vdots \\ \dfrac{\partial f_m}{\partial x_1} & \cdots & \dfrac{\partial f_m}{\partial x_n} \end{bmatrix}$$

where $\partial f_i/\partial x_j$ is evaluated at $\mathbf{x}_0$. In particular, this implies that T is uniquely determined; there is no other matrix satisfying condition (1).

Proof. By Theorem 4(v), condition (2) is the same as

$$\underset{\mathbf{h}\to\mathbf{0}}{\text{limit}} \frac{|f_i(\mathbf{x}_0 + \mathbf{h}) - f_i(\mathbf{x}_0) - (T\mathbf{h})_i|}{\|\mathbf{h}\|} = 0,\ 1 \le i \le m$$

Here $(T\mathbf{h})_i$ stands for the ith component of the column vector $T\mathbf{h}$. Now let $\mathbf{h} = a\mathbf{e}_j = (0, \dots, a, \dots, 0)$, the number a in the jth spot and zeros elsewhere. We get

$$\underset{a\to 0}{\text{limit}} \frac{|f_i(\mathbf{x}_0 + a\mathbf{e}_j) - f_i(\mathbf{x}_0) - a(T\mathbf{e}_j)_i|}{|a|} = 0$$

or in other words

$$\underset{a\to 0}{\text{limit}} \left| \frac{f_i(\mathbf{x}_0 + a\mathbf{e}_j) - f_i(\mathbf{x}_0)}{a} - (T\mathbf{e}_j)_i \right| = 0$$

so that

$$\underset{a\to 0}{\text{limit}} \frac{f_i(\mathbf{x}_0 + a\mathbf{e}_j) - f_i(\mathbf{x}_0)}{a} = (T\mathbf{e}_j)_i$$

But this limit is nothing more than the partial derivative $\partial f_i/\partial x_j$ evaluated at the point $\mathbf{x}_0$. Thus we have proved that $\partial f_i/\partial x_j$ exists and equals $(T\mathbf{e}_j)_i$. But $(T\mathbf{e}_j)_i = t_{ij}$ (see Section 1.4), so our theorem follows. ∎

Our next task is to show that differentiability implies continuity.

Theorem 8. *Let $f\colon U \subset \mathbb{R}^n \to \mathbb{R}^m$ be differentiable at $\mathbf{x}_0$. Then f is continuous at $\mathbf{x}_0$, and furthermore, $\|f(\mathbf{x}) - f(\mathbf{x}_0)\| < M_1 \|\mathbf{x} - \mathbf{x}_0\|$ for some constant M_1 and $\mathbf{x}$ near $\mathbf{x}_0$, $\mathbf{x} \neq \mathbf{x}_0$.*

Proof. We shall need to use the result of Exercise 2, p. 131; namely $\|Df(\mathbf{x}_0) \cdot \mathbf{h}\| \le M\|\mathbf{h}\|$ where M is the square root of the sum of the squares of the matrix elements in $Df(\mathbf{x}_0)$.

Choose $\varepsilon = 1$. Then by the definition of the derivative (see formula (2)) there is a $\delta_1 > 0$ such that $0 < \|\mathbf{h}\| < \delta_1$ implies

$$\|f(\mathbf{x}_0 + \mathbf{h}) - f(\mathbf{x}_0) - Df(\mathbf{x}_0) \cdot \mathbf{h}\| < \varepsilon\|\mathbf{h}\| = \|\mathbf{h}\|$$

Now notice that if $\|\mathbf{h}\| < \delta_1$, then

$$\begin{aligned}\|f(\mathbf{x}_0 + \mathbf{h}) - f(\mathbf{x}_0)\| &= \|f(\mathbf{x}_0 + \mathbf{h}) - f(\mathbf{x}_0) - Df(\mathbf{x}_0)\cdot\mathbf{h} + Df(\mathbf{x}_0)\cdot\mathbf{h}\|\\ &\le \|f(\mathbf{x}_0 + \mathbf{h}) - f(\mathbf{x}_0) - Df(\mathbf{x}_0)\cdot\mathbf{h}\| + \|Df(\mathbf{x}_0)\cdot\mathbf{h}\|\\ &< \|\mathbf{h}\| + M\|\mathbf{h}\| = (1 + M)\|\mathbf{h}\|\end{aligned}$$

(Notice that in this derivation we have used the triangle inequality.) Setting $M_1 = 1 + M$ proves the second assertion of the theorem.

Now let ε' be any positive number, and let δ be the smaller of δ_1 and $\varepsilon'/(1 + M)$. Then $\|\mathbf{h}\| < \delta$ implies

$$\|f(\mathbf{x}_0 + \mathbf{h}) - f(\mathbf{x}_0)\| < (1 + M)\varepsilon'/(1 + M) = \varepsilon'$$

which proves that

$$\operatorname*{limit}_{\mathbf{h}\to\mathbf{0}} f(\mathbf{x}_0 + \mathbf{h}) = f(\mathbf{x}_0)$$

or that f is continuous at $\mathbf{x}_0$. ∎

We asserted in Section 2.3 that an important criterion for differentiability is that the partials exist and are continuous.

Theorem 9. *Let $f\colon U \subset \mathbb{R}^n \to \mathbb{R}^m$. Suppose the partial derivatives $\partial f_i/\partial x_j$ of f all exist and are continuous in a neighborhood of a point $\mathbf{x} \in U$. Then f is differentiable at $\mathbf{x}$.*

Proof. (In this proof we are going to use the Mean Value Theorem from one-variable calculus. See p. 79 above for the statement.)

It suffices to consider the case $m = 1$, for the same reasons as in the proof of Theorem 16, so let us assume $f\colon U \subset \mathbb{R}^n \to \mathbb{R}$. We have to show that

$$\operatorname*{limit}_{\mathbf{h}\to\mathbf{0}} \frac{\left| f(\mathbf{x} + \mathbf{h}) - f(\mathbf{x}) - \sum_{i=1}^{n} \frac{\partial f}{\partial x_i}(\mathbf{x})h_i \right|}{\|\mathbf{h}\|} = 0$$

Write

$$\begin{aligned} &f(x_1 + h_1, \ldots, x_n + h_n) - f(x_1, \ldots, x_n)\\ &= f(x_1 + h_1, \ldots, x_n + h_n) - f(x_1, x_2 + h_2, \ldots, x_n + h_n)\\ &\quad + f(x_1, x_2 + h_2, \ldots, x_n + h_n) - f(x_1, x_2, x_3 + h_3, \ldots, x_n + h_n)\\ &\quad + \cdots\\ &\quad + f(x_1, \ldots, x_{n-1} + h_{n-1}, x_n + h_n) - f(x_1, \ldots, x_{n-1}, x_n + h_n)\\ &\quad + f(x_1, \ldots, x_{n-1}, x_n + h_n) - f(x_1, \ldots, x_n)\end{aligned}$$

(This is called a "telescoping sum" since each term cancels with the succeeding or preceding one, except the first and last.) By the Mean Value Theorem, this expression may be written as

$$f(\mathbf{x} + \mathbf{h}) - f(\mathbf{x}) = \frac{\partial f}{\partial x_1}(\mathbf{y}_1)h_1 + \frac{\partial f}{\partial x_2}(\mathbf{y}_2)h_2 + \cdots + \frac{\partial f}{\partial x_n}(\mathbf{y}_n)h_n$$

where $\mathbf{y}_1 = (c_1, x_2 + h_2, \ldots, x_n + h_n)$, and c_1 lies between x_1 and $x_1 + h_1$;

$\mathbf{y}_2 = (x_1, c_2, x_3 + h_3, \ldots, x_n + h_n)$, and c_2 lies between x_2 and $x_2 + h_2$; and $\mathbf{y}_n = (x_1, \ldots, x_{n-1}, c_n)$ where c_n lies between x_n and $x_n + h_n$. Thus we can write

$$\left| f(\mathbf{x}+\mathbf{h}) - f(\mathbf{x}) - \sum_{i=1}^{n} \frac{\partial f}{\partial x_i}(\mathbf{x})h_i \right|$$
$$= \left| \left(\frac{\partial f}{\partial x_1}(\mathbf{y}_1) - \frac{\partial f}{\partial x_1}(\mathbf{x}) \right) h_1 + \cdots + \left(\frac{\partial f}{\partial x_n}(\mathbf{y}_n) - \frac{\partial f}{\partial x_n}(\mathbf{x}) \right) h_n \right|$$

By the triangle inequality, this expression is less than or equal to

$$\left| \frac{\partial f}{\partial x_1}(\mathbf{y}_1) - \frac{\partial f}{\partial x_1}(\mathbf{x}) \right| |h_1| + \cdots + \left| \frac{\partial f}{\partial x_n}(\mathbf{y}_n) - \frac{\partial f}{\partial x_n}(\mathbf{x}) \right| |h_n|$$
$$\le \left\{ \left| \frac{\partial f}{\partial x_1}(\mathbf{y}_1) - \frac{\partial f}{\partial x_1}(\mathbf{x}) \right| + \cdots + \left| \frac{\partial f}{\partial x_n}(\mathbf{y}_n) - \frac{\partial f}{\partial x_n}(\mathbf{x}) \right| \right\} \|\mathbf{h}\|$$

since $|h_i| \le \|\mathbf{h}\|$. Thus we have proved that

$$\frac{\left| f(\mathbf{x}+\mathbf{h}) - f(\mathbf{x}) - \sum_{i=1}^{n} \frac{\partial f}{\partial x_i}(\mathbf{x})h_i \right|}{\|\mathbf{h}\|}$$
$$\le \left| \frac{\partial f}{\partial x_1}(\mathbf{y}_1) - \frac{\partial f}{\partial x_1}(\mathbf{x}) \right| + \cdots + \left| \frac{\partial f}{\partial x_n}(\mathbf{y}_n) - \frac{\partial f}{\partial x_n}(\mathbf{x}) \right|$$

But since the partials are continuous by assumption, the right side approaches 0 as $\mathbf{h} \to \mathbf{0}$, so that the left side approaches 0 as well. ∎

Finally, we shall prove the all-important Chain Rule.

Theorem 11. *Let $U \subset \mathbb{R}^n$ and $V \subset \mathbb{R}^m$ be open. Let $f\colon U \subset \mathbb{R}^n \to \mathbb{R}^m$ and $g\colon V \subset \mathbb{R}^m \to \mathbb{R}^p$ be given functions such that f maps U into V, so that $g \circ f$ is defined. Suppose f is differentiable at $\mathbf{x}_0$ and g is differentiable at $\mathbf{y}_0 = f(\mathbf{x}_0)$. Then $g \circ f$ is differentiable at $\mathbf{x}_0$ and*

$$D(g \circ f)(\mathbf{x}_0) = Dg(\mathbf{y}_0)Df(\mathbf{x}_0)$$

Proof. According to the definition of the derivative, we must verify that

$$\lim_{\mathbf{x} \to \mathbf{x}_0} \frac{\|g(f(\mathbf{x})) - g(f(\mathbf{x}_0)) - Dg(\mathbf{y}_0)Df(\mathbf{x}_0) \cdot (\mathbf{x} - \mathbf{x}_0)\|}{\|\mathbf{x} - \mathbf{x}_0\|} = 0$$

First rewrite the numerator and apply the triangle inequality as follows:

$$\|g(f(\mathbf{x})) - g(f(\mathbf{x}_0)) - Dg(\mathbf{y}_0) \cdot (f(\mathbf{x}) - f(\mathbf{x}_0))$$
$$+ Dg(\mathbf{y}_0) \cdot (f(\mathbf{x}) - f(\mathbf{x}_0) - Df(\mathbf{x}_0) \cdot (\mathbf{x} - \mathbf{x}_0))\|$$
$$\le \|g(f(\mathbf{x})) - g(f(\mathbf{x}_0)) - Dg(\mathbf{y}_0) \cdot (f(\mathbf{x}) - f(\mathbf{x}_0))\|$$
$$+ \|Dg(\mathbf{y}_0) \cdot [f(\mathbf{x}) - f(\mathbf{x}_0) - Df(\mathbf{x}_0) \cdot (\mathbf{x} - \mathbf{x}_0)]\|$$

As in Theorem 8, $\|Dg(\mathbf{y}_0) \cdot \mathbf{h}\| \le M\|\mathbf{h}\|$ for some constant M. Thus the above expression is less than or equal to

$$\|g(f(\mathbf{x})) - g(f(\mathbf{x}_0)) - Dg(\mathbf{y}_0) \cdot (f(\mathbf{x}) - f(\mathbf{x}_0))\|$$
$$+ M\|f(\mathbf{x}) - f(\mathbf{x}_0) - Df(\mathbf{x}_0) \cdot (\mathbf{x} - \mathbf{x}_0)\| \qquad (3)$$

Since f is differentiable at $\mathbf{x}_0$, given $\varepsilon > 0$, there is a $\delta_1 > 0$ such that $0 < \|\mathbf{x} - \mathbf{x}_0\| < \delta_1$ implies

$$\frac{\|f(\mathbf{x}) - f(\mathbf{x}_0) - Df(\mathbf{x}_0) \cdot (\mathbf{x} - \mathbf{x}_0)\|}{\|\mathbf{x} - \mathbf{x}_0\|} < \frac{\varepsilon}{2M}$$

This makes the second term in Equation (3) less than $\varepsilon\|\mathbf{x} - \mathbf{x}_0\|/2$.

Let us turn to the first term. By Theorem 8, $\|f(\mathbf{x}) - f(\mathbf{x}_0)\| < M_1\|\mathbf{x} - \mathbf{x}_0\|$ for a constánt M_1 if $\mathbf{x}$ is near $\mathbf{x}_0$, say $0 < \|\mathbf{x} - \mathbf{x}_0\| < \delta_2$. Now choose δ_3 such that $0 < \|\mathbf{y} - \mathbf{y}_0\| < \delta_3$ implies

$$\|g(\mathbf{y}) - g(\mathbf{y}_0) - Dg(\mathbf{y}_0) \cdot (\mathbf{y} - \mathbf{y}_0)\| < \varepsilon\|\mathbf{y} - \mathbf{y}_0\|/2M_1$$

Since $\mathbf{y} = f(\mathbf{x})$ and $\mathbf{y}_0 = f(\mathbf{x}_0)$, $\|\mathbf{y} - \mathbf{y}_0\| < \delta_3$ if $\|\mathbf{x} - \mathbf{x}_0\| < \delta_3/M_1$ and $\|\mathbf{x} - \mathbf{x}_0\| < \delta_2$, and so

$$\|g(f(\mathbf{x})) - g(f(\mathbf{x}_0)) - Dg(\mathbf{y}_0) \cdot (f(\mathbf{x}) - f(\mathbf{x}_0))\|$$
$$< \varepsilon\|f(\mathbf{x}) - f(\mathbf{x}_0)\|/2M_1 < \varepsilon\|\mathbf{x} - \mathbf{x}_0\|/2$$

Thus if $\delta = \min(\delta_1, \delta_2, \delta_3/M_1)$, Equation (3) is less than $\varepsilon\|\mathbf{x} - \mathbf{x}_0\|/2 + \varepsilon\|\mathbf{x} - \mathbf{x}_0\|/2$ and so

$$\frac{\|g(f(\mathbf{x})) - g(f(\mathbf{x}_0)) - Dg(\mathbf{y}_0)Df(\mathbf{x}_0)(\mathbf{x} - \mathbf{x}_0)\|}{\|\mathbf{x} - \mathbf{x}_0\|} < \frac{\varepsilon}{2} + \frac{\varepsilon}{2} = \varepsilon$$

for $0 < \|\mathbf{x} - \mathbf{x}_0\| < \delta$. This proves the theorem. ∎

The student has already met with a number of examples illustrating the above theorems. Let us consider two of a more technical nature.

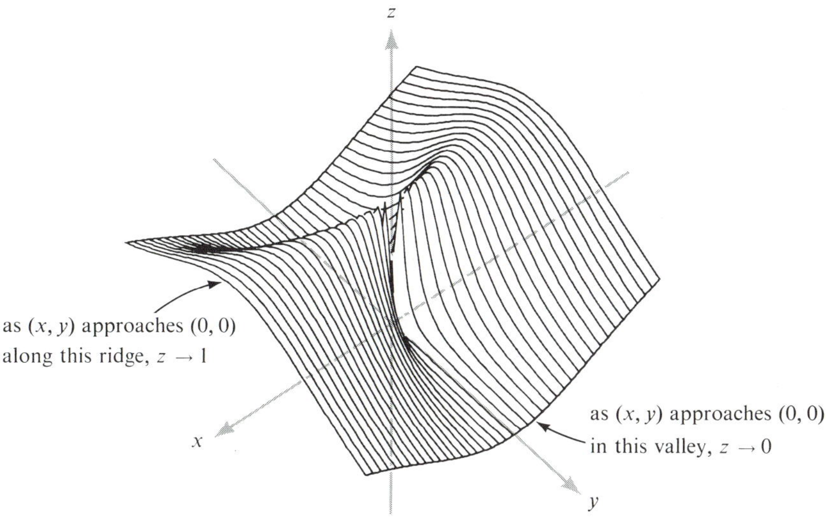

$$z = \frac{x^2}{x^2 + y^2} \qquad -1 \le x \le 1,\ -1 \le y \le 1$$

FIGURE 2.7.1
This function is not differentiable at (0, 0).

EXAMPLE 1 (Figure 2.7.1). Let

$$f\colon \mathbb{R}^2 \to \mathbb{R}, f(x, y) = \begin{cases} \dfrac{x^2}{x^2+y^2} & (x, y) \neq (0, 0) \\ 0 & (x, y) = (0, 0) \end{cases}$$

Is f differentiable?

On $\mathbb{R}^2\backslash\{(0, 0)\}$ f is differentiable, since it has continuous partial derivatives on this set. However, f is not differentiable at $(0, 0)$; in fact, f is not even continuous there. To see this, approach $(0, 0)$ along the line $x = y$. Clearly, $f(x, x) = \frac{1}{2}$; as $x \to 0$ this does not converge to 0 (see Example 10, p. 80).

EXAMPLE 2 (Figure 2.7.2). Let

$$f(x, y) = \begin{cases} \dfrac{xy}{\sqrt{x^2+y^2}} & (x, y) \neq (0, 0) \\ 0 & (x, y) = (0, 0) \end{cases}$$

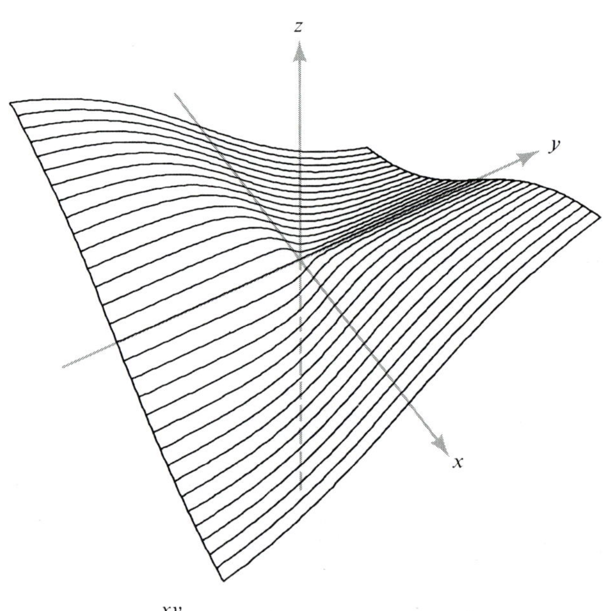

$$z = \frac{xy}{\sqrt{x^2+y^2}} \qquad -3 \leq x \leq 3, \; -3 \leq y \leq 3$$

FIGURE 2.7.2

This function is not differentiable at $(0, 0)$ *because it is "crinkled."*

Is f differentiable at $(0, 0)$? The answer is not immediately clear. However, we note that

$$\begin{aligned}\frac{\partial f}{\partial x}(0, 0) &= \operatorname*{limit}_{x\to 0} \frac{f(x, 0) - f(0, 0)}{x} \\ &= \operatorname*{limit}_{x\to 0} \frac{(x \cdot 0)/\sqrt{x^2+0} - 0}{x} = \operatorname*{limit}_{x\to 0} \frac{0-0}{x} = 0\end{aligned}$$

and similarly, $\partial f/\partial y(0, 0) = 0$. Thus the partials exist at (0, 0). Also, if $(x, y) \neq (0, 0)$, then

$$\frac{\partial f}{\partial x} = \frac{y\sqrt{x^2 + y^2} - 2x(xy)/2\sqrt{x^2 + y^2}}{x^2 + y^2} = \frac{y}{\sqrt{x^2 + y^2}} - \frac{x^2 y}{(x^2 + y^2)^{3/2}}$$

which does not have a limit as $(x, y) \to (0, 0)$, since the second term becomes indefinitely large, as can be seen by letting $x = My$. Thus the partials are not continuous at (0, 0), so we cannot use Theorem 9.

We might now try to show that f is not differentiable (f itself is continuous, so the method of Example 1 will not work). If $Df(0, 0)$ existed, by Theorem 16 it would have to be the zero matrix, since $\partial f/\partial x$ and $\partial f/\partial y$ are zero at (0, 0). Thus, by the definition of differentiability, for any $\varepsilon > 0$ there would be a $\delta > 0$ such that $0 < \|(h_1, h_2)\| < \delta$ implies

$$\frac{|f(h_1, h_2) - f(0, 0)|}{\|(h_1, h_2)\|} < \varepsilon$$

that is

$$|f(h_1, h_2)| < \varepsilon\|(h_1, h_2)\|$$

or

$$|h_1 h_2| < \varepsilon(h_1^2 + h_2^2)$$

But if we choose $h_1 = h_2$, this reads

$$\tfrac{1}{2} < \varepsilon$$

which is untrue if we choose $\varepsilon \leq \frac{1}{2}$. Hence, f is *not* differentiable at (0, 0).

EXERCISES

1. Let $f(x, y, z) = (e^x, \cos y, \sin z)$. Compute Df. In general, when will Df be a diagonal matrix?
2. (a) Let $A\colon \mathbb{R}^n \to \mathbb{R}^m$ be a linear transformation given by the matrix $A = (a_{ij})$ so that $A\mathbf{x}$ has components $y_i = \sum a_{ij} x_j$. Let $M = (\sum a_{ij}^2)^{1/2}$. Use the Schwarz inequality to prove that $\|A\mathbf{x}\| \leq M\|\mathbf{x}\|$
   (b) Use the inequality derived in (a) to show that a linear transformation $T\colon \mathbb{R}^n \to \mathbb{R}^m$ given by a matrix $T = (t_{ij})$ is continuous.
   (c) Let $A\colon \mathbb{R}^n \to \mathbb{R}^m$ be a linear transformation given by a matrix. If
   $$\underset{\mathbf{x}\to\mathbf{0}}{\text{limit}} \frac{A\mathbf{x}}{\|\mathbf{x}\|} = \mathbf{0}$$
   show that $A = 0$.
3. Let $f\colon A \to B$ and $g\colon B \to C$ be maps between open subsets of Euclidean space, and let $\mathbf{x}_0$ be in A or be a boundary point of A and $\mathbf{y}_0$ be in B or be a boundary point of B.
   (a) If $\underset{\mathbf{x}\to\mathbf{x}_0}{\text{limit}} f(\mathbf{x}) = \mathbf{y}_0$ and $\underset{\mathbf{y}\to\mathbf{y}_0}{\text{limit}} g(\mathbf{y}) = \mathbf{w}$, show that $\underset{\mathbf{x}\to\mathbf{x}_0}{\text{limit}} g(f(\mathbf{x}))$ need not equal $\mathbf{w}$.
   (b) If $\mathbf{y}_0 \in B$ and $\mathbf{w} = g(\mathbf{y}_0)$ show that $\underset{\mathbf{x}\to\mathbf{x}_0}{\text{limit}} g(f(\mathbf{x})) = \mathbf{w}$.

4. A function $f: A \subset \mathbb{R}^n \to \mathbb{R}^m$ is called *uniformly continuous* if for every $\varepsilon > 0$ there is a $\delta > 0$ such that for all $\mathbf{p}$ and $\mathbf{q} \in A$, $\|\mathbf{p} - \mathbf{q}\| < \delta$ implies $\|f(\mathbf{p}) - f(\mathbf{q})\| < \varepsilon$. (Note that a uniformly continuous function is continuous; try to describe explicitly the extra property that a uniformly continuous function has.)
   (a) Prove that a linear map $T: \mathbb{R}^n \to \mathbb{R}^m$ is uniformly continuous.
   (b) Prove $x \mapsto 1/x$ on $]0, 1]$ is continuous, but not uniformly continuous.
5. Let $A = (a_{ij})$ be a *symmetric* $n \times n$ matrix (that is, $a_{ij} = a_{ji}$) and define $f(\mathbf{x}) = \langle \mathbf{x}, A\mathbf{x} \rangle$ so $f: \mathbb{R}^n \to \mathbb{R}$. Show that $\nabla f(\mathbf{x})$ is the vector $2A\mathbf{x}$.
6. Let

$$f(x, y) = \begin{cases} \dfrac{2xy^2}{x^2 + y^4} & (x, y) \neq (0, 0) \\ 0 & (x, y) = (0, 0) \end{cases}$$

   Show that $\partial f/\partial x$ and $\partial f/\partial y$ exist everywhere. But show that f is not continuous at $(0, 0)$. Is f differentiable?
7. Let $f(x, y) = g(x) + h(y)$, and suppose g is differentiable at x_0 and h is differentiable at y_0. Prove f is differentiable at (x_0, y_0).
8. Use the Schwarz inequality to prove that for any vector $\mathbf{v} \in \mathbb{R}^n$,

$$\operatorname*{limit}_{\mathbf{x} \to \mathbf{x}_0} \langle \mathbf{v}, \mathbf{x} \rangle = \langle \mathbf{v}, \mathbf{x}_0 \rangle$$

9. Prove that if $\operatorname*{limit}_{\mathbf{x} \to \mathbf{x}_0} f(\mathbf{x}) = b$ for $f: A \subset \mathbb{R}^n \to \mathbb{R}$ then

$$\operatorname*{limit}_{\mathbf{x} \to \mathbf{x}_0} (f(\mathbf{x}))^2 = b^2 \quad \text{and} \quad \operatorname*{limit}_{\mathbf{x} \to \mathbf{x}_0} \sqrt{|f(\mathbf{x})|} = \sqrt{|b|}$$

   (You may use Exercise 3(b).)
10. Show that in Theorem 9 it is enough to assume that $n - 1$ partial derivatives are continuous and merely that the other one exists. Does this agree with what you expect when $n = 1$?
11. Define $f: \mathbb{R}^2 \to \mathbb{R}$ by

$$f(x, y) = \begin{cases} \dfrac{xy}{(x^2 + y^2)^{1/2}} & (x, y) \neq (0, 0) \\ 0 & (x, y) = (0, 0) \end{cases}$$

   Show that f is continuous (see Example 2).
12. (a) Does $\displaystyle\operatorname*{limit}_{(x, y) \to (0, 0)} \frac{x}{x^2 + y^2}$ exist?
    (b) Does $\displaystyle\operatorname*{limit}_{(x, y) \to (0, 0)} \frac{x^3}{x^2 + y^2}$ exist?
13. Find $\displaystyle\operatorname*{limit}_{(x, y) \to (0, 0)} \frac{xy}{\sqrt{x^2 + y^2}}$
14. Prove that $s: \mathbb{R}^2 \to \mathbb{R}^1$, $(x, y) \mapsto x + y$, is continuous.
15. Using the definition of continuity, prove that f is continuous at $\mathbf{x}$ if and only if

$$\operatorname*{limit}_{\mathbf{h} \to \mathbf{0}} f(\mathbf{x} + \mathbf{h}) = f(\mathbf{x})$$

16. (a) A sequence x_n of points in $\mathbb{R}^m$ is said to *converge to* $\mathbf{x}$, written $\mathbf{x}_n \to \mathbf{x}$ as $n \to \infty$, iff for any $\varepsilon > 0$ there is an N such that $n \geq N$ implies $\|\mathbf{x} - \mathbf{x}_n\| < \varepsilon$. Show that $\mathbf{y}$ is a boundary point of an open set A if and only if $\mathbf{y}$ is not in A and there is a sequence of distinct points of A converging to $\mathbf{y}$.
    (b) If $U \subset \mathbb{R}^m$ is open, show that $f\colon U \to \mathbb{R}^p$ is continuous if and only if $\mathbf{x}_n \to \mathbf{x} \in U$ implies $f(x)_n \to f(\mathbf{x})$.
17. If $f(\mathbf{x}) = g(\mathbf{x})$ for all $\mathbf{x} \neq \mathbf{a}$ and if $\lim\limits_{\mathbf{x}\to\mathbf{a}} f(\mathbf{x}) = \mathbf{b}$, then show that $\lim\limits_{\mathbf{x}\to\mathbf{a}} g(\mathbf{x}) = \mathbf{b}$ as well.
18. Let $A \subset \mathbb{R}^n$ and let $\mathbf{x}_0$ be a boundary point of A. Let $f\colon A \to \mathbb{R}$ and $g\colon A \to \mathbb{R}$ be functions defined on A such that $\lim\limits_{\mathbf{x}\to\mathbf{x}_0} f(\mathbf{x})$ and $\lim\limits_{\mathbf{x}\to\mathbf{x}_0} g(\mathbf{x})$ exist, and assume that for all $\mathbf{x}$ in some deleted neighborhood of $\mathbf{x}_0$, $f(\mathbf{x}) \leq g(\mathbf{x})$. (A deleted neighborhood of $\mathbf{x}_0$ is any neighborhood of $\mathbf{x}_0$, less $\mathbf{x}_0$ itself.)
    (a) Prove that $\lim\limits_{\mathbf{x}\to\mathbf{x}_0} f(\mathbf{x}) \leq \lim\limits_{\mathbf{x}\to\mathbf{x}_0} g(\mathbf{x})$.
    (HINT: Consider the function $\phi(\mathbf{x}) = g(\mathbf{x}) - f(\mathbf{x})$; prove that $\lim\limits_{\mathbf{x}\to\mathbf{x}_0} \phi(\mathbf{x}) \geq 0$, and then use the fact that the limit of the sum of two functions is the sum of their limits.)
    (b) If $f(\mathbf{x}) < g(\mathbf{x})$ do we necessarily have strict inequality of the limits?
19. Given $f\colon A \subset \mathbb{R}^n \to \mathbb{R}^m$, we say that "$f$ is $o(\mathbf{x})$ as $\mathbf{x} \to \mathbf{0}$" if $\lim\limits_{\mathbf{x}\to\mathbf{0}} f(\mathbf{x})/\|\mathbf{x}\| = \mathbf{0}$.
    (a) If f_1 and f_2 are $o(\mathbf{x})$ as $\mathbf{x} \to \mathbf{0}$, prove that $f_1 + f_2$ is also $o(\mathbf{x})$ as $\mathbf{x} \to \mathbf{0}$ (where $(f_1 + f_2)(\mathbf{x}) = f_1(\mathbf{x}) + f_2(\mathbf{x})$).
    (b) Let $g\colon A \to \mathbb{R}$ be a function with the property that there is a number $c > 0$ such that $|g(\mathbf{x})| \leq c$ for all $\mathbf{x}$ in A (g is called bounded). If f is $o(\mathbf{x})$ as $\mathbf{x} \to \mathbf{0}$, prove that gf is also $o(\mathbf{x})$ as $\mathbf{x} \to \mathbf{0}$ (where $(gf)(\mathbf{x}) = g(\mathbf{x})f(\mathbf{x})$).
    (c) Show that $f(x) = x^2$ is $o(x)$ as $x \to 0$.
    Is $g(x) = x$ $o(x)$ as $x \to 0$?

REVIEW EXERCISES FOR CHAPTER 2

1. Describe the graphs of:
   (a) $f(x, y) = 3x^2 + y^2$
   (b) $f(x, y) = xy + 3x$
2. Describe some appropriate level surfaces and sections of the graphs of:
   (a) $f(x, y, z) = 2x^2 + y^2 + z^2$
   (b) $f(x, y, z) = x^2$
   (c) $f(x, y, z) = xyz$
3. Compute the derivative $Df(\mathbf{x})$ of each of the following functions.
   (a) $f(x, y) = (x^2y, e^{-xy})$
   (b) $f(x) = (x, x)$
   (c) $f(x, y, z) = e^x + e^y + e^z$
   (d) $f(x, y, z) = (x, y, z)$

4. Suppose $f(x, y) = f(y, x)$ for all (x, y). Prove that
$$(\partial f/\partial x)(x, y) = (\partial f/\partial y)(x, y).$$

5. Let $f(x, y) = (1 - x^2 - y^2)^{1/2}$. Show that the plane tangent to the graph of f at $(x_0, y_0, f(x_0, y_0))$ is orthogonal to the vector $(x_0, y_0, f(x_0, y_0))$. Interpret geometrically.

6. Find the equation of the tangent plane of the graph of f at the point $(x_0, y_0, f(x_0, y_0))$ for:
(a) $f\colon \mathbb{R}^2 \to \mathbb{R}, (x, y) \mapsto x - y + 2,$ $\quad (x_0, y_0) = (1, 1)$
(b) $f\colon \mathbb{R}^2 \to \mathbb{R}, (x, y) \mapsto x^2 + 4y^2,$ $\quad (x_0, y_0) = (2, -1)$
(c) $f\colon \mathbb{R}^2 \to \mathbb{R}, (x, y) \mapsto xy,$ $\quad (x_0, y_0) = (-1, -1)$
(d) $f(x, y) = \log(x + y) + x \cos y + \arctan(x + y),$ $\quad (x_0, y_0) = (1, 0)$
(e) $f(x, y) = \sqrt{x^2 + y^2},$ $\quad (x_0, y_0) = (1, 1)$
(f) $f(x, y) = xy,$ $\quad (x_0, y_0) = (2, 1)$

7. Suppose $F(x, y) = (\partial f/\partial x) - (\partial f/\partial y)$ and f is C^2. Prove that
$$\frac{\partial F}{\partial x} + \frac{\partial F}{\partial y} = \frac{\partial^2 f}{\partial x^2} - \frac{\partial^2 f}{\partial y^2}$$

8. Compute the equations of the tangent planes of the following surfaces at the indicated points.
(a) $x^2 + y^2 + z^2 = 3,$ $\quad (1, 1, 1)$
(b) $x^3 - 2y^3 + z^3 = 0,$ $\quad (1, 1, 1)$
(c) $(\cos x)(\cos y)e^z = 0,$ $\quad (\pi/2, 1, 0)$
(d) $e^{xyz} = 1,$ $\quad (1, 1, 0)$

9. Draw some level curves for the following functions.
(a) $f(x, y) = 1/xy$
(b) $f(x, y) = x^2 - xy - y^2$

10. Consider a temperature function $T(x, y) = x \sin y$. Plot a few level curves. Compute ∇T and explain its meaning.

11. Find the following limits if they exist.
(a) $\displaystyle \lim_{(x, y) \to (0, 0)} \frac{\cos xy - 1}{x}$
(b) $\displaystyle \lim_{(x, y) \to (0, 0)} \sqrt{|(x + y)/(x - y)|}, \; x \neq y$

12. Compute the partial derivatives and gradients of the following functions.
(a) $f(x, y, z) = xe^z + y \cos x$
(b) $f(x, y, z) = (x + y + z)^{10}$
(c) $f(x, y, z) = (x^2 + y)/z$

13. Let $\mathbf{F} = F_1(x, y)\mathbf{i} + F_2(x, y)\mathbf{j}$ be a C^1 vector field. Show that if $\mathbf{F} = \nabla f$ for some f then $\partial F_1/\partial y = \partial F_2/\partial x$. Show that $\mathbf{F} = y(\cos x)\mathbf{i} + x(\sin y)\mathbf{j}$ is not a gradient vector field (see p. 114).

14. Let $y(x)$ be a differentiable function defined implicitly by $F(x, y(x)) = 0$. From problem 11(a), Section 2.4, we know that
$$\frac{dy}{dx} = -\frac{\partial F/\partial x}{\partial F/\partial y}$$
Consider the surface $z = F(x, y)$, and suppose F is increasing as a function of x and as a function of y; i.e., $\partial F/\partial x > 0$ and $\partial F/\partial y > 0$. By con-

sidering the graph and the plane $z = 0$, try to argue that for z fixed at $z = 0$ y should *decrease* as x increases and x should *decrease* as y increases. Does this tally with the minus sign in the formula for dy/dx?

15. (a) Consider the graph of a function $f(x, y)$ (Figure 2. Review 1). Let (x_0, y_0) lie on a level curve C, so $\nabla f(x_0, y_0)$ is perpendicular to this curve. Show that the tangent plane of the graph is the plane that (i)

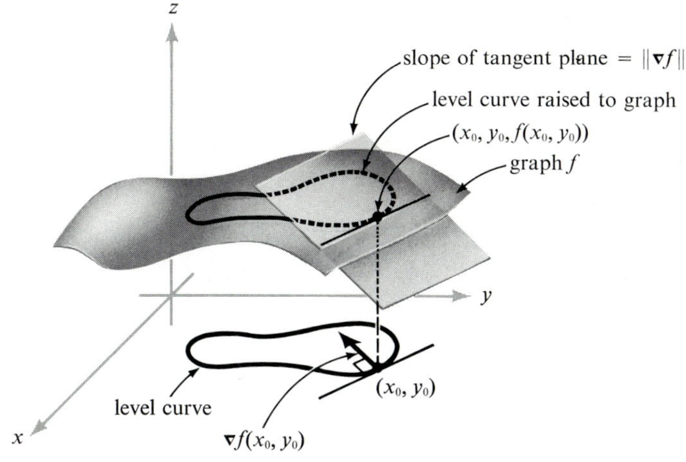

FIGURE 2. REVIEW. 1
Relationship between the gradient of a function and the plane tangent to the function's graph (Exercise 15(a)).

contains the line perpendicular to $\nabla f(x_0, y_0)$ and lying in the horizontal plane $z = f(x_0, y_0)$, and (ii) has slope $\|\nabla f(x_0, y_0)\|$ relative to the xy-plane. By the slope of a plane P relative to the xy-plane we mean the angle θ, $0 \le \theta \le \pi$, between a normal $\mathbf{p}$ to P and the unit vector $\mathbf{k}$.

(b) Use this method to show that the tangent plane of the graph of $f(x, y) = (x + \cos y)x^2$ at $(1, 0, 2)$ is as sketched in Figure 2. Review. 2.

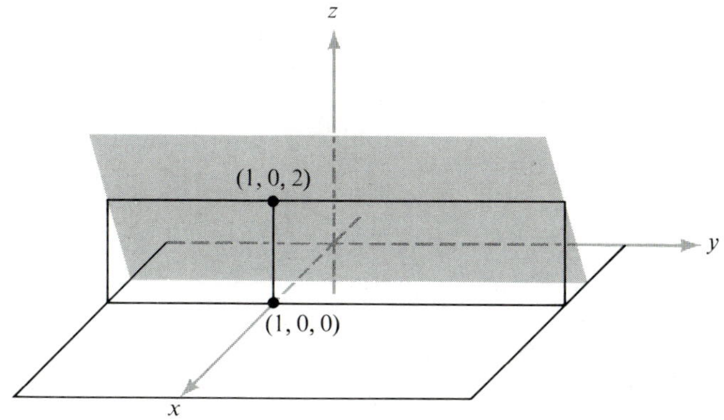

FIGURE 2. REVIEW. 2
The plane referred to in Exercise 15(b).

16. Find the plane tangent to the surface $z = x^2 + y^2$ at the point $(1, -2, 5)$. Explain the geometrical significance, for this surface, of the gradient of $f(x, y) = x^2 + y^2$ (see Exercise 15).

17. In which direction is the directional derivative of $f(x, y) = (x^2 - y^2)/(x^2 + y^2)$ at $(1, 1)$ equal to zero?

18. Find the directional derivative of the given function at the given point and in the direction of the given vector.
(a) $f(x, y, z) = e^x \cos(yz)$, $p_0 = (0, 0, 0)$, $\mathbf{v} = (2, 1, -2)$
(b) $f(x, y, z) = xy + yz + zx$, $p_0 = (1, 1, 2)$, $\mathbf{v} = (10, -1, 2)$

19. Find the tangent plane and normal to the hyperboloid $x^2 + y^2 - z^2 = 18$ at $(3, 5, -4)$.

20. Find the direction in which the function $w = x^2 + xy$ increases most rapidly at the point $(-1, 1)$. What is the magnitude of ∇w in this direction? Interpret this magnitude geometrically.

21. Let f be defined on an open set S in $\mathbb{R}^n$. We say that f is *homogeneous of degree p* over S if $f(\lambda\mathbf{x}) = \lambda^p f(\mathbf{x})$ for every real λ and for every $\mathbf{x}$ in S for which $\lambda\mathbf{x} \in S$. If such a function is differentiable at $\mathbf{x}$, show that $\mathbf{x} \cdot \nabla f(\mathbf{x}) = pf(\mathbf{x})$. This is known as *Euler's Theorem* for homogeneous functions. (HINT: For fixed $\mathbf{x}$, define $g(\lambda) = f(\lambda\mathbf{x})$ and compute $g'(1)$).

22. Prove that if $f(x, y)$ satisfies the Laplace equation

$$\frac{\partial^2 f}{\partial x^2} + \frac{\partial^2 f}{\partial y^2} = 0$$

so does

$$\phi(x, y) = f\left(\frac{x}{x^2 + y^2}, \frac{y}{x^2 + y^2}\right)$$

23. Prove that the functions
(a) $f(x, y) = \log(x^2 + y^2)$
(b) $g(x, y, z) = \dfrac{1}{(x^2 + y^2 + z^2)^{1/2}}$
(c) $h(x, y, z, w) = \dfrac{1}{x^2 + y^2 + z^2 + w^2}$
satisfy the respective Laplace equations:
(a) $f_{xx} + f_{yy} = 0$
(b) $g_{xx} + g_{yy} + g_{zz} = 0$
(c) $h_{xx} + h_{yy} + h_{zz} + h_{ww} = 0$
where $f_{xx} = \partial^2 f/\partial x^2$ etc.

24. If $z = f(x - y)/y$ show that $z + y(\partial z/\partial x) + y(\partial z/\partial y) = 0$

25. Given $w = f(x, y)$ with $x = u + v$, $y = u - v$ show that

$$\frac{\partial^2 w}{\partial u\, \partial v} = \frac{\partial^2 w}{\partial x^2} - \frac{\partial^2 w}{\partial y^2}$$

26. Let f have partial derivatives $\partial f(\mathbf{x})/\partial x_i$, where $i = 1, 2, \ldots, n$, at each point $\mathbf{x}$ of an open set U in $\mathbb{R}^n$. If f has a local maximum or a local minimum at the point $\mathbf{x}_0$ in U, show that $\partial f(\mathbf{x}_0)/\partial x_i = 0$ for each i.

27. Consider the functions defined in $\mathbb{R}^2$ by the following formulas:
    (*i*) $f(x, y) = xy/(x^2 + y^2)$ if $(x, y) \neq (0, 0)$, $f(0, 0) = 0$
    (*ii*) $f(x, y) = x^2y^2/(x^2 + y^4)$ if $(x, y) \neq (0, 0)$, $f(0, 0) = 0$
    (a) In each case show that the partial derivatives $\partial f(x, y)/\partial x$ and $\partial f(x, y)/\partial y$ exist for every (x, y) in $\mathbb{R}^2$ and evaluate these derivatives explicitly in terms of x and y.
    (b) Explain why the functions described in (*i*) and (*ii*) are not differentiable at (0, 0).

28. Compute the gradient vector $\nabla f(x, y)$ at all points (x, y) in $\mathbb{R}^2$ for each of the following functions.
    (a) $f(x, y) = x^2y^2 \log(x^2 + y^2)$ if $(x, y) \neq (0, 0)$, $f(0, 0) = 0$
    (b) $f(x, y) = xy \sin(1/(x^2 + y^2))$ if $(x, y) \neq (0, 0)$, $f(0, 0) = 0$

29. Given a function f defined in $\mathbb{R}^2$, let $F(r, \theta) = f(r \cos \theta, r \sin \theta)$.
    (a) Assume appropriate differentiability properties of f and show that

$$\frac{\partial F}{\partial r}(r, \theta) = \cos \theta \frac{\partial}{\partial x} f(x, y) + \sin \theta \frac{\partial}{\partial y} f(x, y)$$

$$\begin{aligned}\frac{\partial^2}{\partial r^2} F(r, \theta) = \cos^2 \theta \frac{\partial^2}{\partial x^2} f(x, y) \\ + 2 \sin \theta \cos \theta \frac{\partial^2}{\partial x\, \partial y} f(x, y) \\ + \sin^2 \theta \frac{\partial^2}{\partial y^2} f(x, y)\end{aligned}$$

    where $x = r \cos \theta$, $y = r \sin \theta$.
    (b) Verify the formula

$$\|\nabla f(r \cos \theta, r \sin \theta)\|^2 = \left(\frac{\partial}{\partial r} F(r, \theta)\right)^2 + \frac{1}{r^2}\left(\frac{\partial}{\partial \theta} F(r, \theta)\right)^2$$

30. Let $\mathbf{u} = \mathbf{i} - 2\mathbf{j} + 2\mathbf{k}$ and $\mathbf{v} = 2\mathbf{i} + \mathbf{j} - 3\mathbf{k}$. Find: $\|\mathbf{u}\|$, $\mathbf{u} \cdot \mathbf{v}$, $\mathbf{u} \times \mathbf{v}$, and a vector in the same direction as $\mathbf{u}$, but of unit length.

31. Let $h(x, y) = 2e^{-x^2} + e^{-3y^2}$ denote the height on a mountain at position (x, y). In what direction from (1, 0) should one begin walking in order to climb the fastest?

32. Compute the equation of the plane tangent to the graph of

$$f(x, y) = \frac{e^x}{x^2 + y^2}$$

    at $x = 1$, $y = 2$.

33. (a) Give a careful statement of the general form of the Chain Rule.
    (b) Let $f(x, y) = x^2 + y$ and let $h(u) = (\sin 3u, \cos 8u)$. Let $g(u) = f(h(u))$. Compute dg/du at $u = 0$ both directly *and* by using the Chain Rule.

34. (a) Sketch the level curves of $f(x, y) = -x^2 - 9y^2$ for $c = 0, -1, -10$.
    (b) On your sketch, draw in ∇f at (1, 1). Discuss.

CHAPTER 3

VECTOR-VALUED FUNCTIONS

One of our main concerns in Chapter 2 was the study of real-valued functions. This chapter deals with functions whose values are vectors. We shall begin in Section 3.1 with paths, which are maps from $\mathbb{R}$ to $\mathbb{R}^n$. Then we shall go on to study vector fields, and to introduce the main operations of vector differential calculus other than the gradient, namely, the divergence and the curl. We shall consider here some of the geometry associated with these operations, just as we did for the gradient, but the results of the most significance physically will have to wait until we have studied integration theory. Then, in Chapter 7, we shall resume our study of vector calculus.

3.1 PATHS AND VELOCITY

The student undoubtedly thinks of a curve as a line drawn on paper, such as a straight line, a circle, or a sine curve. To deal with such objects effectively, it is convenient to think of a curve in $\mathbb{R}^n$ as the image of a map of an interval $[a, b]$, (or occasionally of all of $\mathbb{R}$) into $\mathbb{R}^n$. We shall call such a map a *path*. A path is in the plane if $n = 2$, and in space if $n = 3$. The image of the path may then correspond to a line we see on paper. See Figure 3.1.1.

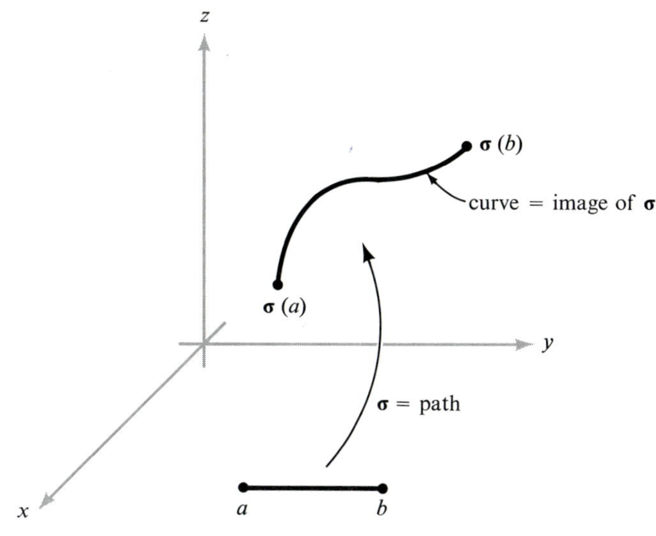

FIGURE 3.1.1
The map $\boldsymbol{\sigma}$ *is the path; its image is the curve we "see."*

Definition. *A **path** in* $\mathbb{R}^n$ *is a map* $\boldsymbol{\sigma}\colon [a, b] \to \mathbb{R}^n$. *If* $\boldsymbol{\sigma}$ *is differentiable (respectively, of class* C^1*) we say* $\boldsymbol{\sigma}$ *is a differentiable (respectively,* C^1*) path. The points* $\boldsymbol{\sigma}(a)$ *and* $\boldsymbol{\sigma}(b)$ *are called the **endpoints** of the path.*

It is useful to write t for the variable and to think of $\boldsymbol{\sigma}(t)$ as tracing out a curve in $\mathbb{R}^n$ as t varies. Often we imagine t to be the time and $\boldsymbol{\sigma}(t)$ as being a moving particle.

If $\boldsymbol{\sigma}$ is a path in $\mathbb{R}^3$, we can write $\boldsymbol{\sigma}(t) = (x(t), y(t), z(t))$ and we call $x(t)$, $y(t)$, and $z(t)$ the *component functions* of $\boldsymbol{\sigma}$. Of course we can similarly form component functions in $\mathbb{R}^2$ or, generally, in $\mathbb{R}^n$.

EXAMPLE 1. The straight line L in $\mathbb{R}^3$ through the point (x_0, y_0, z_0) in the direction of vector $\mathbf{v}$ is the image of the path $\boldsymbol{\sigma}(t) = (x_0, y_0, z_0) + t\mathbf{v}$, for $t \in \mathbb{R}$ (see Figure 3.1.2).

EXAMPLE 2. The unit circle in the plane is represented by the path

$$\boldsymbol{\sigma}\colon \mathbb{R} \to \mathbb{R}^2, \qquad \boldsymbol{\sigma}(t) = (\cos t, \sin t)$$

(see Figure 3.1.3). The image of $\boldsymbol{\sigma}$, that is, the set of points $\boldsymbol{\sigma}(t) \in \mathbb{R}^2$, for $t \in \mathbb{R}$, is the unit circle.

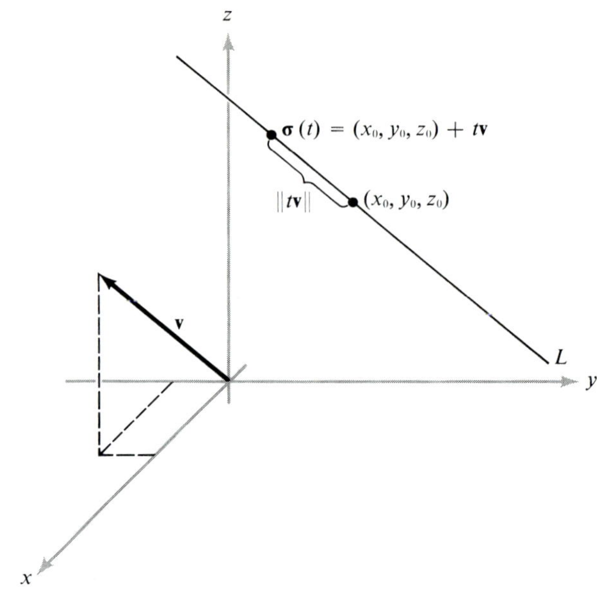

FIGURE 3.1.2
L is the straight line in space through (x_0, y_0, z_0) and in direction $\mathbf{v}$; *its equation is* $\boldsymbol{\sigma}(t) = (x_0, y_0, z_0) + t\mathbf{v}$.

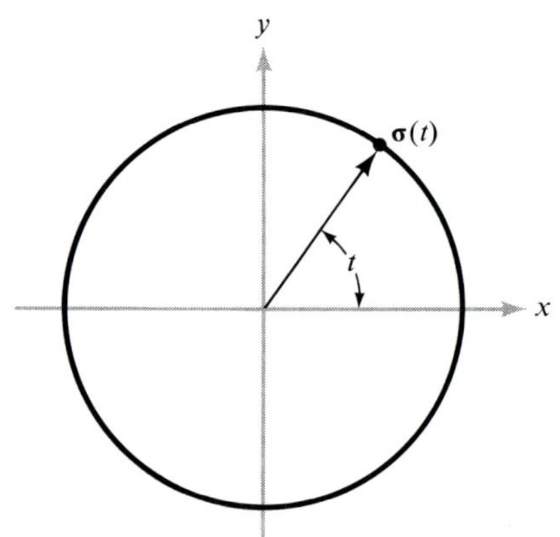

FIGURE 3.1.3
$\boldsymbol{\sigma}(t) = (\cos t, \sin t)$ *is a path whose image is the unit circle.*

EXAMPLE 3. The path $\boldsymbol{\sigma}(t) = (t, t^2)$ has an image that is a parabolic arc. The curve coincides with the graph of $f(x) = x^2$ (see Figure 3.1.4).

EXAMPLE 4. The function $\boldsymbol{\sigma}: t \mapsto (t - \sin t, 1 - \cos t)$ is the position function of a point on a rolling circle of radius 1. The circle lies in the xy-plane and rolls along the x-axis at constant speed; that is, the

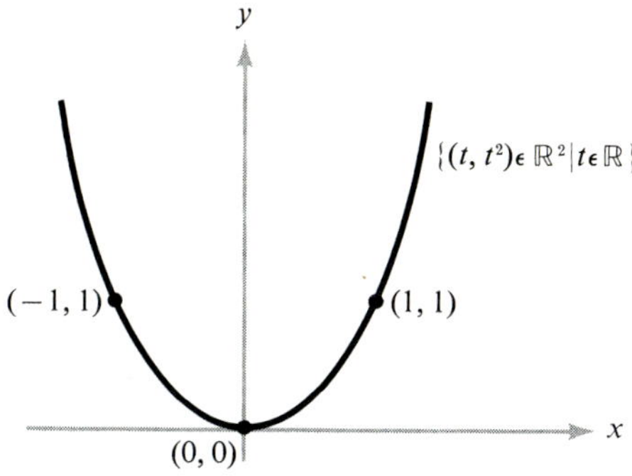

FIGURE 3.1.4
The image of $\boldsymbol{\sigma}(t) = (t, t^2)$ *is the parabola* $y = x^2$.

midpoint of the circle is moving to the right along the line $y = 1$ at a constant speed of 1 radian per unit of time. The motion of the point $\boldsymbol{\sigma}(t)$ is more complicated; its locus is known as the *cycloid* (see Figure 3.1.5).

Usually particles that move in space do so on smooth curves. For example, particles do not usually disappear and spontaneously reappear at another point. Thus we shall restrict our attention to sufficiently smooth paths, say C^1, for the rest of this section without further mention.

From our work in Chapter 2 we know that if $\boldsymbol{\sigma}$ is a path in $\mathbb{R}^3$, the derivative of $\boldsymbol{\sigma}$, $\mathbf{D}\boldsymbol{\sigma}(t)$, is a 3×1 matrix.

$$\mathbf{D}\boldsymbol{\sigma}(t) = \begin{bmatrix} x'(t) \\ y'(t) \\ z'(t) \end{bmatrix}$$

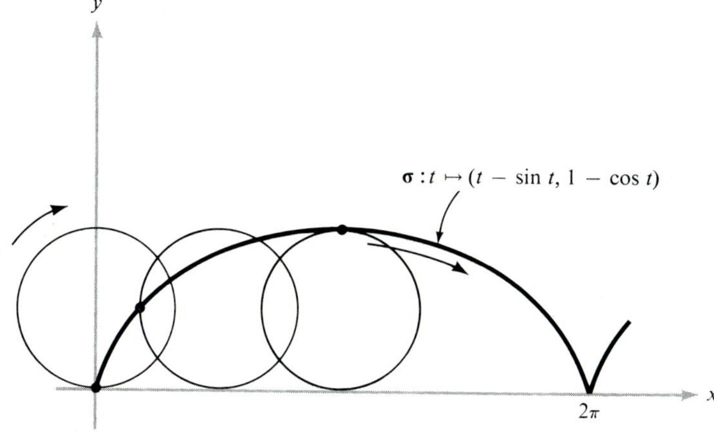

FIGURE 3.1.5
This path $\boldsymbol{\sigma}$ *is called a cycloid. It is the path traced by a point moving on a rolling circle.*

Let us write $\boldsymbol{\sigma}'(t)$ for $D\boldsymbol{\sigma}(t)$. If we adhered religiously to the conventions of linear algebra we would always write $\boldsymbol{\sigma}'$ as a 3×1 matrix. *However for typographical and notational reasons, we often write* $\boldsymbol{\sigma}'(t) = (x'(t), y'(t), z'(t))$. By definition,

$$\boldsymbol{\sigma}'(t_0) = \lim_{h \to 0} \frac{\boldsymbol{\sigma}(t_0 + h) - \boldsymbol{\sigma}(t_0)}{h}$$

Referring to Figure 3.1.6, we can argue intuitively that the vector $\boldsymbol{\sigma}'(t_0)$ ought to be parallel to the line L tangent to the path $\boldsymbol{\sigma}$, and that it ought to represent the velocity of the particle. Indeed, $(\boldsymbol{\sigma}(t_0 + h) - \boldsymbol{\sigma}(t_0))/h$ represents the average directed velocity in the time interval

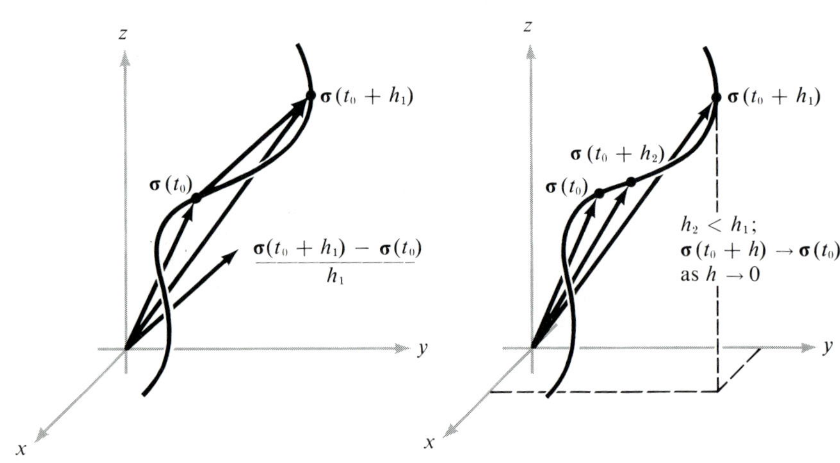

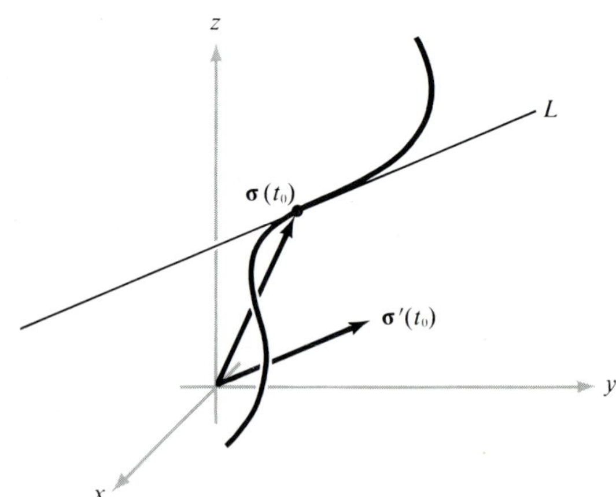

FIGURE 3.1.6
Illustrating the geometry of the definition of the derivative of a path: $\boldsymbol{\sigma}'(t_0) = \lim_{h \to 0} (\boldsymbol{\sigma}(t_0 + h) - \boldsymbol{\sigma}(t_0))/h$.

from t_0 to $t_0 + h$ (that is, total displacement/elapsed time). Hence as $h \to 0$, this expression approaches the instantaneous velocity vector. This leads us to the following definition.

Definition. *Let $\boldsymbol{\sigma}: \mathbb{R} \to \mathbb{R}^3$ be a C^1 path. The* ***velocity vector*** *at $\boldsymbol{\sigma}(t)$ is given by $\mathbf{v}(t) = \boldsymbol{\sigma}'(t) = (x'(t), y'(t), z'(t))$, and the* ***speed*** *of the particle is given by $S(t) = \|\boldsymbol{\sigma}'(t)\|$, the length of the vector $\boldsymbol{\sigma}'(t)$.*

We know that the velocity vector $\boldsymbol{\sigma}'(t_0)$ is parallel to the line L tangent to the path $t \mapsto \boldsymbol{\sigma}(t)$ at the point $\boldsymbol{\sigma}(t_0)$ (Figure 3.1.7). Thus an

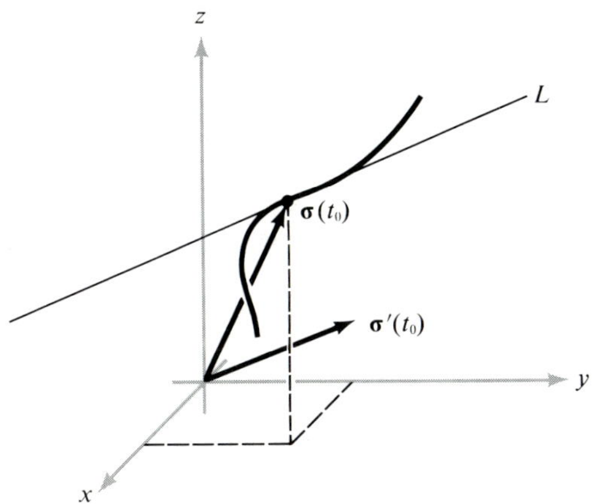

FIGURE 3.1.7
The line L tangent to a path $\boldsymbol{\sigma}$ at $\boldsymbol{\sigma}(t_0)$ has the equation $\mathbf{L}(\lambda) = \boldsymbol{\sigma}(t_0) + \lambda\boldsymbol{\sigma}'(t_0)$.

equation of the line L tangent to $t \mapsto \boldsymbol{\sigma}(t)$ at $\boldsymbol{\sigma}(t_0)$ would be given by the formula $\lambda \mapsto \boldsymbol{\sigma}(t_0) + \lambda\boldsymbol{\sigma}'(t_0)$, where λ ranges over the real numbers (see p. 11).

Definition. *Let $\boldsymbol{\sigma}$ be a C^1 curve in $\mathbb{R}^3$. The* ***tangent line*** *of $\boldsymbol{\sigma}$ at $\boldsymbol{\sigma}(t_0)$ is given in parametric form by*

$$\mathbf{L}(\lambda) = \boldsymbol{\sigma}(t_0) + \lambda\boldsymbol{\sigma}'(t_0)$$

EXAMPLE 5. If $\boldsymbol{\sigma}: t \mapsto (\cos t, \sin t, t)$, then the velocity vector is $\mathbf{v}(t) = \boldsymbol{\sigma}'(t) = (-\sin t, \cos t, 1)$. The speed of a point is the magnitude of the velocity:

$$S(t) = \|\mathbf{v}(t)\| = (\sin^2 t + \cos^2 t + 1)^{1/2} = \sqrt{2}$$

Thus the point moves at constant speed, although its velocity is not constant since it continually changes direction. The trajectory of the point whose motion is given by $\boldsymbol{\sigma}$ is called a (right-circular) helix (see Figure 3.1.8). The helix lies on a right-circular cylinder.

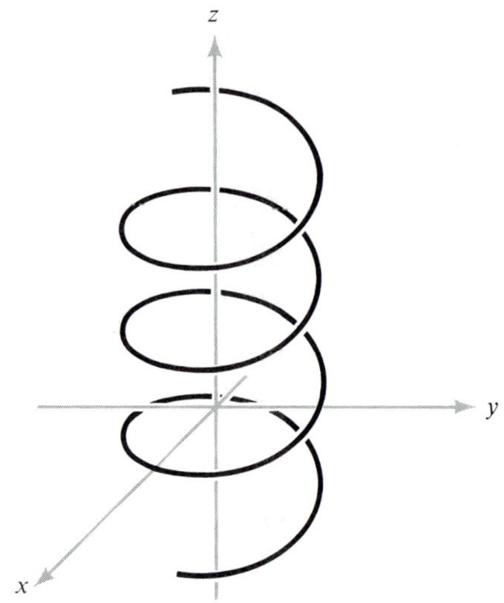

FIGURE 3.1.8
$\boldsymbol{\sigma}(t) = (\cos t, \sin t, t)$ *is a right-circular helix.*

EXAMPLE 6. Consider a particle moving on the path described in Example 5, where t is the time. At time $t = \pi$ the particle leaves the path and flies off on a tangent (as mud leaves a bicycle wheel). Find the location of the particle at time $t = 2\pi$. Assume no forces act on it after it leaves the path (see Figure 3.1.9).

To do this, we note that $\mathbf{v}(\pi) = (0, -1, 1)$ so that the particle, after leaving the first curve, travels in a straight path along a line L that is parallel to the velocity vector $\mathbf{v}(\pi) = \boldsymbol{\sigma}'(\pi)$. If $t \mapsto \mathbf{c}(t)$ represents the path of the particle for $t \geq \pi$, the velocity vector, $\mathbf{c}'(t)$, must be constant, since after the particle leaves the curve no forces act on it. Then $\mathbf{c}'(t) = \boldsymbol{\sigma}'(\pi) = \mathbf{v}(\pi) = (0, -1, 1)$ and $\mathbf{c}(\pi) = \boldsymbol{\sigma}(\pi) = (-1, 0, \pi)$.

Since $t \mapsto \mathbf{c}(t)$ is a straight path parallel to $\mathbf{v}(\pi)$, its equation is given by $t \mapsto \mathbf{w} + t\mathbf{v}(\pi) = \mathbf{w} + t(0, -1, 1)$, where $\mathbf{w}$ is some constant vector. To find $\mathbf{w}$ we note that $\mathbf{c}(\pi) = \mathbf{w} + \pi(0, -1, 1) = \boldsymbol{\sigma}(\pi) = (-1, 0, \pi)$, so $\mathbf{w} = (-1, 0, \pi) - (0, -\pi, \pi) = (-1, \pi, 0)$. Thus $\mathbf{c}(t) = (-1, \pi, 0) + t(0, -1, 1)$. Consequently, $\mathbf{c}(2\pi) = (-1, \pi, 0) + 2\pi(0, -1, 1) = (-1, -\pi, 2\pi)$.

Given a particle moving on a path $\boldsymbol{\sigma}(t)$, it is natural to define the rate of change of the velocity vector to be the *acceleration*. Thus

$$\mathbf{a}(t) = \boldsymbol{\sigma}''(t) = (x''(t), y''(t), z''(t))$$

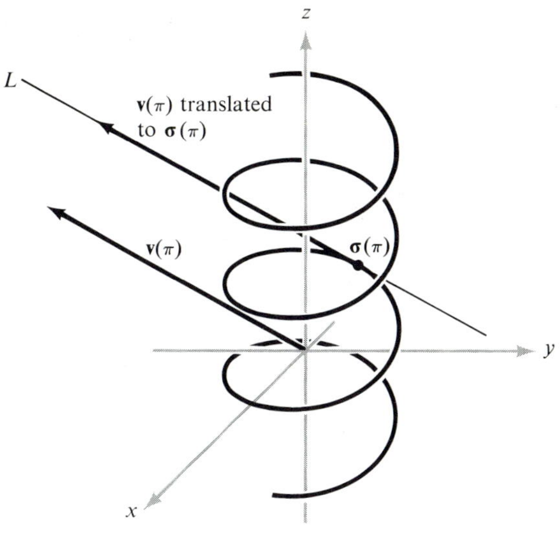

FIGURE 3.1.9
Finding the velocity vector of the right-circular helix. (The drawing is not to scale.)

If a particle of mass m moves in $\mathbb{R}^3$, the force $\mathbf{F}$ acting on it at the point $\boldsymbol{\sigma}(t)$ is related to the acceleration by *Newton's second law*:

$$\mathbf{F}(\boldsymbol{\sigma}(t)) = m\mathbf{a}(t)$$

In the very interesting problem of determining the path $\boldsymbol{\sigma}(t)$ of a particle, given its mass, initial position, and velocity and given a force, Newton's law becomes a differential equation for $\boldsymbol{\sigma}(t)$, and techniques of differential equations can be used to solve it.* For example, a planet moving round the sun (considered to be located at the origin in $\mathbb{R}^3$) in a path $\mathbf{r}(t)$ obeys the law

$$m\mathbf{r}''(t) = -\frac{GmM}{\|\mathbf{r}(t)\|^3}\mathbf{r}(t) = -\frac{GmM}{r^3(t)}\mathbf{r}(t)$$

where M is the mass of the sun, m that of the planet, $r = \|\mathbf{r}\|$, and G is the gravitational constant. The fact that the force is given by $\mathbf{F} = -GmM\mathbf{r}/r^3$ is called *Newton's law of gravitation* (see Figure 3.1.10). We shall not investigate the solution of such equations in this book, but content ourselves with the following special case.

* See, for example, W. E. Boyce and R. C. DiPrima, *Elementary Differential Equations and Boundary Value Problems*, John Wiley & Sons, New York, 1965. Surprisingly, Newton never wrote down his laws as analytical equations. This was first done by Euler around 1750. See C. Truesdell, *Essays in the History of Mechanics*, Springer, New York, 1968.

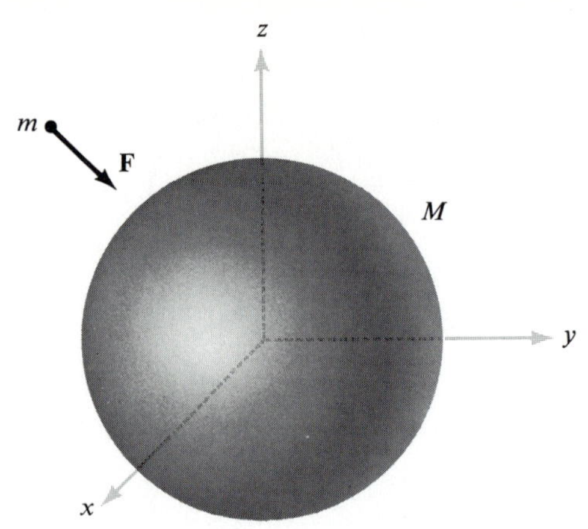

FIGURE 3.1.10
A mass M attracts a mass m with a force **F** *given by Newton's law of gravitation:* $\mathbf{F} = -GmM\mathbf{r}/r^3$.

EXAMPLE 7. Consider a particle of mass m moving at constant speed S in a circular path of radius r_0. Then, supposing it moves in the xy-plane, we can write

$$\mathbf{r}(t) = (r_0 \cos(tS/r_0), r_0 \sin(tS/r_0))$$

since this is a circle of radius r_0 and $\|\mathbf{r}'(t)\| = S$. Then we can see that

$$\mathbf{a}(t) = \mathbf{r}''(t) = \left(-\frac{S^2}{r_0}\cos(tS/r_0), -\frac{S^2}{r_0}\sin(tS/r_0)\right) = -\frac{S^2}{r_0^2}\mathbf{r}(t)$$

Thus, the acceleration is in a direction opposite to $\mathbf{r}(t)$; that is, it is directed towards the center of the circle (see Figure 3.1.11). This acceleration divided by the mass of the particle is called the *centripetal force*. Note that even though the speed is constant, the direction of the velocity is continually changing, which is why an acceleration results.

Now suppose we have a satellite moving with a speed S around a central body with mass M in a circular orbit of radius r_0. Then by Newton's laws,

$$-\frac{S^2 m}{r_0^2}\mathbf{r}(t) = -\frac{GmM}{r_0^3}\mathbf{r}(t)$$

The lengths of the vectors on both sides of this equation must be equal. Hence

$$S^2 = \frac{GM}{r_0}$$

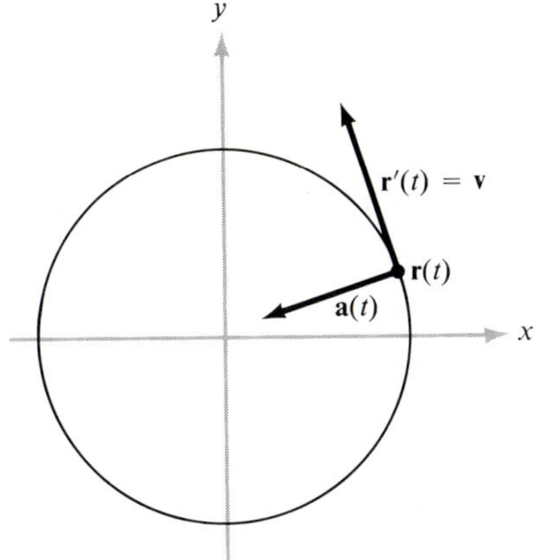

FIGURE 3.1.11
The position of a particle moving with speed S on a circle of radius r_0 is given by the equation $\mathbf{r}(t) = (r_0 \cos(tS/r_0), r_0 \sin(tS/r_0))$, *and its acceleration by* $\mathbf{a}(t) = -S^2\mathbf{r}(t)/r_0^2$.

If T is the period of one revolution, then $2\pi r_0/T = S$; substituting this value for S in the above equation and solving for T, we obtain the rule:

$$T^2 = r_0^3 \frac{(2\pi)^2}{GM}$$

Note that the square of the period is proportional to the cube of the radius. This is one of the three famous laws of Kepler, who observed it before Newton's laws were formulated; it enables one to compute the period of a satellite with a given radius, and vice-versa.

EXERCISES

1. Find $\boldsymbol{\sigma}'(t)$ and $\boldsymbol{\sigma}'(0)$ in each of the following cases.
   (a) $\boldsymbol{\sigma}(t) = (\sin 2\pi t, \cos 2\pi t, 2t - t^2)$
   (b) $\boldsymbol{\sigma}(t) = (e^t, \cos t, \sin t)$
   (c) $\boldsymbol{\sigma}(t) = (t^2, t^3 - 4t, 0)$
   (d) $\boldsymbol{\sigma}(t) = (\sin 2t, \log(1 + t), t)$
2. For each of the following curves, determine the velocity and acceleration vectors, and the equation of the tangent line at the specified value of t.
   (a) $\mathbf{r}(t) = 6t\mathbf{i} + 3t^2\mathbf{j} + t^3\mathbf{k}$, $t = 0$
   (b) $\boldsymbol{\sigma}(t) = (\sin 3t, \cos 3t, 2t^{3/2})$, $t = 1$
   (c) $\boldsymbol{\sigma}(t) = (\cos^2 t, 3t - t^3, t)$, $t = 0$
   (d) $\boldsymbol{\sigma}(t) = (0, 0, t)$, $t = 1$
   (e) $\boldsymbol{\sigma}(t) = (t \sin t, t \cos t, \sqrt{3}t)$, $t = 0$
   (f) $\mathbf{r}(t) = \sqrt{2}\,t\mathbf{i} + e^t\mathbf{j} + e^{-t}\mathbf{k}$, $t = 0$
   (g) $\boldsymbol{\sigma}(t) = t\mathbf{i} + t\mathbf{j} + \frac{2}{3}t^{3/2}\mathbf{k}$, $t = 9$

3. In Exercise 2(a), what force acts on a particle of mass 1 at $t = 0$?
4. Let $\boldsymbol{\sigma}$ be a path in $\mathbb{R}^3$ with zero acceleration. Prove that $\boldsymbol{\sigma}$ is a straight line.
5. Find the path $\boldsymbol{\sigma}$ such that $\boldsymbol{\sigma}(0) = (0, -5, 1)$ and $\boldsymbol{\sigma}'(t) = (t, e^t, t^2)$.
6. Find paths that represent the following curves or trajectories, and sketch.
   (a) $\{(x, y) \mid y = e^x\}$
   (b) $\{(x, y) \mid 4x^2 + y^2 = 1\}$
   (c) A straight line in $\mathbb{R}^3$ passing through the origin and the point (a, b, c).
7. A satellite orbits 500 kilometers above the Earth in a circular orbit. Compute its period. ($G = 6.67 \times 10^{-11}$, $M = 5.98 \times 10^{24}$ kilograms = mass of Earth, radius of Earth $= 6.37 \times 10^3$ kilometers. Here G is given in KMS units—kilograms, meters, seconds.)
8. Suppose a particle follows the path $\boldsymbol{\sigma}(t) = (e^t, e^{-t}, \cos t)$ until it flies off on a tangent at $t = 1$. Where is it at $t = 2$?
9. Let $\boldsymbol{\sigma}(t)$ be a path, $\mathbf{v}(t)$ the velocity, and $\mathbf{a}(t)$ the acceleration. Suppose $\mathbf{F}$ is a vector field, $m > 0$, and $\mathbf{F}(\boldsymbol{\sigma}(t)) = m\mathbf{a}(t)$ (Newton's second law). Prove that

$$\frac{d}{dt}(m\boldsymbol{\sigma}(t) \times \mathbf{v}(t)) = \boldsymbol{\sigma}(t) \times \mathbf{F}(\boldsymbol{\sigma}(t))$$

(i.e., "rate of change of angular momentum = torque"). What can you conclude if $\mathbf{F}(\boldsymbol{\sigma}(t))$ is parallel to $\boldsymbol{\sigma}(t)$? Is this the case in planetary motion?

10. Continue the investigations in Exercise 9 to prove Kepler's law that a planet moving about the sun does so in a fixed plane.

3.2 ARC LENGTH

Consider a given path $\boldsymbol{\sigma}(t)$. The reader should now be accustomed to thinking of $\boldsymbol{\sigma}(t)$ as the path of a particle with speed $S(t) = \|\boldsymbol{\sigma}'(t)\|$. What is the length of this curve as t ranges from, say, a to b? Intuitively, this ought to be nothing more than the total distance travelled, that is, $\int_a^b S(t)\, dt$. This leads us to the following.

Definition. *Let $\boldsymbol{\sigma}\colon [a, b] \to \mathbb{R}^n$ be a C^1 path. The **length** of $\boldsymbol{\sigma}$ is defined to be*

$$l(\boldsymbol{\sigma}) = \int_a^b \|\boldsymbol{\sigma}'(t)\|\, dt$$

In $\mathbb{R}^3$ our formula reads

$$l(\boldsymbol{\sigma}) = \int_a^b \sqrt{(x'(t))^2 + (y'(t))^2 + (z'(t))^2}\, dt$$

and there is a similar formula for curves in $\mathbb{R}^2$.

EXAMPLE 1. The arc length of the curve $\boldsymbol{\sigma}(t) = (r \cos t, r \sin t)$, for $0 \le t \le 2\pi$, is

$$l = \int_0^{2\pi} \sqrt{(-r \sin t)^2 + (r \cos t)^2}\, dt = 2\pi r$$

which is nothing more than the circumference of a circle of radius r. If we allowed $0 \le t \le 4\pi$, we would have gotten $4\pi r$ because the path traverses the same circle twice (Figure 3.2.1).

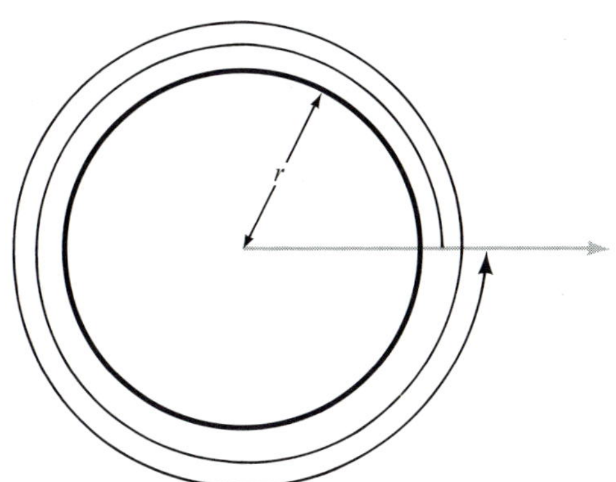

FIGURE 3.2.1
The arc length of a circle traversed twice is $4\pi r$.

In $\mathbb{R}^3$ there is another way to justify the formula for $l(\boldsymbol{\sigma})$ given in the above definition.* This method is based on polygonal approximations and proceeds as follows. We partition the interval $[a, b]$ into N subintervals of equal length:

$$a = t_0 < t_1 < \cdots < t_N = b$$

$$t_{i+1} - t_i = \frac{b-a}{N} \quad \text{for} \quad 0 \le i \le N-1$$

We then consider the polygonal line obtained by joining the successive pairs of points $\boldsymbol{\sigma}(t_i)$, $\boldsymbol{\sigma}(t_{i+1})$ for $0 \le i \le N-1$. This yields a polygonal approximation to $\boldsymbol{\sigma}$ as in Figure 3.2.2. By the formula for distance in $\mathbb{R}^3$, it follows that the line segment from $\boldsymbol{\sigma}(t_i)$ to $\boldsymbol{\sigma}(t_{i+1})$ has length

$$\|\boldsymbol{\sigma}(t_{i+1}) - \boldsymbol{\sigma}(t_i)\| = \sqrt{(x(t_{i+1}) - x(t_i))^2 + (y(t_{i+1}) - y(t_i))^2 + (z(t_{i+1}) - z(t_i))^2}$$

where $\boldsymbol{\sigma}(t) = (x(t), y(t), z(t))$. Applying the Mean Value Theorem to $x(t)$, $y(t)$, and $z(t)$ on $[t_i, t_{i+1}]$ we obtain three points t_i^*, t_i^{**}, and t_i^{***} such that

$$x(t_{i+1}) - x(t_i) = x'(t_i^*)(t_{i+1} - t_i)$$

$$y(t_{i+1}) - y(t_i) = y'(t_i^{**})(t_{i+1} - t_i)$$

and

$$z(t_{i+1}) - z(t_i) = z'(t_i^{***})(t_{i+1} - t_i)$$

* This paragraph assumes an acquaintance with the definite integral defined in terms of Riemann sums. If your background in this is weak, the discussion can be omitted on a first reading (see Chapter 5).

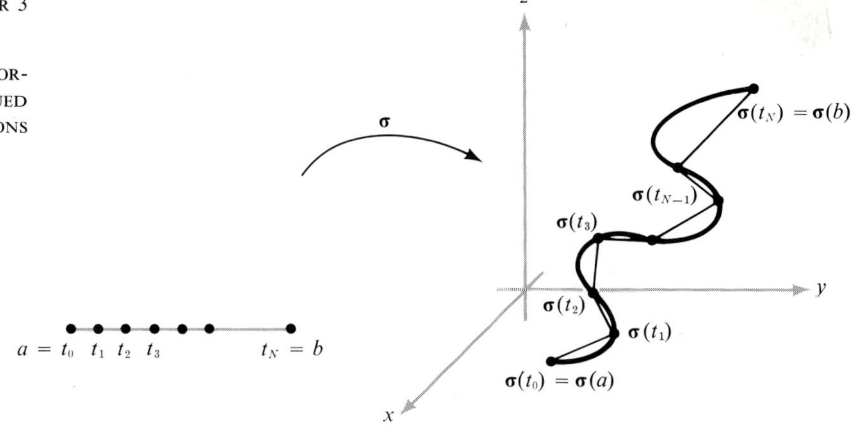

FIGURE 3.2.2
A path $\boldsymbol{\sigma}$ *may be approximated by a polygonal path obtained by joining each* $\boldsymbol{\sigma}(t_i)$ *to* $\boldsymbol{\sigma}(t_{i+1})$ *by a straight line.*

Thus the line segment from $\boldsymbol{\sigma}(t_i)$ to $\boldsymbol{\sigma}(t_{i+1})$ has length

$$\sqrt{(x'(t_i^*))^2 + (y'(t_i^{**}))^2 + (z'(t_i^{***}))^2}\,(t_{i+1} - t_i)$$

Therefore the length of our approximating polygonal line is

$$S_N = \sum_{i=0}^{N-1} \sqrt{(x'(t_i^*))^2 + (y'(t_i^{**}))^2 + (z'(t_i^{***}))^2}\,(t_{i+1} - t_i)$$

As $N \to \infty$, this polygonal line approximates the image of $\boldsymbol{\sigma}$ more closely. Therefore, we define the arc length of $\boldsymbol{\sigma}$ as the limit, if it exists, of the sequence S_N as $N \to \infty$. Since the derivatives x', y', and z' are all assumed to be continuous on $[a, b]$, we can conclude that, in fact, the limit does exist and is given by*

$$\operatorname*{limit}_{N\to\infty} S_N = \int_a^b \sqrt{(x'(t))^2 + (y'(t))^2 + (z'(t))^2}\,dt$$

EXAMPLE 2. Find the arc length of the helix $\boldsymbol{\rho}\colon [0, 4\pi] \to \mathbb{R}^3$, $t \mapsto (\cos 2t, \sin 2t, \sqrt{5}\,t)$. The velocity vector is $\boldsymbol{\rho}'(t) = (-2 \sin 2t, 2 \cos 2t, \sqrt{5})$ which has magnitude

$$\|\boldsymbol{\rho}'(t)\| = \sqrt{4(\sin 2t)^2 + 4(\cos 2t)^2 + 5} = \sqrt{9} = 3$$

* The theory of integration relates the integral to sums by the formula

$$\int_a^b f(t)\,dt = \operatorname*{limit}_{N\to\infty} \sum_{i=0}^{N-1} f(t_i^*)(t_{i+1} - t_i),$$

where $t_0, \ldots, t_N$ is a partition of $[a, b]$, $t_i^* \in [t_i, t_{i+1}]$ is arbitrary, and f is a continuous function. Here we have possibly different points t_i^*, t_i^{**}, and t_i^{***}, so this formula must be extended slightly.

The arc length of $\boldsymbol{\rho}$ is therefore

$$l(\boldsymbol{\rho}) = \int_0^{4\pi} \|\boldsymbol{\rho}'(t)\|\, dt = \int_0^{4\pi} 3\, dt = 12\pi$$

EXAMPLE 3. Consider the point with position function $\boldsymbol{\sigma}\colon t \mapsto (t - \sin t, 1 - \cos t)$ discussed in Example 4, Section 3.1. The velocity vector is $\boldsymbol{\sigma}'(t) = (1 - \cos t, \sin t)$ and the speed of the point $\boldsymbol{\sigma}(t)$ is

$$\|\boldsymbol{\sigma}'(t)\| = \sqrt{(1 - \cos t)^2 + \sin^2 t} = \sqrt{2 - 2\cos t}$$

Hence, $\boldsymbol{\sigma}(t)$ moves at variable speed although, as we discovered earlier, the circle rolls at constant speed. Furthermore, the speed of $\boldsymbol{\sigma}(t)$ is zero when t is an integral multiple of 2π. At these values of t the y-coordinate of the point $\boldsymbol{\sigma}(t)$ is zero and so the point lies on the x-axis.

The image of a C^1 path is not necessarily "very smooth"; indeed it may have sharp bends or changes of direction. For instance, the cycloid in the above example has cusps at all points where $\boldsymbol{\sigma}(t)$ touches the x-axis (that is, where $t = 2\pi n$, $n = 0, \pm 1, \ldots$). Another example is the *hypocycloid of four cusps*, $\boldsymbol{\sigma}\colon [0, 2\pi] \to \mathbb{R}^2$, $t \mapsto (\cos^3 t, \sin^3 t)$, which has cusps at four points (Figure 3.2.3). However, at all such points $\boldsymbol{\sigma}'(t) = \mathbf{0}$, the tangent line is not well defined, and the speed of the point $\boldsymbol{\sigma}(t)$ is zero. Evidently the direction of $\boldsymbol{\sigma}(t)$ may change abruptly at points where it slows to rest.

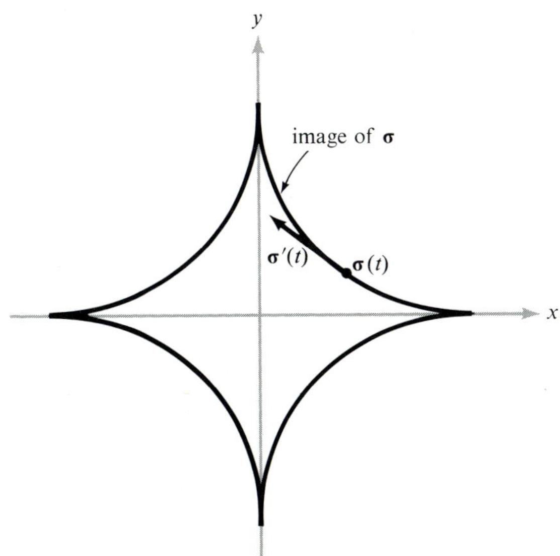

FIGURE 3.2.3
The image of the smooth path $\boldsymbol{\sigma}(t) = (\cos^3 t, \sin^3 t)$, a hypocycloid, does not "look smooth."

If we have a path $\boldsymbol{\sigma}(t) = (x(t), y(t), z(t))$ in $\mathbb{R}^3$, it is conventional to denote it sometimes by $\mathbf{s}(t) = \boldsymbol{\sigma}(t)$, so

$$\boldsymbol{\sigma}'(t) = \frac{d\mathbf{s}}{dt} = \frac{dx}{dt}\mathbf{i} + \frac{dy}{dt}\mathbf{j} + \frac{dz}{dt}\mathbf{k}$$

Thus

$$\|\boldsymbol{\sigma}'(t)\| = \left\|\frac{d\mathbf{s}}{dt}\right\| = \sqrt{\left(\frac{dx}{dt}\right)^2 + \left(\frac{dy}{dt}\right)^2 + \left(\frac{dz}{dt}\right)^2}$$

The arc length of $\boldsymbol{\sigma}$ may be written in this notation as

$$l(\boldsymbol{\sigma}) = \int_a^b \|d\mathbf{s}/dt\| \, dt$$

It is also common to introduce the *arc-length function* $s(t)$ by $s(t) = \int_a^t \|\boldsymbol{\sigma}'(\tau)\| \, d\tau$, so $s'(t) = \|d\mathbf{s}/dt\|$ and $l(\boldsymbol{\sigma}) = \int_a^b s'(t) \, dt$.

EXAMPLE 4. A particle moves along the hypocycloid according to the equations

$$x = \cos^3 t \qquad y = \sin^3 t \qquad a \le t \le b$$

The velocity vector of the particle is

$$\frac{d\mathbf{s}}{dt} = \frac{dx}{dt}\mathbf{i} + \frac{dy}{dt}\mathbf{j} = -(3 \sin t \cos^2 t)\mathbf{i} + (3 \cos t \sin^2 t)\mathbf{j}$$

and its speed is

$$S(t) = s'(t) = \left\|\frac{d\mathbf{s}}{dt}\right\| = (9 \sin^2 t \cos^4 t + 9 \cos^2 t \sin^4 t)^{1/2}$$

$$= 3|\sin t|\,|\cos t|$$

If we differentiate the velocity vector

$$\mathbf{v}(t) = \frac{d\mathbf{s}}{dt} = \frac{dx}{dt}\mathbf{i} + \frac{dy}{dt}\mathbf{j} + \frac{dz}{dt}\mathbf{k}$$

of a moving particle, we obtain its acceleration vector, as we have seen. In this notation the acceleration vector is given by the formula

$$\mathbf{a}(t) = \frac{d\mathbf{v}}{dt} = \frac{d^2x}{dt^2}\mathbf{i} + \frac{d^2y}{dt^2}\mathbf{j} + \frac{d^2z}{dt^2}\mathbf{k}$$

EXAMPLE 5. Let $\boldsymbol{\sigma}(t) = (e^t, t^2, \cos t)$. Then the acceleration vector is $\mathbf{a}(t) = e^t\mathbf{i} + 2\mathbf{j} - (\cos t)\mathbf{k}$.

The definition of arc length can be extended to include paths that are not C^1 but that are formed by piecing together a finite number of C^1

paths. A path $\boldsymbol{\sigma}: [a, b] \to \mathbb{R}^3, t \mapsto (x(t), y(t), z(t))$ is called *piecewise C^1* if there is a partition of $[a, b]$

$$a = t_0 < t_1 < \cdots < t_N = b$$

such that the function $\boldsymbol{\sigma}$ restricted to each interval $[t_i, t_{i+1}]$, $0 \le i \le N - 1$, is continuously differentiable. By this we mean that the derivative exists and is continuous on $[t_i, t_{i+1}]$, computed at the endpoints of each interval by using limits from within the interval (i.e., one-sided limits, as on p. 87). In the case of a path that is piecewise C^1, we define the arc length of the path to be the sum of the arc lengths of the C^1 paths that make it up. That is, if the partition

$$a = t_0 < t_1 < \cdots < t_N = b$$

satisfies the above conditions, we define

$$\text{arc length of } \boldsymbol{\sigma} = \sum_{i=0}^{N-1} (\text{arc length of } \boldsymbol{\sigma} \text{ from } t_i \text{ to } t_{i+1})$$

EXAMPLE 6. The path $\boldsymbol{\tau}: [-1, 1] \to \mathbb{R}^3, t \mapsto (|t|, |t - \frac{1}{2}|, 0)$ is not C^1 because $\tau_1: t \mapsto |t|$ is not differentiable at 0, nor is $\tau_2: t \mapsto |t - \frac{1}{2}|$ differentiable at $\frac{1}{2}$. However, if we take the partition

$$-1 = t_0 < 0 = t_1 < \tfrac{1}{2} = t_2 < 1 = t_3$$

we see that each of the τ_i is continuously differentiable on each of the intervals $[-1, 0]$, $[0, \frac{1}{2}]$, and $[\frac{1}{2}, 1]$, and therefore $\boldsymbol{\tau}$ is continuously differentiable on each interval. Thus, $\boldsymbol{\tau}$ is piecewise C^1.

On $[-1, 0]$, we have $x(t) = -t$, $y(t) = -t + \frac{1}{2}$, $z(t) = 0$, $\|d\mathbf{s}/dt\| = \sqrt{2}$; hence, the arc length of $\boldsymbol{\tau}$ between -1 and 0 is $\int_{-1}^{0} \sqrt{2}\, dt = \sqrt{2}$. Similarly, on $[0, \frac{1}{2}]$, $x(t) = t$, $y(t) = -t + \frac{1}{2}$, $z(t) = 0$, and again $\|d\mathbf{s}/dt\| = \sqrt{2}$, so that the arc length of $\boldsymbol{\tau}$ between 0 and $\frac{1}{2}$ is $\frac{1}{2}\sqrt{2}$. Finally, on $[\frac{1}{2}, 1]$ we have $x(t) = t$, $y(t) = t - \frac{1}{2}$, $z(t) = 0$, and the arc length of $\boldsymbol{\tau}$ between $\frac{1}{2}$ and 1 is $\frac{1}{2}\sqrt{2}$. Thus the total arc length of $\boldsymbol{\tau}$ is $2\sqrt{2}$.

EXERCISES

1. Calculate the arc length of the given curve on the specified interval.*
   (a) The path in Exercise 2(a), Section 3.1, [0, 1]
   (b) The path in Exercise 2(b), Section 3.1, [0, 1]

* Several of these problems make use of the formula

$$\int \sqrt{x^2 + a^2}\, dx = \tfrac{1}{2}(x\sqrt{x^2 + a^2} + a^2 \log(x + \sqrt{x^2 + a^2}))$$

from the table of integrals in the back of the book.

(c) $\mathbf{s}(t) = t\mathbf{i} + t(\sin t)\mathbf{j} + t(\cos t)\mathbf{k}$, $[0, \pi]$
(d) $\mathbf{s}(t) = 2t\mathbf{i} + t\mathbf{j} + t^2\mathbf{k}$, $[0, 2]$
(e) The path in Exercise 2(e), Section 3.1, $[0, 1]$
(f) The path in Exercise 2(f), Section 3.1, $[-1, 1]$
(g) The path in Exercise 2(g), Section 3.1, $[t_0, t_1]$

2. The arc-length function $s(t)$ for a given path $\boldsymbol{\sigma}(t)$, defined by $s(t) = \int_a^t \|\boldsymbol{\sigma}'(\tau)\|\, d\tau$, represents the distance a particle traversing the trajectory of $\boldsymbol{\sigma}$ will have travelled by time t if it starts out at time a; that is, the length of $\boldsymbol{\sigma}$ between $\boldsymbol{\sigma}(a)$ and $\boldsymbol{\sigma}(t)$. Find the arc-length functions for the curves $\boldsymbol{\alpha}(t) = (\cosh t, \sinh t, t)$ and $\boldsymbol{\beta}(t) = (\cos t, \sin t, t)$, with $a = 0$.
3. Let $\boldsymbol{\sigma}$ be the path $\boldsymbol{\sigma}(t) = (2t, t^2, \log t)$, defined for $t > 0$. Find the arc length of $\boldsymbol{\sigma}$ between the points $(2, 1, 0)$ and $(4, 4, \log 2)$.
4. Suppose a particle following the path $\boldsymbol{\sigma}(t) = (t^2, t^3 - 4t, 0)$ flies off on a tangent at $t = 2$. Compute the position of the particle at $t = 3$.
5. Let $\mathbf{c}(t)$ be a given path, $a \le t \le b$. Let $s = \alpha(t)$ be a new variable, where α is a strictly increasing C^1 function given on $[a, b]$. For each s in $[\alpha(a), \alpha(b)]$ there is a unique t with $\alpha(t) = s$. Define the function $\mathbf{d}\colon [\alpha(a), \alpha(b)] \to \mathbb{R}^3$ by $\mathbf{d}(s) = \mathbf{c}(t)$.
   (a) Argue that the image curves of $\mathbf{c}$ and $\mathbf{d}$ are the same.
   (b) Show that $\mathbf{c}$ and $\mathbf{d}$ have the same arc length.

3.3 VECTOR FIELDS

In Chapter 2 we introduced vector fields through the idea of the gradient vector field. In this section we would like to study some general properties of vector fields, including their geometrical and physical significance. A clear understanding of this is important for our work in sections 3.4 and 3.5 and for our studies in Chapter 7.

Definition. *A* ***vector field*** *on $\mathbb{R}^n$ is a map $\mathbf{F}\colon A \subset \mathbb{R}^n \to \mathbb{R}^n$ that assigns to each point $\mathbf{x}$ in its domain A a vector $\mathbf{F}(\mathbf{x})$.*

Graphically we may picture $\mathbf{F}$ as attaching an arrow to each point (Figure 3.3.1). Similarly a map $f\colon A \subset \mathbb{R}^n \to \mathbb{R}$ that assigns a number to each point is called a *scalar field*. For example, a vector field $\mathbf{F}(x, y, z)$ on $\mathbb{R}^3$ has three component scalar fields F_1, F_2, and F_3, so that $\mathbf{F}(x, y, z) = (F_1(x, y, z), F_2(x, y, z), F_3(x, y, z))$.

It is convenient and quite natural to draw the arrow representing $\mathbf{F}(\mathbf{x})$ with its foot at $\mathbf{x}$ rather than at the origin (which is the normal custom for drawing vectors). We regard this displaced vector with foot at $\mathbf{x}$ and the corresponding vector with foot at $\mathbf{0}$ as equivalent. In the remainder of this book we shall be concerned primarily with vector fields on $\mathbb{R}^2$ and $\mathbb{R}^3$, so that the term *vector field* should be taken to mean a vector field on $\mathbb{R}^2$ or $\mathbb{R}^3$ unless otherwise stated.

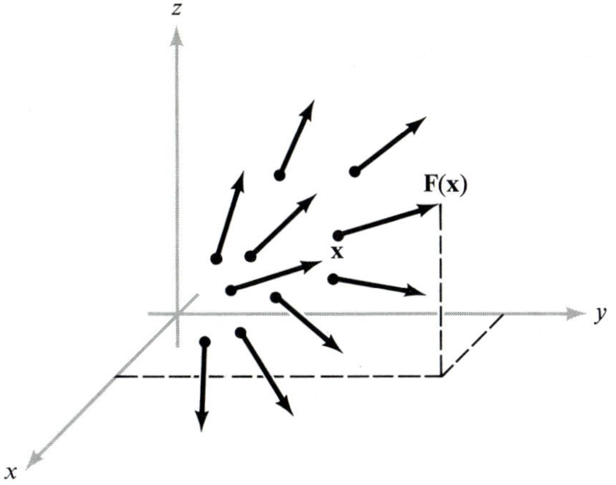

FIGURE 3.3.1
A vector field **F** *assigns a vector (arrow)* **F**(**x**) *to each point* **x** *of its domain.*

EXAMPLE 1. Imagine fluid moving down a pipe in a steady flow. If we attach to each point the fluid velocity at that point we obtain the velocity field **V** of the fluid (see Figure 3.3.2). Notice that the length of the arrows (the speed), as well as the direction of flow, can change from point to point.

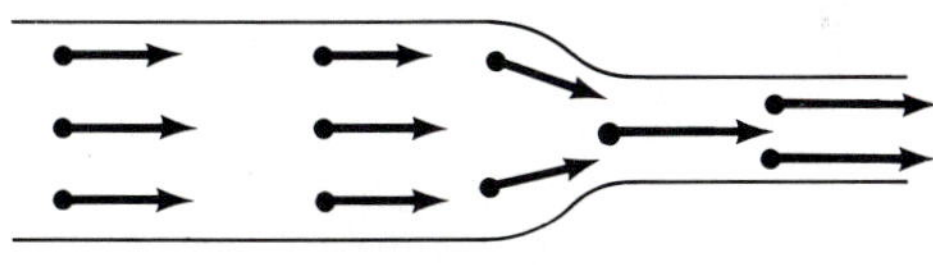

FIGURE 3.3.2
A vector field describing the velocity of flow in a pipe.

EXAMPLE 2. Consider a piece of material that is heated on one side and cooled on another. The temperature at each point within the body yields a scalar field $T(x, y, z)$.

The actual flow of heat may be marked by a field of arrows indicating the direction and magnitude of the flow (Figure 3.3.3). This *energy flux vector field* is given by $\mathbf{J} = -k\,\nabla T$, where $k > 0$ is a constant called the *conductivity*, and ∇T is the gradient of the real-valued function T. Note that the heat flows, as it should, from hot regions toward cold ones, since $-\nabla T$ points in a direction of decreasing T (see Section 2.5).

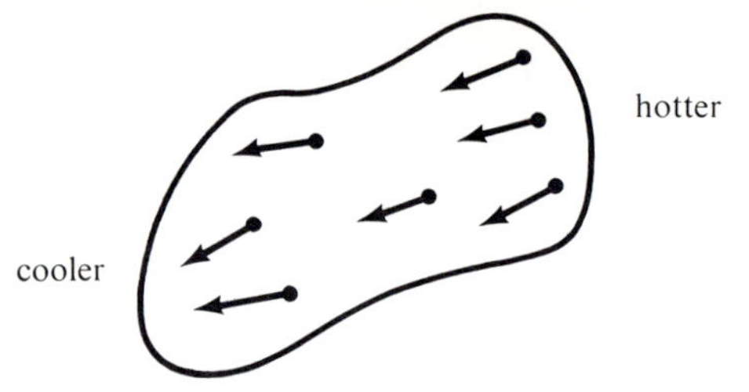

FIGURE 3.3.3
A vector field describing the direction and magnitude of heat flow.

EXAMPLE 3. The force of attraction of the Earth on a mass m can be described by a vector field on $\mathbb{R}^3$, the gravitational force field. According to Newton's law, this field is given by (see sections 2.5 and 3.1).

$$\mathbf{F} = -\frac{mMG}{r^3}\mathbf{r}$$

where $\mathbf{r}(x, y, z) = (x, y, z)$, and $r = \|\mathbf{r}\|$ (see Figure 3.3.4). As we saw earlier (Example 5, Section 2.5), $\mathbf{F}$ is actually a gradient field, $\mathbf{F} = -\nabla V$ where $V = -(mMG)/r$. Note again that $\mathbf{F}$ points in the

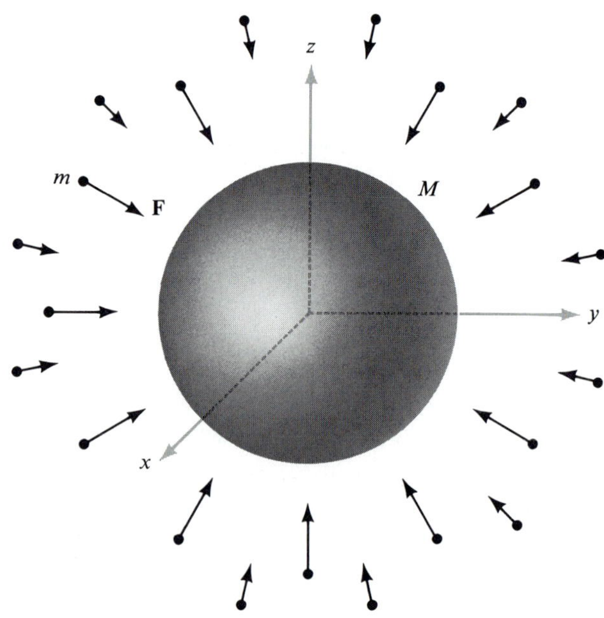

FIGURE 3.3.4
The vector field **F** *given by Newton's law of gravitation.*

direction of decreasing V. Writing $\mathbf{F}$ out in terms of components we see that

$$\mathbf{F}(x, y, z) = \left(\frac{-mMG}{r^3}x, \frac{-mMG}{r^3}y, \frac{-mMG}{r^3}z\right)$$

EXAMPLE 4. In the plane, $\mathbb{R}^2$, the function $\mathbf{V}\colon \mathbb{R}^2\backslash\{\mathbf{0}\} \to \mathbb{R}^2$ defined by

$$\mathbf{V}(x, y) = \frac{y\mathbf{i}}{x^2 + y^2} - \frac{x\mathbf{j}}{x^2 + y^2} = \left(\frac{y}{x^2 + y^2}, -\frac{x}{x^2 + y^2}\right)$$

is a vector field on $\mathbb{R}^2$. This is the velocity field approximating the velocity field of water in "circular" motion such as occurs, for example, when you pull the plug in a tub of water (Figure 3.3.5).

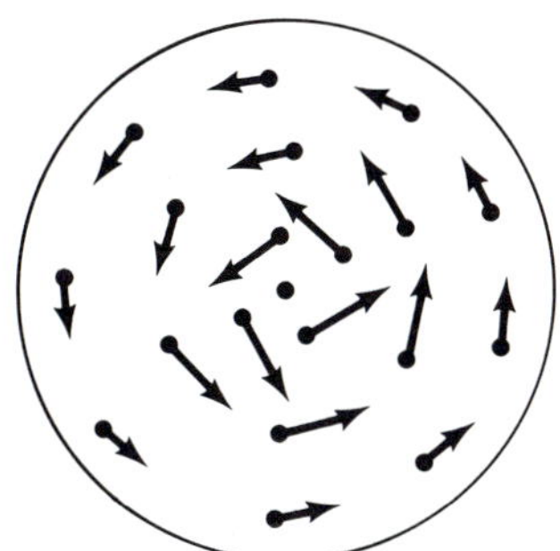

FIGURE 3.3.5
The vector field describing circular flow in a tub.

EXAMPLE 5. According to Coulomb's law, the force acting on a charge e at $\mathbf{r}$ due to a charge Q at the origin is

$$\mathbf{F} = \frac{\varepsilon Qe}{r^3}\mathbf{r} = -\nabla V$$

where $V = \varepsilon Qe/r$ and ε is a constant that depends on the units used. For $Qe > 0$ (like charges) the force is repulsive (Figure 3.3.6(a)) and for $Qe < 0$ (unlike charges) the force is attractive. In this example, as in Example 3, the level surfaces of V are called *equipotential surfaces* since, on them, the potential V is constant. Note that the force field is orthogonal to the equipotential surfaces (the force field is radial and the equipotential surfaces are spheres). This agrees with our general result in Section 2.5. In the case of temperature gradients in which $\mathbf{F} = -k\,\nabla T$, surfaces of constant T are called *isotherms.*

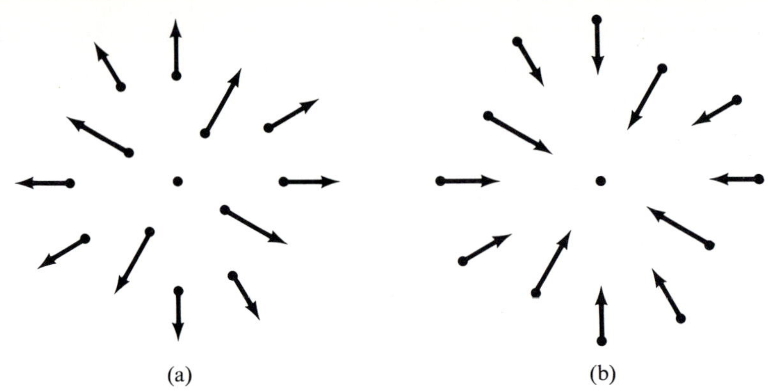

FIGURE 3.3.6
The vector fields associated with (a) like charges ($Qe > 0$) and (b) unlike charges ($Qe < 0$).

We remark that, in general, a vector field need not be a gradient field; that is, a vector field need not be the gradient of a real-valued function. This will become clearer in later chapters. Thus the notion of an equipotential surface only makes sense if the vector field happens to be a gradient field.

Another concept of importance is the notion of a flow line. This idea is easiest to visualize in the context of Example 1. In that case, a flow line is just a path followed by a fluid particle (Figure 3.3.7). These lines are also appropriately called *streamlines* or *integral curves.*

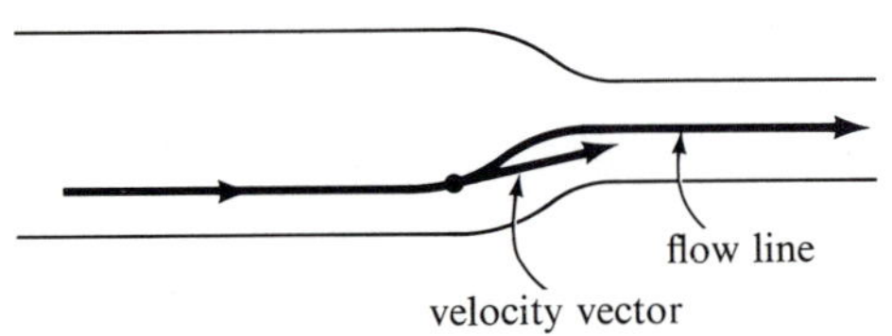

FIGURE 3.3.7
The velocity vector of a fluid is tangent to a flow line.

Definition. *If* $\mathbf{F}$ *is a vector field, a* ***flow line*** *for* $\mathbf{F}$ *is a path* $\boldsymbol{\sigma}(t)$ *such that*

$$\boldsymbol{\sigma}'(t) = \mathbf{F}(\boldsymbol{\sigma}(t))$$

That is, $\mathbf{F}$ *yields the velocity field of the path* $\boldsymbol{\sigma}(t)$.

EXERCISES

1. Let a particle of mass m move on a path $\mathbf{r}(t)$ in a force field $\mathbf{F} = -\nabla V$ on $\mathbb{R}^3$ according to Newton's law.
   (a) Prove that the energy $E = \frac{1}{2}m\|\mathbf{r}'(t)\|^2 + V(\mathbf{r}(t))$ is constant in time. (HINT: Carry out the differentiation dE/dt.)
   (b) If the particle moves on an equipotential surface show that its speed is constant.
2. Sketch a few flow lines of the vector fields
   (a) $\mathbf{F}(x, y) = (y, -x)$
   (b) $\mathbf{F}(x, y) = (x, -y)$
   (c) $\mathbf{F}(x, y) = (x, x^2)$
3. Let $\mathbf{c}(t)$ be a flow line of a gradient field $\mathbf{F} = -\nabla V$. Prove $V(\mathbf{c}(t))$ is a decreasing function of t. Explain.
4. Sketch the gradient field $-\nabla V$ for $V(x, y) = (x + y)/(x^2 + y^2)$. Sketch the equipotential surface $V = 1$.
5. Suppose that the isotherms in a region are all concentric spheres centered at the origin. Prove that the energy flux vector field points either toward or away from the origin.

3.4 DIVERGENCE AND CURL OF A VECTOR FIELD

This section deals with two basic operations that can be performed on vector fields. The full significance of these operations will emerge in Chapter 7, but we can discuss their meaning informally here.

First let us consider the curl of a vector field. This associates with any vector field on $\mathbb{R}^3$, $\mathbf{F} = F_1\mathbf{i} + F_2\mathbf{j} + F_3\mathbf{k} = (F_1, F_2, F_3)$, the vector field curl $\mathbf{F}$ defined by

$$\text{curl }\mathbf{F} = \left(\frac{\partial F_3}{\partial y} - \frac{\partial F_2}{\partial z}\right)\mathbf{i} + \left(\frac{\partial F_1}{\partial z} - \frac{\partial F_3}{\partial x}\right)\mathbf{j} + \left(\frac{\partial F_2}{\partial x} - \frac{\partial F_1}{\partial y}\right)\mathbf{k}$$

The lengthy formula is much easier to remember if we rewrite it using "operator" notation. Let us formally introduce the symbol "del"

$$\nabla = \mathbf{i}\frac{\partial}{\partial x} + \mathbf{j}\frac{\partial}{\partial y} + \mathbf{k}\frac{\partial}{\partial z},$$

∇ is an operator; that is, it makes sense when it acts or operates on real-valued functions. Specifically, ∇f, ∇ operating on f, is given by

$$\nabla f = \mathbf{i}\frac{\partial f}{\partial x} + \mathbf{j}\frac{\partial f}{\partial y} + \mathbf{k}\frac{\partial f}{\partial z},$$

the *gradient* of f. This formal notation is quite useful; if we view ∇ as a vector with components $\partial/\partial x$, $\partial/\partial y$, $\partial/\partial z$, then we can also take the formal cross product

$$\nabla \times \mathbf{F} = \begin{vmatrix} \mathbf{i} & \mathbf{j} & \mathbf{k} \\ \dfrac{\partial}{\partial x} & \dfrac{\partial}{\partial y} & \dfrac{\partial}{\partial z} \\ F_1 & F_2 & F_3 \end{vmatrix}$$

$$= \left(\frac{\partial F_3}{\partial y} - \frac{\partial F_2}{\partial z}\right)\mathbf{i} + \left(\frac{\partial F_1}{\partial z} - \frac{\partial F_3}{\partial x}\right)\mathbf{j} + \left(\frac{\partial F_2}{\partial x} - \frac{\partial F_1}{\partial y}\right)\mathbf{k}$$

$$= \text{curl } \mathbf{F}$$

Thus curl $\mathbf{F} = \nabla \times \mathbf{F}$, and we shall often use the latter expression.

EXAMPLE 1. Let $\mathbf{F}(x, y, z) = x\mathbf{i} + xy\mathbf{j} + \mathbf{k}$. Find $\nabla \times \mathbf{F}$.

We have

$$\nabla \times \mathbf{F} = \begin{vmatrix} \mathbf{i} & \mathbf{j} & \mathbf{k} \\ \dfrac{\partial}{\partial x} & \dfrac{\partial}{\partial y} & \dfrac{\partial}{\partial z} \\ x & xy & 1 \end{vmatrix} = (0 - 0)\mathbf{i} - (0 - 0)\mathbf{j} + (y - 0)\mathbf{k}$$

Thus $\nabla \times \mathbf{F} = y\mathbf{k}$.

The following theorem states a basic property of the curl. The result should be compared with the fact that for any vector $\mathbf{v}$, we have $\mathbf{v} \times \mathbf{v} = \mathbf{0}$.

Theorem 1. *For any C^2 function f we have*

$$\nabla \times (\nabla f) = \mathbf{0}$$

that is, the curl of any gradient is the zero vector.

Proof. Let us write out the components. Since $\nabla f = (\partial f/\partial x, \partial f/\partial y, \partial f/\partial z)$ we have, by definition,

$$\nabla \times \nabla f = \begin{vmatrix} \mathbf{i} & \mathbf{j} & \mathbf{k} \\ \dfrac{\partial}{\partial x} & \dfrac{\partial}{\partial y} & \dfrac{\partial}{\partial z} \\ \dfrac{\partial f}{\partial x} & \dfrac{\partial f}{\partial y} & \dfrac{\partial f}{\partial z} \end{vmatrix} = \left(\frac{\partial^2 f}{\partial y\, \partial z} - \frac{\partial^2 f}{\partial z\, \partial y}\right)\mathbf{i} + \left(\frac{\partial^2 f}{\partial z\, \partial x} - \frac{\partial^2 f}{\partial x\, \partial z}\right)\mathbf{j} + \left(\frac{\partial^2 f}{\partial x\, \partial y} - \frac{\partial^2 f}{\partial y\, \partial x}\right)\mathbf{k}$$

Each component is zero because of the symmetry property of mixed partial derivatives; hence, the desired result follows. ∎

As mentioned above, the full physical significance of the curl will be brought out later, when we study Stokes' Theorem in Chapter 7. However, we can now consider a simple situation that shows why the curl is associated with rotations.

EXAMPLE 2. Suppose we have a rigid body B rotating about an axis L. The rotational motion of the body can be completely described by a vector $\mathbf{w}$ along the axis of rotation, the direction being chosen so that the body rotates about $\mathbf{w}$ as in Figure 3.4.1. Moreover, we take the

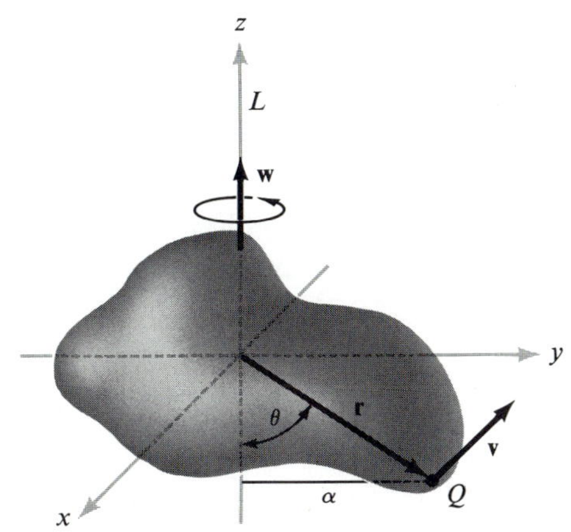

FIGURE 3.4.1
The velocity $\mathbf{v}$ *and angular velocity* $\mathbf{w}$ *of a rotating body are related by* $\mathbf{v} = \mathbf{w} \times \mathbf{r}$.

length $\omega = \|\mathbf{w}\|$ to be the angular speed of the body B—that is, the tangential speed of any point of B divided by its distance to the axis L of rotation.

We also assume that we have selected a coordinate system so that L is the z-axis. Let Q be any point of B and let α be the distance from Q to L. Clearly,

$$\alpha = \|\mathbf{r}\| \sin \theta$$

where $\mathbf{r}$ is the vector whose initial point is the origin and whose terminal point is Q. Then the tangential velocity $\mathbf{v}$ of Q is directed counterclockwise along the tangent to a circle parallel to the xy-plane and with radius α. The magnitude of this velocity is

$$\|\mathbf{v}\| = \omega\alpha = \omega\|\mathbf{r}\| \sin \theta = \|\mathbf{w}\| \, \|\mathbf{r}\| \sin \theta$$

We have seen (p. 28) that the direction and magnitude of $\mathbf{v}$ imply that

$$\mathbf{v} = \mathbf{w} \times \mathbf{r}$$

Because of our choice of axes, we can write $\mathbf{w} = \omega\mathbf{k}$, $\mathbf{r} = x\mathbf{i} + y\mathbf{j} + z\mathbf{k}$, so that

$$\mathbf{v} = \mathbf{w} \times \mathbf{r} = -\omega y\mathbf{i} + \omega x\mathbf{j}$$

and moreover

$$\text{curl } \mathbf{v} = \begin{vmatrix} \mathbf{i} & \mathbf{j} & \mathbf{k} \\ \dfrac{\partial}{\partial x} & \dfrac{\partial}{\partial y} & \dfrac{\partial}{\partial z} \\ -\omega y & \omega x & 0 \end{vmatrix} = 2\omega\mathbf{k} = 2\mathbf{w}$$

Hence, for the rotation of a rigid body, the curl of the velocity vector field is a vector field directed along the axis of rotation with magnitude twice the angular speed.

If a vector field $\mathbf{F}$ represents the flow of a fluid (see Example 1, Section 3.3), then curl $\mathbf{F} = \mathbf{0}$ means physically that the fluid is free from rotations or is irrotational; that is, it is free from whirlpools. Justification of this idea, and therefore of the use of the word irrotational, depends on Stokes' Theorem, but we can say informally that curl $\mathbf{F} = \mathbf{0}$ means that if a small paddle wheel is placed in the fluid, it will move with the fluid, but will not rotate. For example, it is an experimental fact that fluid draining from a tub is usually irrotational except right at the center (see Figure 3.4.2).

EXAMPLE 3. Let us verify that the vector field of Example 4, Section 3.3, is irrotational. Indeed,

$$\nabla \times \mathbf{V} = \begin{vmatrix} \mathbf{i} & \mathbf{j} & \mathbf{k} \\ \dfrac{\partial}{\partial x} & \dfrac{\partial}{\partial y} & \dfrac{\partial}{\partial z} \\ \dfrac{y}{x^2 + y^2} & \dfrac{-x}{x^2 + y^2} & 0 \end{vmatrix}$$

$$= 0\mathbf{i} + 0\mathbf{j} + \left(\frac{\partial}{\partial x}\left(\frac{-x}{x^2 + y^2}\right) - \frac{\partial}{\partial y}\left(\frac{y}{x^2 + y^2}\right)\right)\mathbf{k}$$

$$= \left(\frac{-(x^2 + y^2) + 2x^2}{(x^2 + y^2)^2} + \frac{-(x^2 + y^2) + 2y^2}{(x^2 + y^2)^2}\right)\mathbf{k}$$

$$= \mathbf{0}$$

EXAMPLE 4. Let $\mathbf{V}(x, y, z) = y\mathbf{i} - x\mathbf{j}$. Show that $\mathbf{V}$ is not a gradient field.

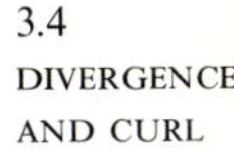

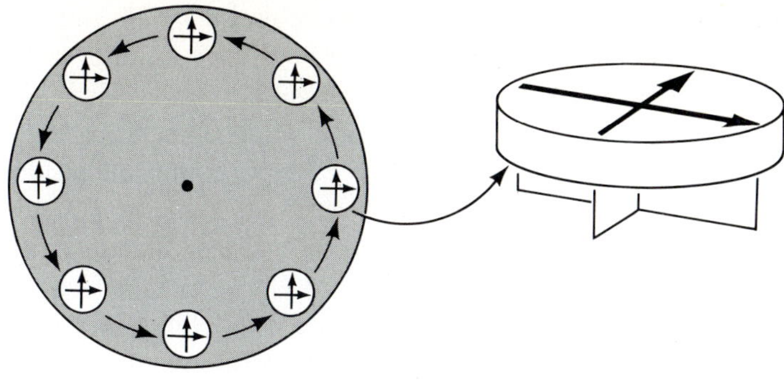

FIGURE 3.4.2
The velocity field $\mathbf{V}(x, y, z) = (y\mathbf{i} - x\mathbf{j})/(x^2 + y^2)$ *is irrotational; a small "paddle wheel" moving in the fluid will not rotate.*

Indeed, if $\mathbf{V}$ were a gradient field, then by Theorem 1 we would have curl $\mathbf{V} = \mathbf{0}$. But

$$\operatorname{curl} \mathbf{V} = \begin{vmatrix} \mathbf{i} & \mathbf{j} & \mathbf{k} \\ \dfrac{\partial}{\partial x} & \dfrac{\partial}{\partial y} & \dfrac{\partial}{\partial z} \\ y & -x & 0 \end{vmatrix} = -2\mathbf{k} \neq \mathbf{0}$$

The flow lines for this vector field, just as for the one in Example 3, are circles about the origin in the xy-plane, but this velocity field has rotation. In such a flow, a small paddle wheel rotates once as it circulates around the origin (Figure 3.4.3).

Another basic operation is the *divergence*, defined by

$$\operatorname{div} \mathbf{F} = \nabla \cdot \mathbf{F} = \frac{\partial F_1}{\partial x} + \frac{\partial F_2}{\partial y} + \frac{\partial F_3}{\partial z}$$

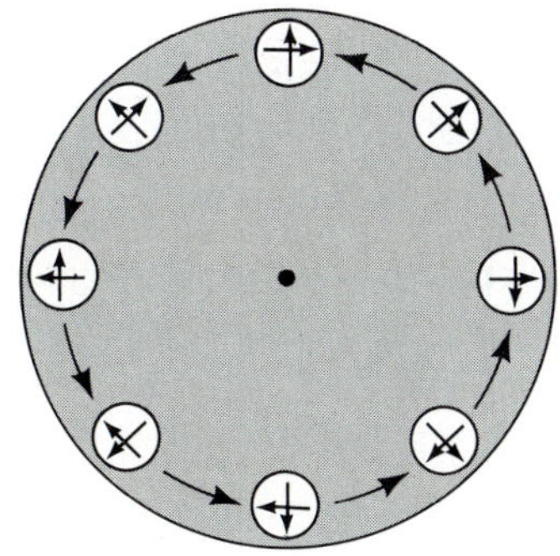

FIGURE 3.4.3
The velocity field $\mathbf{V}(x, y, z) = y\mathbf{i} - x\mathbf{j}$ *is rotational; a small "paddle wheel" moving in the fluid rotates.*

In operator notation, div $\mathbf{F}$ is just the dot product of ∇ and $\mathbf{F}$. Note that $\nabla \times \mathbf{F}$ is a vector field whereas $\nabla \cdot \mathbf{F}: \mathbb{R}^3 \to \mathbb{R}$, so that $\nabla \cdot \mathbf{F}$ is a scalar field. We read $\nabla \cdot \mathbf{F}$ as "the divergence of $\mathbf{F}$."

The full significance of the divergence will become apparent when we discuss Gauss' Theorem, but we can give some of its physical meaning here. If we imagine $\mathbf{F}$ as the velocity field of a gas (or a fluid), then div $\mathbf{F}$ represents the rate of expansion per unit volume of the gas (or fluid). For example, if $\mathbf{F}(x, y, z) = x\mathbf{i} + y\mathbf{j} + z\mathbf{k}$, then div $\mathbf{F} = 3$; this means the gas is expanding at the rate of 3 cubic units per unit of volume per unit of time. This is reasonable, because in this case $\mathbf{F}$ is an outward radial vector, and as the gas moves outward along the flow lines it expands. (See Section 3.3 for a discussion of flow lines).

The next theorem is analogous to Theorem 1.

Theorem 2. *For any C^2 vector field* $\mathbf{F}$,

$$\text{div curl } \mathbf{F} = \nabla \cdot (\nabla \times \mathbf{F}) = 0$$

that is, the divergence of any curl is zero.

The proof is similar to that of Theorem 1, in that it rests on the equality of the mixed partial derivatives. The student should write out the details.

We have seen that $\nabla \times \mathbf{F}$ is related to rotations and $\nabla \cdot \mathbf{F}$ is related to compressions and expansions. This leads to the following common terminology. If $\nabla \cdot \mathbf{F} = 0$ we say $\mathbf{F}$ is *incompressible*, just as we call $\mathbf{F}$ irrotational if $\nabla \times \mathbf{F} = \mathbf{0}$.

EXAMPLE 5. Compute the divergence of

$$\mathbf{F} = x^2y\mathbf{i} + z\mathbf{j} + xyz\mathbf{k}$$

Indeed

$$\text{div } \mathbf{F} = \frac{\partial}{\partial x}(x^2y) + \frac{\partial}{\partial y}(z) + \frac{\partial}{\partial z}(xyz) = 2xy + 0 + xy = 3xy$$

EXAMPLE 6. From Theorem 2 we can conclude that $\mathbf{F}$ in Example 5 cannot be the curl of another vector field, or else it would have zero divergence.

Finally, let us mention the *Laplace operator* ∇^2, which operates on functions f as follows:

$$\nabla^2 f = \nabla \cdot (\nabla f) = \frac{\partial^2 f}{\partial x^2} + \frac{\partial^2 f}{\partial y^2} + \frac{\partial^2 f}{\partial z^2}$$

If $\mathbf{F} = F_1\mathbf{i} + F_2\,\mathbf{j} + F_3\,\mathbf{k}$ is a C^2 vector field we can also define $\nabla^2\mathbf{F}$ in terms of components

$$\nabla^2\mathbf{F} = \nabla^2 F_1\mathbf{i} + \nabla^2 F_2\,\mathbf{j} + \nabla^2 F_3\,\mathbf{k}$$

This operator plays an important role in many physical laws. We shall discuss these in Chapter 7.

EXERCISES

1. Compute the gradient ∇f for the following functions.
   (a) $f(x, y, z) = \sqrt{x^2 + y^2 + z^2}$
   (b) $f(x, y, z) = xy + yz + xz$
   (c) $f(x, y, z) = 1/(x^2 + y^2 + z^2)$
2. Compute the divergence $\nabla \cdot \mathbf{F}$ of the following vector fields.
   (a) $\mathbf{F}(x, y, z) = x\mathbf{i} + y\mathbf{j} + z\mathbf{k}$
   (b) $\mathbf{F}(x, y, z) = yz\mathbf{i} + xz\mathbf{j} + xy\mathbf{k}$
   (c) $\mathbf{F}(x, y, z) = (x^2 + y^2 + z^2)(3\mathbf{i} + 4\mathbf{j} + 5\mathbf{k})$
3. Compute the curl $\nabla \times \mathbf{F}$ of each vector field in Exercise 2.
4. Verify that the vector field of Example 4, Section 3.3, is incompressible. Can you interpret this result physically?
5. Verify that $\mathbf{F} = y\mathbf{i} + x\mathbf{j}$ is incompressible.
6. Let $\mathbf{F}(x, y, z) = 3x^2y\mathbf{i} + (x^3 + y^3)\mathbf{j}$.
   (a) Verify that curl $\mathbf{F} = \mathbf{0}$.
   (b) Find a function* f such that $\mathbf{F} = \nabla f$.
   (c) Is it true that for a vector field $\mathbf{F}$ such a function f can exist only if curl $\mathbf{F} = \mathbf{0}$?
7. Let $f(x, y, z) = x^2y^2 + y^2z^2$. Verify directly that $\nabla \times \nabla f = \mathbf{0}$.
8. Show that the real and imaginary parts of each of the following complex functions form the components of an irrotational and incompressible vector field in the plane.
   (a) $(x - iy)^2$
   (b) $(x - iy)^3$
   (c) $e^{x-iy} = e^x(\cos y - i \sin y)$
9. Show that $\mathbf{F} = y(\cos x)\mathbf{i} + x(\sin y)\mathbf{j}$ is *not* a gradient vector field.
10. Let $\mathbf{r}(x, y, z) = x\mathbf{i} + y\mathbf{j} + z\mathbf{k}$. From Exercise 3(a), we know that $\nabla \times \mathbf{r} = \mathbf{0}$. Can you see why this is so physically, by visualizing $\mathbf{r}$ as the velocity field of a fluid?

3.5 VECTOR DIFFERENTIAL CALCULUS

We now have these basic operations on hand: gradient, divergence, curl, and the Laplace operator. This section develops their properties and the relationships between them a little further.

* Specific techniques for constructing f in general are given in Chapter 7. The one in this problem should be sought directly.

Table 3.1. Some Common Formulas of Vector Analysis

1. $\nabla(f+g) = \nabla f + \nabla g$
2. $\nabla(cf) = c\nabla f$, for a constant c
3. $\nabla(fg) = f\nabla g + g\nabla f$
4. $\nabla(f/g) = (g\nabla f - f\nabla g)/g^2$, at points where $g(\mathbf{x}) \neq 0$
5. $\operatorname{div}(\mathbf{F} + \mathbf{G}) = \operatorname{div} \mathbf{F} + \operatorname{div} \mathbf{G}$
6. $\operatorname{curl}(\mathbf{F} + \mathbf{G}) = \operatorname{curl} \mathbf{F} + \operatorname{curl} \mathbf{G}$
7. $\nabla(\mathbf{F} \cdot \mathbf{G}) = (\mathbf{F} \cdot \nabla)\mathbf{G} + (\mathbf{G} \cdot \nabla)\mathbf{F} + \mathbf{F} \times \operatorname{curl} \mathbf{G} + \mathbf{G} \times \operatorname{curl} \mathbf{F}$
8. $\operatorname{div}(f\mathbf{F}) = f \operatorname{div} \mathbf{F} + \mathbf{F} \cdot \nabla f$
9. $\operatorname{div}(\mathbf{F} \times \mathbf{G}) = \mathbf{G} \cdot \operatorname{curl} \mathbf{F} - \mathbf{F} \cdot \operatorname{curl} \mathbf{G}$
10. $\operatorname{div} \operatorname{curl} \mathbf{F} = 0$
11. $\operatorname{curl}(f\mathbf{F}) = f \operatorname{curl} \mathbf{F} + \nabla f \times \mathbf{F}$
12. $\operatorname{curl}(\mathbf{F} \times \mathbf{G}) = \mathbf{F} \operatorname{div} \mathbf{G} - \mathbf{G} \operatorname{div} \mathbf{F} + (\mathbf{G} \cdot \nabla)\mathbf{F} - (\mathbf{F} \cdot \nabla)\mathbf{G}$
13. $\operatorname{curl} \operatorname{curl} \mathbf{F} = \operatorname{grad} \operatorname{div} \mathbf{F} - \nabla^2\mathbf{F}$
14. $\operatorname{curl} \nabla f = \mathbf{0}$
15. $\nabla(\mathbf{F} \cdot \mathbf{F}) = 2(\mathbf{F} \cdot \nabla)\mathbf{F} + 2\mathbf{F} \times (\operatorname{curl} \mathbf{F})$
16. $\nabla^2(fg) = f\nabla^2 g + g\nabla^2 f + 2(\nabla f \cdot \nabla g)$
17. $\operatorname{div}(\nabla f \times \nabla g) = 0$
18. $\nabla \cdot (f\nabla g - g\nabla f) = f\nabla^2 g - g\nabla^2 f$
19. $\mathbf{H} \cdot (\mathbf{F} \times \mathbf{G}) = \mathbf{G} \cdot (\mathbf{H} \times \mathbf{F}) = \mathbf{F} \cdot (\mathbf{G} \times \mathbf{H})$
20. $\mathbf{H} \cdot ((\mathbf{F} \times \nabla) \times \mathbf{G}) = ((\mathbf{H} \cdot \nabla)\mathbf{G}) \cdot \mathbf{F} - \mathbf{H} \cdot (\mathbf{F} \cdot \nabla)\mathbf{G}$
21. $\mathbf{F} \times (\mathbf{G} \times \mathbf{H}) = (\mathbf{F} \cdot \mathbf{H})\mathbf{G} - \mathbf{H}(\mathbf{F} \cdot \mathbf{G})$

NOTE: f and g denote scalar fields; **F**, **G**, and **H** denote vector fields.

In Table 3.1 above are summarized some basic general formulas that are useful when computing with vector fields. Some of these, such as formulas 10 and 14, were given in Section 3.4. Others are proven in the examples and exercises.

Some expressions in this table require explanation. First,

$$\mathbf{V} = (\mathbf{F} \cdot \nabla)\mathbf{G}$$

has, by definition, components $V_i = \mathbf{F} \cdot (\nabla G_i)$, for $i = 1, 2, 3$, where $\mathbf{G} = (G_1, G_2, G_3)$. Second, $\nabla^2\mathbf{F}$ has components $\nabla^2 F_i$, where $\mathbf{F} = (F_1, F_2, F_3)$. In the expression $(\mathbf{F} \times \nabla) \times \mathbf{G}$, the ∇ is to operate only on **G**.

EXAMPLE 1. Prove Formula 8 in Table 3.1.

$f\mathbf{F}$ has components fF_i, for $i = 1, 2, 3$, so

$$\operatorname{div}(f\mathbf{F}) = \frac{\partial}{\partial x}(fF_1) + \frac{\partial}{\partial y}(fF_2) + \frac{\partial}{\partial z}(fF_3)$$

However, $\partial/\partial x(fF_1) = f\,\partial F_1/\partial x + F_1\,\partial f/\partial x$, with similar expressions for the other terms. Therefore

$$\begin{aligned}\operatorname{div}(f\mathbf{F}) &= f\left(\frac{\partial F_1}{\partial x} + \frac{\partial F_2}{\partial y} + \frac{\partial F_3}{\partial z}\right) + F_1\frac{\partial f}{\partial x} + F_2\frac{\partial f}{\partial y} + F_3\frac{\partial f}{\partial z}\\ &= f\nabla\cdot\mathbf{F} + \mathbf{F}\cdot\nabla f\end{aligned}$$

EXAMPLE 2. Let $\mathbf{r}$ be the vector field $\mathbf{r}(x, y, z) = (x, y, z)$ (the position vector), and let $r = \|\mathbf{r}\|$. Compute ∇r and $\nabla\cdot(r\mathbf{r})$.

We have

$$\nabla r = \left(\frac{\partial r}{\partial x}, \frac{\partial r}{\partial y}, \frac{\partial r}{\partial z}\right)$$

Now $r(x, y, z) = \sqrt{x^2 + y^2 + z^2}$ so for $r \neq 0$ we get

$$\frac{\partial r}{\partial x} = \frac{x}{\sqrt{x^2 + y^2 + z^2}} = \frac{x}{r}$$

Thus

$$\nabla r = \left(\frac{x}{r}, \frac{y}{r}, \frac{z}{r}\right) = \frac{\mathbf{r}}{r}$$

For the second part, use Formula 8 to write

$$\nabla\cdot(r\mathbf{r}) = r\nabla\cdot\mathbf{r} + \mathbf{r}\cdot\nabla r$$

Now $\nabla\cdot\mathbf{r} = \partial x/\partial x + \partial y/\partial y + \partial z/\partial z = 3$, and we have computed $\nabla r = \mathbf{r}/r$ above. Since $\mathbf{r}\cdot\mathbf{r}/r = r^2/r$,

$$\nabla\cdot(r\mathbf{r}) = 3r + \frac{r^2}{r} = 4r$$

EXAMPLE 3. Show that $\nabla f \times \nabla g$ is always incompressible. In fact, deduce Formula 17 of Table 3.1 from Formula 9.

By Formula 9,

$$\operatorname{div}(\nabla f \times \nabla g) = \nabla g\cdot(\nabla\times\nabla f) - \nabla f\cdot(\nabla\times\nabla g)$$

which is zero, since $\nabla\times\nabla f = \mathbf{0}$ and $\nabla\times\nabla g = \mathbf{0}$.

In the following exercises the reader will get practice with this kind of manipulation. Later in the text we shall have use for the identities in Exercise 8.

EXERCISES

1. Suppose $\nabla \cdot \mathbf{F} = 0$ and $\nabla \cdot \mathbf{G} = 0$. Which of the following necessarily have zero divergence?
   (a) $\mathbf{F} + \mathbf{G}$
   (b) $\mathbf{F} \times \mathbf{G}$
   (c) $\mathbf{F}(\mathbf{F} \cdot \mathbf{G})$
2. Prove formulas 1 to 6 of Table 3.1.
3. Prove formulas 7, 9, and 11 of Table 3.1.
4. Prove formulas 12, 13, 15, and 16 of Table 3.1.
5. Prove formulas 18 to 21 of Table 3.1.
6. Let $\mathbf{F} = 2xz^2\mathbf{i} + \mathbf{j} + y^3zx\mathbf{k}$, $\mathbf{G} = x^2\mathbf{i} + y^2\mathbf{j} + z^2\mathbf{k}$ and $f = x^2y$. Compute the following quantities.
   (a) ∇f
   (b) $\nabla \times \mathbf{F}$
   (c) $(\mathbf{F} \cdot \nabla)\mathbf{G}$
   (d) $\mathbf{F} \cdot (\nabla f)$
   (e) $\mathbf{F} \times \nabla f$
7. Does $\nabla \times \mathbf{F}$ have to be perpendicular to $\mathbf{F}$?
8. Let $\mathbf{r}(x, y, z) = (x, y, z)$ and $r = \sqrt{x^2 + y^2 + z^2} = \|\mathbf{r}\|$. Prove the following identities (make use of Table 3.1 as much as possible).
   (a) $\nabla(1/r) = -\mathbf{r}/r^3$, $r \neq 0$; and, in general, $\nabla(r^n) = nr^{n-2}\mathbf{r}$ and $\nabla(\log r) = \mathbf{r}/r^2$
   (b) $\nabla^2(1/r) = 0$, $r \neq 0$; and, in general, $\nabla^2 r^n = n(n+1)r^{n-2}$
   (c) $\nabla \cdot (\mathbf{r}/r^3) = 0$; and, in general, $\nabla \cdot (r^n\mathbf{r}) = (n+3)r^n$
   (d) $\nabla \times \mathbf{r} = \mathbf{0}$; and, in general, $\nabla \times (r^n\mathbf{r}) = \mathbf{0}$.

REVIEW EXERCISES FOR CHAPTER 3

1. Compute the divergence of the following vector fields at the points indicated.
   (a) $\mathbf{F}(x, y, z) = x\mathbf{i} + 3xy\mathbf{j} + z\mathbf{k}$, $\quad (0, 1, 0)$
   (b) $\mathbf{F}(x, y, z) = y\mathbf{i} + z\mathbf{j} + x\mathbf{k}$, $\quad (1, 1, 1)$
   (c) $\mathbf{F}(x, y, z) = (x + y)^3\mathbf{i} + (\sin xy)\mathbf{j} + (\cos xyz)\mathbf{k}$, $\quad (2, 0, 1)$
2. Compute the curl of each vector field in Exercise 1 at the given point.
3. (a) Let $f(x, y, z) = xyz^2$; compute ∇f.
   (b) Let $\mathbf{F}(x, y, z) = xy\mathbf{i} + yz\mathbf{j} + zy\mathbf{k}$; compute $\nabla \times \mathbf{F}$.
   (c) Compute $\nabla \times (f\mathbf{F})$ using Formula 11 of Table 3.1. Compare with a direct computation.
4. Compute $\nabla \cdot \mathbf{F}$ and $\nabla \times \mathbf{F}$ for the following vector fields.
   (a) $\mathbf{F} = 2x\mathbf{i} + 3y\mathbf{j} + 4z\mathbf{k}$
   (b) $\mathbf{F} = x^2\mathbf{i} + y^2\mathbf{j} + z^2\mathbf{k}$
   (c) $\mathbf{F} = (x + y)\mathbf{i} + (y + z)\mathbf{j} + (z + x)\mathbf{k}$
5. Find the equation of the plane tangent to each surface at the indicated point.

(a) $z = x^2 + y^2$, $(0, 0, 0)$
(b) $z = x^2 - y^2 + x$, $(1, 0, 2)$
(c) $z = (x + y)^2$, $(3, 2, 25)$

6. What altitude must a satellite have in order that it appear stationary in the sky when viewed from Earth? (See Exercise 7, Section 3.1, for units.)

7. Let $\boldsymbol{\sigma}: \mathbb{R} \to \mathbb{R}^3$ be a path and $h: \mathbb{R} \to \mathbb{R}$ a strictly increasing differentiable function. The composition $\boldsymbol{\sigma} \circ h: \mathbb{R} \to \mathbb{R}^3$ is called the *reparametrization* of $\boldsymbol{\sigma}$ by h; argue that $\boldsymbol{\sigma} \circ h$ has the same trajectory as $\boldsymbol{\sigma}$, and prove that if $\boldsymbol{\alpha} = \boldsymbol{\sigma} \circ h$ then $\boldsymbol{\alpha}'(t) = h'(t)\boldsymbol{\sigma}'(h(t))$.

8. (a) Let $\boldsymbol{\alpha}$ be any differentiable path whose speed is never zero. Let $s(t)$ be the arc-length function for $\boldsymbol{\alpha}$, $s(t) = \int_a^t \|\boldsymbol{\alpha}'(\tau)\|\, d\tau$. Let $t(s)$ be the inverse function of s. Prove that the curve $\boldsymbol{\beta} = \boldsymbol{\alpha} \circ t$ has unit speed; i.e., $\|\boldsymbol{\beta}'(s)\| = 1$.
   (b) Let $\boldsymbol{\sigma}$ be the path $\boldsymbol{\sigma}(t) = (a \cos t, a \sin t, bt)$. Find a path $\boldsymbol{\alpha}$ that has the same trajectory as $\boldsymbol{\sigma}$, but has unit speed, $\|\boldsymbol{\alpha}'(t)\| = 1$; i.e., find a unit-speed reparametrization of $\boldsymbol{\sigma}$.

9. Let a particle of mass m move on the path $\boldsymbol{\sigma}(t) = (t^2, \sin t, \cos t)$. Compute the force acting on the particle at $t = 0$.

10. (a) Let $\mathbf{c}(t)$ be a path with $\|\mathbf{c}(t)\| = \text{constant}$; i.e., the curve lies on a sphere. Prove that $\mathbf{c}'(t)$ is orthogonal to $\mathbf{c}(t)$.
   (b) Let $\mathbf{c}$ be a path whose speed is never zero. Show that $\mathbf{c}$ has constant speed iff the acceleration vector $\mathbf{c}''$ is always perpendicular to the velocity vector $\mathbf{c}'$.

11. Let a particle travel on the path $\mathbf{c}(t) = (t, t^2, t \cos t)$ and, at $t = \pi$, leave this curve on a tangent. Where is the particle at time $t = 2\pi$?

12. (a) For the following functions $f: \mathbb{R}^3 \to \mathbb{R}$ and $\mathbf{g}: \mathbb{R} \to \mathbb{R}^3$ find ∇f and $\mathbf{g}'$ and evaluate $(f \circ \mathbf{g})'(1)$
      (*i*) $f(x, y, z) = xyz$, $\mathbf{g}(t) = (t \cos t, \sin t)$
      (*ii*) $f(x, y, z) = xyz$, $\mathbf{g}(t) = (6t, 3t^2, t^3)$
   (b) Let $f(x, y) = (e^x, x + y)$ and $g(u, v) = (u, \cos v, v + u)$. Compute the derivative of $g \circ f$ at $(0, 0)$ in two ways.

13. Compute the directional derivative of the following functions in the given direction at the given point.
   (a) $f(x, y, z) = xyz$, $\mathbf{v} = (1/\sqrt{14})(\mathbf{i} + 3\mathbf{j} + 2\mathbf{k})$, $(1, 1, 1)$
   (b) $f(x, y, z) = x^2 + y$, $\mathbf{v} = \frac{12}{13}\mathbf{i} + \frac{3}{13}\mathbf{j} + \frac{4}{13}\mathbf{k}$, $(1, 0, 0)$

14. (a) Review the proof that $\nabla f(x, y, z)$ is perpendicular to the surface $f(x, y, z) = \text{constant}$.
   (b) Find a unit normal to the surface $x^3y + xz = 1$ at the point $(1, 2, -1)$.
   (c) Find an equation for the plane tangent to the surface in (b) at the indicated point.
   (d) Find the angle between the surfaces $x^2 + y^2 + z^2 = 3$ and $x = z^2 + y^2 - 3$ at the point $(-1, 1, -1)$.

15. Let $\mathbf{F}(x, y, z) = (x^2, 0, z(1 + x))$. Show that $\boldsymbol{\sigma}(t) = (1/(1 - t), 0, e^t/(1 - t))$ is a flow line of $\mathbf{F}$.

16. Let $\mathbf{v} = 3\mathbf{i} + 2\mathbf{j} + \mathbf{k}$ and $\mathbf{w} = \mathbf{i} - \mathbf{j}$.
   (a) Compute $\mathbf{v} + 3\mathbf{w}$, $\langle \mathbf{v}, \mathbf{w} \rangle$, and $\mathbf{v} \times \mathbf{w}$.
   (b) Compute the area of the parallelogram spanned by $\mathbf{v}$ and $\mathbf{w}$.

17. (a) Does $\lim_{(x, y)\to(0, 0)} (x^2 - y^2)/(x^2 + y^2)$ exist?
    (b) Explain what it means for $f: \mathbb{R}^n \to \mathbb{R}^m$ to be continuous at $\mathbf{x}_0 \in \mathbb{R}^n$. (State the precise definition.)
    (c) Explain what it means for $f: \mathbb{R}^n \to \mathbb{R}^m$ to be differentiable at $\mathbf{x}_0 \in \mathbb{R}^n$. (State the precise definition.)
18. (a) A bug, finding himself in a toxic environment caused by chemical AJT, decides to move in a direction that will decrease the concentration of AJT the fastest. If the concentration of AJT is given by $\sigma(x, y, z) = e^{-3x} + \sin(yz) + e^{-z2}$, with the bug at (0, 0, 0), in which direction should he swim?
    (b) Discuss briefly the theory behind your answer to (a).
19. Compute the tangent planes of the following surfaces at the indicated points.
    (a) $z = x^2 + 3y^3 + \sin(xy)$; $x = 1$, $y = 0$
    (b) $x^2 + 3y^2 + 4z^2 = 10$; $(0, \sqrt{2}, 1)$
20. (a) Let $\mathbf{F} = 2xye^z\mathbf{i} + e^z x^2\mathbf{j} + (x^2ye^z + z^2)\mathbf{k}$. Compute $\nabla \cdot \mathbf{F}$ and $\nabla \times \mathbf{F}$.
    (b) Find a function $f(x, y, z)$ such that $\mathbf{F} = \nabla f$. Discuss briefly.

CHAPTER 4

HIGHER-ORDER DERIVATIVES; MAXIMA AND MINIMA

In one-variable calculus, to test a function $f(x)$ for a local maximum or minimum we look for critical points x_0, that is, points x_0 for which $f'(x_0) = 0$, and at each such point we check the sign of the second derivative $f''(x_0)$. If $f''(x_0) < 0$, $f(x_0)$ is a local maximum of f; if $f''(x_0) > 0$, $f(x_0)$ is a local minimum of f; if $f''(x_0) = 0$, the test fails.

One of the goals of this chapter is to extend these methods to real-valued functions of several variables. We shall begin in Section 4.1 with a discussion of Taylor's Theorem; this will then be used in Section 4.2 to derive tests for maxima, minima, and saddle points. As with functions of one variable, such methods help one to visualize the shape of a graph.

In Section 4.3 we shall study the problem of maximizing a real-valued function subject to supplementary conditions, also referred to as constraints. For example, we might wish to maximize $f(x, y, z)$ among those (x, y, z) constrained to lie on the unit sphere, $x^2 + y^2 + z^2 = 1$.

In Section 4.4 we shall describe a few physical applications of the preceding material, relating to equilibrium points of physical systems and their stability.

4.1 TAYLOR'S THEOREM

We shall use Taylor's Theorem in several variables to derive a test for different types of extrema, finally obtaining a test much like the second-derivative test that is learned in one-variable calculus. There are other important applications of this theorem as well. Basically, Taylor's Theorem gives us "higher-order" approximations to a function by using more than just the first derivative of the function.

For smooth functions of one variable $x \mapsto f(x)$, Taylor's Theorem asserts that

$$f(x) = f(a) + f'(a)(x-a) + \frac{f''(a)}{2!}(x-a)^2 + \cdots + \frac{f^{(k)}(a)}{k!}(x-a)^k + R_k(x, a) \tag{1}$$

where

$$R_k(x, a) = \int_a^x \frac{(x-t)^k}{k!} f^{(k+1)}(t)\, dt \tag{1'}$$

is the remainder. For x near a this error $R_k(x, a)$ is small to "order k." This means that

$$\frac{R_k(x, a)}{(x-a)^k} \to 0 \text{ as } x \to a \tag{2}$$

In other words, $R_k(x, a)$ is small compared to the (already small) quantity $(x-a)^k$. The proof of (2) follows by using the Mean Value Theorem for Integrals to express the remainder (1′) in Lagrange's form:

$$R_k(x, a) = \frac{f^{(k+1)}(c)(x-a)^{k+1}}{(k+1)!}$$

for some c between x and a.

Our goal in this section is to prove a more general theorem valid for functions of several variables. The above theorem for functions of one variable will then follow as a corollary.

We already know a first-order version, that is, when $k = 1$. Indeed, if $f\colon \mathbb{R}^n \to \mathbb{R}$ is differentiable at $\mathbf{x}_0$, and we define

$$R_1(\mathbf{x}, \mathbf{x}_0) = f(\mathbf{x}) - f(\mathbf{x}_0) - Df(\mathbf{x}_0) \cdot (\mathbf{x} - \mathbf{x}_0)$$

so

$$f(\mathbf{x}) = f(\mathbf{x}_0) + Df(\mathbf{x}_0)(\mathbf{x} - \mathbf{x}_0) + R_1(\mathbf{x}, \mathbf{x}_0)$$

then by the definition of differentiability,

$$\frac{|R_1(\mathbf{x}, \mathbf{x}_0)|}{\|\mathbf{x} - \mathbf{x}_0\|} \to 0 \text{ as } \mathbf{x} \to \mathbf{x}_0$$

that is, $R_1(\mathbf{x}, \mathbf{x}_0)$ vanishes to first order at $\mathbf{x}_0$. Let us summarize, writing $\mathbf{h} = \mathbf{x} - \mathbf{x}_0$, and $R_1(\mathbf{x}, \mathbf{x}_0) = R_1(\mathbf{h}, \mathbf{x}_0)$ (an admitted abuse of notation!).

Theorem 1. *(First-order Taylor's Theorem). Let $f\colon U \subset \mathbb{R}^n \to \mathbb{R}$ be differentiable at $\mathbf{x}_0 \in U$. Then we may write*

$$f(\mathbf{x}_0 + \mathbf{h}) = f(\mathbf{x}_0) + \sum_{i=1}^{n} h_i \frac{\partial f}{\partial x_i}(\mathbf{x}_0) + R_1(\mathbf{h}, \mathbf{x}_0)$$

where $R_1(\mathbf{h}, \mathbf{x}_0)/\|\mathbf{h}\| \to 0$ as $\mathbf{h} \to \mathbf{0}$ in $\mathbb{R}^n$.

The *second-order Taylor formula* is as follows:

Theorem 2. *Let $f\colon U \subset \mathbb{R}^n \to \mathbb{R}$ have continuous partials of third order.* Then we may write*

$$f(\mathbf{x}_0 + \mathbf{h}) = f(\mathbf{x}_0) + \sum_{i=1}^{n} h_i \frac{\partial f}{\partial x_i}(\mathbf{x}_0) + \frac{1}{2} \sum_{i,j=1}^{n} h_i h_j \frac{\partial^2 f}{\partial x_i\, \partial x_j}(\mathbf{x}_0)$$
$$+ R_2(\mathbf{h}, \mathbf{x}_0)$$

where $R_2(\mathbf{h}, \mathbf{x}_0)/\|\mathbf{h}\|^2 \to 0$ as $\mathbf{h} \to \mathbf{0}$.

In the course of the proof, we shall obtain a useful explicit formula (see (5′) below) for the remainder R_2. This formula is a generalization of formula (1′).

Proof of Theorem 2. If we notice that, by the Chain Rule,

$$\frac{d}{dt} f(\mathbf{x}_0 + t\mathbf{h}) = Df(\mathbf{x}_0 + t\mathbf{h}) \cdot \mathbf{h}$$
$$= \sum_{i=1}^{n} \frac{\partial f}{\partial x_i}(\mathbf{x}_0 + t\mathbf{h}) h_i$$

then we can integrate both sides from $t = 0$ to $t = 1$ to obtain

$$f(\mathbf{x}_0 + \mathbf{h}) - f(\mathbf{x}_0) = \int_0^1 \sum_{i=1}^{n} \frac{\partial f}{\partial x_i}(\mathbf{x}_0 + t\mathbf{h}) h_i \, dt$$

* For the statement of the theorem as given here, f actually needs only to be of class C^2, but for a convenient form of the remainder we assume f is C^3. Also, if one assumes the one-variable version, then application of it to $g(t) = f(\mathbf{x}_0 + t\mathbf{h})$ yields the version given here for several variables.

We shall now integrate the expression on the right-hand side by parts. Remember the general formula

$$\int_0^1 u \frac{dv}{dt}\,dt = -\int_0^1 v \frac{du}{dt}\,dt + uv\Big|_0^1$$

In this case, let $u = \partial f/\partial x_i\,(\mathbf{x}_0 + t\mathbf{h})h_i$ and let $v = t - 1$. Therefore

$$\sum_{i=1}^{n} \int_0^1 \frac{\partial f}{\partial x_i}(\mathbf{x}_0 + t\mathbf{h})h_i\,dt$$

$$= \sum_{i,j=1}^{n} \int_0^1 (1-t)\frac{\partial^2 f}{\partial x_i\,\partial x_j}(\mathbf{x}_0 + t\mathbf{h})h_i h_j\,dt + \sum_{i=1}^{n} h_i \frac{\partial f}{\partial x_i}(\mathbf{x}_0)$$

since

$$\frac{du}{dt} = \sum_{j=1}^{n} \frac{\partial^2 f}{\partial x_i\,\partial x_j}(\mathbf{x}_0 + t\mathbf{h})h_i h_j$$

by the Chain Rule, and

$$uv\Big|_0^1 = (t-1)\frac{\partial f}{\partial x_i}(\mathbf{x}_0 + t\mathbf{h})h_i\Big|_{t=0}^{1}$$

$$= \frac{\partial f}{\partial x_i}(\mathbf{x}_0)h_i$$

Thus we have proved the identity

$$\left.\begin{aligned} f(\mathbf{x}_0 + \mathbf{h}) - f(\mathbf{x}_0) &= \sum_{i=1}^{n} \frac{\partial f}{\partial x_i}(\mathbf{x}_0)h_i + R_1(\mathbf{h}, \mathbf{x}_0) \\ \text{where}\qquad & \\ R_1(\mathbf{h}, \mathbf{x}_0) &= \sum_{i,j=1}^{n} \int_0^1 (1-t)\frac{\partial^2 f}{\partial x_i\,\partial x_j}(\mathbf{x}_0 + t\mathbf{h})h_i h_j\,dt \end{aligned}\right\} \tag{3}$$

(Formula (3) gives an explicit formula for the remainder in Theorem 1.)

If we integrate the expression for $R_1(\mathbf{h}, \mathbf{x}_0)$ by parts, with

$$u = \frac{\partial^2 f}{\partial x_i\,\partial x_j}(\mathbf{x}_0 + t\mathbf{h})h_i h_j$$

and

$$v = -(t-1)^2/2$$

we get

$$R_1(\mathbf{h}, \mathbf{x}_0) = \sum_{i,j,k=1}^{n} \int_0^1 \frac{(t-1)^2}{2} \frac{\partial^3 f}{\partial x_i\,\partial x_j\,\partial x_k}(\mathbf{x}_0 + t\mathbf{h})h_i h_j h_k\,dt$$

$$+ \sum_{i,j=1}^{n} \frac{1}{2}\frac{\partial^2 f}{\partial x_i\,\partial x_j}(\mathbf{x}_0)h_i h_j$$

Thus we have proved that

$$\left.\begin{aligned} f(\mathbf{x}_0+\mathbf{h}) &= f(\mathbf{x}_0) + \sum_{i=1}^{n} h_i \frac{\partial f}{\partial x_i}(\mathbf{x}_0) \\ &\quad + \frac{1}{2}\sum_{i,j=1}^{n} h_i h_j \frac{\partial^2 f}{\partial x_i\,\partial x_j}(\mathbf{x}_0) + R_2(\mathbf{h},\mathbf{x}_0) \\ \text{where}\qquad & \\ R_2(\mathbf{h},\mathbf{x}_0) &= \sum_{i,j,k=1}^{n} \int_0^1 \frac{(t-1)^2}{2} \frac{\partial^3 f}{\partial x_i\,\partial x_j\,\partial x_k}(\mathbf{x}_0+t\mathbf{h}) h_i h_j h_k\,dt \end{aligned}\right\} \quad (4)$$

The integrand is a continuous function of t and is therefore bounded on a small neighborhood of $\mathbf{x}_0$ (since it has to be close to its value at $\mathbf{x}_0$). Thus for a constant $M \geq 0$ we get, for $\|\mathbf{h}\|$ small,

$$|R_2(\mathbf{h},\mathbf{x}_0)| \leq \|\mathbf{h}\|^3 M$$

In particular,

$$\frac{|R_2(\mathbf{h},\mathbf{x}_0)|}{\|\mathbf{h}\|^2} \leq \|\mathbf{h}\| M \to 0 \text{ as } \mathbf{h}\to\mathbf{0},$$

as required by the theorem. A similar argument for R_1 shows that $|R_1(\mathbf{h},\mathbf{x}_0)|/\|\mathbf{h}\| \to 0$ as $\mathbf{h}\to\mathbf{0}$, although this also follows from the definition of differentiability, as noted on p. 173. ▮

Corollary (Explicit form of the remainder).

(*i*) *In Theorem 1*

$$\begin{aligned} R_1(\mathbf{h},\mathbf{x}_0) &= \sum_{i,j=1}^{n} \int_0^1 (1-t) \frac{\partial^2 f}{\partial x_i\,\partial x_j}(\mathbf{x}_0+t\mathbf{h}) h_i h_j\,dt \\ &= \sum_{i,j=1}^{n} \frac{1}{2} \frac{\partial^2 f}{\partial x_i\,\partial x_j}(\mathbf{c}_{ij}) h_i h_j \end{aligned} \quad (5)$$

where $\mathbf{c}_{ij}$ *lies somewhere on the line joining* $\mathbf{x}_0$ *to* $\mathbf{x}_0+\mathbf{h}$.

(*ii*) *In Theorem 2*

$$\begin{aligned} R_2(\mathbf{h},\mathbf{x}_0) &= \sum_{i,j,k=1}^{n} \int_0^1 \frac{(t-1)^2}{2} \frac{\partial^3 f}{\partial x_i\,\partial x_j\,\partial x_k}(\mathbf{x}_0+t\mathbf{h}) h_i h_j h_k\,dt \\ &= \sum_{i,j,k=1}^{n} \frac{1}{3!} \frac{\partial^3 f}{\partial x_i\,\partial x_j\,\partial x_k}(\mathbf{c}_{ijk}) h_i h_j h_k \end{aligned} \quad (5')$$

where $\mathbf{c}_{ijk}$ *lies somewhere on the line joining* $\mathbf{x}_0$ *to* $\mathbf{x}_0+\mathbf{h}$.

These formulas were obtained in the course of the proof of Theorem 2 (see formulas (3) and (4)). The formulas involving $\mathbf{c}_{ij}$ and $\mathbf{c}_{ijk}$

(Lagrange's form of the remainder) are obtained by applying the Second Mean Value Theorem for Integrals. Recall that this states*

$$\int_a^b f(x)g(x)\,dx = f(c)\int_a^b g(x)\,dx$$

provided f and g are continuous and $g \geq 0$ on $[a, b]$; here c is some number between a and b. Can you see how this applies to the corollary?

It is not hard to guess the general form of Taylor's Theorem. For example, the third-order Taylor formula is:

$$f(\mathbf{x}_0 + \mathbf{h}) = f(\mathbf{x}_0) + \sum_{i=1}^{n} h_i \frac{\partial f}{\partial x_i}(\mathbf{x}_0) + \frac{1}{2}\sum_{i,\,j=1}^{n} h_i h_j \frac{\partial^2 f}{\partial x_i\,\partial x_j}(\mathbf{x}_0)$$

$$+ \frac{1}{3!}\sum_{i,\,j,\,k=1}^{n} h_i h_j h_k \frac{\partial^3 f}{\partial x_i\,\partial x_j\,\partial x_k}(\mathbf{x}_0) + R_3(\mathbf{h}, \mathbf{x}_0)$$

where $R_3(\mathbf{h}, \mathbf{x}_0)/\|\mathbf{h}\|^3 \to 0$ as $\mathbf{h} \to \mathbf{0}$, and so on. The general formula can be proved by induction, using the method of proof given above.

EXAMPLE 1. Compute the second-order Taylor formula for $f(x, y) = \sin(x + 2y)$, $\mathbf{x}_0 = (0, 0)$.

Notice that

$$f(0, 0) = 0$$

$$\frac{\partial f}{\partial x}(0, 0) = \cos(0 + 2 \cdot 0) = 1$$

$$\frac{\partial f}{\partial y}(0, 0) = 2\cos(0 + 2 \cdot 0) = 2$$

$$\frac{\partial^2 f}{\partial x^2}(0, 0) = 0$$

$$\frac{\partial^2 f}{\partial y^2}(0, 0) = 0$$

$$\frac{\partial^2 f}{\partial x\,\partial y}(0, 0) = 0$$

Thus

$$f(\mathbf{h}) = f(h_1, h_2) = h_1 + 2h_2 + R_2(\mathbf{h}, \mathbf{0})$$

where

$$R_2(\mathbf{h}, \mathbf{0})/\|\mathbf{h}\|^2 \to 0 \text{ as } \mathbf{h} \to \mathbf{0}$$

* See, for instance, McAloon and Tromba, *Calculus*, p. 280. Harcourt, Brace, Jovanovich, New York, 1972.

EXAMPLE 2. Compute the second-order Taylor formula for $f(x, y) = e^x \cos y$, $x_0 = 0$, $y_0 = 0$.

Here

$$f(0, 0) = 1$$

$$\frac{\partial f}{\partial x}(0, 0) = 1$$

$$\frac{\partial f}{\partial y}(0, 0) = 0$$

$$\frac{\partial^2 f}{\partial x^2}(0, 0) = 1$$

$$\frac{\partial^2 f}{\partial y^2}(0, 0) = -1$$

$$\frac{\partial^2 f}{\partial x\, \partial y}(0, 0) = 0$$

so

$$f(\mathbf{h}) = f(h_1, h_2) = 1 + h_1 + \tfrac{1}{2}h_1^2 - \tfrac{1}{2}h_2^2 + R_2(\mathbf{h}, \mathbf{0})$$

where

$$R_2(\mathbf{h}, \mathbf{0})/\|\mathbf{h}\|^2 \to 0 \text{ as } \mathbf{h} \to \mathbf{0}$$

In the case of functions of one variable, one can develop $f(x)$ in an infinite power series, called the *Taylor series*:

$$f(x_0 + h) = f(x_0) + f'(x_0)h + \frac{f''(x_0)h^2}{2} + \cdots + \frac{f^{(k)}(x_0)h^k}{k!} + \cdots$$

provided one can show that $R_k(h, x_0) \to 0$ as $k \to \infty$. Similarly, for functions of several variables the above terms are replaced by the corresponding ones involving partial derivatives, as we have seen in Theorem 2. Again, one can represent such a function by its Taylor series provided one can show that $R_k \to 0$ as $k \to \infty$. This point is examined further in Exercise 7.

EXERCISES

In each of exercises 1 to 6 determine the second-order Taylor formula for the given function about the given point (x_0, y_0).

1. $f(x, y) = (x + y)^2$, $x_0 = 0$, $y_0 = 0$
2. $f(x, y) = e^{x+y}$, $x_0 = 0$, $y_0 = 0$

3. $f(x, y) = 1/(x^2 + y^2 + 1)$, $x_0 = 0$, $y_0 = 0$
4. $f(x, y) = e^{-x^2-y^2} \cos(xy)$, $x_0 = 0$, $y_0 = 0$
5. $f(x, y) = \sin(xy) + \cos(xy)$, $x_0 = 0$, $y_0 = 0$
6. $f(x, y) = e^{(x-1)^2} \cos y$, $x_0 = 1$, $y_0 = 0$
7. A function $f: \mathbb{R} \to \mathbb{R}$ is called *analytic* provided

$$f(x + h) = f(x) + f'(x)h + \cdots + \frac{f^{(k)}(x)}{k!} h^k + \cdots$$

(i.e., the series on the right-hand side converges and equals $f(x + h)$).

(a) Suppose f satisfies the following condition: on any closed interval $[a, b]$ there is a constant M such that for all $k = 1, 2, 3, \ldots$, $|f^{(k)}(x)| \leq M^k$ for all $x \in [a, b]$. Prove that f is analytic.

(b) Let $f(x) = \begin{cases} e^{-1/x} & x > 0 \\ 0 & x \leq 0 \end{cases}$

Show that f is a C^∞ function, but f is not analytic.

(c) Give a definition of analytic functions from $\mathbb{R}^n$ to $\mathbb{R}$. Generalize (a) to this class of functions.

(d) Develop $f(x, y) = e^{x+y}$ in a power series about $x_0 = 0$, $y_0 = 0$.

4.2 EXTREMA OF REAL-VALUED FUNCTIONS

Among the most basic geometric features of the graph of a function are its extremal points, at which the function attains its greatest and least values. In this section, we shall derive a method for determining these points. In fact, the method reveals local extrema as well. These are points at which the function attains a maximum or minimum value relative only to nearby points. Let us begin by defining our terms.

Definition. *If $f: U \subset \mathbb{R}^n \to \mathbb{R}$ is a given scalar function, a point $\mathbf{x}_0 \in U$ is called a **local minimum** of f if there is a neighborhood V of $\mathbf{x}_0$ such that for all points $\mathbf{x} \in V, f(\mathbf{x}) \geq f(\mathbf{x}_0)$. (See Figure 4.2.1.) Similarly, $\mathbf{x}_0 \in U$ is a **local maximum** if there is a neighborhood V of $\mathbf{x}_0$ such that $f(\mathbf{x}) \leq f(\mathbf{x}_0)$ for all $\mathbf{x} \in V$. And $\mathbf{x}_0 \in U$ is said to be a **local, or relative, extremum** if it is either a local minimum or a local maximum. A point $\mathbf{x}_0$ is a **critical point** of f if $Df(\mathbf{x}_0) = 0$. A critical point that is not a local extremum is called a **saddle point**.**

The location of extrema is based on the following fact, which should be familiar from one-variable calculus (the case $n = 1$): every extremum is a critical point.

Theorem 3. *If $f: U \subset \mathbb{R}^n \to \mathbb{R}$ is differentiable and $\mathbf{x}_0 \in U$ is a local extremum, then $Df(\mathbf{x}_0) = 0$; that is, $\mathbf{x}_0$ is a critical point of f.*

* The term "saddle point" is sometimes not used this generally.

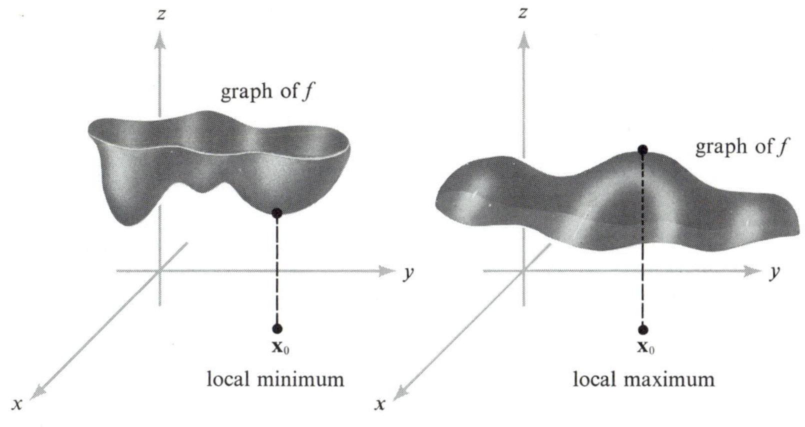

FIGURE 4.2.1
Local maximum and local minimum points for a function of two variables.

Proof. Suppose that f achieves a local maximum at $\mathbf{x}_0$. Then for any $\mathbf{h} \in \mathbb{R}^n$, the function $g(t) = f(\mathbf{x}_0 + t\mathbf{h})$ has a local maximum at $t = 0$. Thus from one-variable calculus $g'(0) = 0$. On the other hand, by the Chain Rule,

$$g'(0) = Df(\mathbf{x}_0) \cdot \mathbf{h}$$

Thus $Df(\mathbf{x}_0) \cdot \mathbf{h} = 0$ for every $\mathbf{h}$, and so $Df(\mathbf{x}_0) = 0$. The case in which f achieves a local minimum at $\mathbf{x}_0$ is entirely analogous. ∎

If we remember that $Df(\mathbf{x}_0) = 0$ means that all the components of $Df(\mathbf{x}_0)$ are zero, we can rephrase the result of Theorem 3: If $\mathbf{x}_0$ is a local extremum, then

$$\frac{\partial f}{\partial x_i}(\mathbf{x}_0) = 0, \qquad i = 1, \ldots, n$$

that is, each partial derivative is zero at $\mathbf{x}_0$. In other words, $\nabla f(\mathbf{x}_0) = \mathbf{0}$, where ∇f is the gradient of f.

If we seek to find the extrema or local extrema of a function, then Theorem 3 states that we should look among the critical points. Sometimes these can be tested by inspection, but usually we use tests (to be developed below) analogous to the second-derivative test in one-variable calculus.

EXAMPLE 1. Consider the function $f\colon \mathbb{R}^2 \to \mathbb{R}$, $(x, y) \mapsto x^2 + y^2$. Of course, we already know that this function has a single minimum at the origin, but let us ignore this and apply the method described above.

We must identify the critical points of f by solving the equations $\partial f(x, y)/\partial x = 0$, $\partial f(x, y)/\partial y = 0$, for x and y. But

$$\frac{\partial}{\partial x} f(x, y) = 2x$$

$$\frac{\partial}{\partial y} f(x, y) = 2y$$

so the only critical point is the origin (0, 0), where the value of the function is zero. As $f(x, y) \geq 0$, this point is a relative minimum—in fact, an absolute minimum—of f.

EXAMPLE 2. Consider the function of Example 4 of Section 2.1, $f: \mathbb{R}^2 \to \mathbb{R}$, $(x, y) \mapsto x^2 - y^2$. Ignoring for the moment that this function is a saddle without extrema, let us apply the method of Theorem 3 for the location of extrema.

As in Example 1, we find that f has only one critical point, at the origin, and the value of f there is zero. Examining values of f directly for points near the origin, we see that $f(x, 0) \geq f(0, 0)$ and $f(0, y) \leq f(0, 0)$. As x or y can be taken arbitrarily small, the origin cannot be either a relative minimum or a relative maximum. Therefore this function can have no relative extrema (see Figure 4.2.2).

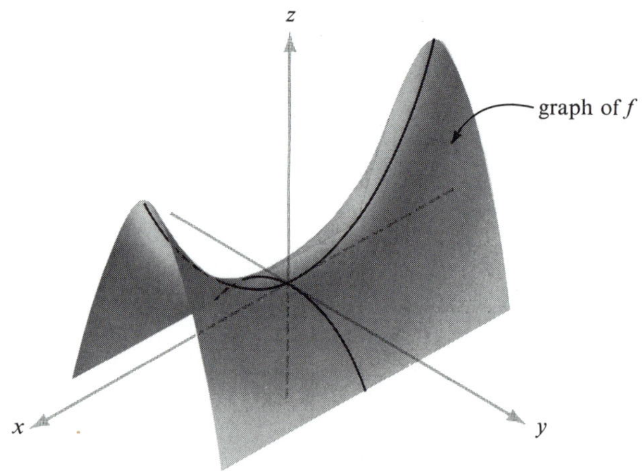

FIGURE 4.2.2
A function of two variables with a saddle point.

The phenomenon in this example, called a critical point of saddle type, or a saddle point, may also occur in case $n = 1$. In that case, when a point is a critical point without being a local extremum, it is called an *inflection point*. For example, $f(x) = x^3$ has a critical point at $x = 0$, but $x = 0$ is not a relative extremum.

EXAMPLE 3. Find all the critical points of $z = x^2y + y^2x$.

Differentiating, we obtain

$$\frac{\partial z}{\partial x} = 2xy + y^2, \qquad \frac{\partial z}{\partial y} = 2xy + x^2$$

Equating the partials to zero yields

$$2xy + y^2 = 0, \qquad 2xy + x^2 = 0$$

Subtracting, we obtain $x^2 = y^2$. Thus $x = \pm y$. Substituting $x = +y$ in the first equation above, we find that

$$2y^2 + y^2 = 3y^2 = 0$$

so $y = 0$ and thus $x = 0$. If $x = -y$ then

$$-2y^2 + y^2 = -y^2 = 0$$

so $y = 0$ and therefore $x = 0$.

Hence the only critical point is (0, 0). However, we can not tell without further analysis whether this point is a local maximum, a local minimum, or neither.

The remainder of this section is devoted to deriving a criterion, depending on the second derivative, for a critical point to be a relative extremum. In the special case $n = 1$, our criterion will reduce to the familiar condition $f''(x) > 0$ for a minimum and $f''(x) < 0$ for a maximum. But in the general context, the second derivative is a fairly complicated mathematical object. So, to state our criterion, we introduce a version of the second derivative called the Hessian.

Definition. *Suppose $f\colon U \subset \mathbb{R}^n \to \mathbb{R}$ has second-order partial derivatives $(\partial^2 f/\partial x_i\, \partial x_j)(\mathbf{x}_0)$, for $i, j = 1, \ldots, n$, at a point $\mathbf{x}_0 \in U$. Then the **Hessian of f at $\mathbf{x}_0$** is the quadratic function* defined by*

$$Hf(\mathbf{x}_0)\colon \mathbb{R}^n \to \mathbb{R},\ (h_1, \ldots, h_n) \mapsto \frac{1}{2} \sum_{i,j=1}^{n} \frac{\partial^2 f}{\partial x_i\, \partial x_j}(\mathbf{x}_0) h_i h_j$$

This function is usually used at critical points $\mathbf{x}_0 \in U$. In this case, $Df(\mathbf{x}_0) = 0$, and the Taylor formula (see Theorem 2, Section 4.1) may be written in the form

$$f(\mathbf{x}_0 + \mathbf{h}) = f(\mathbf{x}_0) + Hf(\mathbf{x}_0)(\mathbf{h}) + R_2(\mathbf{h}, \mathbf{x}_0)$$

Thus at a critical point the Hessian equals the first nonconstant term in the Taylor series of f.

A quadratic function $g\colon \mathbb{R}^n \to \mathbb{R}$ is called *positive definite* if $g(\mathbf{h}) \geq 0$

* A function $g(\mathbf{h})$ is called quadratic if it has the form $g(\mathbf{h}) = \sum_{i,j=1}^{n} a_{ij} h_i h_j$ for an $n \times n$ matrix (a_{ij}).

for all $\mathbf{h} \in \mathbb{R}^n$, and $g(\mathbf{h}) = 0$ only for $\mathbf{h} = \mathbf{0}$. Similarly, g is *negative definite* if $g(\mathbf{h}) \leq 0$, and $g(\mathbf{h}) = 0$ for $\mathbf{h} = \mathbf{0}$ only.

Note that if $n = 1$, $Hf(x_0)(h) = \frac{1}{2}f''(x_0)h^2$, which is positive definite if $f''(x_0) > 0$. We are now ready to state the criterion for relative extrema.

Theorem 4. *If $f: U \subset \mathbb{R}^n \to \mathbb{R}$ is of class C^3, $\mathbf{x}_0 \in U$ is a critical point of f, and the Hessian $Hf(\mathbf{x}_0)$ is positive definite, then $\mathbf{x}_0$ is a relative minimum of f. Similarly, if $Hf(\mathbf{x}_0)$ is negative definite, then $\mathbf{x}_0$ is a relative maximum.*

Actually, we shall prove that the extrema are *strict*. A relative maximum $\mathbf{x}_0$ is called *strict* if $f(\mathbf{x}) < f(\mathbf{x}_0)$ for nearby $\mathbf{x} \neq \mathbf{x}_0$. A strict relative minimum is defined similarly.

The proof of Theorem 4 requires Taylor's Theorem and the following result from linear algebra.

Lemma 1. *If $B = (b_{ij})$ is an $n \times n$ real matrix, and if the associated quadratic function*

$$H: \mathbb{R}^n \to \mathbb{R}, \ (h_1, \ldots, h_n) \mapsto \frac{1}{2} \sum_{i,j=1}^{n} b_{ij} h_i h_j$$

is positive definite, then there exists a constant $M > 0$ such that for all $\mathbf{h} \in \mathbb{R}^n$,

$$H(\mathbf{h}) \geq M\|\mathbf{h}\|^2$$

Proof. For $\|\mathbf{h}\| = 1$ set $g(\mathbf{h}) = H(\mathbf{h})$. Then g is a continuous function of $\mathbf{h}$ for $\|\mathbf{h}\| = 1$ and so achieves a minimum value, say M.* For general $\mathbf{h}$, since H is quadratic, we have

$$H(\mathbf{h}) = H\left(\frac{\mathbf{h}}{\|\mathbf{h}\|}\|\mathbf{h}\|\right) = H\left(\frac{\mathbf{h}}{\|\mathbf{h}\|}\right)\|\mathbf{h}\|^2 \geq M\|\mathbf{h}\|^2 \quad \blacksquare$$

Note that the quadratic function associated with the matrix $(\partial^2 f/\partial x_i\, \partial x_j)$ is exactly the Hessian.

Proof of Theorem 4. Recall that if $f: U \subset \mathbb{R}^n \to \mathbb{R}$ is of class C^3 and $\mathbf{x}_0 \in U$ is a critical point, Taylor's Theorem may be expressed in the form

$$f(\mathbf{x}_0 + \mathbf{h}) - f(\mathbf{x}_0) = Hf(\mathbf{x}_0)(\mathbf{h}) + R_2(\mathbf{h}, \mathbf{x}_0)$$

where

$$R_2(\mathbf{h}, \mathbf{x}_0)/\|\mathbf{h}\|^2 \to 0$$

as $\mathbf{h} \to \mathbf{0}$.

* Here we are using, without proof, a theorem analogous to a theorem in calculus that states that every continuous function on an interval $[a, b]$ achieves a maximum and minimum.

Since $Hf(\mathbf{x}_0)$ is positive definite by hypothesis, there is a constant $M > 0$ such that for all $\mathbf{h} \in \mathbb{R}^n$

$$Hf(\mathbf{x}_0)(\mathbf{h}) \geq M\|\mathbf{h}\|^2$$

Since $R_2(\mathbf{h}, \mathbf{x}_0)/\|\mathbf{h}\|^2 \to 0$ as $\mathbf{h} \to \mathbf{0}$, there is a $\delta > 0$ such that for $0 < \|\mathbf{h}\| < \delta$

$$|R_2(\mathbf{h}, \mathbf{x}_0)| < M\|\mathbf{h}\|^2$$

Thus $0 < Hf(\mathbf{x}_0)(\mathbf{h}) + R_2(\mathbf{h}, \mathbf{x}_0) = f(\mathbf{x}_0 + \mathbf{h}) - f(\mathbf{x}_0)$ for $0 < \|\mathbf{h}\| < \delta$, so $\mathbf{x}_0$ is a relative minimum, in fact, a strict relative minimum.

The proof in the negative-definite case is similar, or else follows by applying the above to $-f$, and is left as an exercise. ∎

EXAMPLE 4. Consider again the function $f\colon \mathbb{R}^2 \to \mathbb{R}$, $(x, y) \mapsto x^2 + y^2$. Then $(0, 0)$ is a critical point, and f is already in the form of Taylor's Theorem

$$f((0, 0) + (h_1, h_2)) = f(0, 0) + (h_1^2 + h_2^2) + 0$$

We can see directly that the Hessian at $(0, 0)$ is

$$Hf(\mathbf{0})(\mathbf{h}) = h_1^2 + h_2^2$$

which is clearly positive definite. Thus $(0, 0)$ is a relative minimum.

For functions of two variables $f(x, y)$, the Hessian may be written as follows:

$$Hf(x, y)(\mathbf{h}) = \tfrac{1}{2}(h_1, h_2)\begin{bmatrix} \dfrac{\partial^2 f}{\partial x^2} & \dfrac{\partial^2 f}{\partial y\,\partial x} \\ \dfrac{\partial^2 f}{\partial x\,\partial y} & \dfrac{\partial^2 f}{\partial y^2} \end{bmatrix}\begin{pmatrix} h_1 \\ h_2 \end{pmatrix}$$

Now we shall give a useful criterion for when a quadratic function defined by such a 2×2 matrix is positive definite. This will then be applied to Theorem 4.

Lemma 2. *Let*

$$B = \begin{bmatrix} a & b \\ b & c \end{bmatrix} \quad \text{and} \quad H(\mathbf{h}) = \tfrac{1}{2}(h_1, h_2)B\begin{pmatrix} h_1 \\ h_2 \end{pmatrix}$$

Then $H(\mathbf{h})$ *is positive definite if and only if* $a > 0$ *and* $\det B = ac - b^2 > 0$.

Proof. We have

$$H(\mathbf{h}) = \tfrac{1}{2}(h_1, h_2)\begin{pmatrix} ah_1 + bh_2 \\ bh_1 + ch_2 \end{pmatrix} = \tfrac{1}{2}(ah_1^2 + 2bh_1h_2 + ch_2^2)$$

Let us complete the square, writing

$$H(\mathbf{h}) = \frac{1}{2}\left(a\left(h_1 + \frac{b}{a}h_2\right)^2 + \left(c - \frac{b^2}{a}\right)h_2^2\right)$$

Suppose H is positive definite. Setting $h_2 = 0$ we see that $a > 0$. Setting $h_1 = -(b/a)h_2$ we get $c - b^2/a > 0$ or $ac - b^2 > 0$. Conversely, if $a > 0$ and $c - b^2/a > 0$, $H(\mathbf{h})$ is a sum of squares, so $H(\mathbf{h}) \geq 0$. If $H(\mathbf{h}) = 0$ then each square must be zero. This implies that both h_1 and h_2 must be zero, so $H(\mathbf{h})$ is positive definite. ∎

For $H(\mathbf{h})$ negative definite the criteria are seen to be $a < 0$, $ac - b^2 > 0$. There are similar criteria for an $n \times n$ symmetric matrix B. Consider the n square submatrices along the diagonal (see Figure 4.2.3). Then B is positive definite (that is, the quadratic function

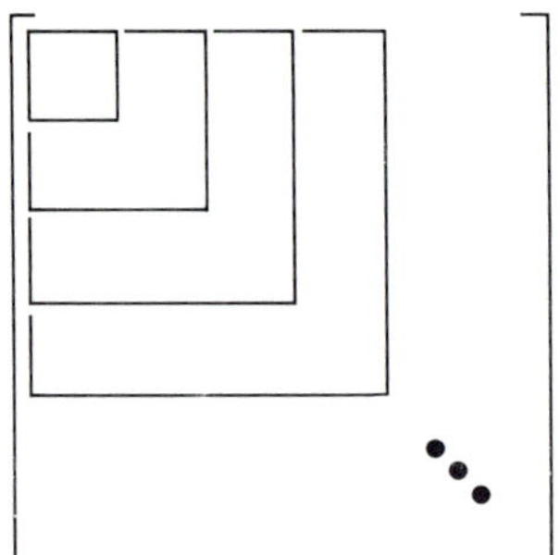

FIGURE 4.2.3
"Diagonal" submatrices are used in the criterion for positive definiteness; they must all have determinant > 0.

associated with B is positive definite) if and only if the determinants of these diagonal submatrices are all greater than zero. For negative definite the signs should be alternately < 0 and > 0. We shall not prove this general case here.

We can use Lemma 2 and Theorem 4 to immediately obtain the following result.

Theorem 5. *Let $f(x, y)$ be of class C^3 on an open set U in $\mathbb{R}^2$. A point (x_0, y_0) is a (strict) local minimum of f provided the following three conditions hold:*

(*i*) $\dfrac{\partial f}{\partial x}(x_0, y_0) = \dfrac{\partial f}{\partial y}(x_0, y_0) = 0$

(*ii*) $\dfrac{\partial^2 f}{\partial x^2}(x_0, y_0) > 0$

(*iii*) $D = \left(\frac{\partial^2 f}{\partial x^2}\right)\left(\frac{\partial^2 f}{\partial y^2}\right) - \left(\frac{\partial^2 f}{\partial x\,\partial y}\right)^2 > 0$ *at* (x_0, y_0)

(D is called the discriminant.) If in (ii) we have < 0 *instead of* > 0 *and condition (iii) is unchanged, then we have a (strict) local maximum.*

EXAMPLE 5. Consider the function $f\colon \mathbb{R}^2 \to \mathbb{R}$, $(x, y) \mapsto x^2 - 2xy + 2y^2$. We compute easily that, as in Example 4, the origin is the only critical point, $f(0, 0) = 0$, and the Hessian is

$$Hf(\mathbf{0})(\mathbf{h}) = h_1^2 - 2h_1h_2 + 2h_2^2 = (h_1 - h_2)^2 + h_2^2$$

which is clearly positive definite, so f has a relative minimum at (0, 0). Alternatively, we can apply Theorem 5. At (0, 0), $\partial^2 f/\partial x^2 = 2$, $\partial^2 f/\partial y^2 = 4$, and $\partial^2 f/\partial x\,\partial y = -2$. Conditions (*i*), (*ii*), and (*iii*) hold, so that f has a relative minimum at (0, 0).

If, in Theorem 5, D is < 0 then we have a saddle point. In fact one can prove that $f(x, y)$ is larger than $f(x_0, y_0)$ as we move away from (x_0, y_0) in some direction and smaller in the orthogonal direction (see Exercise 14). The general appearance is thus similar to that shown in Figure 4.2.2. The appearance of the case $D = 0$ must be examined by further analysis.

In summary: When, in an example dealing with functions of two variables, all critical points have been found and their associated Hessians computed, some of these Hessians may be positive definite, indicating relative minima; some may be negative definite, indicating relative maxima; and some may take both positive and negative values, indicating saddle points. The shape of the graph at a saddle point *where* $D < 0$ is like that in Figure 4.2.2. Critical points where $D \neq 0$ (or the Hessian is positive or negative definite) are called non-degenerate. Thus there are non-degenerate maxima, minima, and saddle points. The remaining critical points, where $D = 0$, may be tested directly, with level sets and sections (Section 2.1) or some other method. Such critical points are called *degenerate*, and the methods developed in this chapter fail to provide a picture of the behavior of a function near such points.

EXAMPLE 6. Locate the maxima, minima, and saddle points of the function

$$f(x, y) = \log(x^2 + y^2 + 1)$$

We must first locate the critical points of this function; so, according to Theorem 3, we calculate

$$\nabla f(x, y) = \frac{2x}{x^2 + y^2 + 1}\mathbf{i} + \frac{2y}{x^2 + y^2 + 1}\mathbf{j}$$

Thus $\nabla f(x, y) = \mathbf{0}$ if and only if $(x, y) = (0, 0)$, and so the only critical point of f is $(0, 0)$. Now we must determine whether this is a maximum, a minimum, or a saddle point. The second partials are

$$\frac{\partial^2 f}{\partial x^2} = \frac{2(x^2 + y^2 + 1) - (2x)(2x)}{(x^2 + y^2 + 1)^2}$$

$$\frac{\partial^2 f}{\partial y^2} = \frac{2(x^2 + y^2 + 1) - (2y)(2y)}{(x^2 + y^2 + 1)^2}$$

and

$$\frac{\partial^2 f}{\partial x\,\partial y} = \frac{-2x(2y)}{(x^2 + y^2 + 1)^2}$$

So

$$\frac{\partial^2 f}{\partial x^2}(0, 0) = 2 = \frac{\partial^2 f}{\partial y^2}(0, 0)$$

and

$$\frac{\partial^2 f}{\partial x\,\partial y}(0, 0) = 0$$

which yields

$$D = 2 \cdot 2 = 4 > 0$$

Since $(\partial^2 f/\partial x^2)(0, 0) > 0$ we conclude by Theorem 5 that $(0, 0)$ is a local minimum.

EXAMPLE 7. The graph of the function $g(x, y) = 1/xy$ is a surface S in $\mathbb{R}^3$. Find the points on S that are closest to the origin $(0, 0, 0)$.

The distance from (x, y, z) to $(0, 0, 0)$ is given by the formula

$$d(x, y, z) = \sqrt{x^2 + y^2 + z^2}$$

If $(x, y, z) \in S$ then d can be expressed as a function $d_*(x, y) = d(x, y, 1/xy)$ of two variables:

$$d_*(x, y) = \sqrt{x^2 + y^2 + \frac{1}{x^2 y^2}}$$

Note that the minimum (if it exists) cannot occur "too near" the

x-axis or y-axis because d_* is not defined if either $x = 0$ or $y = 0$, and d_* gets very large as x or y approaches the x-axis or the y-axis.

Since $d_* > 0$ it will be minimized when $d_*^2(x, y) = x^2 + y^2 + (1/x^2y^2) = f(x, y)$ is minimized. (This function f is much easier to deal with.) We calculate the gradient

$$\nabla f(x, y) = \nabla d_*^2(x, y) = \left(2x - \frac{2}{x^3y^2}\right)\mathbf{i} + \left(2y - \frac{2}{y^3x^2}\right)\mathbf{j}$$

This is $\mathbf{0}$ if and only if

$$\left(2x - \frac{2}{x^3y^2}\right) = 0 = \left(2y - \frac{2}{y^3x^2}\right)$$

that is, $x^4y^2 - 1 = 0$ and $x^2y^4 - 1 = 0$. From the first equation we get $y^2 = 1/x^4$ and substituting this into the second equation we obtain

$$\frac{x^2}{x^8} = 1 = \frac{1}{x^6}$$

Thus $x = \pm 1$ and $y = \pm 1$, and it therefore follows that f has four critical points, namely, $(1, 1)$, $(1, -1)$, $(-1, 1)$, and $(-1, -1)$. To determine whether these are local minima, local maxima, or saddle points we apply Theorem 5:

$$\frac{\partial^2 f}{\partial x^2} = 2 + \frac{6}{x^4y^2}, \qquad \frac{\partial^2 f}{\partial y^2} = 2 + \frac{6}{x^2y^4}, \qquad \frac{\partial^2 f}{\partial y\,\partial x} = \frac{4}{x^3y^3}$$

so

$$\frac{\partial^2 f}{\partial x^2}(a, b) = \frac{\partial^2 f}{\partial y^2}(a, b) = 8$$

where (a, b) is any one of the above four critical points, and $(\partial^2 f/\partial x\,\partial y)(a, b) = \pm 4$.

We see that in any of the above cases $D = 64 - 16 = 48 > 0$ and $(\partial^2 f/\partial x^2)(a, b) > 0$, so each critical point is a local minimum, and these are all the local minima for f.

Finally, note that $d_*^2(a, b) = 3$ for all these critical points and so the points on the surface that are closest to $(0, 0, 0)$ are $(1, 1, 1)$, $(1, -1, -1)$, $(-1, 1, -1)$, and $(-1, -1, 1)$ with $d_* = \sqrt{3}$ at these points. Thus $d_* \geq \sqrt{3}$ and is equal to $\sqrt{3}$ when $(x, y) = (\pm 1, \pm 1)$.

EXAMPLE 8. Analyze the behavior of $z = x^5y + xy^5 + xy$ at its critical points.

The first partial derivatives are

$$\frac{\partial z}{\partial x} = 5x^4y + y^5 + y = y(5x^4 + y^4 + 1)$$

and

$$\frac{\partial z}{\partial y} = x(5y^4 + x^4 + 1)$$

The terms $5x^4 + y^4 + 1$ and $5y^4 + x^4 + 1$ are always greater than or equal to 1 so it follows that the only critical point is (0, 0).

The second partials are

$$\frac{\partial^2 z}{\partial x^2} = 20x^3 y, \qquad \frac{\partial^2 z}{\partial y^2} = 20xy^3$$

and

$$\frac{\partial^2 z}{\partial x\, \partial y} = 5x^4 + 5y^4 + 1$$

Thus at (0, 0), $D = -1$, so (0, 0) is a non-degenerate saddle point and the graph of z near (0, 0) looks like Figure 4.2.2.

EXERCISES

In exercises 1 to 10, find the critical points of the given function and then determine whether they are local maxima, local minima, or saddle points.

1. $f(x, y) = x^2 - y^2 + xy$
2. $f(x, y) = x^2 + y^2 - xy$
3. $f(x, y) = x^2 + y^2 + 2xy$
4. $f(x, y) = x^2 + y^2 + 3xy$
5. $f(x, y) = e^{1 + x^2 - y^2}$
6. $f(x, y) = x^2 - 3xy + 5x - 2y + 6y^2 + 8$
7. $f(x, y) = 3x^2 + 2xy + 2x + y^2 + y + 4$
8. $f(x, y) = \sin(x^2 + y^2)$ (consider only the critical point (0, 0))
9. $f(x, y) = \cos(x^2 + y^2)$ (consider only the critical points (0, 0), $(\sqrt{\pi/2}, \sqrt{\pi/2})$, and $(0, \sqrt{\pi})$)
10. $f(x, y) = \log(2 + \sin xy)$
11. An examination of the function $f\colon \mathbb{R}^2 \to \mathbb{R}$, $(x, y) \mapsto (y - 3x^2)(y - x^2)$, will give an idea of the difficulty of finding conditions that guarantee that a critical point is a relative extremum when Theorem 5 fails. Show that
    (a) the origin is a critical point of f;
    (b) f has a relative minimum at (0, 0) on every straight line through (0, 0), that is, if $g(t) = (at, bt)$, then $f \circ g\colon \mathbb{R} \to \mathbb{R}$ has a relative minimum at 0, for every choice of a and b;
    (c) the origin is not a relative minimum of f.
12. Let $f(x, y) = Ax^2 + E$. What are the critical points of f? Are they local maxima or local minima?
13. Let $f(x, y) = x^2 - 2xy + y^2$. Here $D = 0$. Can you say if the critical points are local minima, local maxima, or saddle points?

*14. Show that if (x_0, y_0) is a saddle point of $f(x, y)$ and $D < 0$ then there are points (x, y) near (x_0, y_0) at which $f(x, y) > f(x_0, y_0)$ and, similarly, points for which $f(x, y) < f(x_0, y_0)$.

15. Determine the nature of the critical points of the function

$$f(x, y, z) = x^2 + y^2 + z^2 + xy$$

4.3 CONSTRAINED EXTREMA AND LAGRANGE MULTIPLIERS

Often in problems we want to maximize a function subject to certain *constraints* or *side conditions*. Such situations arise, for example, in economics. Suppose we are selling two kinds of goods, say I and II; let x and y represent the quantity of each sold. Then let $f(x, y)$ represent the profit we earn when x amount of I and y amount of II is sold. But our production is controlled by our capital, so we are constrained to work subject to a relation, say $g(x, y) = 0$. Thus we want to maximize $f(x, y)$ among those (x, y) satisfying $g(x, y) = 0$. We call the condition $g(x, y) = 0$ the constraint in the problem.

The purpose of this section is to develop some methods for handling this and similar problems.

Theorem 6. *Let $f\colon U \subset \mathbb{R}^n \to \mathbb{R}$ and $g\colon U \subset \mathbb{R}^n \to \mathbb{R}$ be given smooth functions. Let $\mathbf{x}_0 \in U$, $g(\mathbf{x}_0) = c_0$, and let S be the level set for g with value c_0 (recall that this is the set of points $\mathbf{x} \in \mathbb{R}^n$ with $g(\mathbf{x}) = c_0$). Assume $\nabla g(\mathbf{x}_0) \neq \mathbf{0}$.*

If $f \mid S$, which denotes f restricted to S, has a maximum or minimum at $\mathbf{x}_0$, then there is a real number λ such that

$$\boxed{\nabla f(\mathbf{x}_0) = \lambda \nabla g(\mathbf{x}_0)}$$

Proof. Actually, we do not have enough machinery to give a thorough proof, but we can provide the essential points.

Recall that the tangent space of S at $\mathbf{x}_0$ is defined as the space orthogonal to $\nabla g(\mathbf{x}_0)$ (see Section 2.5). We motivated this definition by considering tangents to paths $\mathbf{c}(t)$ that lie in S, as follows: if $\mathbf{c}(t)$ is a path in S and $\mathbf{c}(0) = \mathbf{x}_0$, then $\mathbf{c}'(0)$ is a tangent vector to S at $\mathbf{x}_0$; but

$$\frac{d}{dt} g(\mathbf{c}(t)) = \frac{d}{dt} c_0 = 0$$

and on the other hand, by the Chain Rule,

$$\left. \frac{d}{dt} g(\mathbf{c}(t)) \right|_{t=0} = \nabla g(\mathbf{x}_0) \cdot \mathbf{c}'(0)$$

so $\nabla g(\mathbf{x}_0) \cdot \mathbf{c}'(0) = 0$; that is, $\mathbf{c}'(0)$ is orthogonal to $\nabla g(\mathbf{x}_0)$.

If $f \mid S$ has a maximum at $\mathbf{x}_0$, then certainly $f(\mathbf{c}(t))$ has a maximum at $t = 0$. By one-variable calculus, $df(\mathbf{c}(t))/dt|_{t=0} = 0$. Hence by the Chain Rule

$$0 = \frac{d}{dt} f(\mathbf{c}(t))\bigg|_{t=0} = \nabla f(\mathbf{x}_0) \cdot \mathbf{c}'(0)$$

Thus $\nabla f(\mathbf{x}_0)$ is perpendicular to the tangent of every curve in S and so is also perpendicular to the tangent space of S at $\mathbf{x}_0$. Hence, since the space perpendicular to this tangent space is one-dimensional,* $\nabla f(\mathbf{x}_0)$ and $\nabla g(\mathbf{x}_0)$ are parallel. Since $\nabla g(\mathbf{x}_0) \neq \mathbf{0}$ it follows that $\nabla f(\mathbf{x}_0)$ is a multiple of $\nabla g(\mathbf{x}_0)$, which is exactly the conclusion of the theorem. ▮

Let us extract from this proof the geometry of the situation. We can formulate things in the following way.

Corollary. *If f, when constrained to a surface S, has a maximum or minimum at $\mathbf{x}_0$, then $\nabla f(\mathbf{x}_0)$ is perpendicular to S at $\mathbf{x}_0$ (see Figure 4.3.1).*

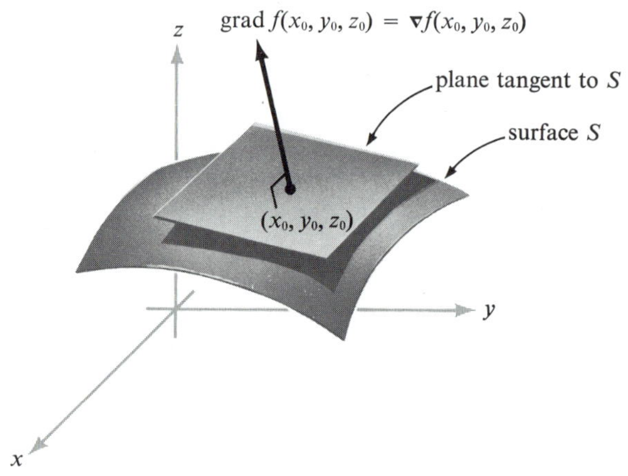

FIGURE 4.3.1
The geometry of constrained extrema.

These results tell us that to find the constrained extrema of f we must look among those $\mathbf{x}_0$ satisfying the conclusions of the theorem or of the corollary. We shall give several illustrations of how to use each.

When the method of Theorem 6 is used we must look for a point $\mathbf{x}_0$ and a constant λ, called a *Lagrange multiplier*, such that $\nabla f(\mathbf{x}_0) = \lambda \nabla g(\mathbf{x}_0)$. This method is more analytical in nature than the method of the corollary to Theorem 6, which is more geometrical.

* This is intuitively obvious but actually requires some care to prove. For details, see J. Marsden, *Elementary Classical Analysis*, W. H. Freeman and Company, San Francisco, 1974, Chapter 7.

Unfortunately, for constrained problems there is no simple test to distinguish maxima from minima as there is for unconstrained problems. Therefore one must examine each $\mathbf{x}_0$ separately using the given data, or apply other geometric arguments.

EXAMPLE 1. Let $S \subset \mathbb{R}^2$ be a line through $(-1, 0)$ inclined at 45°, and let f: $\mathbb{R}^2 \to \mathbb{R}$, $(x, y) \mapsto x^2 + y^2$. Then $S = \{(x, y) \mid y - x - 1 = 0\}$, so here we set $g(x, y) = y - x - 1$ and $c_0 = 0$. The relative extrema of $f \mid S$ must be found among the points at which ∇f is orthogonal to S, that is, inclined at $-45°$. But $\nabla f(x, y) = (2x, 2y)$, which has the desired slope only when $x = -y$, or when (x, y) lies on the line L through the origin inclined at $-45°$. This can occur in the set S only for the single point at which L and S intersect (see Figure 4.3.2). Reference to the level curves

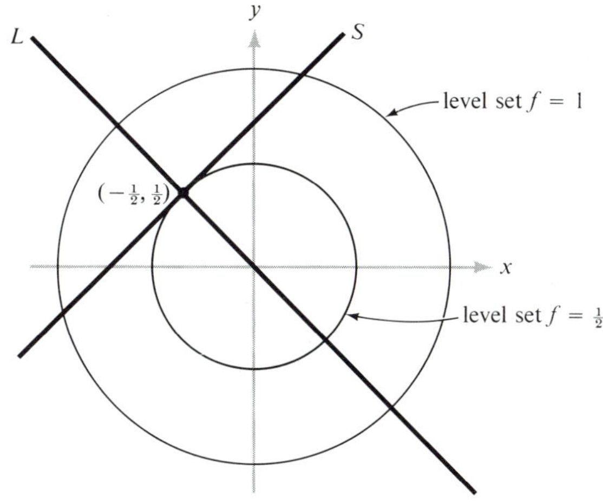

FIGURE 4.3.2
The geometry associated with finding the extrema of $f(x, y) = x^2 + y^2$ restricted to $S = \{(x, y) \mid y - x - 1 = 0\}$.

of f indicates that this point, $(-\frac{1}{2}, \frac{1}{2})$, is a relative minimum of $f \mid S$ (but not of f).

EXAMPLE 2. Let f: $\mathbb{R}^2 \to \mathbb{R}$, $(x, y) \mapsto x^2 - y^2$, and let S be the circle of radius 1 around the origin. Thus S is the level curve for g with value 1, where g: $\mathbb{R}^2 \to \mathbb{R}$, $(x, y) \mapsto x^2 + y^2$. As both of these functions have been studied in previous examples, we know their level curves, and these are shown in Figure 4.3.3. Clearly, the gradient of f is orthogonal to S at the four points $(0, \pm 1)$ and $(\pm 1, 0)$, which are relative minima and maxima, respectively, of $f \mid S$.

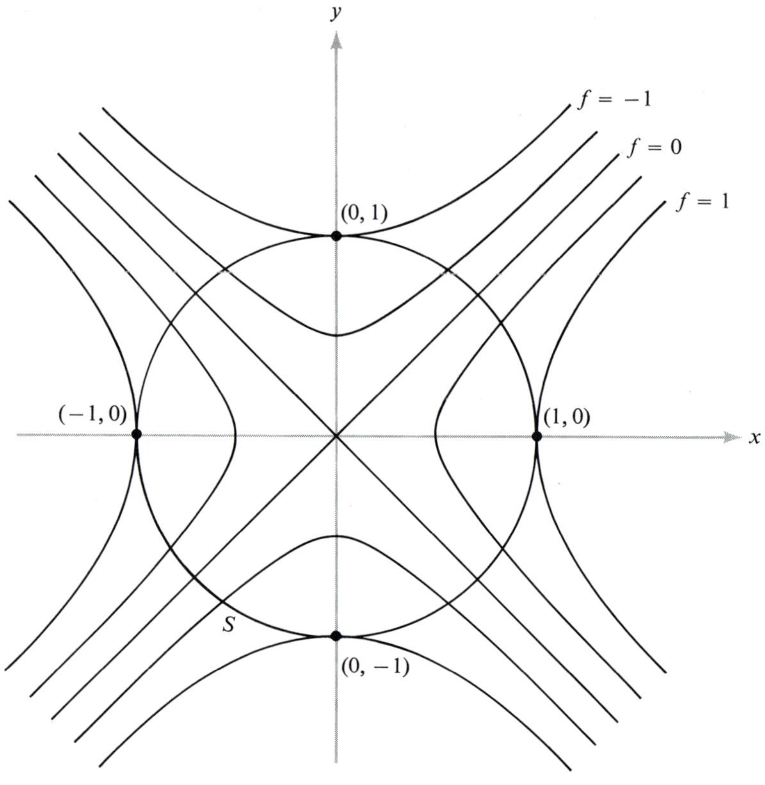

FIGURE 4.3.3
The geometry associated with the problem of finding the extrema of $x^2 - y^2$ *on* $S = \{(x, y) \mid x^2 + y^2 = 1\}$.

Let us also do this problem analytically by the method of Lagrange multipliers. Clearly

$$\nabla f(x, y) = \left(\frac{\partial f}{\partial x}, \frac{\partial f}{\partial y}\right) = (2x, -2y)$$

and

$$\nabla g(x, y) = (2x, 2y)$$

Thus, according to Theorem 6, we must find a λ such that

$$(2x, -2y) = \lambda(2x, 2y)$$

and

$$(x, y) \in S \quad \text{that is} \quad x^2 + y^2 = 1$$

These conditions yield three equations, which can be solved for the three unknowns x, y, and λ. From $2x = \lambda 2x$ we conclude that either

$x = 0$ or $\lambda = 1$. If $x = 0$ then $y = \pm 1$ and $-2y = \lambda 2y$ implies $\lambda = -1$. If $\lambda = 1$, then $y = 0$ and $x = \pm 1$. Thus we get the points $(0, \pm 1)$ and $(\pm 1, 0)$, as before. As we have mentioned, the method only locates potential extrema; whether they are maxima, minima, or neither must be determined by other means.

If a surface S is defined by a number of constraints

$$g_1(x_1, \ldots, x_n) = c_1$$
$$g_2(x_1, \ldots, x_n) = c_2$$
$$\vdots$$
$$g_k(x_1, \ldots, x_n) = c_k$$

then Theorem 6 may be generalized as follows: If f has a maximum or minimum at $\mathbf{x}_0$ on S, there must exist constants $\lambda_1, \ldots, \lambda_k$ such that

$$\nabla f(\mathbf{x}_0) = \lambda_1 \nabla g_1(\mathbf{x}_0) + \cdots + \lambda_k \nabla g_k(\mathbf{x}_0)$$

This case may be proved by generalizing the method used to prove Theorem 6. We leave the argument to the interested reader. Let us give an example of how this more general formulation may be used.

EXAMPLE 3. Find the extreme points of $f(x, y, z) = x + y + z$ subject to the conditions $x^2 + y^2 = 2$ and $x + z = 1$.

Here there are two constraints

$$g_1(x, y, z) = x^2 + y^2 - 2 = 0$$
$$g_2(x, y, z) = x + z - 1 = 0$$

Thus we must find x, y, z, λ_1, and λ_2 such that

$$\nabla f(x, y, z) = \lambda_1 \nabla g_1(x, y, z) + \lambda_2 \nabla g_2(x, y, z)$$

and

$$g_1(x, y, z) = 0$$
$$g_2(x, y, z) = 0$$

that is, computing the gradients and equating components,

$$1 = \lambda_1 \cdot 2x + \lambda_2 \cdot 1$$
$$1 = \lambda_1 \cdot 2y + \lambda_2 \cdot 0$$
$$1 = \lambda_1 \cdot 0 + \lambda_2 \cdot 1$$

and

$$x^2 + y^2 = 2$$
$$x + z = 1$$

These are five equations for x, y, z, λ_1, and λ_2. From the third, $\lambda_2 = 1$ and so $2x\lambda_1 = 0$, $2y\lambda_1 = 1$. Since the second implies $\lambda_1 \neq 0$, we have $x = 0$. Thus $y = \pm\sqrt{2}$ and $z = 1$. Hence our points are $(0, \pm\sqrt{2}, 1)$. By inspection one can show that $(0, \sqrt{2}, 1)$ gives a maximum, and $(0, -\sqrt{2}, 1)$ a minimum.

EXAMPLE 4. Maximize $f(x, y, z) = x + z$ subject to the constraint $x^2 + y^2 + z^2 = 1$.

Here we use Theorem 6. We seek λ and (x, y, z) such that

$$1 = 2x\lambda$$
$$0 = 2y\lambda$$
$$1 = 2z\lambda$$

and

$$x^2 + y^2 + z^2 = 1$$

Since $\lambda \neq 0$, we get $y = 0$. From the first and third equations, $x = z$, and so from the fourth, $x = \pm 1/\sqrt{2} = z$. Hence our points are $(1/\sqrt{2}, 0, 1/\sqrt{2})$ and $(-1/\sqrt{2}, 0, -1/\sqrt{2})$. Comparing the values of f at these points, we can see that the first point yields the maximum of f and the second the minimum.

EXAMPLE 5. Find the largest volume a rectangular box can have subject to the constraint that the surface area be fixed at 10 square meters.

Here, if x, y, z are the lengths of the sides the volume is $f(x, y, z) = xyz$. The constraint is that $2(xy + xz + yz) = 10$; that is, $xy + xz + yz = 5$. Thus our conditions are

$$yz = \lambda(y + z)$$
$$xz = \lambda(x + z)$$
$$xy = \lambda(y + x)$$
$$xy + xz + yz = 5$$

First of all, $x \neq 0$, for $x = 0$ implies $yz = 5$ and $0 = \lambda z$, so $\lambda = 0$ and $yz = 0$. Similarly, $y \neq 0$, $z \neq 0$, $x + y \neq 0$, etc. Elimination of λ from the first two equations gives $yz/(y + z) = xz/(x + z)$, which gives $x = y$; similarly, $y = z$. Substituting these values into the last equation, we obtain $3x^2 = 5$, or $x = \sqrt{5/3}$. Thus $x = y = z = \sqrt{5/3}$, and $xyz = (5/3)^{3/2}$. This is the solution; it should be geometrically clear that the maximum occurs when $x = y = z$.

Some general guidelines may be useful for problems such as these. First of all, if the surface S is bounded (like an ellipsoid) then f must have a maximum and a minimum on S.* In particular, if f has only two points satisfying the conditions of Theorem 6 or its corollary, then one must be a maximum and one must be a minimum. However, if there are more than two such points, some can also be saddle points. Also, if S is not bounded (e.g., a hyperboloid) then f need not have any maxima nor minima.

EXERCISES

In exercises 1 to 5 find the extrema of f subject to the stated constraints.

1. $f(x, y, z) = x - y + z$, $x^2 + y^2 + z^2 = 2$
2. $f(x, y) = x - y$, $x^2 - y^2 = 2$
3. $f(x, y) = x$, $x^2 + 2y^2 = 3$
4. $f(x, y) = 3x + 2y$, $2x^2 + 3y^2 = 3$
5. $f(x, y, z) = x + y + z$, $x^2 - y^2 = 1$, $2x + z = 1$

Find the relative extrema of $f \mid S$ in exercises 6 to 9.

6. $f\colon \mathbb{R}^2 \to \mathbb{R}$, $(x, y) \mapsto x^2 + y^2$, $S = \{(x, 2) \mid x \in \mathbb{R}\}$
7. $f\colon \mathbb{R}^2 \to \mathbb{R}$, $(x, y) \mapsto x^2 + y^2$, $S = \{(x, y) \mid y \geq 2\}$
8. $f\colon \mathbb{R}^2 \to \mathbb{R}$, $(x, y) \mapsto x^2 - y^2$, $S = \{(x, \cos x) \mid x \in \mathbb{R}\}$
9. $f\colon \mathbb{R}^3 \to \mathbb{R}$, $(x, y, z) \mapsto x^2 + y^2 + z^2$, $S = \{(x, y, z) \mid z \geq -2 + x^2 + y^2\}$
10. A rectangular box with no top is to have a surface area of 16 square meters. Find the dimensions that maximize its volume.
11. Design a cylindrical can (with a lid) to contain 1 liter of water, using the minimum amount of metal.

4.4 SOME APPLICATIONS

In this section we shall give some applications of the mathematical methods that we have developed in the preceding sections.

Let $\mathbf{F}$ denote a force field defined on a certain domain U of $\mathbb{R}^3$. Thus $\mathbf{F}\colon U \to \mathbb{R}^3$ is a given vector field. Let us agree that a particle (with mass m) is to move along a path $\boldsymbol{\sigma}(t)$ in such a way that Newton's law holds: mass × acceleration = force; that is, the path $\boldsymbol{\sigma}(t)$ is to satisfy the equation

$$m\boldsymbol{\sigma}''(t) = \mathbf{F}(\boldsymbol{\sigma}(t)) \tag{1}$$

If $\mathbf{F}$ is a potential field with potential V, that is, if $\mathbf{F} = -\text{grad } V$, then

$$\tfrac{1}{2}m\|\boldsymbol{\sigma}'(t)\|^2 + V(\boldsymbol{\sigma}(t)) = \text{constant} \tag{2}$$

* This is proved in more advanced courses. See, for example, J. Marsden, *Elementary Classical Analysis*, W. H. Freeman and Company, San Francisco, 1974, Chapter 4.

(The first term is called the *kinetic energy*.) Indeed, by differentiating with the Chain Rule,

$$\frac{d}{dt}\left\{\frac{1}{2}m\|\boldsymbol{\sigma}'(t)\|^2 + V(\boldsymbol{\sigma}(t))\right\} = m\boldsymbol{\sigma}'(t)\cdot\boldsymbol{\sigma}''(t) + \text{grad } V(\boldsymbol{\sigma}(t))\cdot\boldsymbol{\sigma}'(t)$$

$$= [m\boldsymbol{\sigma}''(t) + \text{grad } V(\boldsymbol{\sigma}(t))]\cdot\boldsymbol{\sigma}'(t) = 0$$

since $m\boldsymbol{\sigma}''(t) = -\text{grad } V(\boldsymbol{\sigma}(t))$. This proves (2).

Definition. *A point $\mathbf{x}_0 \in U$ is called a* ***position of equilibrium*** *if the force at that point is zero: $\mathbf{F}(\mathbf{x}_0) = \mathbf{0}$. A point $\mathbf{x}_0$ that is a position of equilibrium is said to be* ***stable*** *if for every $\rho > 0$ and $\varepsilon > 0$, we can choose numbers $\rho_0 > 0$ and $\varepsilon_0 > 0$ such that a material point situated anywhere at a distance less than ρ_0 from $\mathbf{x}_0$, after initially receiving kinetic energy in amount less than ε_0, will forever remain a distance from $\mathbf{x}_0$ less than ρ and possess kinetic energy less than ε (see Figure 4.4.1).*

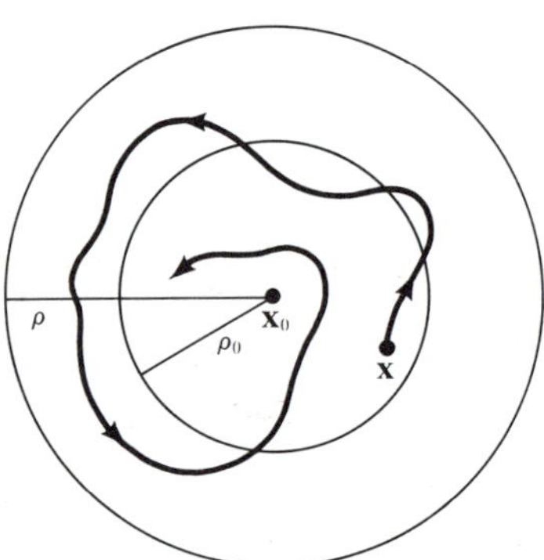

FIGURE 4.4.1
Motion near a stable point $\mathbf{x}_0$.

Thus if we have a position of equilibrium, stability at $\mathbf{x}_0$ means that a slowly moving particle near $\mathbf{x}_0$ will always remain near $\mathbf{x}_0$ and keep moving slowly. If we have an unstable equilibrium point $\mathbf{x}_0$, then $\boldsymbol{\sigma}(t) = \mathbf{x}_0$ solves the equation $m\boldsymbol{\sigma}''(t) = \mathbf{F}(\boldsymbol{\sigma}(t))$, but nearby solutions may move away from $\mathbf{x}_0$ as time progresses. For example a pencil balancing on its tip illustrates an unstable configuration, whereas a ball hanging on a spring illustrates a stable equilibrium.

Theorem 7.

(*i*) *Critical points of a potential are the positions of equilibrium.*

(*ii*) *In a potential field, a point $\mathbf{x}_0$ at which the potential takes a strict local minimum is a position of stable equilibrium.*

Recall that a function f is said to have a *strict* local minimum at the point $\mathbf{x}_0$ if there exists a neighborhood U of $\mathbf{x}_0$ such that $f(\mathbf{x}) > f(\mathbf{x}_0)$ for all $\mathbf{x}$ in U other than $\mathbf{x}_0$.

Proof. The first assertion (*i*) is quite obvious from the definition $\mathbf{F} = -\text{grad } V$; equilibrium points $\mathbf{x}_0$ are exactly critical points of V, at which $\nabla V(\mathbf{x}_0) = \mathbf{0}$.

To prove (*ii*) we shall make use of the law of conservation of energy, equation (2). We have

$$\tfrac{1}{2}m\|\boldsymbol{\sigma}'(t)\|^2 + V(\boldsymbol{\sigma}(t)) = \tfrac{1}{2}m\|\boldsymbol{\sigma}'(0)\|^2 + V(\boldsymbol{\sigma}(0))$$

We shall argue slightly informally to amplify and illuminate the central ideas involved. Let us choose a small neighborhood of $\mathbf{x}_0$ and start our particle with a small kinetic energy. As t increases, the particle moves away from $\mathbf{x}_0$ on a path $\boldsymbol{\sigma}(t)$ and $V(\boldsymbol{\sigma}(t))$ increases (since $V(\boldsymbol{\sigma}(0))$ is a strict minimum), so the kinetic energy must decrease. If the initial kinetic energy is sufficiently small, then in order for the particle to escape from our neighborhood of $\mathbf{x}_0$, outside of which V has increased by a definite amount, the kinetic energy would have to become negative (which is impossible). Thus the particle cannot escape the neighborhood. ∎

EXAMPLE 1. Find the points that are positions of equilibrium, and determine whether or not they are stable, if the force field $\mathbf{F} = F_x\mathbf{i} + F_y\mathbf{j} + F_z\mathbf{k}$ is given by $F_x = -k^2x$, $F_y = -k^2y$, $F_z = -k^2z$ $(k \neq 0)$.*

The field $\mathbf{F}$ is a potential field with potential $V = \frac{1}{2}k^2(x^2 + y^2 + z^2)$. The only critical point of V is the origin. The Hessian of V at the origin is $HV(0, 0, 0)(h_1, h_2, h_3) = \frac{1}{2}k^2(h_1^2 + h_2^2 + h_3^2)$, which is positive definite. It follows that the origin is a strict minimum of V. Thus, by (*i*) and (*ii*) of Theorem 7, we have shown that the origin is a position of stable equilibrium.

Let a material point in a potential field V be constrained to remain on the level surface S given by the equation $\phi(x, y, z) = 0$, with grad $\phi \neq \mathbf{0}$. If in formula (1) we replace $\mathbf{F}$ by the component of $\mathbf{F}$ parallel to S, we ensure that the particle will remain on S.† By analogy with Theorem 7, we have:

Theorem 8.

(*i*) *If at a point P on the surface S the potential $V|S$ has an extreme value, then the point P is a position of equilibrium on the surface.*

(*ii*) *If a point $P \in S$ is a strict local minimum of the potential $V|S$, then the point P is a position of stable equilibrium.*

The proof of this theorem will be omitted. It is similar to the proof of Theorem 7, with the additional fact that the equation of motion uses only the component of $\mathbf{F}$ along the surface.‡

* The force field in this example is that governing the motion of a three-dimensional harmonic oscillator.

† If $\phi(x, y, z) = x^2 + y^2 + z^2 - r^2$, the particle is constrained to move on a sphere; for instance, it may be whirling on a string. The part subtracted from $\mathbf{F}$ to make it parallel to S is normal to S and is called the *centripetal force*.

‡ These ideas can be applied to quite a number of interesting physical situations, such as molecular vibrations. The stability of such systems is an important question. For further information consult the physics literature (e.g., H. Goldstein, *Classical Mechanics*, Chapter 10, Addison-Wesley, Reading, Mass., 1950), and the mathematics literature (e.g., M. Hirsch and S. Smale, *Differential Equations, Dynamical Systems and Linear Algebra*, Academic Press, New York, 1974).

EXAMPLE 2. Let **F** be the gravitational field near the surface of the Earth, that is, $\mathbf{F} = (F_x, F_y, F_z)$ where $F_x = 0$, $F_y = 0$, and $F_z = -mg$. What are the positions of equilibrium, if a material point with mass m is constrained to the sphere $\phi(x, y, z) = x^2 + y^2 + z^2 - r^2 = 0$ $(r > 0)$? Which of these are stable?

Notice that **F** is a potential field with $V = mgz$. Using the method of Lagrange multipliers introduced in the preceding section to locate the possible extrema, we have the equations

$$\nabla V = \lambda \nabla \phi$$
$$\phi = 0$$

or in terms of components

$$0 = 2\lambda x$$
$$0 = 2\lambda y$$
$$mg = 2\lambda z$$
$$x^2 + y^2 + z^2 - r^2 = 0$$

and the solution of the above simultaneous equations is $x = 0$, $y = 0$, $z = \pm r$, $\lambda = \pm mg/2r$. By Theorem 8 it follows that the points $P_1 = (0, 0, -r)$ and $P_2 = (0, 0, r)$ are positions of equilibrium. By observation of the potential function $V = mgz$ and by Theorem 8(*ii*) it follows that P_1 is a strict minimum and hence a stable point, whereas P_2 is not. This conclusion should be physically obvious.

We conclude this section with a geometric problem.

EXAMPLE 3. Suppose we have a curve defined by the equation

$$\phi(x, y) = Ax^2 + 2Bxy + Cy^2 - 1 = 0$$

Find the maximum and minimum distance of the curve to the origin. These are called the lengths of the semi-major and the semi-minor axis.

The problem is equivalent to finding the extreme values of $f(x, y) = x^2 + y^2$ subject to the constraining condition $\phi(x, y) = 0$. Using the Lagrange multiplier method, we have the following equations:

$$2x + \lambda(2Ax + 2By) = 0 \tag{1}$$
$$2y + \lambda(2Bx + 2Cy) = 0 \tag{2}$$
$$Ax^2 + 2Bxy + Cy^2 = 1 \tag{3}$$

Adding $(1) \times x$ to $(2) \times y$, we obtain $2(x^2 + y^2) + 2\lambda(Ax^2 + 2Bxy + Cy^2) = 0$. By (3), it follows that $x^2 + y^2 + \lambda = 0$. Let $t = -1/\lambda = 1/(x^2 + y^2)$ ($\lambda = 0$ is impossible since $(0, 0)$ is not on the curve $\phi(x, y) = 0$). Then (1) and (2) can be written as follows:

$$\left.\begin{aligned} 2(A - t)x + 2By &= 0 \\ 2Bx + 2(C - t)y &= 0 \end{aligned}\right\} \tag{4}$$

If these two equations are to have a nontrivial solution (remember $(x, y) = (0, 0)$ is not on our curve and so is not a solution), it follows from a theorem of linear algebra that their determinant vanishes:*

$$\begin{vmatrix} A - t & B \\ B & C - t \end{vmatrix} = 0$$

Since this equation is quadratic in t, there are two solutions, say t_1 and t_2. Since $-\lambda = x^2 + y^2$, we have $\sqrt{x^2 + y^2} = \sqrt{-\lambda}$. Now $\sqrt{x^2 + y^2}$ is the distance from the point (x, y) to the origin. Therefore, if (x_1, y_1) and (x_2, y_2) denote the nontrivial solutions to (4) corresponding to t_1 and t_2, we have that $\sqrt{x_2^2 + y_2^2} = 1/\sqrt{t_2}$ and $\sqrt{x_1^2 + y_1^2} = 1/\sqrt{t_1}$. Consequently, if $t_1 > t_2$ the lengths of the semi-minor and semi-major axes are $1/\sqrt{t_1}$ and $1/\sqrt{t_2}$, respectively. If the curve is an ellipse, both t_1 and t_2 are real and positive. What happens with a hyperbola or a parabola?

EXERCISES

1. Let a particle move in a potential field in $\mathbb{R}^2$ given by $V(x, y) = 3x^2 + 2xy + 2x + y^2 + y + 4$. Find the stable equilibrium points, if any.
2. Let a particle move in a potential field in $\mathbb{R}^2$ given by $V(x, y) = x^2 - 2xy + y^2 + y^3 + x^4$. Is $(0, 0)$ a position of stable equilibrium?
3. Let a particle be constrained to move on the sphere $x^2 + y^2 + z^2 = 1$, subject to gravitational forces (as in Example 2) as well as an additional potential $V(x, y, z) = x + y$. Find the stable equilibrium points, if any.
4. Attempt to formulate a definition and a theorem saying that if a potential has a maximum at $\mathbf{x}_0$, then $\mathbf{x}_0$ is a position of unstable equilibrium. Watch out for pitfalls in your argument.
5. Try to find the extremes of $xy + yz$ among points satisfying $xz = 1$.

REVIEW EXERCISES FOR CHAPTER 4

1. Find the equation of the plane tangent to the surface at the indicated point.
   (a) $z = 2x^2 + y^2, \quad (x, y) = (0, 0)$
   (b) $z = x^2 - 3y^2 + x, \quad (x, y) = (1, 0)$
   (c) $z = x + 2y, \quad (x, y) = (3, 2)$
2. Analyze the behavior of the following functions at the indicated points.
   (a) $z = x^2 - y^2 + 3xy, \quad (x, y) = (0, 0)$
   (b) $z = x^2 + y^2 + Cxy, \quad (x, y) = (0, 0)$

* The matrix of coefficients of the equations cannot have an inverse, because this would imply that the solution is zero. From Section 1.4 we know that a matrix that does not have an inverse has determinant zero.

3. Find the equation of the plane tangent to the surface S given by the graph of
   (a) $f(x, y) = \sqrt{x^2 + y^2} + (x^2 + y^2)$ at $(1, 0, 2)$;
   (b) $f(x, y) = \sqrt{x^2 + 2xy - y^2 + 1}$ at $(1, 1, \sqrt{3})$.
4. Find and classify the extreme values (if any) of the functions on $\mathbb{R}^2$ defined by the following expressions:
   (a) $y^2 - x^3$
   (b) $(x - 1)^2 + (x - y)^2$
   (c) $x^2 + xy^2 + y^4$
5. (a) Find the minimum distance from the origin in $\mathbb{R}^3$ to the surface $z = \sqrt{x^2 - 1}$.
   (b) Repeat (a) for the surface $z = 6xy + 7$.
6. Let $f\colon \mathbb{R}^3 \to \mathbb{R}$ be C^1 and let $z\colon \mathbb{R}^2 \to \mathbb{R}$ be C^1. Consider the composite function

$$h(x, y) = f(x, y, z(x, y))$$

Show that

$$\frac{\partial h}{\partial x} = \frac{\partial f}{\partial x} + \frac{\partial f}{\partial z}\frac{\partial z}{\partial x}$$

and

$$\frac{\partial h}{\partial y} = \frac{\partial f}{\partial y} + \frac{\partial f}{\partial z}\frac{\partial z}{\partial y}$$

7. (a) Let $f(x, y, z) = xyz^2$; compute ∇f.
   (b) Let $\mathbf{F}(x, y, z) = xy\mathbf{i} + yz\mathbf{j} + zy\mathbf{k}$; compute $\nabla \times \mathbf{F}$.
   Compute $\nabla \cdot \mathbf{F}$ and $\nabla \times \mathbf{F}$ for the following vector fields.
   (c) $\mathbf{F} = 2x\mathbf{i} + 3y\mathbf{j} + 4z\mathbf{k}$
   (d) $\mathbf{F} = x^2\mathbf{i} + y^2\mathbf{j} + z^2\mathbf{k}$
   (e) $\mathbf{F} = (x + y)\mathbf{i} + (y + z)\mathbf{j} + (z + x)\mathbf{k}$
8. Find the first few terms in the Taylor expansion of $f(x, y) = e^{xy} \cos x$ about $x = 0$, $y = 0$.
9. Find the extreme value of $z = xy$, subject to the condition $x + y = 1$.
10. Find the points on the surface $z^2 - xy = 1$ nearest to the origin.
11. Find the extreme value of $z = \cos^2 x + \cos^2 y$ subject to the condition $x + y = \pi/4$.
12. (a) Let $(\partial f/\partial x)(x, y) = (\partial f/\partial y)(x, y) = 0$ for every (x, y) in an open disc D. Show that $f(x, y)$ is a constant in D.
    (b) Use the result of (a) to show that if $\partial f/\partial x = \partial g/\partial x$ and $\partial f/\partial y = \partial g/\partial y$ in D, then f and g differ by a constant.
13. Find the shortest distance from the point $(0, b)$ to the parabola $x^2 - 4y = 0$. Solve this problem using the Lagrange multiplier and also without using Lagrange's method.
14. Solve the following geometric problems by Lagrange's method.
    (a) Find the shortest distance from the point (a_1, a_2, a_3) in $\mathbb{R}^3$ to the plane whose equation is given by $b_1x_1 + b_2x_2 + b_3x_3 + b_0 = 0$, where $(b_1, b_2, b_3) \neq (0, 0, 0)$.

(b) Find the point on the line of intersection of the two planes $a_1 x_1 + a_2 x_2 + a_3 x_3 + a_0 = 0$ and $b_1 x_1 + b_2 x_2 + b_3 x_3 + b_0 = 0$ that is nearest to the origin.

(c) Show that the volume of the largest rectangular parallelepiped that can be inscribed in the ellipsoid

$$\frac{x^2}{a^2} + \frac{y^2}{b^2} + \frac{z^2}{c^2} = 1$$

is $8abc/3\sqrt{3}$.

15. A particle moves in a potential $V(x, y) = x^3 - y^2 + x^2 + 3xy$. Determine if (0, 0) is a stable equilibrium point.

16. Study the nature of the function $f(x, y) = x^3 - 3xy^2$ near (0, 0). Show that the point (0, 0) is a degenerate critical point, that is, $D = 0$. This surface is called a "monkey saddle."

CHAPTER 5

INTEGRATION

In this chapter we shall study the integration of real-valued functions of several variables; we are especially interested in integrals of functions of two variables, or *double integrals*, as they are called. The double integral has a basic geometric interpretation as volume, and can be defined rigorously as a limit of approximating sums. We shall present several techniques for evaluating double integrals, consider some applications, and then discuss improper integrals. Finally, we shall introduce integrals of functions of three variables, or *triple integrals*.

5.1 INTRODUCTION

In this introductory section, we shall briefly discuss some of the geometric aspects of the double integral, deferring a more rigorous discussion in terms of Riemann sums until Section 5.2.

Let us consider, then, a continuous function of two variables f: $R \subset \mathbb{R}^2 \to \mathbb{R}$ whose domain R is a rectangle with sides parallel to the coordinate axes. The rectangle R can be described in terms of the two closed intervals $[a, b]$ and $[c, d]$, representing the sides of R along the

x- and y-axes respectively, as in Figure 5.1.1. In this case, we say that R is the *Cartesian product* of $[a, b]$ and $[c, d]$ and write $R = [a, b] \times [c, d]$.

Let us assume that $f(x, y) \geq 0$ on R. The graph of $z = f(x, y)$ is then a surface lying above the rectangle R. This surface, the rectangle R, and the four planes $x = a$, $x = b$, $y = c$, and $y = d$ form the boundary of a region V in space (see Figure 5.1.1). The problem of how to rigorously

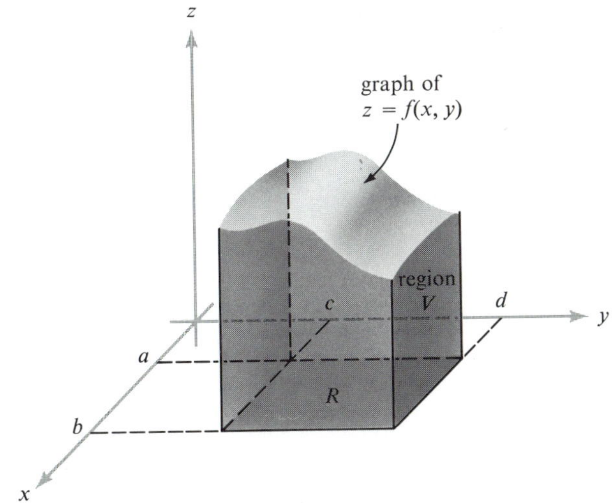

FIGURE 5.1.1
The region V in space is bounded by the graph of f, the rectangle R, and the four vertical sides indicated.

define the volume of V has to be faced, and we shall solve it by the classical method of exhaustion, or in more modern terms, the method of Riemann sums, in Section 5.2. However in order to gain an intuitive grasp of this method, let us provisionally assume that the volume of a region has been defined. This is not unreasonable since we feel intuitively that every region has a volume. Then the volume of the region above R and under the graph of f is called the (*double*) *integral* of f over R and is denoted by

$$\int_R f, \quad \int_R f(x, y)\, dA, \quad \int_R f(x, y)\, dx\, dy, \quad \text{or} \quad \iint_R f(x, y)\, dx\, dy$$

EXAMPLE 1. (a) If $f(x, y) = k$, where k is a positive constant, then $\int_R f(x, y)\, dA = k(b - a)(d - c)$, since the integral is equal to the volume of a rectangular box with base R and height k.

(b) If $f(x, y) = 1 - x$ and $R = [0, 1] \times [0, 1]$, then $\int_R f(x, y)\, dA = \frac{1}{2}$ since the integral is equal to the volume of the triangular solid shown in Figure 5.1.2.

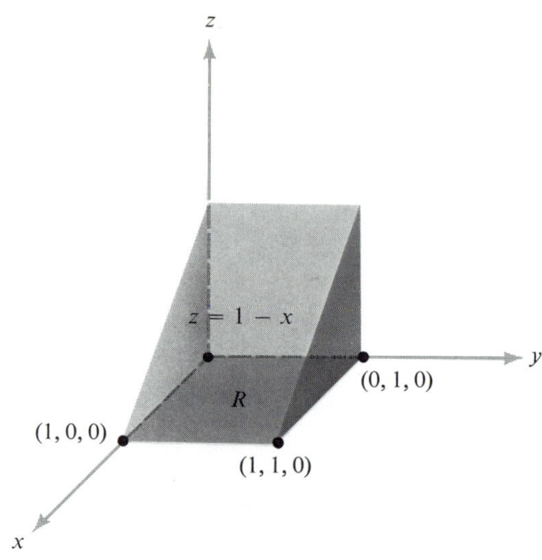

FIGURE 5.1.2
Volume under the graph $z = 1 - x$ *over* $R = [0, 1] \times [0, 1]$.

EXAMPLE 2. Suppose $z = f(x, y) = x^2 + y^2$ and $R = [-1, 1] \times [0, 1]$. Then the integral $\int_R f = \int_R (x^2 + y^2)\, dx\, dy$ is equal to the volume of the solid sketched in Figure 5.1.3. We shall compute this integral in Example 3.

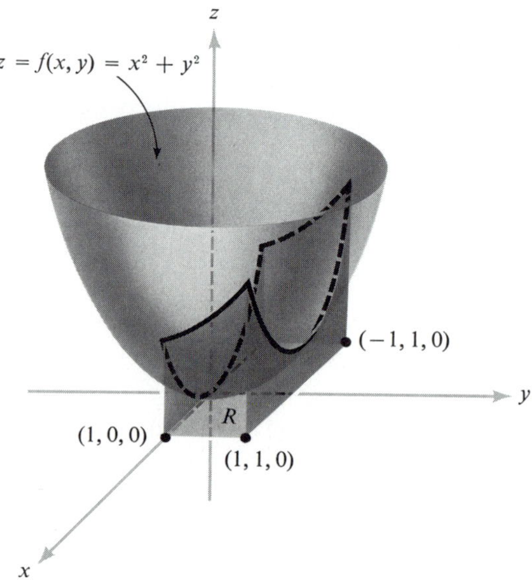

FIGURE 5.1.3
Volume under $z = x^2 + y^2$ *over* $R = [-1, 1] \times [0, 1]$.

These ideas are similar to the idea of a single integral $\int_a^b f(x)\,dx$, which represents the area under the graph of f if f is ≥ 0 and, say, continuous; see Figure 5.1.4.* We should also recall that $\int_a^b f(x)\,dx$

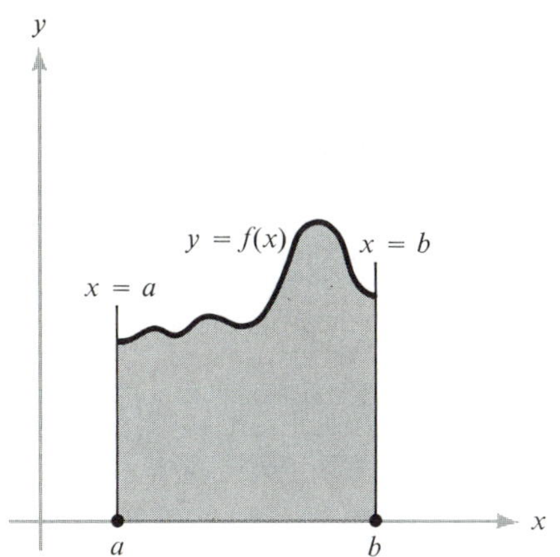

FIGURE 5.1.4
Area under the graph of a nonnegative continuous function f from x = a to x = b is $\int_a^b f(x)\,dx$.

can be rigorously defined, without recourse to the area concept, as a limit of Riemann sums. Thus we can approximate $\int_a^b f(x)\,dx$ by choosing a partition $a = x_0 < x_1 < \cdots < x_n = b$ of $[a, b]$, selecting points $c_i \in [x_i, x_{i+1}]$, and forming the Riemann sum

$$\sum_{i=0}^{n-1} f(c_i)(x_{i+1} - x_i) \approx \int_a^b f(x)\,dx$$

(see Figure 5.1.5). We shall examine the analogous process for double integrals in the next section.

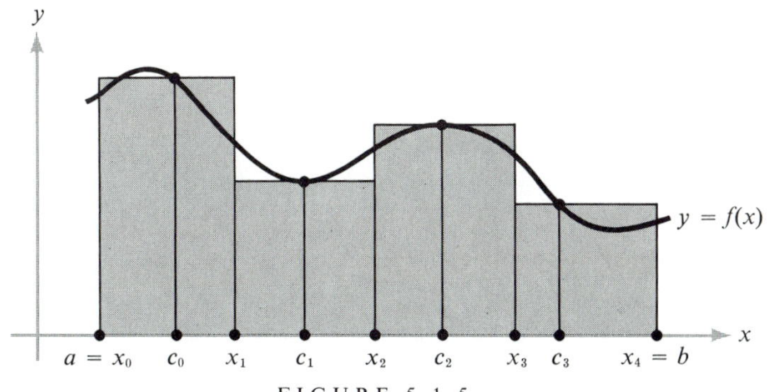

FIGURE 5.1.5
The sum of the areas of the shaded rectangles is a Riemann sum approximating the area under f from x = a to x = b.

* The reader not already familiar with this idea should review the appropriate sections of their introductory calculus text.

There is a method for computing volumes known as *Cavalieri's Principle.** Suppose we have a solid body and we let $A(x)$ denote its cross-sectional area measured at a distance x from a reference plane (Figure 5.1.6). According to Cavalieri's Principle, the volume of the body is given by

$$\text{volume} = \int_a^b A(x)\, dx$$

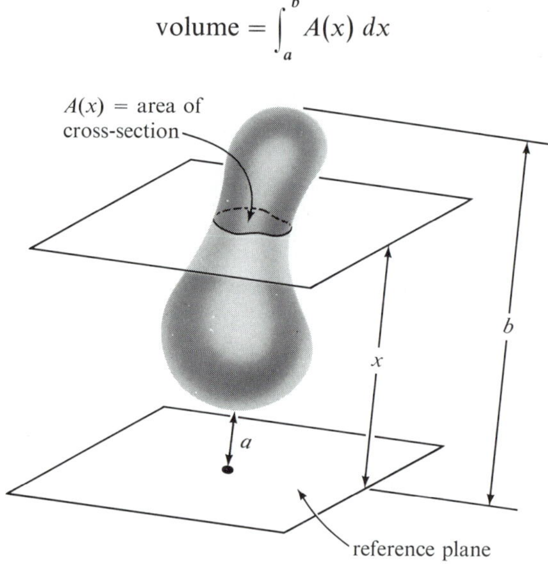

FIGURE 5.1.6
A solid body with cross-sectional area $A(x)$ at distance x from a reference plane.

where a and b are the minimum and maximum distances from the reference plane. This can be made intuitively clear. If we partition $[a, b]$ into $x_0 = a < x_1 < \cdots < x_n = b$, then an approximating Riemann sum for the above integral is

$$\sum_{i=0}^{n-1} A(c_i)(x_{i+1} - x_i)$$

But this sum also approximates the volume of the body, since $A(x)\Delta x$ is the volume of a slab with cross-sectional area $A(x)$ and thickness Δx (Figure 5.1.7). Therefore it is reasonable to accept the above formula for the volume. A more careful justification of this method is given below. For now, let us see how this result provides a method for evaluating double integrals.

* Bonaventura Cavalieri (1598–1647) was a pupil of Galileo and a professor in Bologna. His investigations into area and volume were important building blocks of the foundations of calculus. Although his methods were criticized by his contemporaries, similar ideas had been used by Archimedes in antiquity, and were later taken up by the "fathers" of calculus, Newton and Leibnitz.

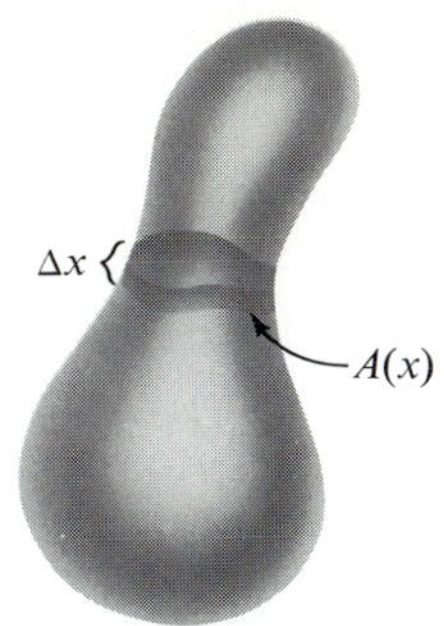

FIGURE 5.1.7
Volume of a slab with cross-sectional area $A(x)$ and thickness Δx equals $A(x)\,\Delta x$. The total volume of the body is $\int_a^b A(x)\,dx$.

Let us consider the solid region under a graph $z = f(x, y)$ defined on the region $[a, b] \times [c, d]$, where f is continuous and greater than zero. There are two natural cross-sectional area functions: one obtained by using cutting planes perpendicular to the x-axis, and the other obtained by using cutting planes perpendicular to the y-axis. The cross section determined by a cutting plane $x = x_0$, of the first sort, is the plane region under the graph of $z = f(x_0, y)$ from $y = c$ to $y = d$ (Figure 5.1.8). When we fix $x = x_0$ we have the function $y \mapsto f(x_0, y)$, which is continuous on $[c, d]$. The cross-sectional area $A(x_0)$ is, therefore, equal to the integral $\int_c^d f(x_0, y)\,dy$. Thus the cross-sectional area function A has domain $[a, b]$, and $A: x \mapsto \int_c^d f(x, y)\,dy$. By Cavalieri's

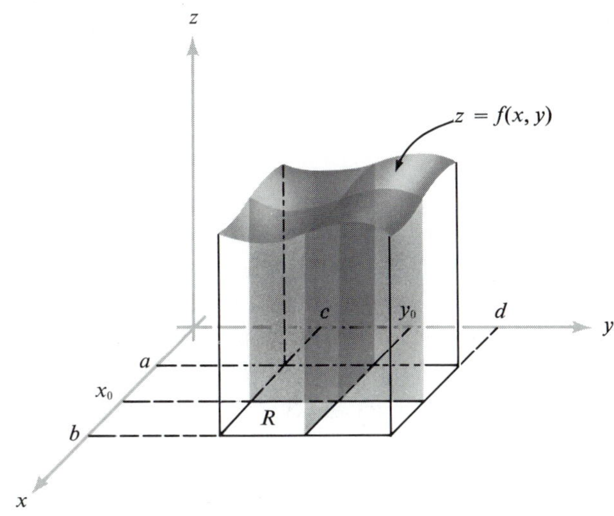

FIGURE 5.1.8
Two different cross sections sweeping out the volume under $z = f(x, y)$.

Principle, the volume V of the region under $z = f(x, y)$ must be equal to

$$V = \int_a^b A(x)\,dx = \int_a^b \left[\int_c^d f(x, y)\,dy\right] dx$$

The integral $\int_a^b [\int_c^d f(x, y)\,dy]\,dx$ is known as an *iterated integral*, because it is obtained by integrating with respect to y and then integrating the result with respect to x. Since $\int_R f(x, y)\,dA$ is equal to the volume V,

$$\int_R f(x, y)\,dA = \int_a^b \left[\int_c^d f(x, y)\,dy\right] dx \tag{1}$$

If we reverse the roles of x and y in the above discussion and use cutting planes perpendicular to the y-axis, we obtain

$$\int_R f(x, y)\,dA = \int_c^d \left[\int_a^b f(x, y)\,dx\right] dy \tag{2}$$

The expression on the right of (2) is the iterated integral obtained by integrating with respect to x and then integrating the result with respect to y.

Thus, if our intuition about volumes is correct, formulas (1) and (2) ought to be valid and equal. The equality of the integrals (1) and (2) and the double integral also holds when the concepts we are discussing are defined rigorously. This result is known as Fubini's Theorem. We shall give a proof of this theorem in the next section.

As the following examples illustrate, the notion of the iterated integral and equations (1) and (2) provide a powerful method for computing the double integral of a function of two variables.

EXAMPLE 3. Let $z = f(x, y) = x^2 + y^2$ and let $R = [-1, 1] \times [0, 1]$. Let us evaluate the integral $\int_R (x^2 + y^2)\,dx\,dy$. By equation (2), we have

$$\int_R (x^2 + y^2)\,dx\,dy = \int_0^1 \left[\int_{-1}^1 (x^2 + y^2)\,dx\right] dy$$

To find $\int_{-1}^1 (x^2 + y^2)\,dx$, we treat y as a constant and integrate with respect to x. Since $x \mapsto x^3/3 + y^2x$ is an antiderivative of $x \mapsto x^2 + y^2$, we can integrate, using methods of elementary calculus, to obtain

$$\int_{-1}^1 (x^2 + y^2)\,dx = \left[\frac{x^3}{3} + y^2x\right]_{x=-1}^1 = \frac{2}{3} + 2y^2$$

Next we integrate $y \mapsto \frac{2}{3} + 2y^2$ with respect to y from 0 to 1, to obtain

$$\int_0^1 \left(\frac{2}{3} + 2y^2\right) dy = \left[\frac{2}{3}y + \frac{2}{3}y^3\right]_{y=0}^1 = \frac{4}{3}$$

Hence the volume of the solid in Figure 5.1.3 is $\frac{4}{3}$. For completeness, let us evaluate $\int_R (x^2 + y^2)\, dx\, dy$ using (1)—that is, integrating with respect to y and then with respect to x. We have

$$\int_R (x^2 + y^2)\, dx\, dy = \int_{-1}^{1} \left[\int_0^1 (x^2 + y^2)\, dy\right] dx$$

Treating x as a constant in the y-integration, we obtain

$$\int_0^1 (x^2 + y^2)\, dy = \left[x^2 y + \frac{y^3}{3}\right]_{y=0}^{1} = x^2 + \frac{1}{3}$$

Next we evaluate $\int_{-1}^{1} (x^2 + \frac{1}{3})\, dx$ to obtain

$$\int_{-1}^{1} \left(x^2 + \frac{1}{3}\right) dx = \left[\frac{x^3}{3} + \frac{x}{3}\right]_{x=-1}^{1} = \frac{4}{3}$$

which agrees with our previous answer.

EXAMPLE 4. Compute $\int_S \cos x \sin y\, dx\, dy$ where S is the square $[0, \pi/2] \times [0, \pi/2]$ (see Figure 5.1.9). We have by equation (2)

$$\int_S \cos x \sin y\, dx\, dy = \int_0^{\pi/2} \left[\int_0^{\pi/2} \cos x \sin y\, dx\right] dy$$
$$= \int_0^{\pi/2} \sin y\, dy = 1$$

In the next section, we shall use Riemann sums to rigorously define the double integral for a wide class of functions of two variables

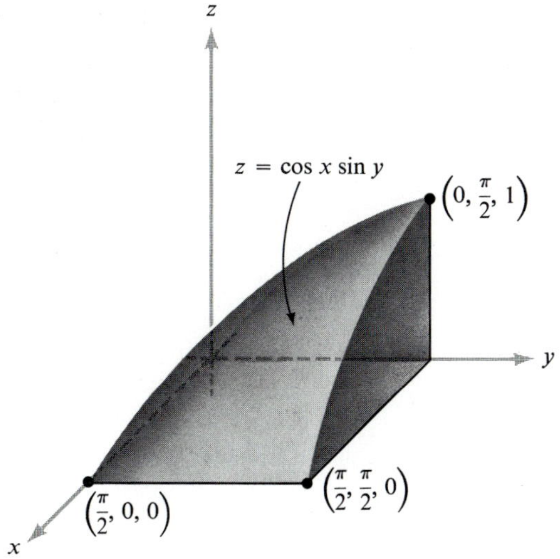

FIGURE 5.1.9
Volume under $z = \cos x \sin y$ *over the rectangle* $[0, \pi/2] \times [0, \pi/2]$.

without recourse to the notion of volume. Although we shall drop the requirement that $f(x, y) \geq 0$, equations (1) and (2) will remain valid. Therefore, the iterated integral will again provide the key to computing the double integral. In Section 5.3 we shall treat double integrals over regions more general than rectangles.

Finally, we remark that it is common to delete the brackets in iterated integrals such as (1) and (2) above, so

$$\int_a^b \int_c^d f(x, y)\, dy\, dx = \int_a^b \left[\int_c^d f(x, y)\, dy\right] dx$$

and

$$\int_c^d \int_a^b f(x, y)\, dx\, dy = \int_c^d \left[\int_a^b f(x, y)\, dx\right] dy$$

EXERCISES

1. Evaluate the following iterated integrals.
   (a) $\int_{-1}^{1} \int_0^1 (x^4 y + y^2)\, dy\, dx$
   (b) $\int_0^{\pi/2} \int_0^1 (y \cos x + 2)\, dy\, dx$
   (c) $\int_0^1 \int_0^1 (xye^{x+y})\, dy\, dx$
   (d) $\int_{-1}^{0} \int_1^2 (-x \log y)\, dy\, dx$
2. Evaluate the integrals in Exercise 1 by integrating with respect to x and then with respect to y.
3. Use Cavalieri's formula to show that the volumes of two cylinders with the same base and height are equal (see Figure 5.1.10).

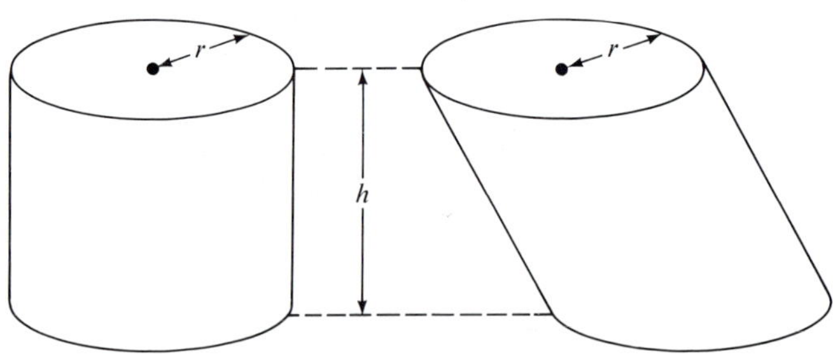

FIGURE 5.1.10
Two cylinders with same base and height have the same volume.

4. (a) Argue that the volume of revolution shown in Figure 5.1.11 is

$$\pi \int_a^b [f(x)]^2\, dx$$

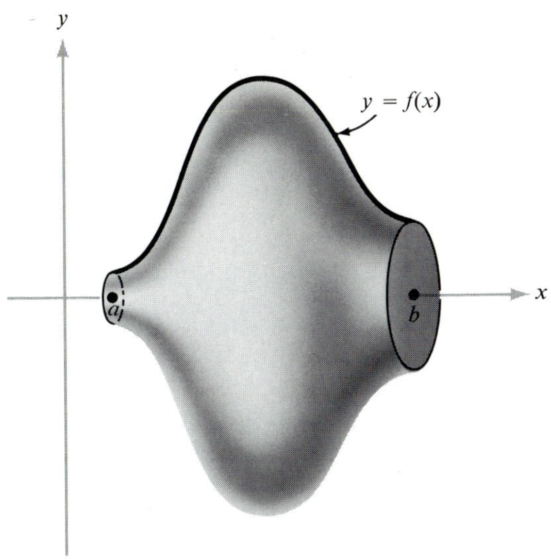

FIGURE 5.1.11
This solid of revolution has volume $\pi \int_a^b [f(x)]^2\, dx$.

(b) Show that the volume of the region obtained by rotating the graph of the parabola $y = -x^2 + 2x + 3$, $-1 \le x \le 3$, about the x-axis is $512\pi/15$ (see Figure 5.1.12).

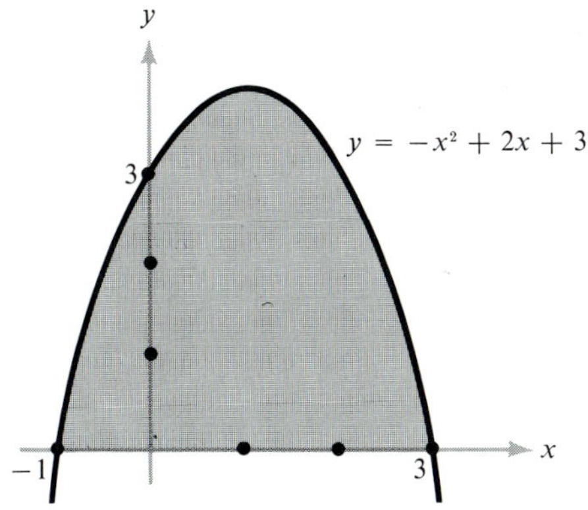

FIGURE 5.1.12
Region between the graph of $y = -x^2 + 2x + 3$ *and the x-axis.*

5. Evaluate the following double integrals where R is the rectangle $[0, 2] \times [-1, 0]$.
(a) $\int_R (x^2y^2 + x)\, dy\, dx$
(b) $\int_R (|y| \cos \frac{1}{4}\pi x)\, dy\, dx$
(c) $\int_R (-xe^x \sin \frac{1}{2}\pi y)\, dy\, dx$

6. Find the volume bounded by the graph of $f(x, y) = 1 + 2x + 3y$, the rectangle $[1, 2] \times [0, 1]$, and the four vertical sides of the rectangle R as in Figure 5.1.1.
7. Repeat Exercise 6 for the surface $f(x, y) = x^4 + y^2$ and the rectangle $[-1, 1] \times [-3, -2]$.

5.2 THE DOUBLE INTEGRAL OVER A RECTANGLE*

We are now ready to give a rigorous definition of the double integral as the limit of a sequence of sums. This will then be used to *define* the volume of the region under the graph of a function $f(x, y)$. We shall not require that $f(x, y) \geq 0$, but if $f(x, y)$ assumes negative values we shall not interpret the integral as a volume, just as we make a similar reservation for the area under the graph of a function of one variable. In addition, we shall discuss some of the fundamental algebraic properties of the double integral and prove Fubini's Theorem, which states that the double integral can be calculated as an iterated integral. To begin, let us establish some notation for partitions and sums.

Consider a rectangle $R \subset \mathbb{R}^2$ that is the Cartesian product $R = [a, b] \times [c, d]$. By the *regular partition* of R of order n we mean the pair of regular partitions of $[a, b]$ and of $[c, d]$ of order n, that is, the two collections of $n + 1$ equally spaced points $\{x_j\}_{j=0}^n$ and $\{y_k\}_{k=0}^n$ with

$$a = x_0 < x_1 < \cdots < x_n = b, \qquad c = y_0 < y_1 < \cdots < y_n = d$$

and

$$x_{j+1} - x_j = \frac{b-a}{n}, \qquad y_{k+1} - y_k = \frac{d-c}{n}$$

(see Figure 5.2.1).

Let R_{jk} be the rectangle $[x_j, x_{j+1}] \times [y_k, y_{k+1}]$, and let c_{jk} be any point in R_{jk}. Suppose $f\colon R \to \mathbb{R}$ is a bounded real-valued function. Form the sum

$$S_n = \sum_{j,k=0}^{n-1} f(c_{jk})\, \Delta x\, \Delta y = \sum_{j,k=0}^{n-1} f(c_{jk})\, \Delta A \tag{1}$$

where

$$\Delta x = x_{j+1} - x_j = \frac{b-a}{n}, \qquad \Delta y = y_{k+1} - y_k = \frac{d-c}{n}$$

* This section is intended to show how the theory of integration goes, without providing the detailed proofs. In a course such as this, the material should be treated lightly, to be taken up more seriously in a later course.

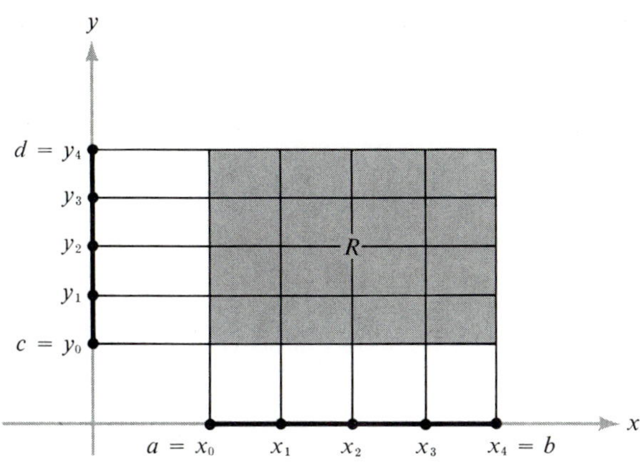

FIGURE 5.2.1
A regular partition of a rectangle R.

and

$$\Delta A = \Delta x \, \Delta y$$

This sum is taken over all j's and k's from 0 to $n - 1$, so there are n^2 terms.

Definition. *If the sequence $\{S_n\}$ converges to a limit S as $n \to \infty$, with the limit S being the same for any choice of points c_{jk} in the rectangles R_{jk}, then we say that f is **integrable** over R and we write*

$$\int_R f, \quad \int_R f(x, y)\, dA, \quad \int_R f(x, y)\, dx\, dy, \quad \text{or} \quad \iint_R f(x, y)\, dx\, dy$$

for the limit S.

*A sum S_n is called a **Riemann sum** for f.*

Thus we can rewrite integrability in the following way:

$$\lim_{n \to \infty} \sum_{j,k=0}^{n-1} f(c_{jk})\, \Delta x\, \Delta y = \int_R f$$

for any choice of $c_{jk} \in R_{jk}$.

The proof of the following basic theorem is not difficult, but requires an idea, namely the concept of uniform continuity, that we shall not develop in this course. We therefore state the theorem without proof.*

Theorem 1. *Any continuous function defined on a rectangle R is integrable.*

* For a proof see, for example, J. Marsden, *Elementary Classical Analysis*, W. H. Freeman and Company, San Francisco, 1974, Chapter 8.

If $f(x, y) \geq 0$, the existence of $\lim_{n\to\infty} S_n$ has a straightforward geometric meaning. Consider the graph of $z = f(x, y)$ as the top of a solid whose base is the rectangle R. If we take each c_{jk} to be a point where $f(x, y)$ has its minimum value* on R_{jk}, then $f(c_{jk})\,\Delta x\,\Delta y$ represents the volume of a rectangular box with base R_{jk}. The sum $\sum_{j,k=0}^{n-1} f(c_{jk})\,\Delta x\,\Delta y$ equals the volume of an inscribed solid part of which is shown in Figure 5.2.2. Similarly, if c_{jk} is a point where $f(x, y)$

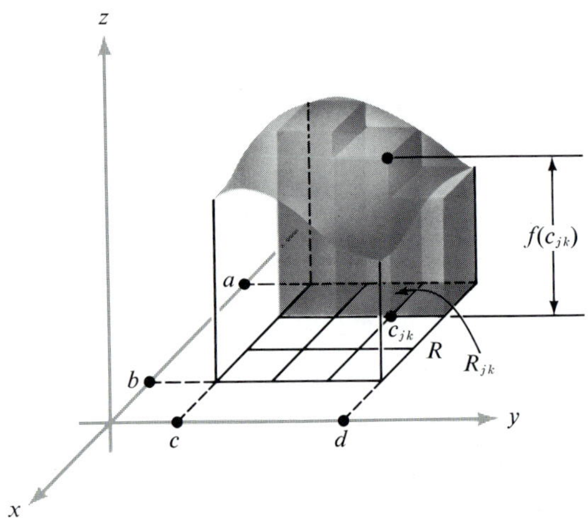

FIGURE 5.2.2
The sum of inscribed boxes approximates the volume under the graph of $z = f(x, y)$.

has its maximum on R_{jk}, then the sum $\sum_{j,k=0}^{n-1} f(c_{jk})\,\Delta x\,\Delta y$ is equal to the volume of a circumscribed solid; see Figure 5.2.3. Therefore, if $\lim_{n\to\infty} S_n$ exists and is independent of $c_{jk} \in R_{jk}$, it follows that the volumes of the inscribed and circumscribed solids approach the same limit as $n \to \infty$. It is therefore reasonable to call this limit the exact volume of the solid under the graph of f. Thus the method of Riemann sums supports the concepts introduced on an intuitive basis in Section 5.1.

There is a theorem guaranteeing the existence of the integral of certain discontinuous functions as well. We shall need this result in the next section in order to discuss the integral of functions over regions more general than rectangles. The statement of this theorem requires the following definition.

* Such c_{jk} exist by virtue of the continuity of f on R, but we shall not prove this fact.

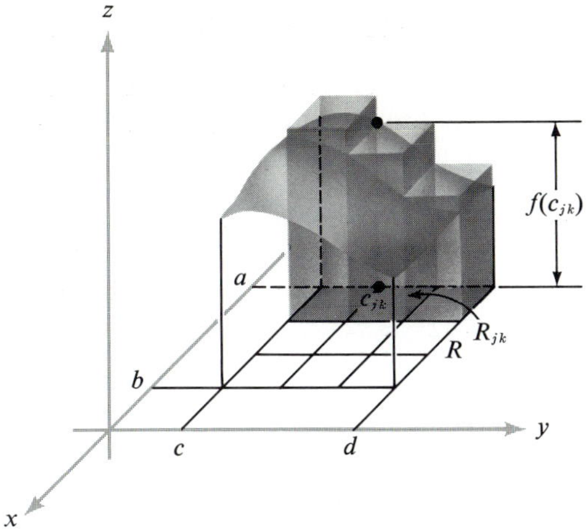

FIGURE 5.2.3
The volume of circumscribed boxes also approximates the volume under $z = f(x, y)$.

Definition. *Let $B \subset R$ be a subset of R (see Figure 5.2.4). In the nth regular partition of R, let b_n be the sum of the areas of those rectangles in the partition that intersect B, that is, those rectangles containing at least one point of B. Then B is said to have **area zero** if $\lim_{n \to \infty} b_n = 0$.*

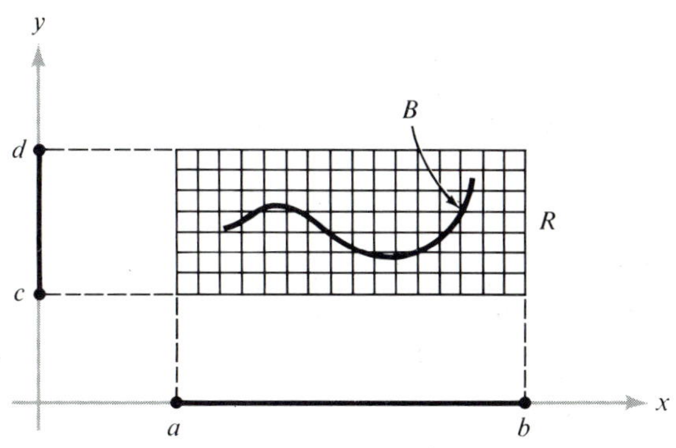

FIGURE 5.2.4
A rectangle R containing a subset B of area zero.

This definition is reasonable, for b_n represents the area of circumscribing rectangles. In the following examples we measure the areas of sets that should, intuitively, be of area zero.

EXAMPLE 1. Let B be a subset of R consisting of a finite number of points. Show that B has area zero.

Suppose B consists of m points $p_1, p_2, \ldots, p_m$. Then each point p_j, $j = 1, \ldots, m$, can be in at most four rectangles of the subdivision (see

Figure 5.2.5). Since the area of each rectangle in the nth regular partition is $(b-a)(d-c)/n^2$, we have

$$0 < b_n \leq \frac{4m}{n^2}(b-a)(d-c)$$

Thus $\lim_{n\to\infty} b_n = 0$, and B has area zero.

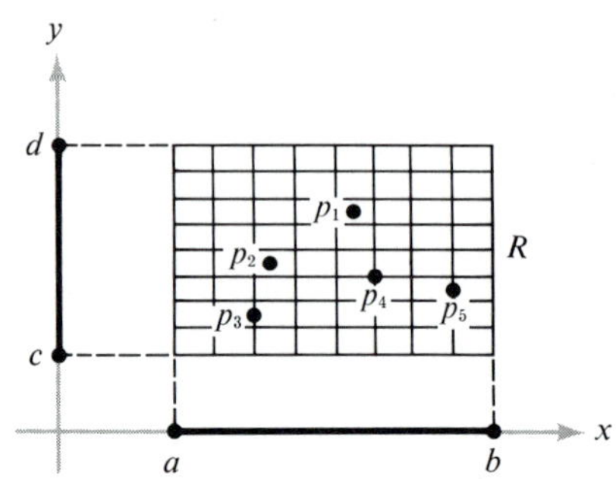

FIGURE 5.2.5
Each point of R lies in at most four rectangles of the partition.

EXAMPLE 2. Let B be the boundary of R (the set consisting of the four edges of R). Show that B has area zero.

For the nth regular partition, the total area b_n of those $4n-4$ rectangles containing points of the boundary is given explicitly by

$$b_n = \left\{\left(\frac{b-a}{n}\right)\left(\frac{d-c}{n}\right)\right\}(4n-4) = \frac{4(n-1)}{n}\frac{(b-a)(d-c)}{n}$$

(see Figure 5.2.6). Thus

$$0 < b_n < \frac{4(b-a)(d-c)}{n} \quad \text{since} \quad \frac{n-1}{n} < 1$$

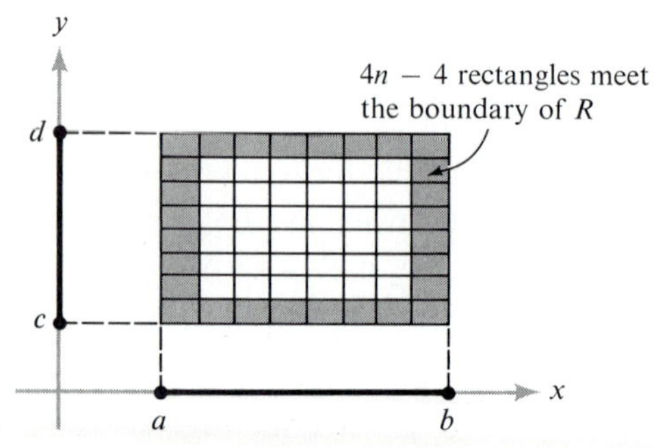

FIGURE 5.2.6
The boundary of a rectangle has area zero.

and hence
$$\lim_{n\to\infty} b_n = 0$$

The result of Example 2 extends to an important class of curves in the plane, namely the graphs of continuous functions $y = \phi(x)$ and $x = \psi(y)$. In the next section we shall use without proof the fact that the graph of a continuous real function $\phi\colon [a, b] \to \mathbb{R}$ has area zero. In fact, any set composed of a finite number of such graphs has area zero.

Thus the curves in Figure 5.2.7 are all sets of area zero. We shall use

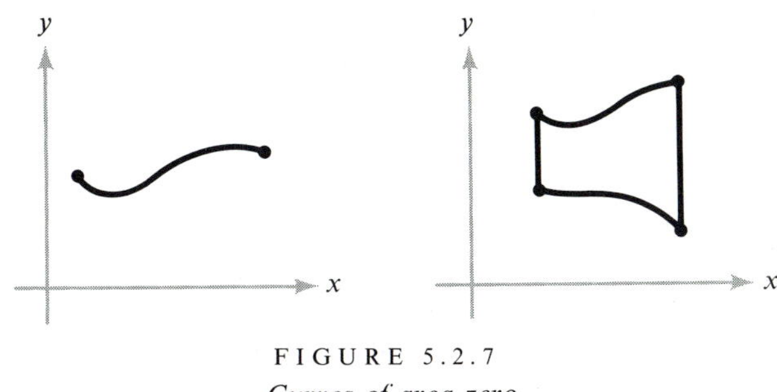

FIGURE 5.2.7
Curves of area zero.

the notion of zero area in the next section when we define the integral over regions bounded by such curves.

The next result, also stated without proof, provides an important criterion for determining if a function is integrable.

Theorem 2. *Let $f\colon R \to \mathbb{R}$ be a bounded real-valued function on the rectangle R, and suppose that the set of points where f is discontinuous has area zero. Then f is integrable over R.*

Recall that a function is *bounded* if there is a number $M > 0$ such that $-M \le f(x, y) \le M$ for all (x, y) in the domain of f. A continuous function on a closed rectangle is always bounded, but, for example $f(x, y) = 1/x$ on $]0, 1] \times [0, 1]$ is not bounded, because $1/x$ becomes arbitrarily large for x near 0.

Using Theorem 2 and the remarks preceding it, we see that the functions sketched in Figure 5.2.8 are integrable over R, since these functions are continuous except on sets of area zero.

Geometrically, Theorem 2 means that if a non-negative function f is not "too badly behaved" (that is, if its discontinuities form a set of area zero), then the volumes of the circumscribed and inscribed solids will approximate the "true" volume under its graph (see Figure 5.2.9).

From the definition of the integral as a limit of sums and the limit theorems, we can quickly deduce some fundamental properties of the

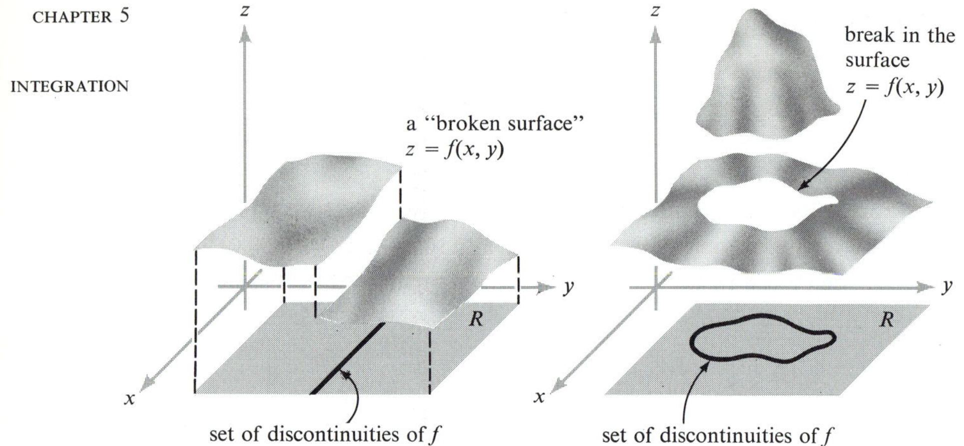

FIGURE 5.2.8
What the graphs of discontinuous functions of two variables might look like.

integral $\int_R f(x, y)\,dA$; these properties are essentially the same as for the integral of a real-valued function of a single variable.

Let f and g be integrable functions on the rectangle R, and let c be a constant. Then $f + g$ and cf are integrable, and

(i) (*linearity*)
$$\int_R [f(x, y) + g(x, y)]\,dA = \int_R f(x, y)\,dA + \int_R g(x, y)\,dA$$

(ii) (*homogeneity*) $\int_R cf(x, y)\,dA = c\int_R f(x, y)\,dA$

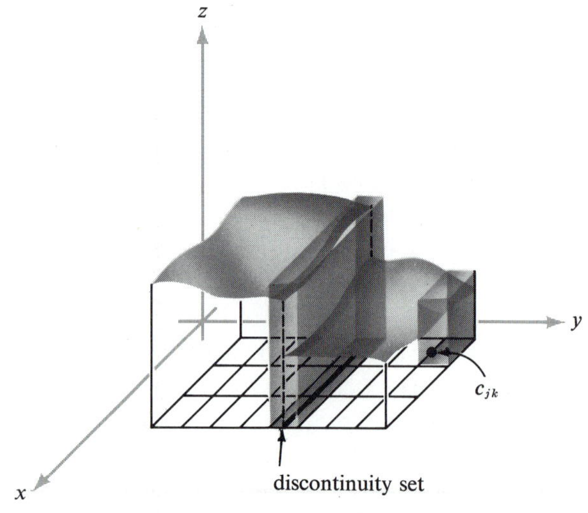

FIGURE 5.2.9
Graph of a discontinuous function and two circumscribing boxes.

(*iii*) (*monotonicity*) If $f(x, y) \geq g(x, y)$, then

$$\int_R f(x, y)\, dA \geq \int_R g(x, y)\, dA$$

(*iv*) (*additivity*) If R_i, $i = 1, \ldots, m$, are pairwise disjoint rectangles such that f is integrable over each R_i and if $Q = R_1 \cup R_2 \cup \cdots \cup R_m$ is a rectangle, then $f: Q \to \mathbb{R}$ is integrable over Q and

$$\int_Q f(x, y)\, dA = \sum_{i=1}^{m} \int_{R_i} f(x, y)\, dA$$

Properties (*i*) and (*ii*) are a consequence of the definition of the integral as a limit of a sum and the following facts for convergent sequences $\{S_n\}$ and $\{T_n\}$

$$\operatorname*{limit}_{n\to\infty} (T_n + S_n) = \operatorname*{limit}_{n\to\infty} T_n + \operatorname*{limit}_{n\to\infty} S_n$$

$$\operatorname*{limit}_{n\to\infty} (cS_n) = c \operatorname*{limit}_{n\to\infty} S_n$$

To demonstrate monotonicity we first observe that if $h(x, y) \geq 0$ and $\{S_n\}$ is a sequence of Riemann sums that converges to $\int_R h(x, y)\, dA$, then $S_n \geq 0$ for all n, so that $\int_R h(x, y)\, dA = \operatorname*{limit}_{n\to\infty} S_n \geq 0$. If $f(x, y) \geq g(x, y)$ for all $(x, y) \in R$, then $(f - g)(x, y) \geq 0$ for all (x, y) and using (*i*) and (*ii*), we have

$$\int_R f(x, y)\, dA - \int_R g(x, y)\, dA = \int_R [f(x, y) - g(x, y)]\, dA \geq 0$$

This proves (*iii*). The proof of (*iv*) is left as an exercise.

Another important result is the inequality

$$\left| \int_R f \right| \leq \int_R |f| \tag{2}$$

To see why (2) is true, note that, by the definition of absolute value, $-|f| \leq f \leq |f|$; so from the monotonicity and homogeneity of integration (with $c = -1$)

$$-\int_R |f| \leq \int_R f \leq \int_R |f|$$

which is equivalent to formula (2).

Although we have noted the integrability of a variety of functions, we have not yet established rigorously a general method of computing integrals. In the case of one variable we avoid computing $\int_a^b f(x)\,dx$ from its definition as a limit of a sum by using the Fundamental Theorem of Integral Calculus.

Let us recall that this important theorem tells us that *if f is continuous, then*

$$\int_a^b f(x)\,dx = F(b) - F(a)$$

where F is an antiderivative of f; that is, $F' = f$.

This technique won't work as stated for functions $f(x, y)$ of two variables. However, as we indicated in Section 5.1, we can often reduce a double integral over a rectangle to iterated single integrals; the Fundamental Theorem then applies to these single integrals. Fubini's Theorem, which was mentioned in the last section, establishes this reduction to iterated integrals rigorously, by using Riemann sums. As we saw in Section 5.1 this reduction

$$\begin{aligned}\int_R f(x, y)\,dA &= \int_a^b \left[\int_c^d f(x, y)\,dy\right] dx \\ &= \int_c^d \left[\int_a^b f(x, y)\,dx\right] dy\end{aligned}$$

is a consequence of Cavalieri's Principle, at least if $f(x, y) \geq 0$. In terms of Riemann sums, it corresponds to the following equality

$$\begin{aligned}\sum_{j,\,k=0}^{n-1} f(c_{jk})\,\Delta x\,\Delta y &= \sum_{j=0}^{n-1}\left(\sum_{k=0}^{n-1} f(c_{jk})\,\Delta y\right)\Delta x \\ &= \sum_{k=0}^{n-1}\left(\sum_{j=0}^{n-1} f(c_{jk})\,\Delta x\right)\Delta y\end{aligned}$$

which may be proved more generally as follows: *Let $[a_{jk}]$ be an $n \times n$ matrix, $0 \leq j \leq n-1$, $0 \leq k \leq n-1$. Let $\sum_{j,\,k=0}^{n-1} a_{jk}$ be the sum of the n^2 matrix entries. Then*

$$\begin{aligned}\sum_{j,\,k=0}^{n-1} a_{jk} &= \sum_{j=0}^{n-1}\left(\sum_{k=0}^{n-1} a_{jk}\right) \\ &= \sum_{k=0}^{n-1}\left(\sum_{j=0}^{n-1} a_{jk}\right)\end{aligned} \tag{3}$$

In the first equality, the right-hand side represents summing the matrix entries by rows:

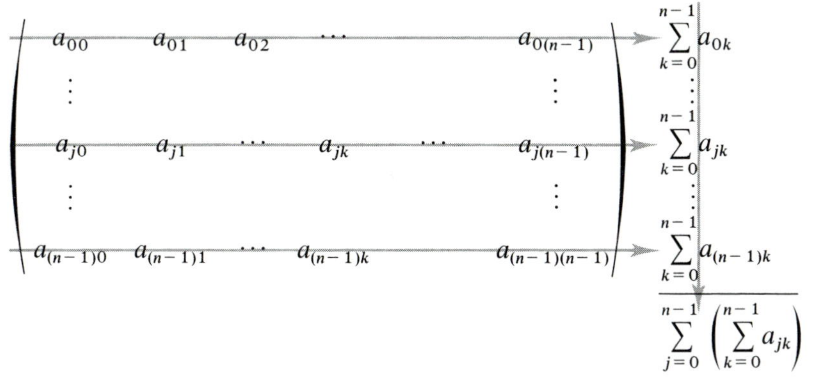

Clearly this is equal to $\sum_{j,k=0}^{n-1} a_{jk}$, that is, the sum of all the a_{jk}'s. Similarly, $\sum_{k=0}^{n-1} (\sum_{j=0}^{n-1} a_{jk})$ represents a summing of the matrix entries by columns. This establishes (3) and makes the reduction to iterated integrals quite plausible if we remember that integrals can be approximated by the corresponding Riemann sums. The actual proof of Fubini's Theorem exploits this idea.

Before we proceed to the proof, it may be helpful to recall how Cavalieri's Principle makes plausible the formula

$$\int_R f(x, y)\, dA = \int_a^b \left[\int_c^d f(x, y)\, dy\right] dx = \int_c^d \left[\int_a^b f(x, y)\, dx\right] dy \quad (4)$$

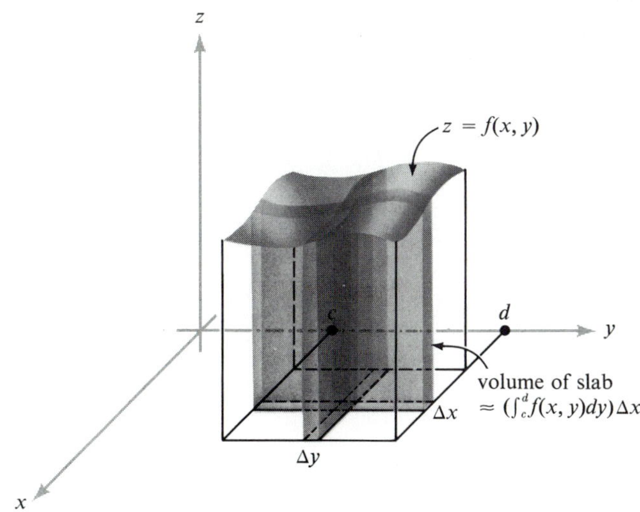

FIGURE 5.2.10
Geometric interpretation of the iterated integral.

If we slice up the volume under the graph of f into slabs parallel to the y-axis, then we can see that the total volume under the graph is approximately equal to the sum of the quantities $\left[\int_c^d f(x, y)\,dy\right]\Delta x$; that is, we have $\int_R f(x, y)\,dA = \int_a^b\left[\int_c^d f(x, y)\,dy\right]dx$. Similarly, the second equality above is proved by slicing the volume into slabs parallel to the x-axis (see Figure 5.2.10).

Theorem 3 (Fubini's Theorem). *Let f be a continuous function with domain a rectangle $R = [a, b] \times [c, d]$. Then*

$$\int_a^b \int_c^d f(x, y)\,dy\,dx = \int_c^d \int_a^b f(x, y)\,dx\,dy = \int_R f(x, y)\,dA \qquad (4')$$

Proof. We shall first show that

$$\int_a^b \int_c^d f(x, y)\,dy\,dx = \int_R f(x, y)\,dA$$

Let $c = y_0 < y_1 < \cdots < y_n = d$ be a partition of $[c, d]$ into n equal parts. Define

$$F(x) = \int_c^d f(x, y)\,dy$$

then

$$F(x) = \sum_{k=0}^{n-1} \int_{y_k}^{y_{k+1}} f(x, y)\,dy$$

Using the integral version of the Mean Value Theorem,* for each fixed x and for each k we have (see Figure 5.2.11)

$$\int_{y_k}^{y_{k+1}} f(x, y)\,dy = f(x, Y_k(x))(y_{k+1} - y_k)$$

where the point $Y_k(x)$ belongs to $[y_k, y_{k+1}]$ and may depend on x and n. We have thus shown that

$$F(x) = \sum_{k=0}^{n-1} f(x, Y_k(x))(y_{k+1} - y_k)$$

Now by the definition of the integral in one variable as a limit of Riemann sums

$$\int_a^b F(x)\,dx = \int_a^b \left[\int_c^d f(x, y)\,dy\right] dx$$

$$= \operatorname*{limit}_{n\to\infty} \sum_{j=0}^{n-1} F(p_j)(x_{j+1} - x_j)$$

where $a = x_0 < x_1 < \cdots < x_n = b$ is a partition of the interval $[a, b]$ into n

* This states that if $g(x)$ is continuous on $[a, b]$, $\int_a^b g(x)\,dx = g(c)(b - a)$ for some point $c \in [a, b]$. We also used this in Section 4.1.

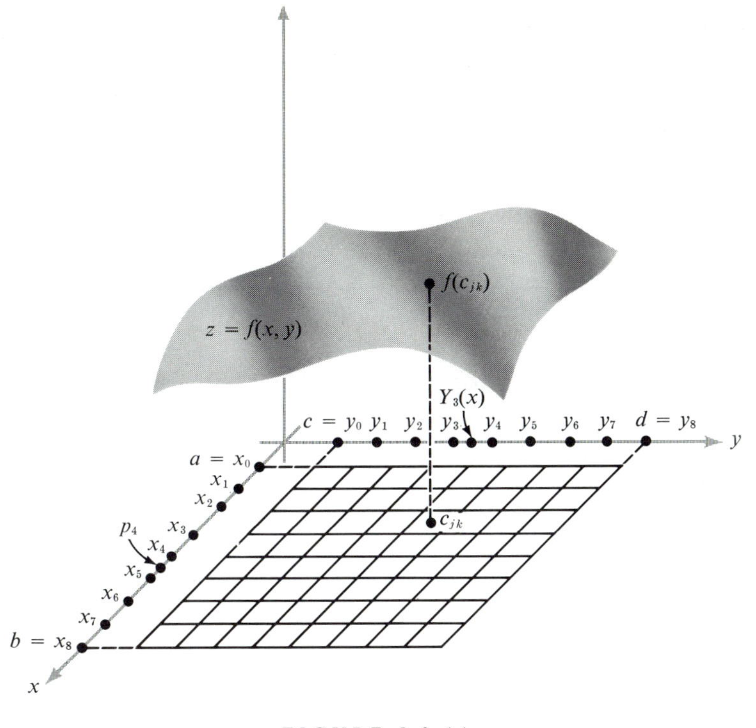

FIGURE 5.2.11
The notation needed in the proof of Fubini's Theorem; $n = 8$.

equal parts and p_j is any point in $[x_j, x_{j+1}]$. Setting $c_{jk} = (p_j, Y_k(p_j)) \in R_{jk}$, we have (substituting p_j for x above)

$$F(p_j) = \sum_{k=0}^{n-1} f(c_{jk})(y_{k+1} - y_k)$$

Therefore

$$\begin{aligned}
\int_a^b \int_c^d f(x, y)\, dy\, dx &= \int_a^b F(x)\, dx \\
&= \lim_{n\to\infty} \sum_{j=0}^{n-1} F(p_j)(x_{j+1} - x_j) \\
&= \lim_{n\to\infty} \sum_{j=0}^{n-1} \sum_{k=0}^{n-1} f(c_{jk})(y_{k+1} - y_k)(x_{j+1} - x_j) \\
&= \int_R f(x, y)\, dA
\end{aligned}$$

Thus we have proved that

$$\int_a^b \int_c^d f(x, y)\, dy\, dx = \int_R f(x, y)\, dA$$

In exactly the same fashion we can show that

$$\int_c^d \int_a^b f(x, y)\, dx\, dy = \int_R f(x, y)\, dA$$

These two conclusions are exactly what we wanted to prove. ▮

Fubini's Theorem can be generalized to the case where f is not necessarily continuous. Although we shall nct present a proof, we state here this more general version.

***Theorem 3′* (*Fubini's Theorem*).** *Let f be a bounded function with domain a rectangle $R = [a, b] \times [c, d]$, and suppose the discontinuities of f form a set of area zero. If*

$$\int_c^d f(x, y)\, dy \quad \textit{exists for each} \quad x \in [a, b]$$

then

$$\int_a^b \left[\int_c^d f(x, y)\, dy\right] dx \quad \textit{exists}$$

and

$$\int_a^b \int_c^d f(x, y)\, dy\, dx = \int_R f(x, y)\, dA$$

Similarly, if

$$\int_a^b f(x, y)\, dx \quad \textit{exists for each} \quad y \in [c, d]$$

then

$$\int_c^d \left[\int_a^b f(x, y)\, dx\right] dy \quad \textit{exists}$$

and

$$\int_c^d \int_a^b f(x, y)\, dx\, dy = \int_R f(x, y)\, dA$$

Thus, if all these conditions hold simultaneously,

$$\int_a^b \int_c^d f(x, y)\, dy\, dx = \int_c^d \int_a^b f(x, y)\, dx\, dy = \int_R f(x, y)\, dA$$

REMARK: The assumptions made for this version of Fubini's Theorem are more complicated than those we made in Theorem 3. They are necessary because if f is not continuous everywhere, for example, there is no guarantee that $\int_c^d f(x, y)\,dy$ will exist for each x.

EXAMPLE 3. Compute $\int_R (x^2 + y)\,dA$ where R is the square $[0, 1] \times [0, 1]$. By Fubini's Theorem

$$\int_R (x^2 + y)\,dA = \int_0^1 \int_0^1 (x^2 + y)\,dx\,dy = \int_0^1 \left[\int_0^1 (x^2 + y)\,dx\right] dy$$

By the Fundamental Theorem of Integral Calculus the x-integration may be performed:

$$\int_0^1 (x^2 + y)\,dx = \left[\frac{x^3}{3} + yx\right]_{x=0}^1 = \frac{1}{3} + y$$

Thus

$$\int_R (x^2 + y)\,dA = \int_0^1 \left[\frac{1}{3} + y\right] dy = \left[\frac{1}{3}y + \frac{y^2}{2}\right]_0^1 = \frac{5}{6}$$

What we have done is hold y fixed, integrate with respect to x, and then evaluate the result between the given limits for the x variable. Next we integrated the remaining function (of y alone) with respect to y to obtain the final answer.

A consequence of Fubini's Theorem is that interchanging the order of integration in the iterated integrals does not change the answer. Let us verify this for the above example. We have

$$\int_0^1 \int_0^1 (x^2 + y)\,dy\,dx = \int_0^1 \left[x^2 y + \frac{y^2}{2}\right]_{y=0}^1 dx = \int_0^1 \left[x^2 + \frac{1}{2}\right] dx$$

$$= \left[\frac{x^3}{3} + \frac{x}{2}\right]_0^1 = \frac{5}{6}$$

We have seen that when $f(x, y) \geq 0$ on $R = [a, b] \times [c, d]$, the integral $\int_R f(x, y)\,dA$ can be interpreted as a volume. If the function also takes on negative values, then the double integral can be thought of as the sum of all volumes lying between the surface $z = f(x, y)$ and the plane $z = 0$, bounded by the planes $x = a$, $x = b$, $y = c$ and $y = d$; here the volumes above $z = 0$ are counted as positive and those below as negative. However, Fubini's Theorem as stated remains valid in the case where $f(x, y)$ is negative or changes sign on R; that is, there is no restriction on the sign of f in the hypotheses of the theorem.

EXAMPLE 4. Let R be the rectangle $[-2, 1] \times [0, 1]$ and let $f(x, y) = y(x^3 - 12x)$; $f(x, y)$ takes on both positive and negative values on R. Evaluate the integral $\int_R f(x, y)\,dx\,dy = \int_R y(x^3 - 12x)\,dx\,dy$.

By Fubini's Theorem, we may write

$$\int_R y(x^3 - 12x)\,dx\,dy = \int_0^1 \left[\int_{-2}^1 y(x^3 - 12x)\,dx\right] dy = \frac{57}{4}\int_0^1 y\,dy = \frac{57}{8}$$

Alternatively, integrating first with respect to y, we find

$$\begin{aligned}\int_R y(x^3 - 12x)\,dy\,dx &= \int_{-2}^1 \left[\int_0^1 (x^3 - 12x)y\,dy\right] dx \\ &= \frac{1}{2}\int_{-2}^1 (x^3 - 12x)\,dx \\ &= \frac{1}{2}\left[\frac{x^4}{4} - 6x^2\right]_{-2}^1 = \frac{57}{8}\end{aligned}$$

EXERCISES

1. Let A and B be two subsets of a rectangle R and suppose each of A and B have area zero. Prove that $A \cup B$ and $A \cap B$ have area zero.
2. Let f be continuous, $f \geq 0$, on the rectangle R. If $\int_R f\,dA = 0$, prove that $f = 0$ on R.
3. Evaluate each of the following integrals if $R = [0, 1] \times [0, 1]$.
   (a) $\int_R (x^3 + y^2)\,dA$
   (b) $\int_R ye^{xy}\,dA$
   (c) $\int_R (xy)^2 \cos x^3\,dA$
   (d) $\int_R (x^m y^n)\,dx\,dy$
   (e) $\int_R (ax + by + c)\,dx\,dy$
   (f) $\int_R \sin(x + y)\,dx\,dy$
4. Compute the volume of the solid bounded by the xz-plane, the yz-plane, the xy-plane, the planes $x = 1$ and $y = 1$, and the surface $z = x^2 + y^4$.
5. Let f be continuous on $[a, b]$ and g continuous on $[c, d]$. Show that $\int_R [f(x)g(y)]\,dx\,dy = \left[\int_a^b f(x)\,dx\right]\left[\int_c^d g(y)\,dy\right]$, where $R = [a, b] \times [c, d]$.
6. Compute the volume of the solid bounded by $z = \sin y$, $0 \leq y \leq \pi/2$, $0 \leq x \leq 1$, and the xy-plane.
7. Compute the volume of the solid bounded by $z = x^2 + y$, the rectangle $R = [0, 1] \times [1, 2]$, and the "vertical sides" of R.

*8. Let $f\colon [0, 1] \times [0, 1] \to \mathbb{R}$ be defined by

$$f(x, y) = \begin{cases} 1 & x \text{ rational} \\ 2y & x \text{ irrational} \end{cases}$$

Show that the iterated integral $\int_0^1 \left[\int_0^1 f(x, y)\, dy\right] dx$ exists but f is not integrable.

9. Let f be continuous on $R = [a, b] \times [c, d]$; for $a < x < b$, $c < y < d$, define

$$F(x, y) = \int_a^x \int_c^y f(u, v)\, dv\, du$$

Show that $\partial^2 F/\partial x\, \partial y = \partial^2 F/\partial y\, \partial x = f(x, y)$. Use this example to discuss the relationship between Fubini's Theorem and the equality of mixed partial derivatives (see Section 2.6).

10. Let B be a subset of a rectangle R and define

$$f(x, y) = \begin{cases} 0 & (x, y) \notin B \\ 1 & (x, y) \in B \end{cases}$$

(a) Indicate why $\int_R f\, dA$ should be interpreted as the area of B.
(b) Prove that if B has area zero then $\int_R f\, dA = 0$.

11. Compute $\int_R \cosh xy\, dx\, dy$ where $R = [0, 1] \times [0, 1]$.

*12. Although Fubini's Theorem holds for most functions met in practice, one must still exercise some caution. It certainly does not hold for all functions. For example, one could divide the unit square into infinitely many rectangles of the form $[1/(m+1), 1/m] \times [1/(n+1), 1/n]$, as in Figure 5.2.12.

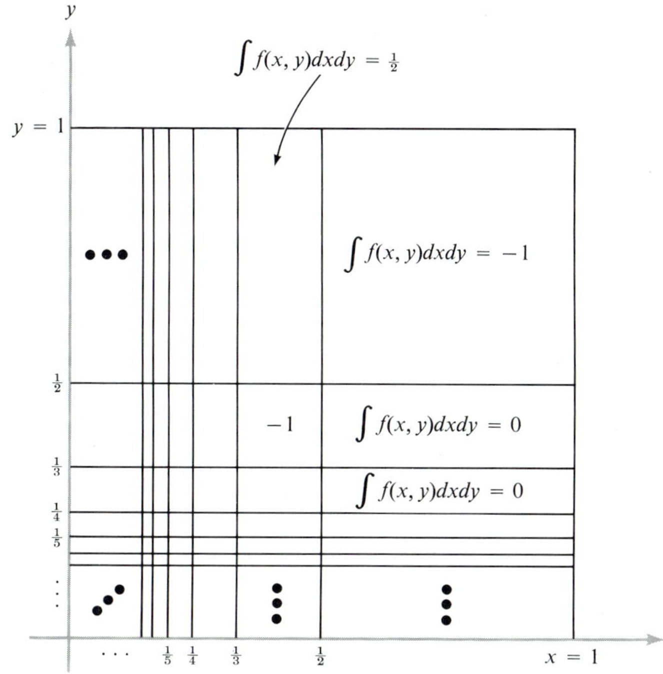

FIGURE 5.2.12
Construction of a function that does not satisfy Fubini's Theorem (Exercise 12).

Define f in such a way that the volume under the graph of f over each rectangle takes on values according to the following table:

$$\begin{array}{ccccccc}
\cdots & \frac{1}{32} & \frac{1}{16} & \frac{1}{8} & \frac{1}{4} & \frac{1}{2} & -1 \\
\cdots & \frac{1}{16} & \frac{1}{8} & \frac{1}{4} & \frac{1}{2} & -1 & 0 \\
\cdots & \frac{1}{8} & \frac{1}{4} & \frac{1}{2} & -1 & 0 & 0 \\
\cdots & \frac{1}{4} & \frac{1}{2} & -1 & 0 & 0 & 0 \\
 & \vdots & \vdots & \vdots & \vdots & \vdots & \vdots
\end{array}$$

Define f to be zero at $(0, 0)$. Each row adds to zero, so adding rows and then columns gives a result of zero. On the other hand the columns add to

$$\cdots \quad -\tfrac{1}{32} \quad -\tfrac{1}{16} \quad -\tfrac{1}{8} \quad -\tfrac{1}{4} \quad -\tfrac{1}{2} \quad -1$$

so adding columns and then rows gives a result of -2. Do the discontinuities have area zero? Why doesn't Fubini's Theorem hold for this function?

5.3 THE DOUBLE INTEGRAL OVER MORE GENERAL REGIONS

Our goal in this section is twofold; first we wish to define the integral $\int_D f(x, y)\, dA$ on regions D more general than rectangles, and second, we want to develop a technique for evaluating this type of integral. To accomplish this, we shall define three special types of subsets of the xy-plane, and then extend the notion of the double integral to them.

Suppose we are given two continuous real-valued functions ϕ_1: $[a, b] \to \mathbb{R}$, ϕ_2: $[a, b] \to \mathbb{R}$ that satisfy $\phi_2(t) \leq \phi_1(t)$ for all $t \in [a, b]$. Let D be the set of all points (x, y) such that

$$x \in [a, b], \qquad \phi_2(x) \leq y \leq \phi_1(x)$$

This region D is said to be of *type 1*. Figure 5.3.1 shows various examples of regions of type 1. The curves and straight line segments that bound the region taken together constitute the *boundary* of D, denoted ∂D.

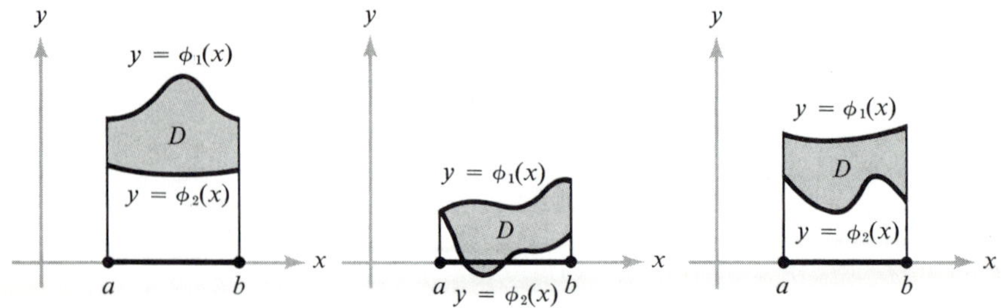

FIGURE 5.3.1
Some regions of type 1.

We say that a region D is of *type 2* if there are continuous functions $\phi_1, \phi_2\colon [c, d] \to \mathbb{R}$ such that D is the set of points (x, y) satisfying

$$y \in [c, d], \quad \phi_2(y) \le x \le \phi_1(y)$$

where $\phi_2(t) \le \phi_1(t)$, $t \in [c, d]$. Again, the curves that bound the region D constitute its boundary ∂D. Some examples of type 2 regions are shown in Figure 5.3.2.

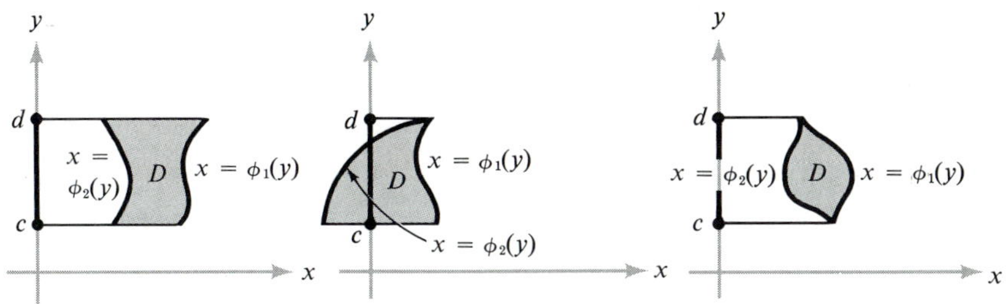

FIGURE 5.3.2
Some regions of type 2.

Finally, a region of *type 3* is one that is both type 1 and type 2; an example of a type 3 region is the unit disc (Figure 5.3.3).

Sometimes we shall refer to regions of types 1, 2, and 3 as *elementary regions*. Note that the boundary ∂D of an elementary region has area zero.

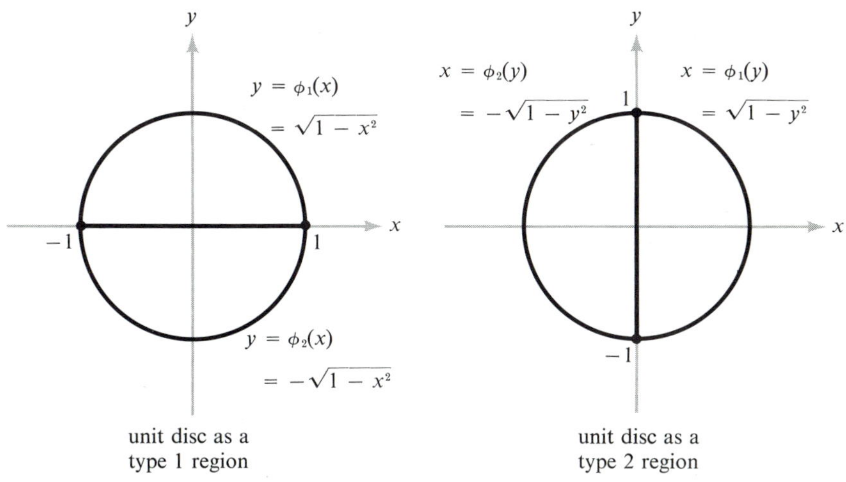

FIGURE 5.3.3
The unit disc, a region of type 3.

Definition. *If D is an elementary region in the plane, we can find a rectangle R that contains D. Assume that we have chosen such an R.*

Given $f\colon D \to \mathbb{R}$, where f is continuous (and hence bounded), we would like to define $\int_D f(x, y)\, dA$, the ***integral of f over the set D****. To do this we "extend" f to a function f^* defined on all of R by*

$$f^*(x, y) = \begin{cases} f(x, y) & (x, y) \in D \\ 0 & (x, y) \notin D \quad \text{and} \quad (x, y) \in R \end{cases}$$

Now f^* is bounded (since f is), and continuous except possibly on the boundary of D (see Figure 5.3.4). The boundary of D has area zero, so f^*

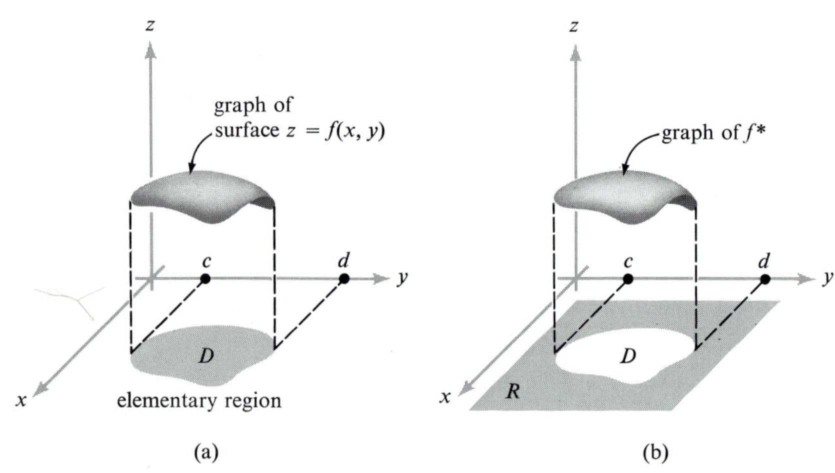

FIGURE 5.3.4
(a) Graph of surface $z = f(x, y)$ over elementary region D. (b) Shaded region shows graph of $z = f^*(x, y)$ on some rectangle R containing D. From this picture we see that boundary points of D may be points of discontinuity of f^*, since the graph of $z = f^*(x, y)$ can be broken at these points.

is integrable over R by Theorem 2, Section 5.2. Therefore we can define

$$\int_D f(x, y)\, dA = \int_R f^*(x, y)\, dA$$

When $f(x, y) \geq 0$ on D, we can interpret the integral $\int_D f(x, y)\, dA$ as the volume of the three-dimensional region between the graph of f and D, as is evident from Figure 5.3.4.

We have defined $\int_D f(x, y)\, dx\, dy$ by choosing a rectangle R that encloses D. It should be intuitively clear that the value of $\int_D f(x, y)\, dx\, dy$ does not depend on the particular R we select; we shall demonstrate this fact at the end of this section.

If $R = [a, b] \times [c, d]$ is a rectangle containing D, we can use the results on iterated integrals in Section 5.2 to obtain

$$\int_D f(x, y)\, dA = \int_R f^*(x, y)\, dA = \int_a^b \int_c^d f^*(x, y)\, dy\, dx$$
$$= \int_c^d \int_a^b f^*(x, y)\, dx\, dy$$

where f^* equals f in D and is zero outside D, as above. Assume D is a region of type 1 determined by functions $\phi_1\colon [a, b] \to \mathbb{R}$, and $\phi_2\colon [a, b] \to \mathbb{R}$. Consider the iterated integral

$$\int_a^b \int_c^d f^*(x, y)\, dy\, dx$$

and, in particular, the inner integral $\int_c^d f^*(x, y)\, dy$ for some fixed x (Figure 5.3.5). Since by definition, $f^*(x, y) = 0$ if $y > \phi_1(x)$ or $y < \phi_2(x)$ we obtain

$$\int_c^d f^*(x, y)\, dy = \int_{\phi_2(x)}^{\phi_1(x)} f^*(x, y)\, dy = \int_{\phi_2(x)}^{\phi_1(x)} f(x, y)\, dy$$

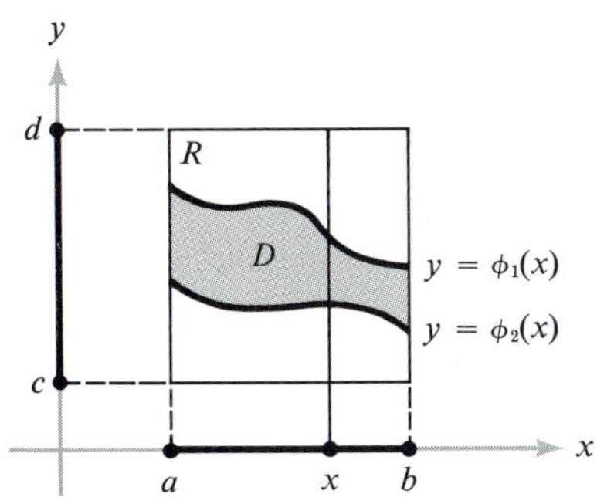

FIGURE 5.3.5
The region between two graphs—a type 1 region.

Thus if D is a region of type 1

$$\int_D f(x, y)\, dA = \int_a^b \int_{\phi_2(x)}^{\phi_1(x)} f(x, y)\, dy\, dx \tag{1}$$

In the case $f(x, y) = 1$ for all $(x, y) \in D$, $\int_D f(x, y)\, dA$ is the area of D. We can check this for formula (1) as follows:

$$\int_a^b \int_{\phi_2(x)}^{\phi_1(x)} f(x, y)\, dy\, dx = \int_a^b [\phi_1(x) - \phi_2(x)]\, dx = A(D)$$

which is the formula for the area of D learned in elementary calculus.

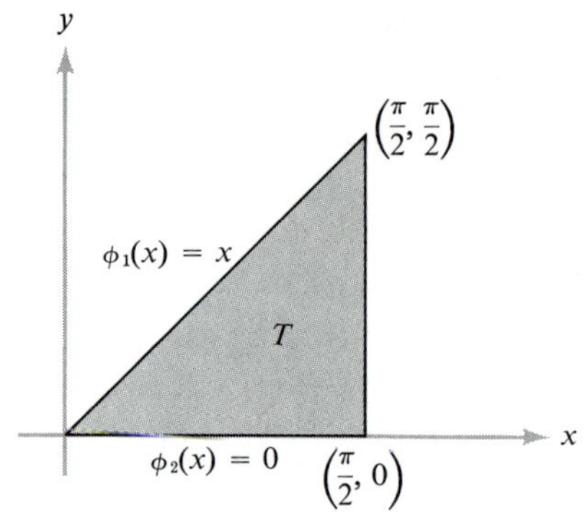

FIGURE 5.3.6
A triangle T represented as a region of type 1.

EXAMPLE 1. Find $\int_T (x^3y + \cos x)\, dA$, where T is the triangle consisting of all points (x, y) such that $0 \le x \le \pi/2$, $0 \le y \le x$. Referring to Figure 5.3.6 and formula (1), we have

$$\begin{aligned}
\int_T (x^3y + \cos x)\, dA &= \int_0^{\pi/2} \int_0^x (x^3y + \cos x)\, dy\, dx \\
&= \int_0^{\pi/2} \left[\frac{x^3y^2}{2} + y \cos x\right]_{y=0}^{x} dx \\
&= \int_0^{\pi/2} \left(\frac{x^5}{2} + x \cos x\right) dx \\
&= \left[\frac{x^6}{12}\right]_0^{\pi/2} + \int_0^{\pi/2} (x \cos x)\, dx \\
&= \frac{\pi^6}{(12)(64)} + [x \sin x + \cos x]_0^{\pi/2} \\
&= \frac{\pi^6}{768} + \frac{\pi}{2} - 1
\end{aligned}$$

In the next example we shall use formula (1) to find the volume of a solid whose base is a nonrectangular region D.

EXAMPLE 2. Find the volume of the tetrahedron bounded by the planes $y = 0$, $z = 0$, $x = 0$, and the plane $y - x + z = 1$ (Figure 5.3.7).

We first note that the given tetrahedron has a triangular base D whose points (x, y) satisfy $-1 \le x \le 0$ and $0 \le y \le 1 + x$; hence D is a region of type 1. (In fact, D is type 3; see Figure 5.3.8.)

For any point (x, y) in D, the height of the surface z above (x, y) is $1 - y + x$. Thus, the volume we seek is given by the integral

$$\int_D (1 - y + x)\, dA$$

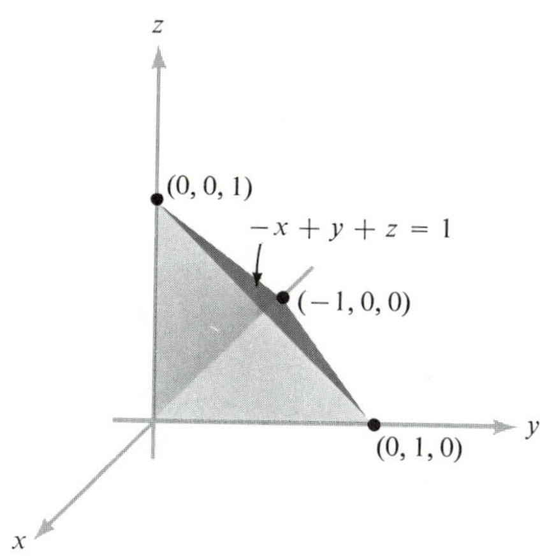

FIGURE 5.3.7
A tetrahedron bounded by the planes $y = 0$, $z = 0$, $x = 0$, and $y - x + z = 1$.

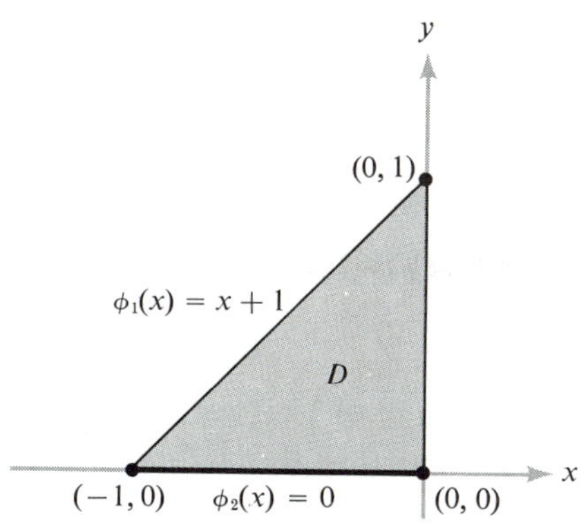

FIGURE 5.3.8
The base of the tetrahedron in Figure 5.3.7 represented as a region of type 1.

Using formula (1) with $\phi_1(x) = x + 1$ and $\phi_2(x) = 0$, we have

$$\begin{aligned}\int_D (1 - y + x)\,dA &= \int_{-1}^{0} \int_{0}^{1+x} (1 - y + x)\,dy\,dx \\ &= \int_{-1}^{0} \left[(1 + x)y - \frac{y^2}{2}\right]_{y=0}^{1+x} dx \\ &= \int_{-1}^{0} \left[\frac{(1 + x)^2}{2}\right] dx \\ &= \left[\frac{(1 + x)^3}{6}\right]_{-1}^{0} = \frac{1}{6}\end{aligned}$$

EXAMPLE 3. Let D be a region of type 1. Describe its area $A(D)$ as a limit of Riemann sums.

If we recall the definition, $A(D) = \int_D dx\, dy$ is the integral over a containing rectangle R of the function $f = 1$. A Riemann sum S_n for this integral is obtained by dividing R into subrectangles and forming the sum $S_n = \sum_{j,k=0}^{n-1} f^*(c_{jk})\, \Delta x\, \Delta y$ as in (1), Section 5.2. Now $f^*(c_{jk})$ is 1 or 0 depending on whether or not c_{jk} is in D. Consider those subrectangles R_{jk} that meet D, and choose c_{jk} in $D \cap R_{jk}$. Thus S_n is the sum of the areas of the subrectangles that meet D and $A(D)$ is the limit of these as $n \to \infty$. Thus $A(D)$ is the limit of the areas of the rectangles "circumscribing" D. The reader should draw a figure to accompany this discussion.

The methods for treating regions of type 2 are entirely analogous. Specifically, if D is the set of points (x, y) such that $y \in [c, d]$, $\phi_2(y) \le x \le \phi_1(y)$, then for f continuous, we have

$$\int_D f(x, y)\, dA = \int_c^d \left[\int_{\phi_2(y)}^{\phi_1(y)} f(x, y)\, dx\right] dy \tag{2}$$

To find the area of D we substitute $f = 1$ in formula (2); this yields

$$\int_D dA = \int_c^d (\phi_1(y) - \phi_2(y))\, dy$$

Note again that this result for area agrees with the results of single-variable calculus for the area of a region between two curves.

Either the method for type 1 or the method for type 2 regions can be used for integrals over regions of type 3.

It also follows from formulas (1) and (2) that $\int_D f$ is independent of the choice of the rectangle R enclosing D used in the definition of $\int_D f$. To see this let us consider the case when D is of type 1. Then formula (1) holds; moreover, on the right side of this formula R does not appear, and thus $\int_D f$ is independent of R.

EXERCISES

1. Evaluate the following iterated integrals and draw the regions D determined by the limits. State whether the regions are of type 1, type 2, or both.
   (a) $\int_0^1 \int_0^{x^2} dy\, dx$
   (b) $\int_1^2 \int_{2x}^{3x+1} dy\, dx$
   (c) $\int_0^1 \int_1^{e^x} (x + y)\, dy\, dx$
   (d) $\int_0^1 \int_{x^3}^{x^2} y\, dy\, dx$

2. Repeat Exercise 1 for the following iterated integrals
   (a) $\int_{-3}^{2} \int_{0}^{y^2} (x^2 + y)\, dx\, dy$
   (b) $\int_{-1}^{1} \int_{-2|x|}^{|x|} e^{x+y}\, dy\, dx$
   (c) $\int_{0}^{1} \int_{0}^{(1-x^2)^{1/2}} dy\, dx$
   (d) $\int_{0}^{\pi/2} \int_{0}^{\cos x} y \sin x\, dy\, dx$
   (e) $\int_{0}^{1} \int_{y^2}^{y} (x^n + y^m)\, dx\, dy, \qquad m, n > 0$
   (f) $\int_{-1}^{0} \int_{0}^{2(1-x^2)^{1/2}} x\, dy\, dx$
3. Use double integrals to compute the area of a circle of radius r.
4. Using double integrals, determine the area of an ellipse with semiaxes of length a and b.
5. What is the volume of a barn that has a rectangular base 20 ft by 40 ft, and vertical walls 30 ft high at the front (which we assume is on the 20-ft side of the barn) and 40 ft high at the rear? The barn has a flat roof.
6. Let D be the region bounded by the positive x- and y-axes and the line $3x + 4y = 10$. Compute

$$\int_D (x^2 + y^2)\, dA$$

7. Let D be the region bounded by the y-axis and the parabola $x = -4y^2 + 3$. Compute

$$\int_D x^3 y\, dx\, dy$$

8. Evaluate $\int_0^1 \int_0^{x^2} (x^2 + xy - y^2)\, dy\, dx$. Describe this iterated integral as a double integral over a certain region D.
9. Let D be the region given as the set of (x, y) where $1 \le x^2 + y^2 \le 2$ and $y \ge 0$. Is D an elementary region? Evaluate $\int_D f(x, y)\, dA$ where $f(x, y) = 1 + xy$.
10. Find the volume of the region inside the surface $z = x^2 + y^2$ and between $z = 0$ and $z = 10$.
11. Find the area enclosed by one period of the sine function $\sin x$, for $0 \le x \le 2\pi$ and the x-axis.
12. Compute the volume of a cone of base radius r and height h.
13. Evaluate $\int_D y\, dA$ where D is the set of points (x, y) such that $0 \le 2x/\pi \le y$, $y \le \sin x$.
14. From Exercise 5, Section 5.2, we know that $\int_a^b \int_c^d f(x)g(y)\, dy\, dx = (\int_a^b f(x)\, dx)(\int_c^d g(y)\, dy)$. Is this true if we integrate $f(x)g(y)$ over any region D (for example, a region of type 1)?
15. Let D be a region given as the set of (x, y) with $-\phi(x) \le y \le \phi(x)$ and $a \le x \le b$, where ϕ is a non-negative continuous function on the interval $[a, b]$. Let $f(x, y)$ be a function on D such that $f(x, y) = -f(x, -y)$ for all $(x, y) \in D$. Argue that $\int_D f(x, y)\, dA = 0$.
16. Use the methods of this section to show that the area of the parallelogram D determined by vectors $\mathbf{a}$ and $\mathbf{b}$ is $|a_1 b_2 - a_2 b_1|$, where $\mathbf{a} = a_1\mathbf{i} + a_2\mathbf{j}$, $\mathbf{b} = b_1\mathbf{i} + b_2\mathbf{j}$.
17. Describe the area $A(D)$ of a region as a limit of areas of inscribed rectangles, as in Example 3.

5.4 CHANGING THE ORDER OF INTEGRATION

Suppose that D is a region of type 3. Thus, being of types 1 and 2, it can be given as the set of points (x, y) such that

$$a \le x \le b, \qquad \phi_2(x) \le y \le \phi_1(x)$$

and also as the set of points (x, y) such that

$$c \le y \le d, \qquad \psi_2(y) \le x \le \psi_1(y)$$

Hence we have the formulas

$$\begin{aligned}\int_D f(x, y)\, dA &= \int_a^b \int_{\phi_2(x)}^{\phi_1(x)} f(x, y)\, dy\, dx \\ &= \int_c^d \int_{\psi_2(y)}^{\psi_1(y)} f(x, y)\, dx\, dy\end{aligned}$$

If we are to compute one of the iterated integrals above, we may do so by evaluating the other iterated integral; this technique is called *changing the order of integration.* It is often useful to make such a change when evaluating iterated integrals, since one of the iterated integrals may be more difficult to compute than the other.

EXAMPLE 1. By changing the order of integration, evaluate

$$\int_0^a \int_0^{(a^2-x^2)^{1/2}} (a^2 - y^2)^{1/2}\, dy\, dx$$

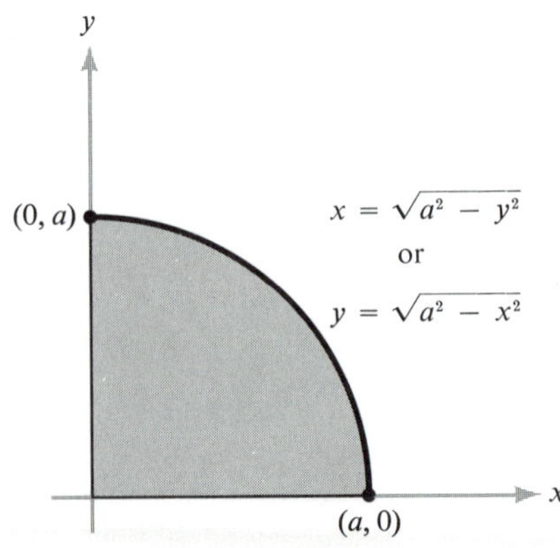

FIGURE 5.4.1
The positive-quadrant portion of a disc of radius a.

Note that x varies between 0 and a, and for fixed x, $0 \le y \le (a^2 - x^2)^{1/2}$. Thus the iterated integral is equivalent to the double integral

$$\int_D (a^2 - y^2)^{1/2}\, dy\, dx$$

where D is the set of points (x, y) such that $0 \le x \le a$ and $0 \le y \le (a^2 - x^2)^{1/2}$. But this is the representation of one quarter (the positive quadrant portion) of the disc of radius a; hence D can also be described as the set of points (x, y) where

$$0 \le y \le a, \qquad 0 \le x \le (a^2 - y^2)^{1/2}$$

(see Figure 5.4.1). Thus

$$\begin{aligned}
&\int_0^a \int_0^{(a^2-x^2)^{1/2}} (a^2 - y^2)^{1/2}\, dy\, dx \\
&= \int_0^a \int_0^{(a^2-y^2)^{1/2}} (a^2 - y^2)^{1/2}\, dx\, dy \\
&= \int_0^a [x(a^2 - y^2)^{1/2}]_{x=0}^{(a^2-y^2)^{1/2}}\, dy \\
&= \int_0^a (a^2 - y^2)\, dy = \left[a^2 y - \frac{y^3}{3}\right]_0^a \\
&= \frac{2a^3}{3}
\end{aligned}$$

We could have evaluated the initial iterated integral directly but, as the reader can easily verify for himself, changing the order of integration makes the problem much simpler computationally.

The next example shows that it may even be "impossible" to evaluate an iterated integral and yet be possible to evaluate the iterated integral obtained by changing the order of integration.

EXAMPLE 2. Evaluate

$$\int_1^2 \int_0^{\log x} (x - 1)\sqrt{1 + e^{2y}}\, dy\, dx$$

First observe that we could not compute this integral in the order given by using the Fundamental Theorem. However, the integral is equal to $\int_D (x - 1)\sqrt{1 + e^{2y}}\, dA$, where D is the set of (x, y) such that

$$1 \le x \le 2 \quad \text{and} \quad 0 \le y \le \log x$$

The region D is of type 3 (see Figure 5.4.2) and so it can be described by

$$0 \le y \le \log 2 \quad \text{and} \quad e^y \le x \le 2$$

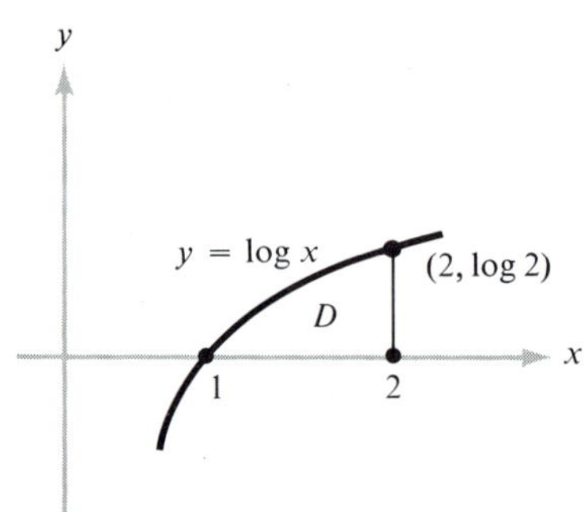

FIGURE 5.4.2
D is the region of integration for Example 2.

Thus the given iterated integral is equal to

$$\begin{aligned}
&\int_0^{\log 2} \int_{e^y}^{2} (x-1)\sqrt{1+e^{2y}}\,dx\,dy \\
&\quad = \int_0^{\log 2} \sqrt{1+e^{2y}} \left[\int_{e^y}^{2} (x-1)\,dx\right] dy \\
&\quad = \int_0^{\log 2} \sqrt{1+e^{2y}} \left[\frac{x^2}{2} - x\right]_{e^y}^{2} dy \\
&\quad = -\int_0^{\log 2} \left(\frac{e^{2y}}{2} - e^y\right)\sqrt{1+e^{2y}}\,dy \\
&\quad = -\tfrac{1}{2}\int_0^{\log 2} e^{2y}\sqrt{1+e^{2y}}\,dy + \int_0^{\log 2} e^y\sqrt{1+e^{2y}}\,dy \qquad (1)
\end{aligned}$$

In the first integral in expression (1) we substitute $u = e^{2y}$ and in the second, $v = e^y$. Hence we obtain

$$-\tfrac{1}{4}\int_1^4 \sqrt{1+u}\,du + \int_1^2 \sqrt{1+v^2}\,dv \qquad (2)$$

Both integrals in (2) are easily found with techniques of calculus (or by consulting the table of integrals at the back of the book). For the first integral we get

$$\begin{aligned}
\tfrac{1}{4}\int_1^4 \sqrt{1+u}\,du &= [\tfrac{1}{6}(1+u)^{3/2}]_1^4 = \tfrac{1}{6}[(1+4)^{3/2} - 2^{3/2}] \\
&= \tfrac{1}{6}[5^{3/2} - 2^{3/2}] \qquad (3)
\end{aligned}$$

The second integral is

$$\begin{aligned}\int_1^2 \sqrt{1+v^2}\,dv &= \tfrac{1}{2}[v\sqrt{1+v^2} + \log(\sqrt{1+v^2}+v)]_1^2 \\ &= \tfrac{1}{2}[2\sqrt{5} + \log(\sqrt{5}+2)] \\ &\quad - \tfrac{1}{2}[\sqrt{2} + \log(\sqrt{2}+1)] \end{aligned} \tag{4}$$

Finally, we subtract (3) from (4) to obtain the answer

$$\frac{1}{2}\left(2\sqrt{5} - \sqrt{2} + \log\frac{\sqrt{5}+2}{\sqrt{2}+1}\right) - \frac{1}{6}[5^{3/2} - 2^{3/2}]$$

To conclude this section we mention an important analog of the Mean Value Theorem of integral calculus.

***Theorem 4* (*Mean Value Theorem for Double Integrals*).** *Suppose $f\colon D \to \mathbb{R}$ is continuous and D is an elementary region. Then for some point (x_0, y_0) in D we have*

$$\int_D f(x, y)\,dA = f(x_0, y_0)\cdot A(D)$$

where $A(D)$ denotes the area of D.

Proof. We cannot prove this theorem with complete rigor because it requires some concepts about continuous functions not proved in this course, but we can sketch the main ideas that underlie the proof.

Since f is continuous on D it has a maximum value M and a minimum value m (proved in advanced calculus). Thus

$$m \le f(x, y) \le M \tag{5}$$

for all $(x, y) \in D$. Furthermore, $f(x_1, y_1) = m$ and $f(x_2, y_2) = M$ for some pairs (x_1, y_1) and (x_2, y_2) in D. From inequality (5) it follows that

$$mA(D) = \int_D m\,dA \le \int_D f(x, y)\,dA \le \int_D M\,dA = MA(D)$$

Therefore dividing through by $A(D)$ we get

$$m \le \frac{1}{A(D)}\int_D f(x, y)\,dA \le M \tag{6}$$

Since a continuous function on D takes on every value between its maximum and minimum values (this is the Intermediate Value Theorem proved in advanced calculus), and the number $(1/A(D))\int_D f(x, y)\,dA$ is by inequality (6) between these values, there must be a point $(x_0, y_0) \in D$ with

$$f(x_0, y_0) = \frac{1}{A(D)}\int_D f(x, y)\,dA$$

But this is precisely the conclusion of Theorem 4. ∎

EXERCISES

1. In the following integrals, change the order of integration, sketch the corresponding regions, and evaluate the integral both ways.
   (a) $\int_0^1 \int_x^1 xy\, dy\, dx$
   (b) $\int_0^{\pi/2} \int_0^{\cos\theta} \cos\theta\, dr\, d\theta$
   (c) $\int_0^1 \int_{2-y}^1 (x+y)^2\, dx\, dy$
   (d) $\int_a^b \int_a^y f(x, y)\, dx\, dy$
2. Find
   (a) $\int_{-1}^1 \int_{|y|}^1 (x+y)^2\, dx\, dy$
   (b) $\int_{-3}^3 \int_{-\sqrt{(9-y^2)}}^{\sqrt{(9-y^2)}} x^2\, dx\, dy$
3. Prove that

$$2\int_a^b \int_x^b f(x)f(y)\, dy\, dx = \left(\int_a^b f(x)\, dx\right)^2$$

4. If $f(x, y) = e^{\sin(x+y)}$ and $D = [-\pi, \pi] \times [-\pi, \pi]$ show that

$$\frac{1}{e} \le \frac{1}{4\pi^2} \int_D f(x, y)\, dA \le e$$

5. Compute $\int_D f(x, y)\, dA$ where $f(x, y) = y^2\sqrt{x}$ and D is the set of (x, y) where $x > 0$, $y > x^2$, $y < 10 - x^2$.
6. Compute the volume of an ellipsoid with semiaxes a, b, and c. (HINT: Use symmetry and first find the volume of one half of the ellipsoid.)
7. Find the volume of the region determined by $x^2 + y^2 + z^2 \le 10$, $z \ge 2$.

5.5 IMPROPER INTEGRALS

In the previous sections we defined the notion of the integral for functions of two variables, and stated criteria guaranteeing that f was indeed integrable over a set D. Recall that one of the hypotheses of Theorem 2 (Section 5.2) was that f be bounded. The following example shows how the sum S_n may fail to converge if f is not bounded.

Let R be the unit square $[0, 1] \times [0, 1]$ and let $f\colon R \to \mathbb{R}$ be defined by

$$f(x, y) = \begin{cases} \dfrac{1}{\sqrt{x}} & x \ne 0 \\ 0 & x = 0 \end{cases}$$

Clearly, f is not bounded on R since, as x gets close to zero, f gets arbitrarily large. Let R_{ij} be a regular partition of R and form the sum (1) of Section 5.2

$$S_n = \sum_{i=0}^{n-1} \sum_{j=0}^{n-1} f(c_{ij})\, \Delta x\, \Delta y$$

Let R_{11} be the subrectangle that contains $(0, 0)$ (see Figure 5.5.1) and choose some $c_{11} \in R_{11}$. For a fixed n, we can make S_n as large as we please by picking c_{11} closer and closer to $(0, 0)$; hence $\lim_{n\to\infty} S_n$ cannot be *independent* of the choice of the c_{ij}.

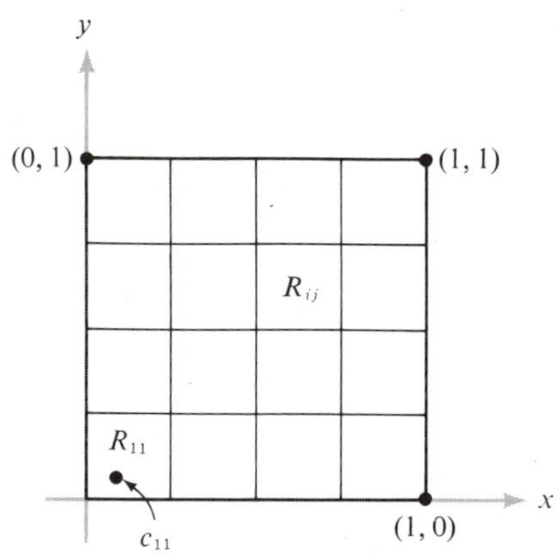

FIGURE 5.5:1
Location of R_{11} *in a partition of* $[0, 1] \times [0, 1]$.

However, let us formally evaluate the iterated integral of f, following the rules for integrating a function of a single variable. We have

$$\int_0^1 \int_0^1 f(x, y)\, dx\, dy = \int_0^1 \int_0^1 \frac{dx}{\sqrt{x}}\, dy = \int_0^1 [2\sqrt{x}]_0^1\, dy$$

$$= \int_0^1 2\, dy = 2$$

Moreover, if we reverse the order of integration we also obtain

$$\int_0^1 \int_0^1 \frac{dy}{\sqrt{x}}\, dx = 2$$

So in some sense this function is integrable. The question is, in what sense?

Recall from one-variable calculus how the improper integral $\int_0^1 dx/\sqrt{x}$ is treated: $1/\sqrt{x}$ is unbounded on the interval $]0, 1]$, yet $\lim_{\delta \to 0} \int_\delta^1 (dx/\sqrt{x}) = 2$, and we *define* $\int_0^1 (dx/\sqrt{x})$ to be this limit. Similarly, for the two-variable case we shall allow the function to be unbounded at certain points on the boundary of its domain and define the improper integral through a limiting process.

More specifically, suppose the region D is of type 1 and $f: D \to \mathbb{R}$ is continuous and bounded except at certain points on the boundary. Assume that D is described by $a \leq x \leq b$, $\phi_2(x) \leq y \leq \phi_1(x)$. Choose numbers $\delta, \eta > 0$ such that $D_{\eta, \delta}$ is the subset of D consisting of points (x, y) with $a + \eta \leq x \leq b - \eta$, $\phi_2(x) + \delta \leq y \leq \phi_1(x) - \delta$ (Figure 5.5.2), where η and δ are chosen small enough so that $D_{\eta, \delta} \subset D$. (If either $\phi_2(a) = \phi_1(a)$ or $\phi_2(b) = \phi_1(b)$ we must modify this slightly because in this case $D_{\eta, \delta}$ may not be a subset of D (see Example 2).) Since f is continuous and bounded on $D_{\eta, \delta}$ the integral $\int_{D_{\eta,\delta}} f$ exists. We can now ask what happens as the region $D_{\eta, \delta}$ expands to fill the region D, that is, as $(\eta, \delta) \to (0, 0)$.

If

$$\lim_{(\eta, \delta) \to (0, 0)} \int_{D_{\eta,\delta}} f$$

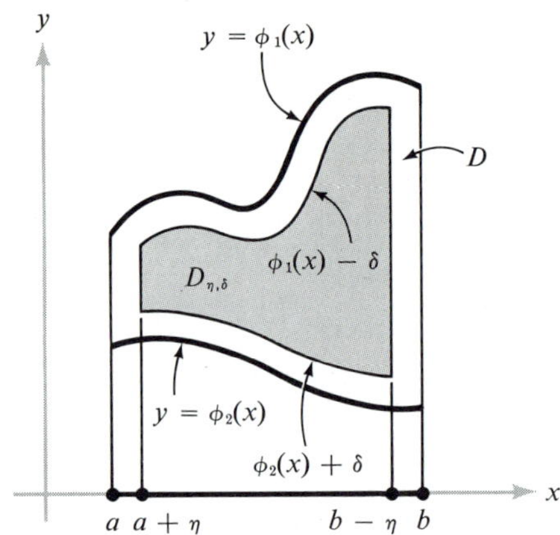

FIGURE 5.5.2
A shrunken domain $D_{\eta,\delta}$ for improper integrals.

exists we define $\int_D f$ to be equal to this limit and say that it is the *improper integral* of f over D. This definition is exactly analogous to the definition of improper integral for a function of one variable.

Since f is integrable over $D_{\eta,\delta}$ we can apply Fubini's Theorem to obtain

$$\int_{D_{\eta,\delta}} f = \int_{a+\eta}^{b-\eta} \int_{\phi_2(x)+\delta}^{\phi_1(x)-\delta} f(x, y)\, dy\, dx \tag{1}$$

So if f is integrable over D

$$\int_D f = \lim_{(\eta,\delta)\to(0,0)} \int_{a+\eta}^{b-\eta} \int_{\phi_2(x)+\delta}^{\phi_1(x)-\delta} f(x, y)\, dy\, dx$$

It may be convenient to work with the iterated limits

$$\lim_{\eta\to 0} \int_{a+\eta}^{b-\eta} \left[\lim_{\delta\to 0} \int_{\phi_2(x)+\delta}^{\phi_1(x)-\delta} f(x, y)\, dy \right] dx \tag{2}$$

if these limits exist. If the limits do exist, then we denote (2) by

$$\int_a^b \int_{\phi_2(x)}^{\phi_1(x)} f(x, y)\, dy\, dx$$

and call it the *iterated improper integral* of f over D. Using more advanced techniques it is possible to show that if $|f|$ is integrable, then if the iterated improper integral exists, it equals $\int_D f$; that is, formula (2) may be used to evaluate the improper integral. Also, if $f \geq 0$, the existence of the limits (2) implies the existence of the double limit defining $\int_D f$; so (2) equals $\int_D f$ in this case. The definition for D a region of type 2 is analogous.

Finally, let us consider the case where D is a region of type 3 and f is unbounded at points on ∂D. For example, suppose D is the set of points (x, y) with

$$a \leq x \leq b, \qquad \phi_2(x) \leq y \leq \phi_1(x)$$

and

$$c \leq y \leq d, \qquad \psi_2(y) \leq x \leq \psi_1(y)$$

If $|f|$ is integrable, and

$$\int_c^d \int_{\psi_2(y)}^{\psi_1(y)} f(x, y)\, dx\, dy \quad \text{and} \quad \int_a^b \int_{\phi_2(x)}^{\phi_1(x)} f(x, y)\, dy\, dx$$

exist, then it can be shown that both iterated integrals are equal and their common value is $\int_D f$. This is Fubini's Theorem for improper integrals.

EXAMPLE 1. Evaluate $\int_D f(x, y)\, dy\, dx$ where $f(x, y) = 1/\sqrt{1 - x^2 - y^2}$ and D is the unit disc $x^2 + y^2 \le 1$.

We can describe D as the set of points (x, y) with $-1 \le x \le 1$, $-\sqrt{1 - x^2} \le y \le \sqrt{1 - x^2}$. Now since ∂D is the set of points (x, y) with $x^2 + y^2 = 1$, f is undefined at every point on ∂D, since at such points the denominator of f is 0. We calculate the iterated improper integrals and obtain

$$\begin{aligned}\int_{-1}^{1} \int_{-\sqrt{(1-x^2)}}^{\sqrt{(1-x^2)}} \frac{dy\, dx}{\sqrt{1 - x^2 - y^2}} &= \int_{-1}^{1} \left[\sin^{-1}\left(\frac{y}{\sqrt{1 - x^2}}\right)\right]_{-\sqrt{(1-x^2)}}^{\sqrt{(1-x^2)}} dx \\ &= \int_{-1}^{1} [\sin^{-1}(1) - \sin^{-1}(-1)]\, dx \\ &= \pi \int_{-1}^{1} dx = 2\pi\end{aligned}$$

In this example we used the fact, stated above, that the iterated improper integral is equal to the improper integral of $1/(\sqrt{1 - x^2 - y^2})$ over the unit disc.

EXAMPLE 2. Let $f(x, y) = 1/(x - y)$ and let D be the set of (x, y) with $0 \le x \le 1$ and $0 \le y \le x$. Show that f is not integrable over D.

Since the denominator of f is zero on the line $y = x$, f is unbounded on part of the boundary of D. Let $0 < \eta < 1$ and $0 < \delta < \eta$ and let $D_{\eta,\delta}$ be the set of (x, y) with $\eta \le x \le 1 - \eta$ and $\delta \le y \le x - \delta$ (Figure 5.5.3). We choose $\delta < \eta$ to guarantee that $D_{\eta,\delta}$ is contained in D. Consider

$$\begin{aligned}\int_{D_{\eta,\delta}} f &= \int_{0+\eta}^{1-\eta} \int_{\delta}^{x-\delta} \frac{1}{x - y}\, dy\, dx \\ &= \int_{\eta}^{1-\eta} [-\log(x - y)]_{\delta}^{x-\delta}\, dx \\ &= \int_{\eta}^{1-\eta} [-\log(\delta) + \log(x - \delta)]\, dx \\ &= [-\log \delta] \int_{\eta}^{1-\eta} dx + \int_{\eta}^{1-\eta} \log(x - \delta)\, dx \\ &= -(1 - 2\eta)\log \delta + [(x - \delta)\log(x - \delta) - (x - \delta)]_{\eta}^{1-\eta}\end{aligned}$$

In the last step we used the fact that $\int \log u\, du = u \log u - u$. Continuing the above set of equalities, we have

$$\begin{aligned}&= -(1 - 2\eta)\log \delta + (1 - \eta - \delta)\log(1 - \eta - \delta) \\ &\quad - (1 - \eta - \delta) - (\eta - \delta)\log(\eta - \delta) + (\eta - \delta)\end{aligned}$$

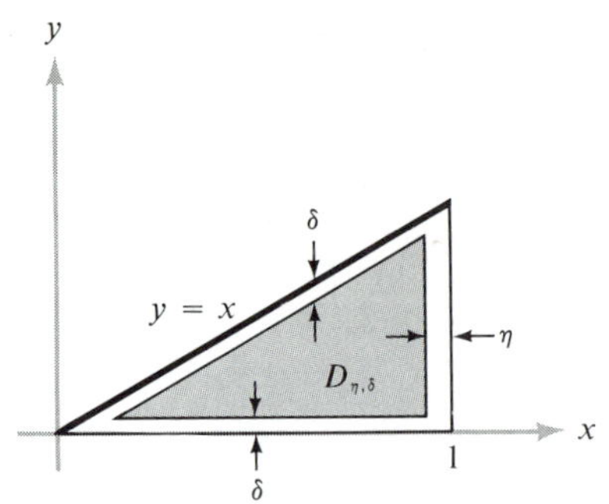

FIGURE 5.5.3
The shrunken domain $D_{\eta,\delta}$ for a triangular domain D.

As $(\eta, \delta) \to (0, 0)$ the second term converges to $1 \log 1 = 0$, and the third and fifth terms converge to 1 and 0, respectively. Let $v = \eta - \delta$. Since $v \log v \to 0$ as $v \to 0$ (a limit established in calculus) we see that the fourth term goes to zero as $(\eta, \delta) \to (0, 0)$. It is the first term which will give us trouble. Now

$$-(1 - 2\eta)\log \delta = -\log \delta + 2\eta \log \delta \tag{3}$$

and it is not hard to see that this does not converge as $(\eta, \delta) \to (0, 0)$. For example, let $\eta = 2\delta$; then (3) becomes

$$-\log \delta + 4\delta \log \delta$$

As before, $4\delta \log \delta \to 0$ as $\delta \to 0$ but $-\log \delta \to +\infty$ as $\delta \to 0$, which shows that (3) does not converge. Hence $\lim_{(\eta, \delta)\to(0, 0)} \int_{D_{\eta,\delta}} f$ does not exist and f is not integrable.

It is important to consider improper integrals because they do arise from natural problems. For example, as we shall see later, one of the formulas for computing the surface area of a hemisphere forces us to consider the improper integral of Example 1.

EXERCISES

1. Evaluate the following integrals if they exist.

   (a) $\displaystyle\int_D \frac{1}{\sqrt{xy}}\, dA$, $D = [0, 1] \times [0, 1]$

   (b) $\displaystyle\int_D \frac{1}{\sqrt{|x - y|}}\, dx\, dy$, $D = [0, 1] \times [0, 1]$

   (HINT: Divide D into two pieces.)

   (c) $\int_D y/x\, dx\, dy$, D bounded by $x = 1$, $x = y$, and $x = 2y$

   (d) $\int_0^1 \int_0^{e^y} \log x\, dx\, dy$

2. (a) Discuss how you would define $\int_D f$ if D is an unbounded region, for example, the set of (x, y) such that $a \le x < \infty$ and $\phi_2(x) \le y \le \phi_1(x)$, where $\phi_2 \le \phi_1$ are given (Figure 5.5.4).
   (b) Evaluate $\int_D xye^{-(x^2+y^2)}\,dx\,dy$ if $x \ge 0$, $0 \le y \le 1$.

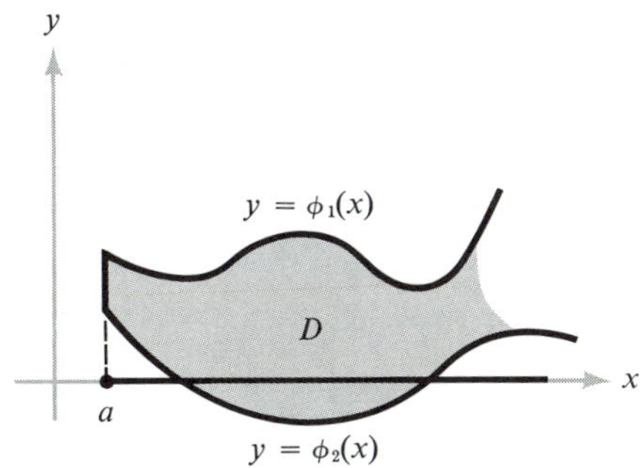

FIGURE 5.5.4
An unbounded region D.

3. Use Exercise 2 to integrate e^{-xy} for $x \ge 0$, $1 \le y \le 2$ in two ways (assume Fubini's Theorem) to show

$$\int_0^\infty \frac{e^{-x} - e^{-2x}}{x}\,dx = \log 2$$

4. Show that the integral $\int_0^1 \int_0^a (x/\sqrt{a^2 - y^2})\,dy\,dx$ exists, and compute its value.

*5. Discuss whether the integral

$$\int_D \frac{x + y}{x^2 + 2xy + y^2}\,dx\,dy$$

exists where $D = [0, 1] \times [0, 1]$. If it exists compute its value.

6. Let f be a non-negative function that may be unbounded and discontinuous on the boundary of an elementary region D. Let g be a similar function such that $f(x, y) \le g(x, y)$ whenever both are defined. Suppose $\int_D g(x, y)\,dA$ exists. Argue informally that this implies the existence of $\int_D f(x, y)\,dA$.

7. Use Exercise 6 to show that

$$\int_D \frac{\sin^2(x - y)}{\sqrt{1 - x^2 - y^2}}\,dy\,dx$$

exists where D is the unit disc $x^2 + y^2 \le 1$.

8. Let f be as in Exercise 6 and let g be a function such that $0 \le g(x, y) \le f(x, y)$ whenever both are defined. Suppose that $\int_D g(x, y)\,dA$ does not exist. Argue informally that $\int_D f(x, y)\,dA$ cannot exist.

9. Use Exercise 8 to show that

$$\int_D \frac{e^{x^2+y^2}}{x - y}\,dy\,dx$$

does not exist, where D is the set of (x, y) with $0 \le x \le 1$ and $0 \le y \le x$.

5.6 THE TRIPLE INTEGRAL

Given a continuous function $f: C \to \mathbb{R}$, where C is some rectangular parallelepiped in $\mathbb{R}^3$, we can define the integral of f over C as a limit of sums just as we did for a function of two variables. Briefly, we partition the three sides of C into n equal parts and form the sum

$$S_n = \sum_{i=0}^{n-1} \sum_{j=0}^{n-1} \sum_{k=0}^{n-1} f(c_{ijk})\, \Delta V$$

where $c_{ijk} \in C_{ijk}$, the ijkth rectangular parallelepiped (or box) in the partition of C, and ΔV is the volume of C_{ijk} (see Figure 5.6.1).

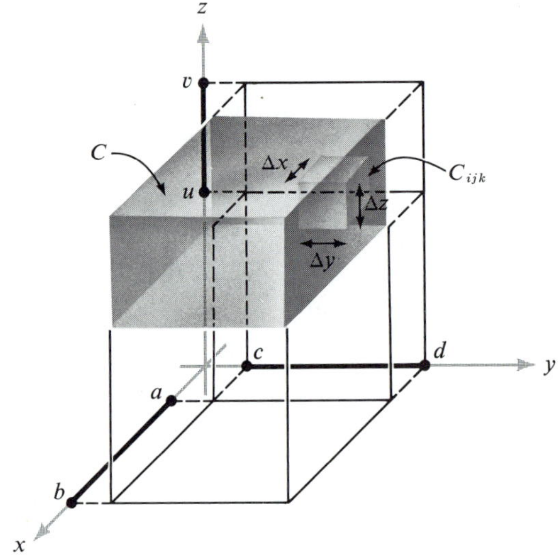

FIGURE 5.6.1
A partition of a box C into n^3 sub-boxes C_{ijk}.

Definition. *Let f be a bounded function of three variables defined on C. If* $\lim_{n\to\infty} S_n$ *exists and the limit is independent of the points c_{ijk}, we call the limit of S_n the* ***triple integral*** *(or simply the integral) of f over C and denote it by*

$$\int_C f, \qquad \int_C f(x, y, z)\, dV, \qquad \int_C f(x, y, z)\, dx\, dy\, dz,$$

$$\text{or} \quad \iiint_C f(x, y, z)\, dx\, dy\, dz$$

As before, one can prove that continuous functions defined on C are integrable.

Suppose the rectangular parallelepiped C is the Cartesian product $[a, b] \times [c, d] \times [u, v]$. Then by analogy with functions of two variables, there are various iterated integrals we can consider, namely

$$\int_u^v \int_c^d \int_a^b f(x, y, z)\, dx\, dy\, dz, \qquad \int_u^v \int_a^b \int_c^d f(x, y, z)\, dy\, dx\, dz,$$

$$\int_a^b \int_u^v \int_c^d f(x, y, z)\, dy\, dz\, dx, \text{ etc.}$$

The order of dx, dy, and dz indicates how the integration is carried out. For example, the first integral above stands for

$$\int_u^v \left[\int_c^d \left(\int_a^b f(x, y, z)\, dx\right) dy\right] dz$$

As in the two-variable case, Fubini's Theorem is valid: if f is continuous then the six possible iterated integrals are all equal. In other words, a triple integral may be reduced to a threefold iterated integration.

To complete the analogy with the double integral, consider the problem of evaluating triple integrals over general regions. For bounded sets $W \subset R^3$ (that is, those sets which can be contained in some box) such that ∂W has "volume zero" (we shall leave the definition of this to the student; see Exercise 12), every continuous function f: $W \to \mathbb{R}$ is integrable via the same type of construction used in the two-dimensional case. That is, extend f to a function f^* that agrees with f on W and is zero outside of W. If B is a box containing W we define

$$\int_W f(x, y, z)\, dV = \int_B f^*(x, y, z)\, dV$$

As in the two-dimensional case, this integral is independent of the choice of B.

As in the two-variable case, we shall restrict our attention to regions of special types. A region W is of type 1 if it can be described as the set of all (x, y, z) such that

$$a \le x \le b, \quad \phi_2(x) \le y \le \phi_1(x) \quad \text{and} \quad \gamma_2(x, y) \le z \le \gamma_1(x, y) \qquad (1)$$

In this definition, γ_i: $D \to \mathbb{R}$, $i = 1, 2$, are continuous functions, D is a region of type 1, and $\gamma_1(x, y) = \gamma_2(x, y)$ implies $(x, y) \in \partial D$. The last condition means that the surfaces $z = \gamma_1(x, y)$ and $z = \gamma_2(x, y)$, if they intersect at all, do so only for $(x, y) \in \partial D$.

A three-dimensional region will also be said to be of type 1 if it can be expressed as the set of all (x, y, z) such that

$$c \le y \le d, \quad \psi_2(y) \le x \le \psi_1(y) \quad \text{and} \quad \gamma_2(x, y) \le z \le \gamma_1(x, y) \quad (2)$$

where $\gamma_i\colon D \to \mathbb{R}$ are as above and D is a two-dimensional region of type 2. Figure 5.6.2 shows two regions of type 1 that are described by conditions (1) and (2) respectively.

A region W is of type 2 if it can be expressed in the form (1) or (2) with the roles of x and z interchanged, and W is of type 3 if it can be

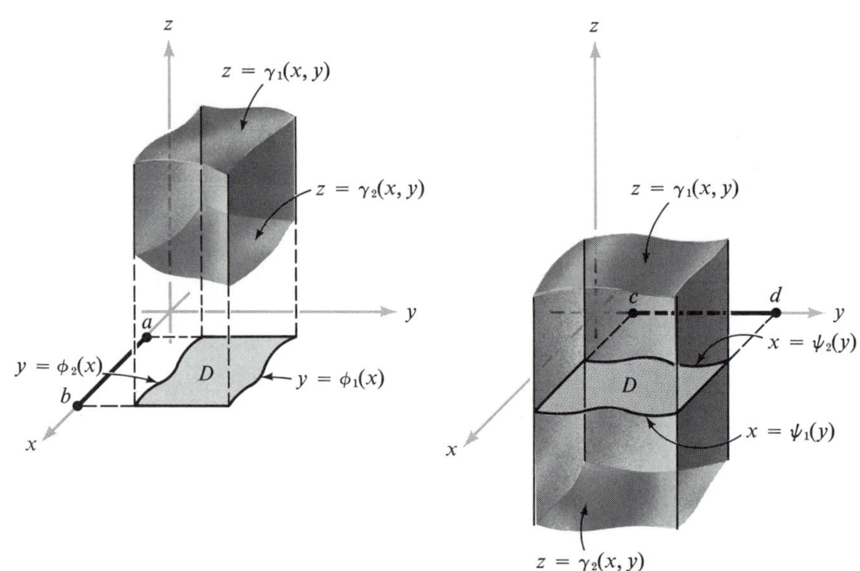

FIGURE 5.6.2
Some regions of type 1 in space.

expressed in the form (1) or (2) with y and z interchanged. A region W that is of type 1, 2, and 3 is said to be of type 4 (Figure 5.6.3). An example of a type 4 region is the ball of radius r, $x^2 + y^2 + z^2 \le r^2$.

Suppose W is of type 1. Then either

$$\begin{aligned}\int_W f(x, y, z)\, dV &= \int_a^b \int_{\phi_2(x)}^{\phi_1(x)} \int_{\gamma_2(x, y)}^{\gamma_1(x, y)} f(x, y, z)\, dz\, dy\, dx \\ &= \int_D \left[\int_{\gamma_2(x, y)}^{\gamma_1(x, y)} f(x, y, z)\, dz\right] dy\, dx \qquad (3)\end{aligned}$$

or

$$\begin{aligned}\int_W f(x, y, z)\, dV &= \int_c^d \int_{\psi_2(y)}^{\psi_1(y)} \int_{\gamma_2(x, y)}^{\gamma_1(x, y)} f(x, y, z)\, dz\, dx\, dy \\ &= \int_D \left[\int_{\gamma_2(x, y)}^{\gamma_1(x, y)} f(x, y, z)\, dz\right] dx\, dy \qquad (4)\end{aligned}$$

according to whether W is defined by (1) or by (2). The proofs of (3) and (4) by Fubini's Theorem are the same as for the two-dimensional case. The student should sketch, or at least try to visualize, the figures associated with (3) and (4). Once all the terms are understood, the formulas are easily remembered. It may help to recall the intuition Cavalieri's Principle provided of Fubini's Theorem.

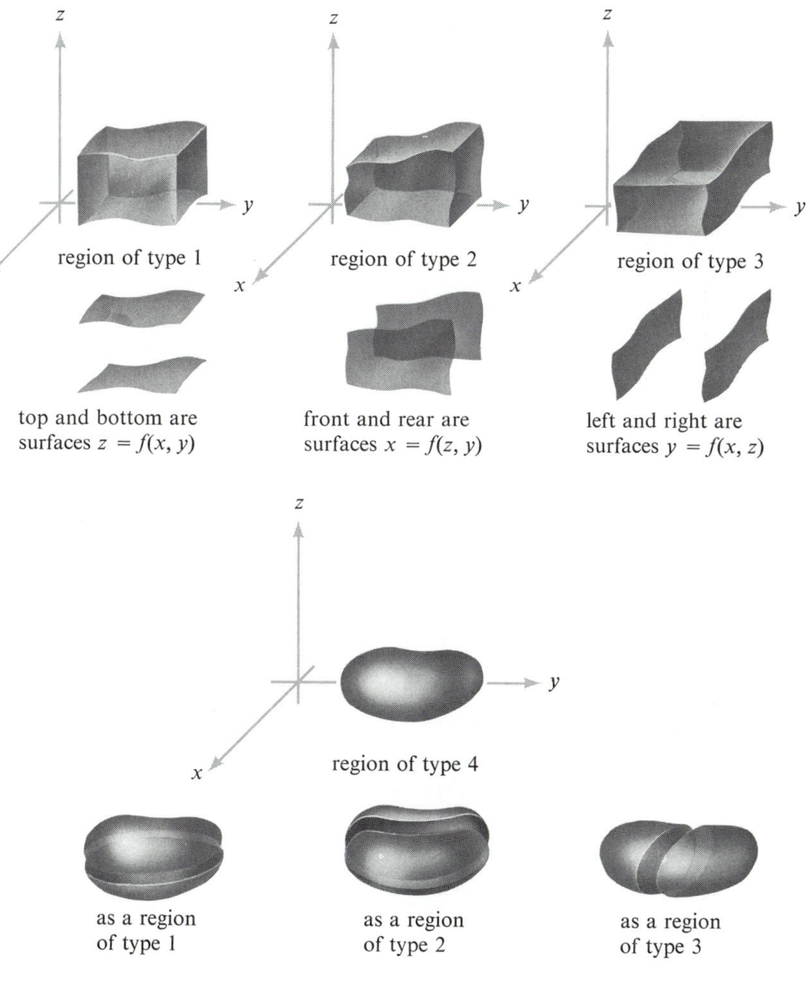

FIGURE 5.6.3
The four possible types of regions in space.

If $f(x, y, z) = 1$ for all $(x, y, z) \in W$, then we obtain

$$\int_W f(x, y, z)\, dV = \int_W 1\, dV = \text{volume } (W)$$

In case W is of type 1 and formula (3) is applicable, we get the formula

$$\text{volume } (W) = \int_a^b \int_{\phi_2(x)}^{\phi_1(x)} \int_{\gamma_2(x,y)}^{\gamma_1(x,y)} dz\, dy\, dx$$

$$= \int_a^b \int_{\phi_2(x)}^{\phi_1(x)} [\gamma_1(x, y) - \gamma_2(x, y)]\, dy\, dx$$

Can you see how to prove this formula from Cavalieri's Principle?

EXAMPLE 1. Verify the formula for the volume of a ball: $\int_W dV = \frac{4}{3}\pi$, where W is the unit ball $x^2 + y^2 + z^2 \leq 1$.

The region W is of type 1; we can describe it as the set of (x, y, z) satisfying

$$-1 \leq x \leq 1, \qquad -\sqrt{1 - x^2} \leq y \leq \sqrt{1 - x^2},$$

and

$$-\sqrt{1 - x^2 - y^2} \leq z \leq \sqrt{1 - x^2 - y^2}$$

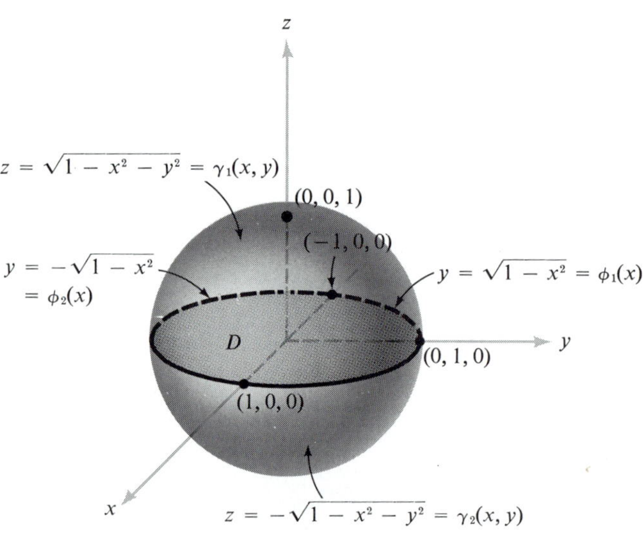

FIGURE 5.6.4
The unit ball expressed as a region of type 1.

(see Figure 5.6.4). Describing W this way is often the most difficult step in the evaluation of a triple integral. Once this has been done appropriately, it remains only to evaluate the given triple integral by using an equivalent iterated integral. In this case we may apply (3) to obtain

$$\int_W dV = \int_{-1}^{1} \int_{-(1-x^2)^{1/2}}^{(1-x^2)^{1/2}} \int_{-(1-x^2-y^2)^{1/2}}^{(1-x^2-y^2)^{1/2}} dz\, dy\, dx$$

Holding y and x fixed and integrating with respect to z yields

$$\int_{-1}^{1}\int_{-(1-x^2)^{1/2}}^{(1-x^2)^{1/2}}\left[z\Big|_{-(1-x^2-y^2)^{1/2}}^{(1-x^2-y^2)^{1/2}}\right] dy\,dx$$

$$= 2\int_{-1}^{1}\left[\int_{-(1-x^2)^{1/2}}^{(1-x^2)^{1/2}}(1-x^2-y^2)^{1/2}\,dy\right] dx$$

Now since x is fixed in the dy-integral, this integral can be expressed as $\int_{-a}^{a}(a^2-y^2)^{1/2}\,dy$, where $a=(1-x^2)^{1/2}$. This integral represents the area of a semicircular region of radius a, so that

$$\int_{-a}^{a}(a^2-y^2)^{1/2}\,dy = \frac{a^2}{2}\pi$$

(Of course we could have evaluated the integral directly by using the table of integrals in Appendix A, but this trick saves quite a bit of effort.) Thus

$$\int_{-(1-x^2)^{1/2}}^{(1-x^2)^{1/2}}(1-x^2-y^2)\,dy = \frac{1-x^2}{2}\pi$$

and so

$$2\int_{-1}^{1}\int_{-(1-x^2)^{1/2}}^{(1-x^2)^{1/2}}(1-x^2-y^2)^{1/2}\,dy\,dx$$

$$= 2\int_{-1}^{1}\pi\frac{1-x^2}{2}\,dx = \pi\int_{-1}^{1}(1-x^2)\,dx$$

$$= \pi\left[x-\frac{x^3}{3}\right]_{-1}^{1} = \frac{4}{3}\pi$$

EXAMPLE 2. Let W be the region bounded by the planes $x=0$, $y=0$, $z=2$ and the surface $z=x^2+y^2$, $x\geq 0$, $y\geq 0$. Compute $\int_W x\,dx\,dy\,dz$.

The region W is sketched in Figure 5.6.5. To write this as a region of type 1, let $\gamma_1(x,y)=2$, $\gamma_2(x,y)=x^2+y^2$, $\phi_1(x)=\sqrt{2-x^2}$, $\phi_2(x)=0$, $a=0$, and $b=\sqrt{2}$. Thus, by formula (3), p. 248,

$$\int_W x\,dx\,dy\,dz = \int_0^{\sqrt{2}}\left[\int_0^{(2-x^2)^{1/2}}\left(\int_{x^2+y^2}^{2} x\,dz\right)dy\right]dx$$

$$= \int_0^{\sqrt{2}}\int_0^{(2-x^2)^{1/2}} x(2-x^2-y^2)\,dy\,dx$$

$$= \int_0^{\sqrt{2}} x\left[(2-x^2)^{3/2}-\frac{(2-x^2)^{3/2}}{3}\right]dx$$

$$= \int_0^{\sqrt{2}}\frac{2x}{3}(2-x^2)^{3/2}\,dx = \frac{-2(2-x^2)^{5/2}}{15}\Bigg|_0^{\sqrt{2}}$$

$$= 2\cdot\frac{2^{5/2}}{15} = \frac{8\sqrt{2}}{15}$$

We can also evaluate the integral by writing W as a region of type 2. We see that W can be expressed as the set of (x, y, z) with $\rho_2(z, y) = 0 \le x \le (z - y^2)^{1/2} = \rho_1(z, y)$ and $(z, y) \in D$, where D is the subset of the yz-plane with $0 \le z \le 2$ and $0 \le y \le z^{1/2}$ (see Figure 5.6.5b).

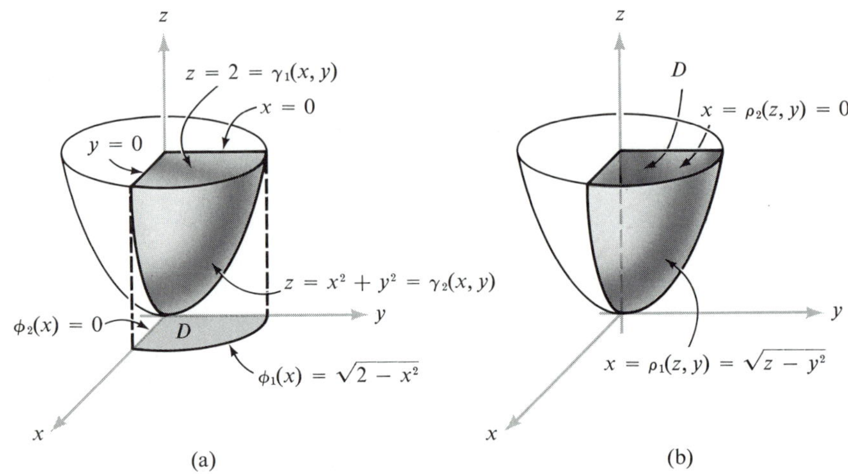

FIGURE 5.6.5
(a) A region of type 1 in $\mathbb{R}^3$ over a domain D. (b) Here D, denoted by the shaded area, is the set of (z, y) with $0 \le z \le 2$ and $0 \le y \le \sqrt{z}$.

Therefore

$$\begin{aligned}
\int_W x\,dx\,dy\,dz &= \int_D \left[\int_{\rho_2(z,y)}^{\rho_1(z,y)} x\,dx\right] dy\,dz \\
&= \int_0^2 \left[\int_0^{z^{1/2}} \left(\int_0^{(z-y^2)^{1/2}} x\,dx\right) dy\right] dz \\
&= \int_0^2 \int_0^{z^{1/2}} \left(\frac{z - y^2}{2}\right) dy\,dz \\
&= \frac{1}{2}\int_0^2 \left(z^{3/2} - \frac{z^{3/2}}{3}\right) dz \\
&= \frac{1}{2}\int_0^2 \frac{2}{3} z^{3/2}\,dz \\
&= \left[\frac{2}{15} z^{5/2}\right]_0^2 = \frac{2}{15}\cdot 2^{5/2} = \frac{8\sqrt{2}}{15}
\end{aligned}$$

EXERCISES

1. Evaluate $\int_W x^2\,dV$ where $W = [0, 1] \times [0, 1] \times [0, 1]$.
2. Evaluate $\int_W e^{-xy} y\,dV$ where $W = [0, 1] \times [0, 1] \times [0, 1]$.

3. Evaluate $\int_W x^2 \cos z \, dV$ where W is the region bounded by the planes $z = 0$, $z = \pi$, $y = 0$, $y = \pi$, $x = 0$, and $x + y = 1$.
4. Find the volume of the region bounded by $z = x^2 + 3y^2$ and $z = 9 - x^2$.
5. Evaluate $\int_0^1 \int_0^{2x} \int_{x+y}^{x^2+y^2} dz \, dy \, dx$ and sketch the region of integration.
6. Find the volume of the solid bounded by the surfaces $x^2 + 2y^2 = 2$, $z = 0$, and $x + y + 2z = 2$.
7. Change the order of integration in

$$\int_0^1 \int_0^x \int_0^y f(x, y, z) \, dz \, dy \, dx$$

to obtain five other forms of the answer. Sketch the region.

8. Find the volume of the region bounded by the surface $x^2 + y^2 + z^2 = 1$ and the region $z^2 \geq x^2 + y^2$.
9. Let f be continuous and let B_ε be the ball of radius ε centered at the point (x_0, y_0, z_0). Let $|B_\varepsilon|$ be the volume of B_ε. Prove that

$$\lim_{\varepsilon \to 0} \frac{1}{|B_\varepsilon|} \int_{B_\varepsilon} f(x, y, z) \, dV = f(x_0, y_0, z_0)$$

10. Find the volume of the region bounded by the surfaces $z = x^2 + y^2$ and $z = 10 - x^2 - 2y^2$. Sketch.
11. Let W be symmetric in the xy-plane: $(x, y, z) \in W$ implies $(x, y, -z) \in W$. Suppose $f(x, y, z) = -f(x, y, -z)$. Prove that

$$\int_W f(x, y, z) \, dV = 0$$

12. Define volume zero by following the definition for area zero. Show that if A and B have volume zero, then $A \cup B$ and $A \cap B$ have volume zero.
13. Use the result of Exercise 11 to prove that $\int_W (1 + x + y) \, dV = 4\pi/3$, where W, the unit ball, is the set of (x, y, z) with $x^2 + y^2 + z^2 \leq 1$.

5.7 THE GEOMETRY OF MAPS FROM $\mathbb{R}^2$ TO $\mathbb{R}^2$

In Chapter 3 we studied vector fields on $\mathbb{R}^2$ and on $\mathbb{R}^3$. Now we wish to investigate these from a somewhat different point of view. We shall be interested in what maps from $\mathbb{R}^2$ to $\mathbb{R}^2$ and $\mathbb{R}^3$ to $\mathbb{R}^3$ do to subsets of these spaces. This geometric understanding will be useful in the next section when we discuss the Change of Variables formula for multiple integrals.

Let D^* be a subset of $\mathbb{R}^2$; suppose we apply a C^1 function or vector field $T: \mathbb{R}^2 \to \mathbb{R}^2$ to D^*. Then T takes points in D^* to points in $\mathbb{R}^2$. We denote this image set of points by D or by $T(D^*)$; hence, $D = T(D^*)$ is the set of all points $(x, y) \in \mathbb{R}^2$ such that

$$(x, y) = T(x^*, y^*) \text{ for some } (x^*, y^*) \in D^*$$

One way to understand the geometry of the map T is to see how it

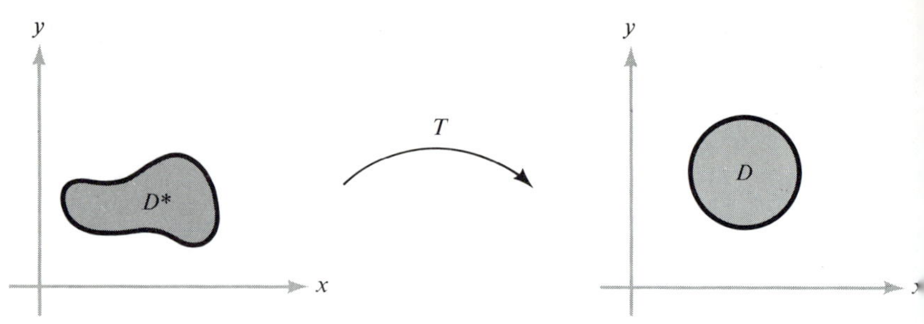

FIGURE 5.7.1
A function T from a domain D^* to a domain D.

deforms or changes D^*. For example, Figure 5.7.1 illustrates a map T that takes a slightly twisted region into a disc.

EXAMPLE 1. Let $D^* \subset \mathbb{R}^2$ be the rectangle $D^* = [0, 1] \times [0, 2\pi]$. Then all points in D^* are of the form (r, θ) where $0 \leq \theta \leq 2\pi$, $0 \leq r \leq 1$. Let T be defined by $T(r, \theta) = (r \cos \theta, r \sin \theta)$. Find the image set D.

We set $(x, y) = (r \cos \theta, r \sin \theta)$. Since $x^2 + y^2 = r^2 \cos^2 \theta + r^2 \sin^2 \theta = r^2 \leq 1$, the set of points $(x, y) \in \mathbb{R}^2$ such that $(x, y) \in D$ has the property that $x^2 + y^2 \leq 1$, and so D is contained in the unit disc. In addition, any point (x, y) in the unit disc can be written as $(r \cos \theta, r \sin \theta)$ for $0 \leq r \leq 1$ and $0 \leq \theta \leq 2\pi$. Thus D is the unit disc (see Figure 5.7.2).

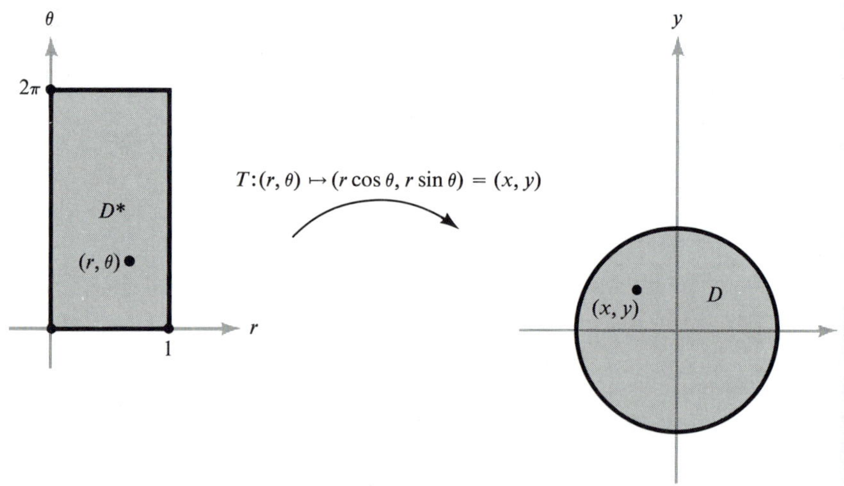

FIGURE 5.7.2
T gives a change of variables to polar coordinates. The unit circle is the image of a rectangle.

EXAMPLE 2. Let T be defined by $T(x, y) = ((x + y)/2, (x - y)/2)$. Let $D^* = [-1, 1] \times [-1, 1] \subset \mathbb{R}^2$ be a square with side of length 2 centered at the origin. Determine the image D obtained by applying T to D^*.

Let us first determine the effect of T on the line $\boldsymbol{\sigma}_1(t) = (t, 1)$, where $-1 \le t \le 1$. We have $T(\boldsymbol{\sigma}_1(t)) = ((t+1)/2, (t-1)/2)$. The map $t \mapsto T(\boldsymbol{\sigma}_1(t))$ is a parametrization of the line $y = x - 1, 0 \le x \le 1$, since $(t-1)/2 = (t+1)/2 - 1$ (see Figure 5.7.3). Let

$$\boldsymbol{\sigma}_2(t) = (1, t), \qquad -1 \le t \le 1$$

$$\boldsymbol{\sigma}_3(t) = (t, -1), \qquad -1 \le t \le 1$$

$$\boldsymbol{\sigma}_4(t) = (-1, t), \qquad -1 \le t \le 1$$

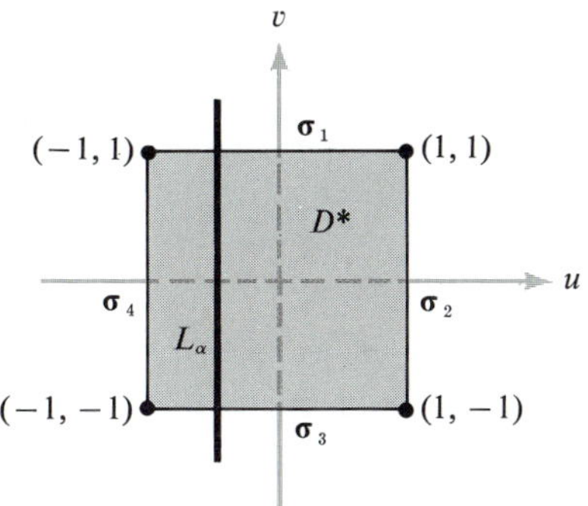

FIGURE 5.7.3
Domain for the transformation T of Example 2.

be parametrizations of the other edges of the square D^*. Using the same argument as above, we see that $T \circ \boldsymbol{\sigma}_2$ is a parametrization of the line $y = 1 - x$, $0 \le x \le 1$, $T \circ \boldsymbol{\sigma}_3$ the line $y = x + 1$, $-1 \le x \le 0$, and $T \circ \boldsymbol{\sigma}_4$ the line $y = -x - 1$, $-1 \le x \le 0$. By this time it seems reasonable to guess that T "flips" the square D^* over and takes it to the square D whose vertices are $(1, 0)$, $(0, 1)$, $(-1, 0)$, $(0, -1)$ (Figure 5.7.4). To prove that this is indeed the case, let $-1 \le \alpha \le 1$ and let L_α

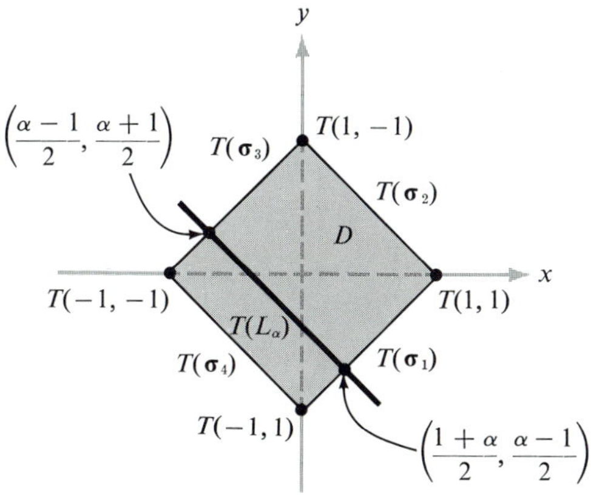

FIGURE 5.7.4
The effect of T on the region D^*.

(Figure 5.7.3) be a fixed line parametrized by $\boldsymbol{\sigma}(t) = (\alpha, t)$, $-1 \le t \le 1$; then $T(\boldsymbol{\sigma}(t)) = ((\alpha + t)/2, (\alpha - t)/2)$ is a parametrization of the line $y = -x + \alpha$, $(\alpha - 1)/2 \le x \le (\alpha + 1)/2$. This line begins, for $t = -1$, at the point $((\alpha - 1)/2, (1 + \alpha)/2)$ and ends up at the point $((1 + \alpha)/2, (\alpha - 1)/2)$; as is easily checked, these points lie on the lines $T \circ \boldsymbol{\sigma}_3$ and $T \circ \boldsymbol{\sigma}_1$, respectively. Thus as α varies between -1 and 1, L_α sweeps out the square D^* while $T(L_\alpha)$ sweeps out the square D determined by the vertices $(-1, 0)$, $(0, 1)$, $(1, 0)$, and $(0, -1)$.

Although we cannot visualize the graph of a function $T\colon \mathbb{R}^2 \to \mathbb{R}^2$, it does help to consider how the function deforms subsets. However, simply looking at these deformations does not give us a complete picture of the behavior of T. We may characterize T further using the notion of a one-to-one correspondence.

Definition. *The function T is **one-to-one** on D^* if $T(u, v) = T(u', v')$ for (u, v) and $(u', v') \in D^*$ implies that $u = u'$ and $v = v'$.*

Geometrically, this statement means that two different points of D^* do not get sent into the same point of D by T. For example, the function $T(x, y) = (x^2 + y^2, y^4)$ is not one-to-one because $T(1, -1) = (2, 1) = T(1, 1)$ yet $(1, -1) \ne (1, 1)$. In other words, a function is one-to-one when it does not collapse two different points together.

EXAMPLE 3. Consider the function $T\colon \mathbb{R}^2 \to \mathbb{R}^2$ of Example 1, $T(r, \theta) = (r \cos \theta, r \sin \theta)$. Show that T is not one-to-one if its domain is all of $\mathbb{R}^2$.

If $\theta_1 \ne \theta_2$, then $T(0, \theta_1) = T(0, \theta_2)$ and so T cannot be one-to-one. This observation implies that if L is the side of the rectangle $D^* = [0, 1] \times [0, 2\pi]$ (Figure 5.7.5), where $0 \le \theta \le 2\pi$ and $r = 0$, then T maps all of L into a single point, the center of the unit disc D.

However, if we consider the set $S^* = \,]0, 1] \times \,]0, 2\pi]$ then $T : S^* \to S$ is one-to-one (see Exercise 1). Evidently, in determining whether a function is one-to-one, the domain chosen must be carefully considered.

EXAMPLE 4. Show that the function $T\colon \mathbb{R}^2 \to \mathbb{R}^2$ of Example 2 is one-to-one.

Suppose $T(x, y) = T(x', y')$; then

$$((x + y)/2, (x - y)/2) = ((x' + y')/2, (x' - y')/2)$$

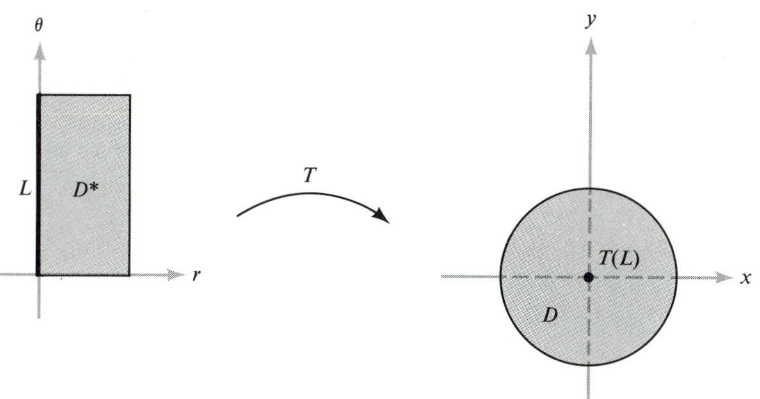

FIGURE 5.7.5
The polar-coordinate transformation T takes the line L to the point (0, 0).

and we have

$$x + y = x' + y'$$
$$x - y = x' - y'$$

Adding, we have

$$2x = 2x'$$

Thus $x = x'$ and hence $y = y'$, which shows that T is one-to-one (with domain all of $\mathbb{R}^2$).

EXERCISES

1. Let $S^* =]0, 1] \times]0, 2\pi]$ and define $T(r, \theta) = (r \cos \theta, r \sin \theta)$. Determine the image set S. Show that T is one-to-one on S^*.
2. Define
$$T(x^*, y^*) = \left(\frac{x^* - y^*}{2}, \frac{x^* + y^*}{2}\right).$$
Show that T rotates the unit square, $D^* = [0, 1] \times [0, 1]$.
3. Let $D^* = [0, 1] \times [0, 1]$ and define T on D^* by $T(u, v) = (-u^2 + 4u, v)$. Find D. Is T one-to-one?
4. Let D^* be the parallelogram bounded by the lines $y = 3x - 4$, $y = 3x$, $y = \frac{1}{2}x$, and $y = \frac{1}{2}(x + 4)$. Let $D = [0, 1] \times [0, 1]$. Find a T such that D is the image of D^* under T.
5. Let $T: \mathbb{R}^3 \to \mathbb{R}^3$ be defined by $(\rho, \phi, \theta) \mapsto (x, y, z)$, where
$$x = \rho \sin \phi \cos \theta, \qquad y = \rho \sin \phi \sin \theta, \qquad z = \rho \cos \phi$$

Let D^* be the set of points (ρ, ϕ, θ) such that $\phi \in [0, \pi]$, $\theta \in [0, 2\pi]$, $\rho \in [0, 1]$. Find $D = T(D^*)$. Is T one-to-one? If not, can we eliminate some subset of D^* (as we did, in Exercise 1, to D^* in Example 1) so that, on the remainder, T will be one-to-one?

6. Let $T(\mathbf{x}) = A\mathbf{x}$, where A is a 2×2 matrix. Show T is one-to-one if and only if the determinant of A is not zero.

5.8 CHANGING VARIABLES IN THE DOUBLE INTEGRAL

Given two regions D and D^* of type 1 or 2 in $\mathbb{R}^2$, a differentiable map T on D^* with image D, that is, $T(D^*) = D$, and any real-valued integrable function $f: D \to \mathbb{R}$, we should like to express $\int_D f(x, y)\, dA$ as an integral over D^* of the composite function $f \circ T$. In this section we shall see how to do this.

Assume the region D^* is a subset of $\mathbb{R}^2$ of type 1 with the coordinate variables designated by (u, v). Furthermore, assume that D is a subset of type 1 of the xy-plane. The map T is given by two coordinate functions

$$T(u, v) = (x(u, v), y(u, v)) \quad \text{for} \quad (u, v) \in D^*$$

Now, as a first guess one might conjecture that

$$\int_D f(x, y)\, dx\, dy \overset{?}{=} \int_{D^*} f(x(u, v), y(u, v))\, du\, dv \tag{1}$$

where $f \circ T(u, v) = f(x(u, v), y(u, v))$ is the composite function defined on D^*. However, if we consider the function $f: D \to \mathbb{R}^2$ where $f(x, y) = 1$, then equation (1) would imply

$$A(D) = \int_D dx\, dy \overset{?}{=} \int_{D^*} du\, dv = A(D^*) \tag{2}$$

It is easy to see that (2) will hold only for a few special cases and not for a general map T. For example, define T by $T(u, v) = (-u^2 + 4u, v)$. Restrict T to the unit square $D^* = [0, 1] \times [0, 1]$ in the uv-plane (see Figure 5.8.1). Then, as in Exercise 3, Section 5.7, T takes D^* onto $D = [0, 3] \times [0, 1]$. Clearly $A(D) \neq A(D^*)$, and so formula (2) is not valid.

What is needed here is a measure of how a transformation $T: \mathbb{R}^2 \to \mathbb{R}^2$ distorts the area of a region. This is given by the *Jacobian determinant*, which is defined as follows.

Definition. *Let $T: D^* \subset \mathbb{R}^2 \to \mathbb{R}^2$ be a C^1 transformation given by*

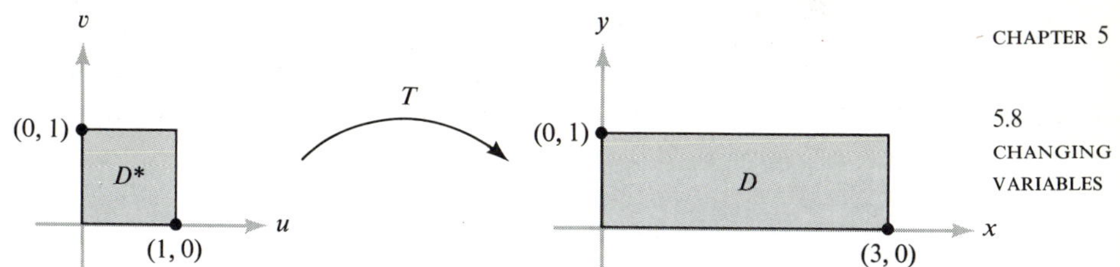

FIGURE 5.8.1
The map T: $(u, v) \mapsto (-u^2 + 4u, v)$ takes the square D^* onto the rectangle D.

*$x = x(u, v)$ and $y = y(u, v)$. The **Jacobian** of T, written $\partial(x, y)/\partial(u, v)$, is the determinant*

$$\begin{vmatrix} \dfrac{\partial x}{\partial u} & \dfrac{\partial x}{\partial v} \\[2ex] \dfrac{\partial y}{\partial u} & \dfrac{\partial y}{\partial v} \end{vmatrix}$$

EXAMPLE 1. The function from $\mathbb{R}^2$ to $\mathbb{R}^2$ that transforms polar coordinates into Cartesian coordinates is given by

$$x = r \cos \theta, \qquad y = r \sin \theta$$

and its Jacobian is

$$\frac{\partial(x, y)}{\partial(r, \theta)} = \begin{vmatrix} \cos \theta & -r \sin \theta \\ \sin \theta & r \cos \theta \end{vmatrix} = r(\cos^2 \theta + \sin^2 \theta) = r$$

If we make suitable restrictions on the function T, we can show that the area of $D = T(D^*)$ is obtained by integrating the absolute value of the Jacobian $\partial(x, y)/\partial(u, v)$ over D^*; that is, we have the equations

$$A(D) = \int_D dx\, dy = \int_{D^*} \left| \frac{\partial(x, y)}{\partial(u, v)} \right| du\, dv \tag{3}$$

To illustrate, from Example 1 in Section 5.7 take $T: D^* \to D$, where $D = T(D^*)$ is the set of (x, y) with $x^2 + y^2 \le 1$ and $D^* = [0, 1] \times [0, 2\pi]$, and $T(r, \theta) = (r \cos \theta, r \sin \theta)$. By formula (3) we have that

$$A(D) = \int_{D^*} \left| \frac{\partial(x, y)}{\partial(r, \theta)} \right| dr\, d\theta = \int_{D^*} r\, dr\, d\theta \tag{4}$$

(here r and θ play the role of u and v). From the above computation it follows that

$$\int_{D^*} r\, dr\, d\theta = \int_0^{2\pi} \int_0^1 r\, dr\, d\theta = \int_0^{2\pi} \left[\frac{r^2}{2} \bigg|_0^1 \right] d\theta = \frac{1}{2} \int_0^{2\pi} d\theta = \pi$$

is the area of D, confirming formula (3) in this case. In fact, we may recall from first-year calculus that (4) is the correct formula for the area of a region in polar coordinates.

It is not easy to prove rigorously assertion (3), that the Jacobian determinant is a measure of how a transformation distorts area. However, looked at in the proper way, it becomes quite plausible.

Recall that $A(D) = \int_D dx\, dy$ was obtained by dividing up D into little rectangles, summing their areas and then taking the limit of this sum as the size of the subrectangles tended to zero (see Example 3, Section 5.3). The problem is that T may map rectangles into regions whose area is not easy to compute. The solution is to approximate these images by simpler regions whose area we can compute. A useful tool for doing this is the derivative of T, which we know (from Chapter 2) gives the best linear approximation to T.

Consider a small rectangle D^* in the uv-plane as shown in Figure 5.8.2. Let T' denote the derivative of T evaluated at (u_0, v_0), so T' is a

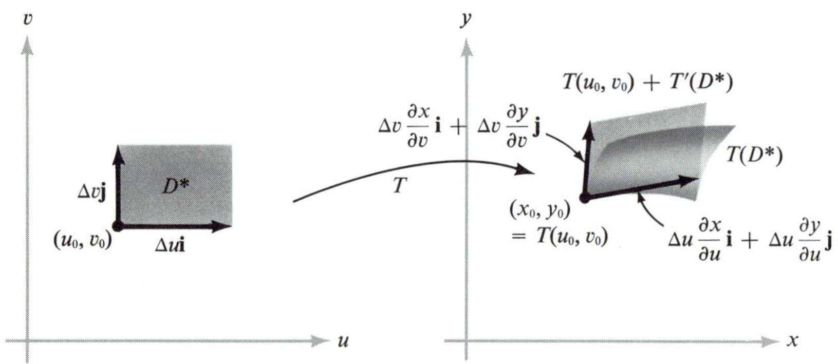

FIGURE 5.8.2
Effect of the transformation T on a small rectangle D^*.

2×2 matrix. From our work in Chapter 2 (see p. 94) we know that a good approximation to $T(u, v)$ is given by

$$T(u_0, v_0) + T' \cdot \begin{pmatrix} \Delta u \\ \Delta v \end{pmatrix}$$

where $\Delta u = u - u_0$ and $\Delta v = v - v_0$. But this mapping takes D^* into a parallelogram with vertex at $T(u_0, v_0)$ and with adjacent sides given by the vectors

$$T'(\Delta u\mathbf{i}) = \begin{bmatrix} \dfrac{\partial x}{\partial u} & \dfrac{\partial x}{\partial v} \\ \dfrac{\partial y}{\partial u} & \dfrac{\partial y}{\partial v} \end{bmatrix} \begin{pmatrix} \Delta u \\ 0 \end{pmatrix} = \Delta u \begin{pmatrix} \dfrac{\partial x}{\partial u} \\ \dfrac{\partial y}{\partial u} \end{pmatrix} = \Delta u \mathbf{T}_u$$

$$T'(\Delta v\mathbf{j}) = \begin{bmatrix} \dfrac{\partial x}{\partial u} & \dfrac{\partial x}{\partial v} \\[2ex] \dfrac{\partial y}{\partial u} & \dfrac{\partial y}{\partial v} \end{bmatrix} \begin{pmatrix} 0 \\ \Delta v \end{pmatrix} = \Delta v \begin{pmatrix} \dfrac{\partial x}{\partial v} \\[2ex] \dfrac{\partial y}{\partial v} \end{pmatrix} = \Delta v \mathbf{T}_v$$

where

$$\mathbf{T}_u = \frac{\partial x}{\partial u}\mathbf{i} + \frac{\partial y}{\partial u}\mathbf{j}$$

$$\mathbf{T}_v = \frac{\partial x}{\partial v}\mathbf{i} + \frac{\partial y}{\partial v}\mathbf{j}$$

are evaluated at (u_0, v_0).

We know from Section 1.3 that the area of the parallelogram with sides equal to the vectors $a\mathbf{i} + b\mathbf{j}$ and $c\mathbf{i} + d\mathbf{j}$ is equal to the absolute value of the determinant

$$\begin{vmatrix} a & b \\ c & d \end{vmatrix} = \begin{vmatrix} a & c \\ b & d \end{vmatrix}$$

Thus the area of $T(D^*)$ is approximately equal to the absolute value of

$$\begin{vmatrix} \dfrac{\partial x}{\partial u}\Delta u & \dfrac{\partial x}{\partial v}\Delta v \\[2ex] \dfrac{\partial y}{\partial u}\Delta u & \dfrac{\partial y}{\partial v}\Delta v \end{vmatrix} = \begin{vmatrix} \dfrac{\partial x}{\partial u} & \dfrac{\partial x}{\partial v} \\[2ex] \dfrac{\partial y}{\partial u} & \dfrac{\partial y}{\partial v} \end{vmatrix} \Delta u\, \Delta v = \frac{\partial(x, y)}{\partial(u, v)}\Delta u\, \Delta v$$

evaluated at (u_0, v_0). But the absolute value of this is just $|\partial(x, y)/\partial(u, v)|\ \Delta u\, \Delta v$.

This fact and a partitioning argument should make formula (3) plausible. Indeed, if we partition D^* into small rectangles with sides of length Δu and Δv, the images of these rectangles are approximated by parallelograms with sides $\mathbf{T}_u\, \Delta u$ and $\mathbf{T}_v\, \Delta v$ and hence with area $|\partial(x, y)/\partial(u, v)|\ \Delta u\, \Delta v$. Thus the area of D^* is approximately $\sum \Delta u\, \Delta v$, where the sum is taken over all the rectangles R inside D^* (see Figure 5.8.3). Hence the area of $T(D^*)$ is approximately the sum $\sum |\partial(x, y)/\partial(u, v)|\ \Delta u\, \Delta v$. In the limit, this sum becomes $\int_{D^*} |\partial(x, y)/\partial(u, v)|\ du\, dv$. The precise technical theorem justifying these manipulations will be given later.

Let us give another argument for the special case (4) of formula (3); that is, the case of polar coordinates. Consider a region D in the xy-plane and a grid corresponding to a partition of the r and θ

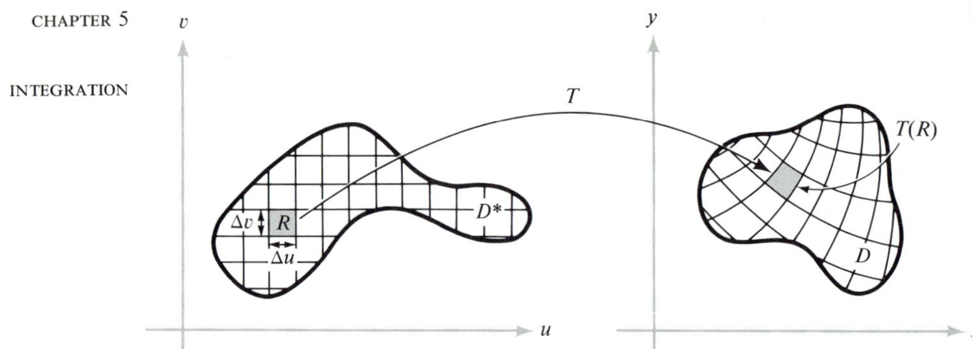

FIGURE 5.8.3
The area of the little rectangle R is $\Delta u\ \Delta v$. The area of $T(R)$ is approximately $|\partial(x, y)/\partial(u, v)|\ \Delta u\ \Delta v$.

variables (Figure 5.8.4). The area of the shaded region shown is approximately $(\Delta r)(r_{jk}\ \Delta\theta)$, since the arc length of a segment of a circle of radius R subtending an angle ϕ is $R\phi$. The total area is then the limit of $\sum r_{jk}\ \Delta r\ \Delta\theta$; that is, $\int_{D*} r\ dr\ d\theta$. The key idea is thus that the jkth "polar rectangle" in the grid has area approximately equal to $r_{jk}\ \Delta r\ \Delta\theta$. (For n large, the jkth polar rectangle will look like a rectangle with sides of length $r_{jk}\ \Delta\theta$ and Δr.) This should provide some

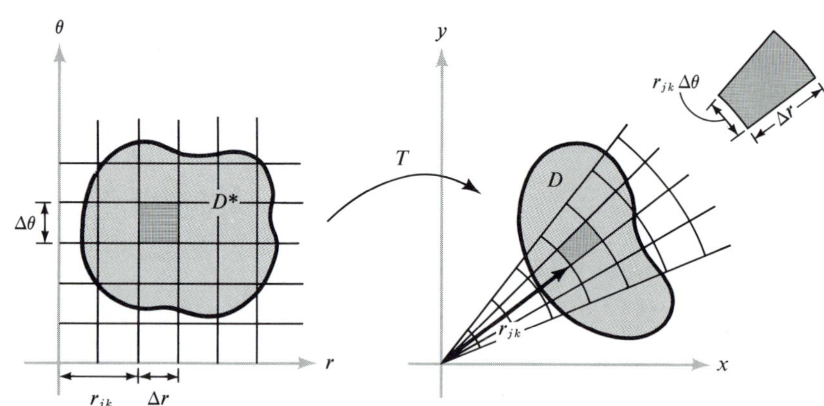

FIGURE 5.8.4
D gets mapped to D under the polar-coordinate mapping T.*

insight into why we say the "area element $\Delta x\ \Delta y$" is transformed into the "area element $r\ \Delta r\ \Delta\theta$." The following example explicates these ideas for a special case.

EXAMPLE 2. Let the elementary region D in the xy-plane be bounded by the graph of a polar equation $r = f(\theta)$ where $\theta_0 \leq \theta \leq \theta_1$ and

$f(\theta) \geq 0$ (see Figure 5.8.5). In the $r\theta$-plane we consider the type-2 region D^* where $\theta_0 \leq \theta \leq \theta_1$ and $0 \leq r \leq f(\theta)$. Under the transformation $x = r \cos \theta$, $y = r \sin \theta$, the region D^* is carried onto the region D.

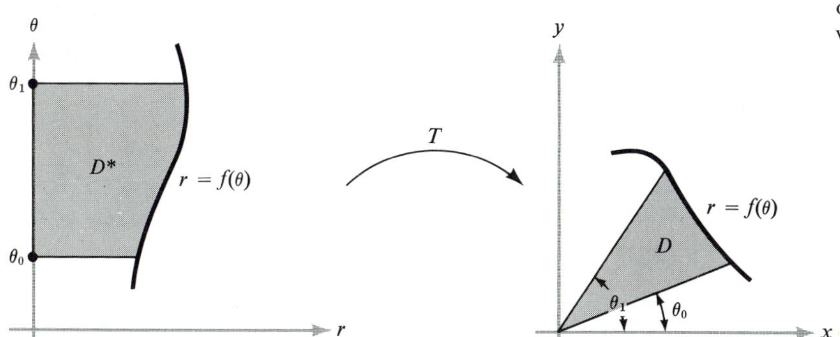

FIGURE 5.8.5
Effect on the region D^* of the polar coordinate mapping.

We have

$$A(D) = \int_D dx\, dy = \int_{D^*} \left| \frac{\partial(x, y)}{\partial(r, \theta)} \right| dr\, d\theta$$

$$= \int_{D^*} r\, dr\, d\theta = \int_{\theta_0}^{\theta_1} \left[\int_0^{f(\theta)} r\, dr \right] d\theta$$

$$= \int_{\theta_0}^{\theta_1} \left[\frac{r^2}{2} \right]_0^{f(\theta)} d\theta = \int_{\theta_0}^{\theta_1} \frac{[f(\theta)^2]}{2}\, d\theta$$

This formula for $A(D)$ may be familiar from one-variable calculus.

Before stating the Change of Variables formula, which is the culmination of the above discussion, let us recall the corresponding result from one-variable calculus. We have

$$\int_a^b f(x(u)) \frac{dx}{du}\, du = \int_{x(a)}^{x(b)} f(x)\, dx \tag{5}$$

where f is continuous and $u \mapsto x(u)$ is continuously differentiable on $[a, b]$.

Proof. Let F be an antiderivative of f; that is, $F' = f$, which is possible by the Fundamental Theorem of Calculus. The right-hand side of (5) becomes

$$\int_{x(a)}^{x(b)} f(x)\, dx = F(x(b)) - F(x(a))$$

To evaluate the left hand side of (5), let $G(u) = F(x(u))$. By the Chain Rule, $G'(u) = F'(x(u))x'(u) = f(x(u))x'(u)$. Hence, again by the Fundamental Theorem,

$$\int_a^b f(x(u))x'(u)\,du = \int_a^b G'(u)\,du = G(b) - G(a) = F(x(b)) - F(x(a))$$

as required. ▮

Suppose now that the C^1 function $u \mapsto x(u)$ is one-to-one on $[a, b]$. Thus, we must have either $dx/du \geq 0$ on $[a, b]$ or $dx/du \leq 0$ on $[a, b]$. (If dx/du is positive and then negative, the function $x = x(u)$ rises and then falls, and thus is not one-to-one; a similar statement applies if dx/du is negative then positive.) Let I^* denote the interval $[a, b]$, and let I denote the closed interval with endpoints $x(a)$ and $x(b)$. (Thus, $I = [x(a), x(b)]$ if $u \mapsto x(u)$ is increasing and $I = [x(b), x(a)]$ if $u \mapsto x(u)$ is decreasing.) With these notations we can rewrite formula (5) as

$$\int_{I^*} f(x(u)) \left| \frac{dx}{du} \right| du = \int_I f(x)\,dx$$

This is the formula that generalizes to double integrals: I^* becomes D^*, I becomes D, and $|dx/du|$ is replaced by $|\partial(x, y)/\partial(u, v)|$. Let us state the result formally (the technical proof is omitted).

Theorem 5 (Change of Variables for double integrals). *Let D and D^* be elementary regions in the plane and let $T: D^* \to D$ be C^1; suppose that T is one-to-one on D^*. Furthermore, suppose that $D = T(D^*)$. Then for any integrable function $f: D \to \mathbb{R}$, we have*

$$\int_D f(x, y)\,dx\,dy = \int_{D^*} f(x(u, v), y(u, v)) \left| \frac{\partial(x, y)}{\partial(u, v)} \right| du\,dv \qquad (6)$$

EXAMPLE 3. Let P be the parallelogram bounded by $y = 2x$, $y = 2x - 2$, $y = x$, and $y = x + 1$ (see Figure 5.8.6). Evaluate $\int_P xy\,dx\,dy$ by making the change of variables

$$x = u - v, \qquad y = 2u - v$$

that is, $T(u, v) = (u - v, 2u - v)$.

The transformation T is one-to-one (see Exercise 6, Section 5.7) and is designed so that it takes the *rectangle* P^* bounded by $v = 0$, $v = -2$, $u = 0$, $u = 1$ onto P. Moreover,

$$\left| \frac{\partial(x, y)}{\partial(u, v)} \right| = \left| \det \begin{bmatrix} 1 & -1 \\ 2 & -1 \end{bmatrix} \right| = 1$$

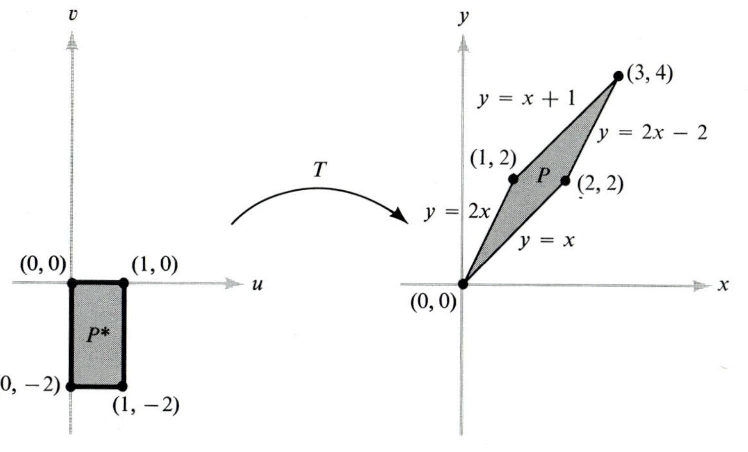

FIGURE 5.8.6
The effect of $T(u, v) = (u - v, 2u - v)$ on the rectangle P^*.

Therefore

$$\begin{aligned}
\int_P xy\,dx\,dy &= \int_{P^*} (u-v)(2u-v)\,du\,dv \\
&= \int_{-2}^{0} \int_{0}^{1} (2u^2 - 3vu + v^2)\,du\,dv \\
&= \int_{-2}^{0} \left[\frac{2}{3}u^3 - \frac{3u^2 v}{2} + v^2 u\right]_0^1 dv \\
&= \int_{-2}^{0} \left[\frac{2}{3} - \frac{3}{2}v + v^2\right] dv \\
&= \left[\frac{2}{3}v - \frac{3}{4}v^2 + \frac{v^3}{3}\right]_{-2}^{0} = -\left[\frac{2}{3}(-2) - 3 - \frac{8}{3}\right] \\
&= -\left[-\frac{12}{3} - 3\right] = 7
\end{aligned}$$

EXAMPLE 4. Evaluate $\int_D \log(x^2 + y^2)\,dx\,dy$, where D is the region in the first quadrant lying between the arcs of the circles

$$x^2 + y^2 = a^2, \qquad x^2 + y^2 = b^2 \qquad (0 < a < b)$$

(see Figure 5.8.7). These circles have equations $r = a$ and $r = b$ in polar coordinates. Therefore the transformation

$$x = r \cos\theta, \qquad y = r \sin\theta$$

sends the rectangle D^* given by $a \le r \le b$, $0 \le \theta \le \frac{1}{2}\pi$ onto the region

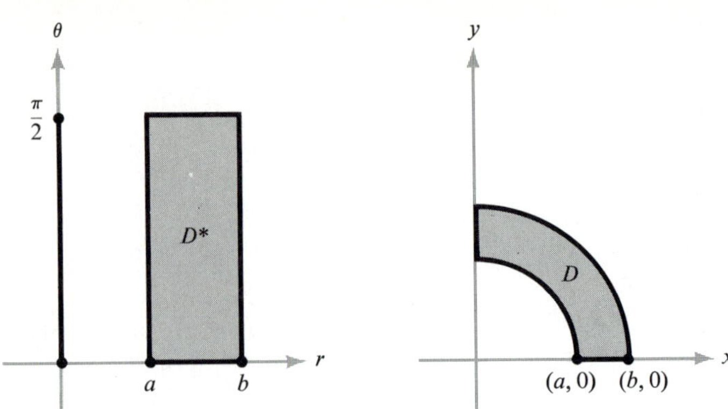

FIGURE 5.8.7
The polar-coordinate mapping takes a rectangle D^* onto part of an annulus D.

D. This transformation is one-to-one on D^* and so, by Theorem 5, we have

$$\int_D \log(x^2 + y^2)\, dx\, dy = \int_{D^*} \log r^2 \left| \frac{\partial(x, y)}{\partial(r, \theta)} \right| dr\, d\theta$$

Now $|\partial(x, y)/\partial(r, \theta)| = r$, as we have seen before; hence the right-hand integral becomes

$$\int_a^b \int_0^{\pi/2} r \log r^2\, d\theta\, dr = \frac{\pi}{2} \int_a^b r \log r^2\, dr = \frac{\pi}{2} \int_a^b 2r \log r\, dr$$

Applying integration by parts, or using the formula

$$\int x \log x\, dx = \frac{x^2}{2} \log x - \frac{x^2}{4}$$

from the table of integrals in Appendix A, we obtain the result

$$\frac{\pi}{2} \int_a^b 2r \log r\, dr = \frac{\pi}{2} \left(b^2 \log b - a^2 \log a - \frac{1}{2}(b^2 - a^2) \right)$$

Suppose we consider the rectangle D^* defined by $0 \le \theta \le 2\pi$, $0 \le r \le a$ in the $r\theta$-plane. Then the transformation T given by $T(r, \theta) = (r \cos \theta, r \sin \theta)$ takes D^* onto the disc D with equation $x^2 + y^2 \le a^2$ in the xy-plane. This transformation represents the change from Cartesian coordinates to polar coordinates. However, T does not satisfy the requirements of the Change of Variables theorem since it is not one-to-one on D^*: in particular, T sends all points with $r = 0$ to $(0, 0)$ (see Figure 5.8.8 and Example 3 of Section 5.7). However, the Change of Variables theorem is still valid in this case. Basically, the reason for this is that the set of points where T is not one-to-one is on

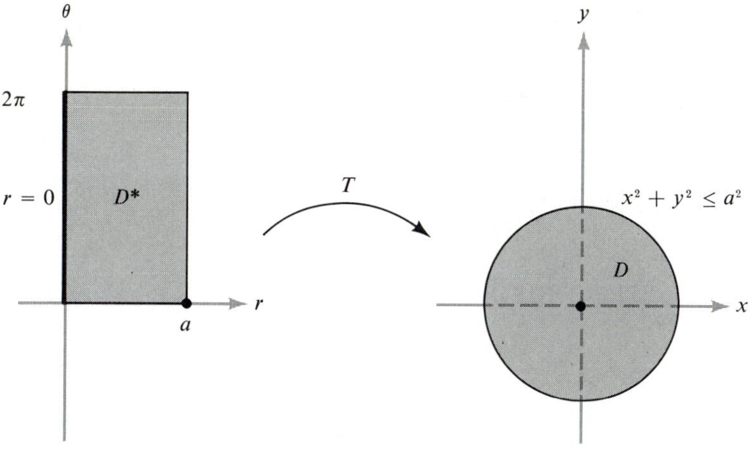

FIGURE 5.8.8
If the rectangle stands against the θ-axis between 0 and 2π, the annulus becomes a disc.

an edge of D^*; this set is of area zero and therefore can be neglected. In summary, the formula

$$\int_D f(x, y)\, dx\, dy = \int_{D^*} f(r \cos \theta, r \sin \theta) r\, dr\, d\theta \tag{7}$$

is valid when T sends D^* onto D in a one-to-one fashion except possibly for points on the edges of D^*. Example 2 provided a simple example of this for the case where $f(x, y)$ is constantly 1. Now we shall consider a more challenging example.

EXAMPLE 5. Evaluate $\int_R \sqrt{x^2 + y^2}\, dx\, dy$ where $R = [0, 1] \times [0, 1]$.

This double integral is equal to the volume of the three-dimensional region shown in Figure 5.8.9. To apply Theorem 5 with polar coordinates, we refer to Figure 5.8.10. We see that R is the image under $T(r, \theta) = (r \cos \theta, r \sin \theta)$ of the region $D^* = D_1^* \cup D_2^*$ where for D_1^* we have $0 \leq \theta \leq \frac{1}{4}\pi$ and $0 \leq r \leq \sec \theta$; for D_2^* we have $\frac{1}{4}\pi \leq \theta \leq \frac{1}{2}\pi$, $0 \leq r \leq \csc \theta$. The transformation T sends D_1^* onto a triangle T_1 and D_2^* onto a triangle T_2. The transformation T is one-to-one except when $r = 0$, so we can apply Theorem 5. From the symmetry of $z = \sqrt{x^2 + y^2}$ on R, we can see that

$$\int_R \sqrt{x^2 + y^2}\, dx\, dy = 2 \int_{T_1} \sqrt{x^2 + y^2}\, dx\, dy$$

Applying formula (7), we obtain

$$\int_{T_1} \sqrt{x^2 + y^2}\, dx\, dy = \int_{D_1*} \sqrt{r^2}\, r\, dr\, d\theta = \int_{D_1*} r^2\, dr\, d\theta$$

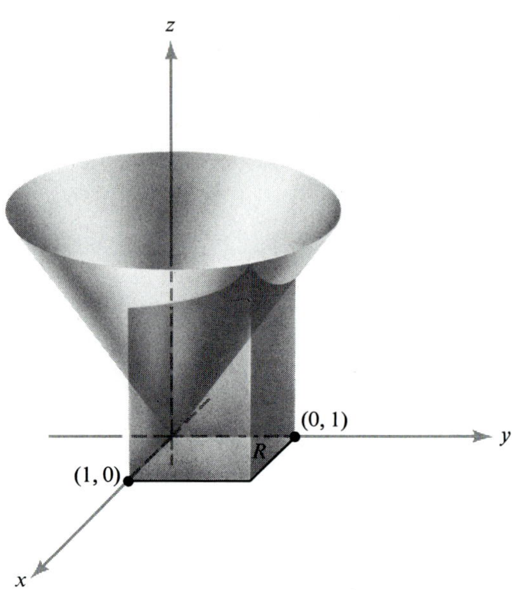

FIGURE 5.8.9
Volume under $z = \sqrt{x^2 + y^2}$ *over* $R = [0, 1] \times [0, 1]$.

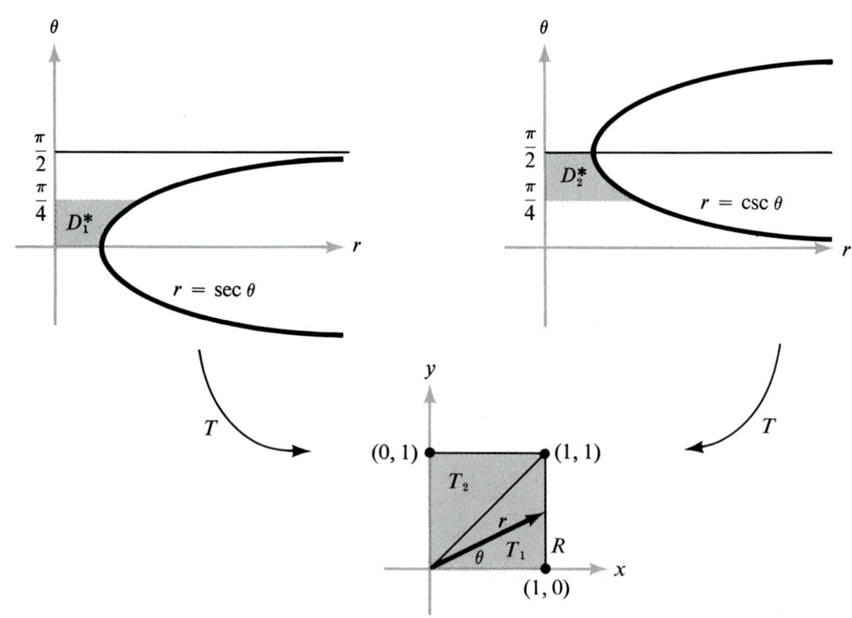

FIGURE 5.8.10
The polar-coordinate transformation takes D_1^* *to the triangle* T_1 *and* D_2^* *to* T_2.

Next we use iterated integration to obtain

$$\int_{D_1*} r^2\, dr\, d\theta = \int_0^{\pi/4} \left[\int_0^{\sec\theta} r^2\, dr \right] d\theta = \frac{1}{3}\int_0^{\pi/4} \sec^3\theta\, d\theta$$

Consulting a table of integrals (see the back of the book) to find $\int \sec^3 x\, dx$, we have

$$\int_0^{\pi/4} \sec^3\theta\, d\theta = \left[\frac{\sec\theta\tan\theta}{2}\right]_0^{\pi/4} + \frac{1}{2}\int_0^{\pi/4} \sec\theta\, d\theta$$

$$= \frac{\sqrt{2}}{2} + \frac{1}{2}\int_0^{\pi/4} \sec\theta\, d\theta$$

Consulting the table again for $\int \sec x\, dx$, we find

$$\frac{1}{2}\int_0^{\pi/4} \sec\theta\, d\theta = \frac{1}{2}[\log|\sec\theta + \tan\theta|]_0^{\pi/4} = \frac{1}{2}\log(1+\sqrt{2})$$

Combining these results and recalling the factor $\frac{1}{3}$, we obtain

$$\int_{D_1*} r^2\, dr\, d\theta = \frac{1}{3}\left(\frac{\sqrt{2}}{2} + \frac{1}{2}\log(1+\sqrt{2})\right) = \frac{1}{6}(\sqrt{2} + \log(1+\sqrt{2}))$$

Multiplying by 2, we obtain the answer

$$\int_R \sqrt{x^2+y^2}\, dx\, dy = \tfrac{1}{3}(\sqrt{2} + \log(1+\sqrt{2}))$$

EXERCISES

1. Let D be the unit circle. Evaluate

$$\int_D \exp(x^2+y^2)\, dx\, dy$$

by making a change of variables to polar coordinates.

2. Let D be the region $0 \le y \le x$ and $0 \le x \le 1$. Evaluate

$$\int_D (x+y)\, dx\, dy$$

by making the change of variables $x = u + v$, $y = u - v$. Check your answer by evaluating the integral directly by using an iterated integral.

3. Let $T(u, v) = (x(u, v), y(u, v))$ be the mapping defined by $T(u, v) = (4u, 2u + 3v)$. Let D^* be the rectangle $[0, 1] \times [1, 2]$. Find $D = T(D^*)$ and evaluate
   (a) $\int_D xy\, dx\, dy$
   (b) $\int_D (x - y)\, dx\, dy$
   by making a change of variables to evaluate them as integrals over D^*.

4. Repeat Exercise 3 for $T(u, v) = (u, v(1 + u))$.

5. Evaluate

$$\int_D \frac{dx\, dy}{\sqrt{1 + x + 2y}}$$

where $D = [0, 1] \times [0, 1]$, by setting $T(u, v) = (u, v/2)$ and evaluating an integral over D^*, where $T(D^*) = D$.

6. Define $T(u, v) = (u^2 - v^2, 2uv)$. Let D^* be the set of (u, v) with $u^2 + v^2 \leq 1$, $u \geq 0$, $v \geq 0$. Find $T(D^*) = D$. Evaluate $\int_D dx\, dy$.

7. Let $T(u, v)$ be as in Exercise 6. By making this change of variables evaluate

$$\int_D \frac{dx\, dy}{\sqrt{x^2 + y^2}}$$

8. Let D^* be a region of type 1 in the uv-plane bounded by

$$v = g(u), \qquad v = h(u) \leq g(u)$$

for $a \leq u \leq b$. Let $T\colon \mathbb{R}^2 \to \mathbb{R}^2$ be the transformation given by

$$x = u, \qquad y = \psi(u, v)$$

Assume $T(D^*) = D$ is a region of type 1; show that if $f\colon D^* \to \mathbb{R}$ is continuous then

$$\int_D f(x, y)\, dx\, dy = \int_{D^*} f(u, \psi(u, v)) \left| \frac{\partial \psi}{\partial v} \right| du\, dv$$

9. Find the area inside the curve $r = 1 + \sin\theta$.

10. (a) Express $\int_0^1 \int_0^{x^2} xy\, dy\, dx$ as an integral over the triangle D^*, which is the set of (u, v) where $0 \leq u \leq 1$, $0 \leq v \leq u$. (HINT: Find a one-to-one mapping T of D^* onto the given region of integration.)
    (b) Evaluate this integral directly and over D^*.

*11. Let D be the region bounded by $x^{3/2} + y^{3/2} = a^{3/2}$, for $x \geq 0$, $y \geq 0$, and the coordinate axes $x = 0$, $y = 0$. Express $\int_D f(x, y)\, dx\, dy$ as an integral over the triangle D^*, which is the set of points $0 \leq u \leq a$, $0 \leq v \leq a - u$.

12. Let D be the unit disc. Express $\int_D (1 + x^2 + y^2)^{3/2}\, dx\, dy$ as an integral over the rectangle $[0, 1] \times [0, 2\pi]$ and evaluate.

13. Using polar coordinates find the area bounded by the *lemniscate* $(x^2 + y^2)^2 = 2a^2(x^2 - y^2)$.

*14. A change of variables can help to find the value of an improper integral over the unbounded region $\mathbb{R}^2$. Evaluate

$$\int_{\mathbb{R}^2} e^{-x^2 - y^2}\, dx\, dy$$

by changing to polar coordinates. Could you have evaluated this integral directly (see Exercise 2, Section 5.5)?

15. Redo Exercise 10 of Section 5.3 using a change of variables and compare the efforts involved in each method.

5.9 CYLINDRICAL AND SPHERICAL COORDINATES

The standard way to represent a point in $\mathbb{R}^3$ is by rectangular coordinates (x, y, z). However, we have seen that both the Cartesian and polar-coordinate systems are useful in working out integration problems. In many instances, a change of coordinates can greatly simplify the form of a problem. With this in mind, we introduce two new coordinate systems for $\mathbb{R}^3$ that are particularly well suited for certain types of problems.

Definition. *The **cylindrical coordinates** (r, θ, z) of a point (x, y, z) are defined by*

$$x = r \cos \theta, \qquad y = r \sin \theta, \qquad z = z \tag{1}$$

or explicitly

$$r^2 = x^2 + y^2, \qquad \theta = \tan^{-1} y/x, \qquad z = z$$

(see Figure 5.9.1).

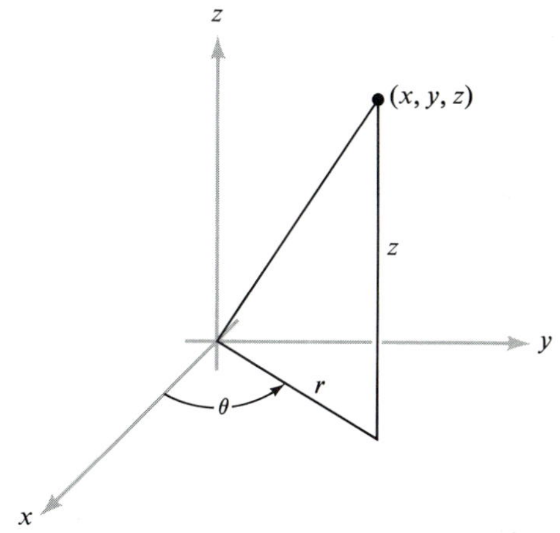

FIGURE 5.9.1
Representing a point (x, y, z) in terms of its cylindrical coordinates r, θ, and z.

In other words, for any point (x, y, z) we represent its first and second coordinates in terms of polar coordinates and leave the third coordinate unchanged. The formula (1) shows that given (r, θ, z) the triple (x, y, z) is completely determined and vice versa if we restrict θ to the interval $[0, 2\pi[$ (sometimes the range $]-\pi, \pi]$ is convenient) and require that $r > 0$.

To see why we use the term cylindrical coordinates, note that if $0 \le \theta < 2\pi$, $-\infty < z < \infty$, and $r = a$ is some positive constant, then the locus of these points is a cylinder of radius a (see Figure 5.9.2).

Cylindrical coordinates are not the only possible generalization of polar coordinates to three dimensions. Recall that in two dimensions

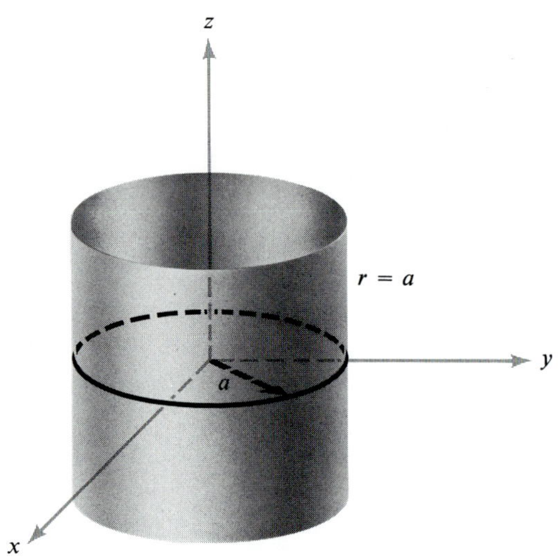

FIGURE 5.9.2
The graph of the points whose cylindrical coordinates satisfy $r = a$ is a cylinder.

the magnitude of the vector $x\mathbf{i} + y\mathbf{j}$ (that is, $\sqrt{x^2 + y^2}$) is the r in the polar coordinate system. For cylindrical coordinates, the length of the vector $x\mathbf{i} + y\mathbf{j} + z\mathbf{k}$,

$$\rho = \sqrt{x^2 + y^2 + z^2}$$

is not one of the coordinates of the system—we use only the magnitude $r = \sqrt{x^2 + y^2}$, the angle θ, and the "height" z.

We now modify this by introducing the *spherical coordinate* system, which uses ρ as a coordinate. Spherical coordinates are often useful for problems that possess spherical symmetry (symmetry about a point) while cylindrical coordinates can be applied when cylindrical symmetry (symmetry about a line) is involved.

Given a point $(x, y, z) \in \mathbb{R}^3$, let

$$\rho = \sqrt{x^2 + y^2 + z^2}$$

and represent x and y by polar coordinates in the xy-plane

$$x = r\cos\theta, \qquad y = r\sin\theta \tag{2}$$

where $r = \sqrt{x^2 + y^2}$ and $\theta = \tan^{-1} y/x$. The coordinate z is given by

$$z = \rho\cos\phi$$

where ϕ is the angle (between 0 and π) the radius vector $\mathbf{v} = x\mathbf{i} + y\mathbf{j} + z\mathbf{k}$ makes with the z-axis, in the plane containing the vector $\mathbf{v}$ and the z-axis (see Figure 5.9.3). Using the dot product we can express ϕ by

$$\cos\phi = \frac{\mathbf{v}\cdot\mathbf{k}}{\|\mathbf{v}\|}$$

or

$$\phi = \cos^{-1}\frac{\mathbf{v}\cdot\mathbf{k}}{\|\mathbf{v}\|}$$

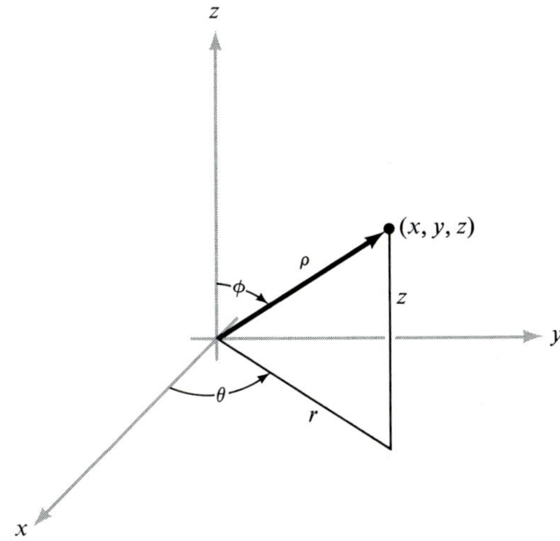

FIGURE 5.9.3
Spherical coordinates (ρ, θ, ϕ); the graph of points satisfying $\rho = a$ is a sphere.

We take as our coordinates the quantities ρ, θ, ϕ. Since

$$r = \rho\sin\phi$$

we can use (2) to find x, y, z in terms of the spherical coordinates ρ, θ, ϕ.

Definition. *The spherical change of coordinates* $S\colon (\rho, \theta, \phi) \mapsto (x, y, z)$ *is given by*

$$x = \rho \sin \phi \cos \theta, \qquad y = \rho \sin \phi \sin \theta, \qquad z = \rho \cos \phi \tag{3}$$

where

$$\rho \geq 0, \qquad 0 \leq \theta < 2\pi, \qquad 0 \leq \phi \leq \pi$$

define the $\rho\theta\phi$*-domain.*

Note that in spherical coordinates the equation of the sphere of radius a takes on the particularly simple form

$$\rho = a$$

EXAMPLE 1. Express (a) the surface $xz = 1$ and (b) the surface $x^2 + y^2 - z^2 = 1$ in spherical coordinates.

We have from (3), $x = \rho \sin \phi \cos \theta$, $z = \rho \cos \phi$, so the surface (a) consists of all (ρ, θ, ϕ) such that

$$\rho^2 \sin \phi \cos \theta \cos \phi = 1$$

For part (b) we can write

$$x^2 + y^2 - z^2 = x^2 + y^2 + z^2 - 2z^2 = \rho^2 - 2\rho^2 \cos^2 \phi$$

so the surface is $\rho^2(1 - 2\cos^2 \phi) = -\rho^2 \cos(2\phi) = 1$.

Spherical and cylindrical coordinates are very useful in the Change of Variables formula for triple integrals, which we state below. First we must define the Jacobian of a transformation from $\mathbb{R}^3$ to $\mathbb{R}^3$—it is a simple extension of the two-variable case.

Definition. *Let* $T\colon W \subset \mathbb{R}^3 \to \mathbb{R}^3$ *be a* C^1 *function defined by* $x = x(u, v, w)$, $y = y(u, v, w)$, $z = z(u, v, w)$. *Then the* ***Jacobian*** *of* T, *written* $\partial(x, y, z)/\partial(u, v, w)$, *is the determinant*

$$\begin{vmatrix} \dfrac{\partial x}{\partial u} & \dfrac{\partial x}{\partial v} & \dfrac{\partial x}{\partial w} \\[2ex] \dfrac{\partial y}{\partial u} & \dfrac{\partial y}{\partial v} & \dfrac{\partial y}{\partial w} \\[2ex] \dfrac{\partial z}{\partial u} & \dfrac{\partial z}{\partial v} & \dfrac{\partial z}{\partial w} \end{vmatrix}$$

The absolute value of this determinant is equal to the volume of the parallelepiped determined by the vectors

$$\mathbf{T}_u = \frac{\partial x}{\partial u}\mathbf{i} + \frac{\partial y}{\partial u}\mathbf{j} + \frac{\partial z}{\partial u}\mathbf{k}$$

$$\mathbf{T}_v = \frac{\partial x}{\partial v}\mathbf{i} + \frac{\partial y}{\partial v}\mathbf{j} + \frac{\partial z}{\partial v}\mathbf{k}$$

$$\mathbf{T}_w = \frac{\partial x}{\partial w}\mathbf{i} + \frac{\partial y}{\partial w}\mathbf{j} + \frac{\partial z}{\partial w}\mathbf{k}$$

(see Section 1.3). Just as in the two-variable case, the Jacobian measures how the transformation T distorts its domain. Hence for volume (triple) integrals, we have the Change of Variables formula

$$\begin{aligned}&\int_D f(x, y, z)\,dx\,dy\,dz\\ &= \int_{D^*} f(x(u, v, w), y(u, v, w), z(u, v, w))\left|\frac{\partial(x, y, z)}{\partial(u, v, w)}\right| du\,dv\,dw \qquad (4)\end{aligned}$$

where D^* is an elementary region in uvw-space corresponding to D in xyz-space, under a coordinate change T: $(u, v, w) \mapsto (x(u, v, w), y(u, v, w), z(u, v, w))$, provided T is C^1 and one-to-one except on a set of volume zero.

Let us apply formula (4) to cylindrical and spherical coordinates. First we compute the Jacobian for the map defining the change to cylindrical coordinates. Since

$$x = r\cos\theta, \qquad y = r\sin\theta, \qquad z = z$$

we have

$$\frac{\partial(x, y, z)}{\partial(r, \theta, z)} = \begin{vmatrix} \cos\theta & -r\sin\theta & 0 \\ \sin\theta & r\cos\theta & 0 \\ 0 & 0 & 1 \end{vmatrix} = r$$

Thus, we obtain the formula

$$\int_D f(x, y, z)\,dx\,dy\,dz = \int_{D^*} f(r\cos\theta, r\sin\theta, z)r\,dr\,d\theta\,dz \qquad (5)$$

Next let us consider the spherical coordinate system. Since

$$x = \rho\sin\phi\cos\theta, \qquad y = \rho\sin\phi\sin\theta, \qquad z = \rho\cos\phi$$

we have

$$\frac{\partial(x, y, z)}{\partial(\rho, \theta, \phi)} = \begin{vmatrix} \sin\phi\cos\theta & -\rho\sin\phi\sin\theta & \rho\cos\phi\cos\theta \\ \sin\phi\sin\theta & \rho\sin\phi\cos\theta & \rho\cos\phi\sin\theta \\ \cos\phi & 0 & -\rho\sin\phi \end{vmatrix}$$

$$= \sin\phi\cos\theta \begin{vmatrix} \rho\sin\phi\cos\theta & \rho\cos\phi\sin\theta \\ 0 & -\rho\sin\phi \end{vmatrix}$$

$$+ \rho\sin\phi\sin\theta \begin{vmatrix} \sin\phi\sin\theta & \rho\cos\phi\sin\theta \\ \cos\phi & -\rho\sin\phi \end{vmatrix}$$

$$+ \rho\cos\phi\cos\theta \begin{vmatrix} \sin\phi\sin\theta & \rho\sin\phi\cos\theta \\ \cos\phi & 0 \end{vmatrix}$$

$$= \sin\phi\cos\theta(-\rho^2\sin^2\phi\cos\theta)$$

$$+ \rho\sin\phi\sin\theta(-\rho\sin^2\phi\sin\theta - \rho\cos^2\phi\sin\theta)$$

$$+ \rho\cos\phi\cos\theta(-\rho\sin\phi\cos\phi\cos\theta)$$

$$= -\rho^2\sin^3\phi\cos^2\theta - \rho^2\sin^3\phi\sin^2\theta$$

$$- \rho^2\cos^2\phi\sin\phi\sin^2\theta - \rho^2\sin\phi\cos^2\phi\cos^2\theta$$

$$= -\rho^2\sin^3\phi(\cos^2\theta + \sin^2\theta)$$

$$- \rho^2\sin\phi\cos^2\phi(\sin^2\theta + \cos^2\theta)$$

$$= -\rho^2\sin^3\phi - \rho^2\sin\phi\cos^2\phi$$

$$= -\rho^2\sin\phi(\sin^2\phi + \cos^2\phi) = -\rho^2\sin\phi$$

Thus we arrive at the formula

$$\int_D f(x, y, z)\,dx\,dy\,dz$$

$$= \int_{D*} f(\rho\sin\phi\cos\theta, \rho\sin\phi\sin\theta, \rho\cos\phi)\rho^2\sin\phi\,d\rho\,d\theta\,d\phi \tag{6}$$

In order to prove the validity of formula (6), we must show that the transformation S on the set D^* is one-to-one except on a set of volume zero. We shall leave this verification as an exercise (see Exercise 8).

EXAMPLE 2. Evaluate $\int_D \exp(x^2 + y^2 + z^2)^{3/2}\,dV$ where D is the unit ball in $\mathbb{R}^3$.

First note that we cannot easily integrate this function using iterated integrals (try it!). Hence, let us try a change of variables. The transformation S into spherical coordinates seems appropriate, since then the entire quantity $x^2 + y^2 + z^2$ can be replaced by one variable, namely, ρ^2. If D^* is the region such that

$$0 \le \rho \le 1, \qquad 0 \le \theta \le 2\pi, \qquad 0 \le \phi \le \pi$$

we may apply formula (6) and write

$$\int_D \exp(x^2 + y^2 + z^2)^{3/2}\, dV = \int_{D^*} \rho^2 e^{\rho^3} \sin\phi\, dV$$

This integral equals the iterated integral

$$\begin{aligned}
&\int_0^1 \int_0^\pi \int_0^{2\pi} e^{\rho^3}\rho^2 \sin\phi\, d\theta\, d\phi\, d\rho \\
&\qquad = 2\pi \int_0^1 \int_0^\pi e^{\rho^3}\rho^2 \sin\phi\, d\phi\, d\rho \\
&\qquad = -2\pi \int_0^1 \rho^2 e^{\rho^3}[\cos\phi]_0^\pi\, d\rho = 4\pi \int_0^1 e^{\rho^3}\rho^2\, d\rho \\
&\qquad = \tfrac{4}{3}\pi \int_0^1 e^{\rho^3}(3\rho^2)\, d\rho = [\tfrac{4}{3}\pi e^{\rho^3}]_0^1 = \tfrac{4}{3}\pi(e-1)
\end{aligned}$$

EXAMPLE 3. Let D be the sphere of radius R and center $(0, 0, 0)$ in $\mathbb{R}^3$. Then $\int_D dx\, dy\, dz$ is the volume of D. This integral may be evaluated by reducing it to iterated integrals (Example 1, Section 5.6) or by regarding D as a volume of revolution, but let us evaluate it here by using spherical coordinates. We get

$$\begin{aligned}
\int_D dx\, dy\, dz &= \int_0^R \int_0^\pi \int_0^{2\pi} \rho^2 \sin\phi\, d\theta\, d\phi\, d\rho \\
&= \frac{R^3}{3} \int_0^\pi \int_0^{2\pi} \sin\phi\, d\theta\, d\phi \\
&= \frac{2\pi R^3}{3} \int_0^\pi \sin\phi\, d\phi \\
&= \frac{2\pi R^3}{3} \{-(\cos(\pi) - \cos(0))\} \\
&= \frac{4\pi R^3}{3}
\end{aligned}$$

which is the familiar formula for the volume of a sphere.

EXERCISES

1. (a) The following points are given in cylindrical coordinates; express each in rectangular coordinates and spherical coordinates: $(1, 45°, 1)$, $(2, \pi/2, -4)$, $(0, 45°, 10)$, $(3, \pi/6, 4)$.
   (b) Change each of the following points from rectangular coordinates to spherical coordinates and to cylindrical coordinates: $(2, 1, -2)$, $(0, 3, 4)$, $(\sqrt{2}, 1, 1)$, $(-2\sqrt{3}, -2, 3)$.
2. Describe the geometric meaning of the following mappings in cylindrical coordinates.
   (a) $(r, \theta, z) \mapsto (r, \theta, -z)$
   (b) $(r, \theta, z) \mapsto (r, \theta + \pi, -z)$
3. Describe the geometric meaning of the following mappings in spherical coordinates.
   (a) $(\rho, \theta, \phi) \mapsto (\rho, \theta + \pi, \phi)$
   (b) $(\rho, \theta, \phi) \mapsto (\rho, \theta, \phi + \pi/2)$
4. (a) Describe the surfaces $r =$ constant, $\theta =$ constant, and $z =$ constant in the cylindrical coordinate system.
   (b) Describe the surfaces $\rho =$ constant, $\theta =$ constant, and $\phi =$ constant in the spherical coordinate system.
5. Show that in order to represent each point in $\mathbb{R}^3$ by spherical coordinates it is only necessary to take values of θ between 0 and 2π, values of ϕ between 0 and π, and values of $\rho \geq 0$. Are coordinates unique if we allow $\rho \leq 0$?
6. Let $T: \mathbb{R}^3 \to \mathbb{R}^3$ be defined by

$$T(u, v, w) = (u \cos v \cos w, u \sin v \cos w, u \sin w)$$

   (a) Show that T is onto the unit sphere; that is, every (x, y, z) with $x^2 + y^2 + z^2 = 1$ can be written as $(x, y, z) = T(u, v, w)$ for some (u, v, w).
   (b) Show that T is not one-to-one.
7. Determine the equations of the following curves and surfaces in spherical and cylindrical coordinates.

   (a) $\dfrac{x^2}{a^2} + \dfrac{y^2}{b^2} + \dfrac{z^2}{c^2} = 1$

   (b) $z^2 = x^2 + y^2$
   (c) the line $y = x = z$

   (d) $z = \tan^{-1} \dfrac{y}{x}, \; x^2 + y^2 = 1$

8. Show that the spherical change-of-coordinate mapping (formula (3)), $S(\rho, \theta, \phi) = (\rho \sin \phi \cos \theta, \rho \sin \phi \sin \theta, \rho \cos \phi)$ is one-to-one except on a set of volume zero.
9. Describe the surface $\sqrt{x^2 + y^2 + z^2} = \rho = \theta$. What is the Cartesian representation of this surface?
10. Describe the surface $r = z \cos \theta$ in rectangular coordinates.

11. Let D be the unit ball. Evaluate

$$\int_D \frac{dx\,dy\,dz}{\sqrt{2 + x^2 + y^2 + z^2}}$$

by making the appropriate change of variables.

12. Let D be the first octant of the ball $x^2 + y^2 + z^2 \le a^2$, where $x \ge 0$, $y \ge 0$, $z \ge 0$. Evaluate

$$\int_D \sqrt{x^2 + y^2 + z^2}/\sqrt{z + (x^2 + y^2 + z^2)^2}\,dx\,dy\,dz$$

by changing variables.

13. Let D be the unbounded region defined as the set of (x, y, z) with $x^2 + y^2 + z^2 \ge 1$. By making a change of variables evaluate the improper integral

$$\int_D \frac{dx\,dy\,dz}{(x^2 + y^2 + z^2)^2}$$

REVIEW EXERCISES FOR CHAPTER 5

1. Evaluate each of the following integrals and describe the region determined by the limits.
   (a) $\int_0^3 \int_{-x^2+1}^{x^2+1} xy\,dy\,dx$
   (b) $\int_0^1 \int_{\sqrt{x}}^1 (x + y)^2\,dy\,dx$
   (c) $\int_0^1 \int_{e^x}^{e^{2x}} x \ln y\,dy\,dx$
2. Reverse the order of integration of the integrals in Exercise 1 and evaluate.
3. Find the volume enclosed by the cone $x^2 + y^2 = z^2$ and the plane $z - y - 1 = 0$.
4. Find the volume between the surfaces $x^2 + y^2 = z$ and $x^2 + y^2 + z^2 = 2$.
5. In Exercise 2, Section 5.5, we discussed integrals over unbounded regions. Use the polar change of coordinates to show $\int_{-\infty}^{\infty} e^{-x^2}\,dx = \sqrt{\pi}$. (HINT: use Fubini's Theorem (you may assume its validity) to show that

$$\left(\int_{-\infty}^{\infty} e^{-x^2}\,dx\right)^2 = \int_{-\infty}^{\infty}\int_{-\infty}^{\infty} e^{-x^2-y^2}\,dx\,dy$$

   and use Exercise 14, Section 5.8.)
6. Use the ideas in Exercise 5 to evaluate $\int_{\mathbb{R}^2} f(x, y)\,dx\,dy$ where $f(x, y) = 1/(1 + x^2 + y^2)^{3/2}$.
7. Find $\int_{\mathbb{R}^3} f(x, y, z)\,dx\,dy\,dz$ where $f(x, y, z) = \exp(-(x^2 + y^2 + z^2)^{3/2})$.
8. Find $\int_{\mathbb{R}^3} f(x, y, z)\,dx\,dy\,dz$ where

$$f(x, y, z) = \frac{1}{(1 + (x^2 + y^2 + z^2)^{3/2})^{3/2}}$$

9. A cylindrical hole of diameter 1 is bored through a sphere of radius 2. Assuming that the axis of the cylinder is the same as the axis of the sphere, find the volume of the solid that remains.

10. Let C_1 and C_2 be two cylinders of infinite extent, of diameter 2, and with axes on the x- and y-axes respectively. Find the volume of $C_1 \cap C_2$.

11. Write the iterated integral $\int_0^1 \int_1^{1-x} \int_x^1 f(x, y, z)\, dz\, dy\, dx$ as an integral over a region in $\mathbb{R}^3$ and then rewrite it in five other possible orders of integration.

*12. Suppose D is the unbounded region on $\mathbb{R}^2$ given by the set of (x, y) with $0 \le x < \infty$, $0 \le y \le x$. Let $f(x, y) = x^{-3/2}e^{y-x}$. Does the improper integral $\int_D f$ exist?

13. Evaluate each of the following iterated integrals.
(a) $\int_0^\infty \int_0^y xe^{-y^3}\, dx\, dy$
(b) $\int_0^1 \int_{y^{1/2}}^{y^3} e^{x/y}\, dx\, dy$
(c) $\int_0^{\pi/2} \int_0^{(\arcsin y)/y} y \cos xy\, dx\, dy$

14. Evaluate each of the following iterated integrals.
(a) $\int_0^1 \int_0^z \int_0^y xy^2z^3\, dx\, dy\, dz$

(b) $$\int_0^1 \int_0^y \int_0^{x-\sqrt{3}} \frac{x}{x^2 + z^2}\, dz\, dx\, dy$$

(c) $\int_1^2 \int_1^z \int_{1/y}^2 yz^2\, dx\, dy\, dz$

15. Find the volume bounded by $x/a + y/b + z/c = 1$ and the coordinate planes.

16. Find the volume bounded by $f(x, y) = 6 - x^2 - y^2$ and $z^2 = x^2 + y^2$.

17. In (a) to (d) below, make the indicated change of variables. (Do not evaluate.)
(a) $\int_0^1 \int_{-1}^1 \int_{-\sqrt{(1-y^2)}}^{\sqrt{(1-y^2)}} (x^2 + y^2)^{1/2}\, dx\, dy\, dz$, cylindrical coordinates
(b) $\int_{-1}^1 \int_{-\sqrt{(1-y^2)}}^{\sqrt{(1-y^2)}} \int_{-\sqrt{(4-x^2-y^2)}}^{\sqrt{(4-x^2-y^2)}} xyz\, dz\, dx\, dy$, cylindrical coordinates
(c) $\int_{-\sqrt{2}}^{\sqrt{2}} \int_{-\sqrt{(2-y^2)}}^{\sqrt{(2-y^2)}} \int_{\sqrt{(x^2+y^2)}}^{\sqrt{(4-x^2-y^2)}} z^2\, dz\, dx\, dy$, spherical coordinates
(d) $\int_0^1 \int_0^{\pi/4} \int_0^{2\pi} \rho^3 \sin 2\phi\, d\theta\, d\phi\, d\rho$, rectangular coordinates

18. (a) Let $\mathbf{F} = 3x^2y\mathbf{i} + zx\mathbf{j} + e^{xy}\mathbf{k}$. Compute $\nabla \cdot \mathbf{F}$ and $\nabla \times \mathbf{F}$.
(b) Is it always true that $\nabla \times (\nabla \times \mathbf{F}) = \mathbf{0}$?

19. Evaluate $\iint_B (x^4 + 2x^2y^2 + y^4)\, dx\, dy$, where B is the portion of the disc of radius 2 (centered at (0, 0)) in the first quadrant.

20. Interchange the order of integration and evaluate

$$\int_0^2 \int_{y/2}^1 (x + y)^2\, dx\, dy$$

21. Evaluate $\iint_B e^{-x^2-y^2}\, dx\, dy$, where B consists of those (x, y) satisfying $x^2 + y^2 \le 1$ and $y \le 0$. Discuss the geometrical meaning of your answer.

22. Change the order of integration and evaluate

$$\int_0^1 \int_{y^{1/2}}^1 (x^2 + y^3x)\, dx\, dy$$

CHAPTER 6

INTEGRALS OVER PATHS AND SURFACES

In Chapter 5 we studied integration over regions in $\mathbb{R}^2$ and $\mathbb{R}^3$. For example, we learned how to evaluate integrals like

$$\int_D f(x, y)\, dA$$

where D is a region in $\mathbb{R}^2$. In this chapter we shall discuss integration over paths and surfaces. This is basic to an understanding of Chapter 7. Indeed, that chapter will relate our results on vector differential calculus (Chapter 3) and vector integral calculus (this chapter) by proving the profound theorems of Green, Gauss, and Stokes. In that chapter we shall also examine some significant physical applications.

6.1 THE PATH INTEGRAL

In this section we shall introduce the concept of a path integral; this is one of the several ways in which integrals of functions of one variable can be generalized to functions of several variables. Besides those in Chapter 5, there are other generalizations, to be discussed in later sections.

Suppose we are given a scalar function $f: \mathbb{R}^3 \to \mathbb{R}$, so f sends points in $\mathbb{R}^3$ to real numbers. It can be extremely useful to be able to integrate the function f along a path $\boldsymbol{\sigma}: I = [a, b] \to \mathbb{R}^3$, where $\boldsymbol{\sigma}(t) = (x(t), y(t), z(t))$. To motivate this notion, let us suppose that the image of $\boldsymbol{\sigma}$ represents a wire. If $f(x, y, z)$ denotes the mass density at (x, y, z), we might want to know the total mass of the wire. If $f(x, y, z)$ indicates temperature, we might want to know the average temperature along the wire. Both types of problems require integrating $f(x, y, z)$ over $\boldsymbol{\sigma}$.

Definition. *The **path integral** or the **integral of** $f(x, y, z)$ **along the path** $\boldsymbol{\sigma}$ is defined when $\boldsymbol{\sigma}: I = [a, b] \to \mathbb{R}^3$ is C^1 and when the composite function $t \mapsto f(x(t), y(t), z(t))$ is continuous on I. We define this integral by the equation*

$$\int_{\boldsymbol{\sigma}} f = \int_a^b f(x(t), y(t), z(t)) \|\boldsymbol{\sigma}'(t)\| \, dt$$

Sometimes $\int_{\boldsymbol{\sigma}} f$ is denoted $\int_{\boldsymbol{\sigma}} f(x, y, z) \, ds$. Notice that another way to write the definition is $\int_{\boldsymbol{\sigma}} f = \int_a^b f(\boldsymbol{\sigma}(t)) \|\boldsymbol{\sigma}'(t)\| \, dt$.

If $\boldsymbol{\sigma}(t)$ is only piecewise C^1 or $f(\boldsymbol{\sigma}(t))$ is piecewise continuous, we can still form $\int_{\boldsymbol{\sigma}} f$ by breaking $[a, b]$ into pieces over which $f(\boldsymbol{\sigma}(t)) \|\boldsymbol{\sigma}'(t)\|$ is continuous, and summing the integrals over the pieces.

Note that we recover the definition of the arc length of $\boldsymbol{\sigma}$ when $f = 1$ (see Section 3.2).

EXAMPLE 1. Let $\boldsymbol{\sigma}$ be the helix $\boldsymbol{\sigma}: [0, 2\pi] \to \mathbb{R}^3$, $t \mapsto (\cos t, \sin t, t)$ (see Figure 3.1.8), and let $f(x, y, z) = x^2 + y^2 + z^2$. To evaluate the integral $\int_{\boldsymbol{\sigma}} f(x, y, z) \, ds$, we first find

$$\begin{aligned} \|\boldsymbol{\sigma}'(t)\| &= \sqrt{\left[\frac{d(\cos t)}{dt}\right]^2 + \left[\frac{d(\sin t)}{dt}\right]^2 + \left[\frac{dt}{dt}\right]^2} \\ &= \sqrt{\sin^2 t + \cos^2 t + 1} = \sqrt{2} \end{aligned}$$

We substitute for x, y, and z to obtain

$$f(x, y, z) = x^2 + y^2 + z^2 = \cos^2 t + \sin^2 t + t^2 = 1 + t^2$$

along $\boldsymbol{\sigma}$. This yields

$$\begin{aligned} \int_{\boldsymbol{\sigma}} f(x, y, z) \, ds &= \int_0^{2\pi} (1 + t^2)\sqrt{2} \, dt = \sqrt{2}\left[t + \frac{t^3}{3}\right]_0^{2\pi} \\ &= \frac{2\sqrt{2}\pi}{3}(3 + 4\pi^2) \end{aligned}$$

If we think of the helix as a wire and $f(x, y, z) = x^2 + y^2 + z^2$ as the mass density, then the total mass of the wire is $(2\sqrt{2}\pi)(3 + 4\pi^2)/3$.

To motivate the definition of the path integral we shall consider "Riemann-like" sums S_N in the same general way as we have done before to define arc length. For simplicity, let $\boldsymbol{\sigma}$ be C^1 on I. Subdivide the interval $I = [a, b]$ by means of a partition

$$a = t_0 < t_1 < \cdots < t_N = b$$

This leads to a decomposition of $\boldsymbol{\sigma}$ into paths $\boldsymbol{\sigma}_i$ (Figure 6.1.1) defined on $[t_i, t_{i+1}]$ for $0 \le i \le N - 1$. Denote the arc length of $\boldsymbol{\sigma}_i$ by Δs_i, thus

$$\Delta s_i = \int_{t_i}^{t_{i+1}} \|\boldsymbol{\sigma}'(t)\| \, dt$$

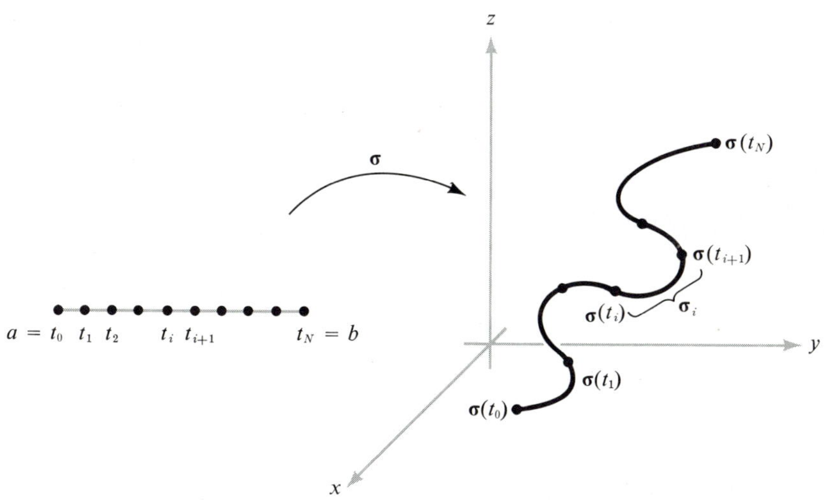

FIGURE 6.1.1
Breaking $\boldsymbol{\sigma}$ into smaller paths $\boldsymbol{\sigma}_i$.

When N is large, the arc length Δs_i is small and $f(x, y, z)$ is approximately constant for points on $\boldsymbol{\sigma}_i$. We consider the sums

$$S_N = \sum_{i=0}^{N-1} f(x_i, y_i, z_i) \, \Delta s_i$$

where $(x_i, y_i, z_i) = \boldsymbol{\sigma}(t)$ for some $t \in [t_i, t_{i+1}]$. These sums are basically Riemann sums, and from their theory it can be shown that

$$\lim_{N \to \infty} S_N = \int_I f(x(t), y(t), z(t)) \|\boldsymbol{\sigma}'(t)\| \, dt = \int_{\boldsymbol{\sigma}} f(x, y, z) \, ds$$

Thus the path integral can be expressed as a limit of Riemann sums.

An interesting application of the path integral is its use in defining the average value of a scalar function along a path. We define the number

$$\frac{\int_{\boldsymbol{\sigma}} f(x, y, z) \, ds}{l(\boldsymbol{\sigma})}$$

as the *average value* of f along $\boldsymbol{\sigma}$. We can justify our definition of average value by considering Riemann sums. The quotient

$$Q_N = \frac{\sum_{i=0}^{N-1} f(x_i, y_i, z_i)\,\Delta s_i}{\sum_{i=0}^{N-1} \Delta s_i} = \frac{\sum_{i=0}^{N-1} f(x_i, y_i, z_i)\,\Delta s_i}{l(\boldsymbol{\sigma})}$$

is the approximate average one obtains by considering $f(x, y, z)$ to be constant along each of the arcs $\boldsymbol{\sigma}_i$. The limit as $N \to \infty$ of the sequence $\{Q_N\}$ is the average value of $f(x, y, z)$ along $\boldsymbol{\sigma}$.

If we think of the helix in Example 1 as a heated wire and $f(x, y, z) = x^2 + y^2 + z^2$ as the temperature, then the average temperature along the wire is

$$\frac{\int_{\boldsymbol{\sigma}} f(x, y, z)\,ds}{l(\boldsymbol{\sigma})} = \frac{2\sqrt{2}\,\pi(3 + 4\pi^2)}{(3)2\sqrt{2}\,\pi} = \frac{1}{3}(3 + 4\pi^2) \text{ degrees}$$

EXAMPLE 2. Find the average y-coordinate of the points on the semicircle parametrized by $\boldsymbol{\rho}\colon [0, \pi] \to \mathbb{R}^3$, $\theta \mapsto (0, a \sin \theta, a \cos \theta)$.

In this example we have $f(x, y, z) = y$. We compute

$$\int_{\boldsymbol{\rho}} f(x, y, z)\,ds = \int_0^{\pi} a \sin \theta (a^2 \cos^2 \theta + a^2 \sin^2 \theta)^{1/2}\,d\theta = 2a^2$$

Since $l(\boldsymbol{\rho}) = \pi a$, the average y-coordinate of points on the semicircle is $2a/\pi$.

The average x-coordinate is clearly zero and the average z-coordinate is also easily computed to be zero. If we let the semicircle represent a wire of uniform density, then the point $(x^*, y^*, z^*) = (0, 2a/\pi, 0)$, whose coordinates are the average x-, y-, and z-coordinates of points on the wire, is the center of gravity of the wire.

An important special case of the path integral occurs when the path $\boldsymbol{\sigma}$ describes a plane curve. Let us examine this case in detail. Suppose then that all points $\boldsymbol{\sigma}(t)$ lie in the xy-plane. Let f be a real-valued function of two variables. The path integral of f along $\boldsymbol{\sigma}$ is

$$\int_{\boldsymbol{\sigma}} f(x, y)\,ds = \int_a^b f(x(t), y(t))\sqrt{x'(t)^2 + y'(t)^2}\,dt$$

When $f(x, y) \geq 0$, this integral has a natural geometric interpretation as the "area of a fence." We can construct a "fence" with base the image of $\boldsymbol{\sigma}$ and with height $f(x, y)$ at (x, y) (Figure 6.1.2). If $\boldsymbol{\sigma}$ winds only once around the image of $\boldsymbol{\sigma}$, the integral $\int_{\boldsymbol{\sigma}} f(x, y)\,ds$ represents the area of a side of this fence. The reader should try to justify this interpretation for himself, using an argument like the one used to justify the arc-length formula.

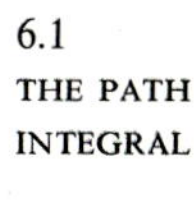

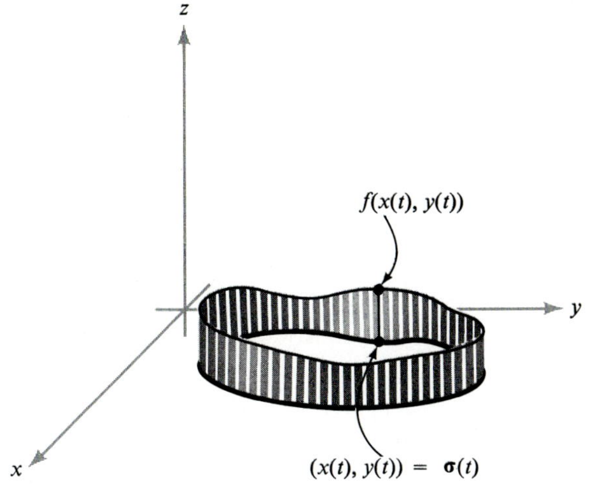

FIGURE 6.1.2
The path integral as the area of a fence.

EXAMPLE 3. Tom Sawyer's aunt has asked him to whitewash both sides of the old fence shown in Figure 6.1.3. Tom estimates that for each 25 square feet of whitewashing he lets someone do for him, the willing victim will pay 5 cents. How much can Tom hope to earn, assuming his aunt will provide whitewash free of charge?

From Figure 6.1.3, the base of the fence in the first quadrant is the path $\boldsymbol{\rho}\colon [0, \pi/2] \to \mathbb{R}^2$, $t \mapsto (30\cos^3 t, 30\sin^3 t)$, and the height of the fence at (x, y) is $f(x, y) = 1 + y/3$. The area of one side of the fence is then

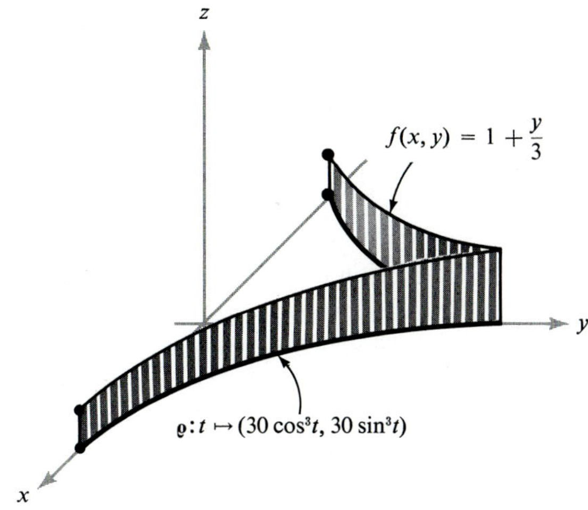

FIGURE 6.1.3
Tom Sawyer's fence.

equal to the path integral $\int_{\boldsymbol{\rho}} f(x, y)\, ds = \int_{\boldsymbol{\rho}} (1 + y/3)\, ds$. Since $\boldsymbol{\rho}'(t) = (-90 \cos^2 t \sin t, 90 \sin^2 t \cos t)$, we have $\|\boldsymbol{\rho}'(t)\| = 90 \sin t \cos t$. So the integral is

$$\int_{\boldsymbol{\rho}} \left(1 + \frac{y}{3}\right) ds = \int_0^{\pi/2} \left(1 + \frac{30 \sin^3 t}{3}\right) 90 \sin t \cos t\, dt$$

$$= 90 \int_0^{\pi/2} (\sin t + 10 \sin^4 t) \cos t\, dt$$

$$= 90 \left[\frac{\sin^2 t}{2} + 2 \sin^5 t\right]_0^{\pi/2}$$

$$= 90(\tfrac{1}{2} + 2) = 225$$

Hence, the area of one side of the fence is 450 square feet. Since both sides are to be whitewashed we must multiply by 2 to find the total area, which is 900 square feet. Dividing by 25 and then multiplying by 5, we find that Tom could realize as much as \$1.80 for the job.

EXAMPLE 4. Since the area of the fence in the above example is 225 square feet on each side and the length of the base is 45 feet, its average height is 225 feet^2/45 feet = 5 feet.

We shall see many further applications of the path integral in Chapter 7, when we study vector analysis.

This concludes our study of integration of *scalar* functions over paths. In the next section we shall turn our attention to the integration of *vector fields* over paths.

EXERCISES

1. Let $f(x, y, z) = y$ and $\boldsymbol{\sigma}(t) = (0, 0, t)$, $0 \le t \le 1$. Prove that $\int_{\boldsymbol{\sigma}} f = 0$.
2. Evaluate the following path integrals $\int_{\boldsymbol{\sigma}} f(x, y, z)\, ds$, where
   (a) $f(x, y, z) = x + y + z$ and $\boldsymbol{\sigma}\colon t \mapsto (\sin t, \cos t, t)$, $t \in [0, 2\pi]$;
   (b) $f(x, y, z) = \cos z$, $\boldsymbol{\sigma}$ as in (a);
   (c) $f(x, y, z) = x \cos z$, $\boldsymbol{\sigma}\colon t \mapsto t\mathbf{i} + t^2\mathbf{j}$, $t \in [0, 1]$.
3. Evaluate the following path integrals $\int_{\boldsymbol{\sigma}} f(x, y, z)\, ds$, where
   (a) $f(x, y, z) = \exp \sqrt{z}$, and $\boldsymbol{\sigma}\colon t \mapsto (1, 2, t^2)$, $t \in [0, 1]$;
   (b) $f(x, y, z) = yz$, and $\boldsymbol{\sigma}\colon t \mapsto (t, 3t, 2t)$, $t \in [1, 3]$;
   (c) $f(x, y, z) = (x + y)/(y + z)$, and $\boldsymbol{\sigma}\colon t \mapsto (t, \frac{2}{3}t^{3/2}, t)$, $t \in [1, 2]$.
4. Let $f\colon \mathbb{R}^3 \backslash \{xz\text{-plane}\} \to \mathbb{R}$ be defined by $f(x, y, z) = 1/y^3$. Evaluate $\int_{\boldsymbol{\sigma}} f(x, y, z)\, ds$ where $\boldsymbol{\sigma}\colon [1, e] \to \mathbb{R}^3$ is given by $\boldsymbol{\sigma}(t) = (\log t)\mathbf{i} + t\mathbf{j} + 2\mathbf{k}$.
5. In exercises 2(a) and 2(b) above find the average value of f over the given curves.
6. Write down a formula for the arc length of the graph of a function $f\colon [a, b] \to \mathbb{R}$. What is the average of the y-coordinate on this graph?

7. Find the average y-coordinate for the path $\boldsymbol{\sigma}: t \mapsto (t^2, t, 3)$, $t \in [0, 1]$.
8. (a) Show that the path integral of $f(x, y)$ along a path given in polar coordinates by $r = r(\theta)$, $\theta_1 \le \theta \le \theta_2$, is

$$\int_{\theta_1}^{\theta_2} f(r\cos\theta, r\sin\theta)\sqrt{r^2 + \left(\frac{dr}{d\theta}\right)^2}\, d\theta$$

   (b) Compute the arc length of $r = 1 + \cos\theta$, $0 \le \theta \le 2\pi$.
9. Write the following limit as a path integral ($t_i \le t_i^* \le t_{i+1}$, other notations as in the text).

$$\lim_{N\to\infty} \sum_{i=1}^{N-1} f(\boldsymbol{\sigma}(t_i^*))[g(\boldsymbol{\sigma}(t_{i+1})) - g(\boldsymbol{\sigma}(t_i))]$$

10. Let $f(x, y) = x - y$, $x = t^4$, $y = t^4$, $-1 \le t \le 1$.
   (a) Compute the integral of f along this path and interpret the answer geometrically. Sketch.
   (b) Evaluate the arc-length function $s(t)$ and redo (a) in terms of s (you may wish to consult Exercise 2, Section 3.2).

6.2 LINE INTEGRALS

If **F** is a force field in space, then a test particle (for example, a small unit charge in an electric force field or a unit mass in a gravitational field) will experience the force **F**. Suppose the particle moves along the image of a path $\boldsymbol{\sigma}$ while being acted upon by **F**. One of the fundamental concepts in physics is the *work done* by **F** on the particle as it traces out the path $\boldsymbol{\sigma}$. If $\boldsymbol{\sigma}$ is a straight-line displacement given by the vector **d** and **F** is a constant force, then the work done by **F** in moving the particle along the path is $\mathbf{F}\cdot\mathbf{d}$.

$$\mathbf{F}\cdot\mathbf{d} = (\text{force}) \times (\text{displacement in direction of force})$$

More generally, if the path is curved we can imagine that it is made up of a succession of infinitesimal straight-line displacements or that it is approximated by a finite number of straight-line displacements. Then (as in our derivation of the formulas for arc length, in Section 3.2, and the path integral, in Section 6.1) we are led to the following formula for the work done by the force field **F** on a particle moving along a path $\boldsymbol{\sigma}: [a, b] \to \mathbb{R}^3$.

$$\text{work done by } \mathbf{F} = \int_a^b \mathbf{F}(\boldsymbol{\sigma}(t)) \cdot \boldsymbol{\sigma}'(t)\, dt$$

Without giving a full proof, we can justify this derivation in some detail. If t ranges over a small interval t to $t + \Delta t$ the particle moves from $\boldsymbol{\sigma}(t)$ to $\boldsymbol{\sigma}(t + \Delta t)$, a vector displacement of $\Delta\mathbf{d} = \boldsymbol{\sigma}(t + \Delta t) - \boldsymbol{\sigma}(t)$ (see Figure 6.2.1).

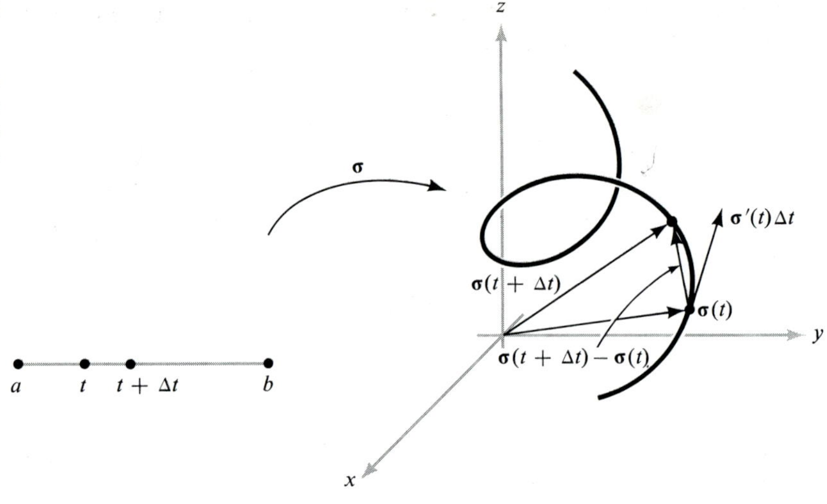

FIGURE 6.2.1
For small Δt, $\boldsymbol{\sigma}(t+\Delta t)-\boldsymbol{\sigma}(t) \approx \boldsymbol{\sigma}'(t)\,\Delta t$.

Now from the definition of the derivative we have the approximation $\Delta\mathbf{d} \approx \boldsymbol{\sigma}'(t)\,\Delta t$. The work done in going from $\boldsymbol{\sigma}(t)$ to $\boldsymbol{\sigma}(t+\Delta t)$ is therefore approximately

$$\mathbf{F}(\boldsymbol{\sigma}(t)) \cdot \Delta\mathbf{d} \approx \mathbf{F}(\boldsymbol{\sigma}(t)) \cdot \boldsymbol{\sigma}'(t)\,\Delta t$$

If we subdivide the interval $[a, b]$ into n equal parts $a = t_0 < t_1 < \cdots < t_n = b$, with $\Delta t = t_{i+1} - t_i$, then the work done by $\mathbf{F}$ is approximately

$$\sum_{i=0}^{n-1} \mathbf{F}(\boldsymbol{\sigma}(t_i)) \cdot \Delta\mathbf{d} \approx \sum_{i=0}^{n-1} \mathbf{F}(\boldsymbol{\sigma}(t_i)) \cdot \boldsymbol{\sigma}'(t_i)\,\Delta t$$

As $n \to \infty$ this approximation becomes better and better, so it is reasonable to take as our definition of work the limit of the above sum as $n \to \infty$. But this limit is given by the integral

$$\int_a^b \mathbf{F}(\boldsymbol{\sigma}(t)) \cdot \boldsymbol{\sigma}'(t)\,dt$$

This fundamental physical notion of work leads us to make the following mathematical definition.

Definition. *Let* $\mathbf{F}$ *be a vector field on* $\mathbb{R}^3$, *continuous on the* C^1 *path* $\boldsymbol{\sigma}\colon [a, b] \to \mathbb{R}^3$. *We define* $\int_{\boldsymbol{\sigma}} \mathbf{F}$, *the* ***line integral*** *of* $\mathbf{F}$ *along* $\boldsymbol{\sigma}$, *by the formula*

$$\int_{\boldsymbol{\sigma}} \mathbf{F} = \int_a^b \mathbf{F}(\boldsymbol{\sigma}(t)) \cdot \boldsymbol{\sigma}'(t)\,dt$$

that is, we integrate the dot product of $\mathbf{F}$ *with* $\boldsymbol{\sigma}'$ *over the interval* $[a, b]$.

Just as with scalar functions, we can also form $\int_{\boldsymbol{\sigma}} \mathbf{F}$ *if* $\mathbf{F}(\boldsymbol{\sigma}(t)) \cdot \boldsymbol{\sigma}'(t)$ *is only piecewise continuous.*

Recall that in Section 6.1 we defined the integral of a real-valued function f along the path $\boldsymbol{\sigma}$ as

$$\int_{\boldsymbol{\sigma}} f = \int_a^b f(\boldsymbol{\sigma}(t))\|\boldsymbol{\sigma}'(t)\|\ dt$$

where $\|\boldsymbol{\sigma}'(t)\|$ is the length of $\boldsymbol{\sigma}'(t)$. The line integral of $\mathbf{F}$ is actually an integral of a particular real-valued function f along $\boldsymbol{\sigma}$. To see this, choose

$$f(\boldsymbol{\sigma}(t)) = \mathbf{F}(\boldsymbol{\sigma}(t)) \cdot \mathbf{T}(t)$$

where

$$\mathbf{T}(t) = \boldsymbol{\sigma}'(t)/\|\boldsymbol{\sigma}'(t)\|$$

is the unit tangent vector (assume $\boldsymbol{\sigma}'(t) \neq \mathbf{0}$ so that $\mathbf{T}$ is defined). But $\mathbf{F} \cdot \mathbf{T}$ is exactly the component of $\mathbf{F}$ tangent to the curve (since $\mathbf{T}$ is a unit vector).

Consequently the line integral of a vector field $\mathbf{F}$ along a path $\boldsymbol{\sigma}(t)$ is equal to the path integral of the real-valued function that is the tangential component of $\mathbf{F}$, which is the projection of $\mathbf{F}$ onto the unit tangent $\mathbf{T}(t)$ (see Figure 1.2.9). Thus we have

$$\begin{aligned}\int_{\boldsymbol{\sigma}} \mathbf{F} &= \int_a^b \left[\mathbf{F}(\boldsymbol{\sigma}(t)) \cdot \frac{\boldsymbol{\sigma}'(t)}{\|\boldsymbol{\sigma}'(t)\|}\right]\|\boldsymbol{\sigma}'(t)\|\ dt \\ &= \int_a^b [\mathbf{F}(\boldsymbol{\sigma}(t)) \cdot \mathbf{T}(t)]\|\boldsymbol{\sigma}'(t)\|\ dt = \int_{\boldsymbol{\sigma}} f\end{aligned}$$

To compute a line integral in any particular case, one can either use the original definition or integrate the tangential component of $\mathbf{F}$ along $\boldsymbol{\sigma}$, whichever is easier or more appropriate.

EXAMPLE 1. Let $\boldsymbol{\sigma}(t) = (\sin t, \cos t, t)$, with $0 \le t \le 2\pi$. Let $\mathbf{F}(x, y, z) = x\mathbf{i} + y\mathbf{j} + z\mathbf{k}$. Then $\mathbf{F}(\boldsymbol{\sigma}(t)) = \mathbf{F}(\sin t, \cos t, t) = (\sin t)\mathbf{i} + (\cos t)\mathbf{j} + t\mathbf{k}$, and $\boldsymbol{\sigma}'(t) = (\cos t)\mathbf{i} - (\sin t)\mathbf{j} + \mathbf{k}$. Therefore,

$$\mathbf{F}(\boldsymbol{\sigma}(t)) \cdot \boldsymbol{\sigma}'(t) = \sin t \cos t - \cos t \sin t + t = t$$

and so

$$\int_{\boldsymbol{\sigma}} \mathbf{F} = \int_0^{2\pi} t\ dt = 2\pi^2$$

There are other common ways of writing line integrals, namely

$$\int_{\boldsymbol{\sigma}} \mathbf{F} \cdot d\mathbf{s} \qquad \text{or} \qquad \int_{\boldsymbol{\sigma}} F_1\ dx + F_2\ dy + F_3\ dz$$

where F_1, F_2, and F_3 are the components of the vector field $\mathbf{F}$. We call the expression $F_1\,dx + F_2\,dy + F_3\,dz$ a *differential form.** By *definition* the integral of a differential form is

$$\int_{\sigma} F_1\,dx + F_2\,dy + F_3\,dz = \int_a^b \left(F_1\frac{dx}{dt} + F_2\frac{dy}{dt} + F_3\frac{dz}{dt}\right)dt = \int_{\sigma}\mathbf{F}$$

EXAMPLE 2. Evaluate $\int_{\sigma} x^2\,dx + xy\,dy + dz$, where $\boldsymbol{\sigma}\colon [0, 1] \to \mathbb{R}^3$ is given by $\boldsymbol{\sigma}(t) = (t, t^2, 1) = (x(t), y(t), z(t))$.

We compute $dx/dt = 1$, $dy/dt = 2t$, $dz/dt = 0$; therefore,

$$\begin{aligned}\int_{\sigma} x^2\,dx + xy\,dy + dz &= \int_0^1 \left([x(t)]^2\frac{dx}{dt} + [x(t)y(t)]\frac{dy}{dt}\right)dt\\ &= \int_0^1 (t^2 + 2t^4)\,dt\\ &= \left[\frac{1}{3}t^3 + \frac{2}{5}t^5\right]_0^1 = \frac{11}{15}\end{aligned}$$

EXAMPLE 3. Evaluate $\int_{\sigma} \cos z\,dx + e^x\,dy + e^y\,dz$, where $\boldsymbol{\sigma}(t) = (1, t, e^t)$ and $0 \le t \le 2$.

We compute $dx/dt = 0$, $dy/dt = 1$, $dz/dt = e^t$, and so

$$\begin{aligned}\int_{\sigma} \cos z\,dx + e^x\,dy + e^y\,dz &= \int_0^2 (0 + e + e^{2t})\,dt\\ &= [et + \tfrac{1}{2}e^{2t}]_0^2 = 2e + \tfrac{1}{2}e^4 - \tfrac{1}{2}\end{aligned}$$

EXAMPLE 4. Let $\boldsymbol{\sigma}$ be the path

$$x = \cos^3\theta, \qquad y = \sin^3\theta, \qquad z = \theta, \qquad 0 \le \theta \le \frac{7\pi}{2}$$

(see Figure 6.2.2). Evaluate the integral $\int_{\sigma} (\sin z\,dx + \cos z\,dy - (xy)^{1/3})\,dz$.

In this case we have

$$\frac{dx}{d\theta} = -3\cos^2\theta\sin\theta, \qquad \frac{dy}{d\theta} = 3\sin^2\theta\cos\theta, \qquad \frac{dz}{d\theta} = 1$$

and so the integral is

$$\begin{aligned}&\int_{\sigma} \sin z\,dx + \cos z\,dy - (xy)^{1/3}\,dz\\ &= \int_0^{7\pi/2} (-3\cos^2\theta\sin^2\theta + 3\sin^2\theta\cos^2\theta - \cos\theta\sin\theta)\,d\theta\\ &= -\int_0^{7\pi/2} \cos\theta\sin\theta\,d\theta = -[\tfrac{1}{2}\sin^2\theta]_0^{7\pi/2} = -\tfrac{1}{2}\end{aligned}$$

* See Section 7.6 for a brief discussion of the theory of differential forms.

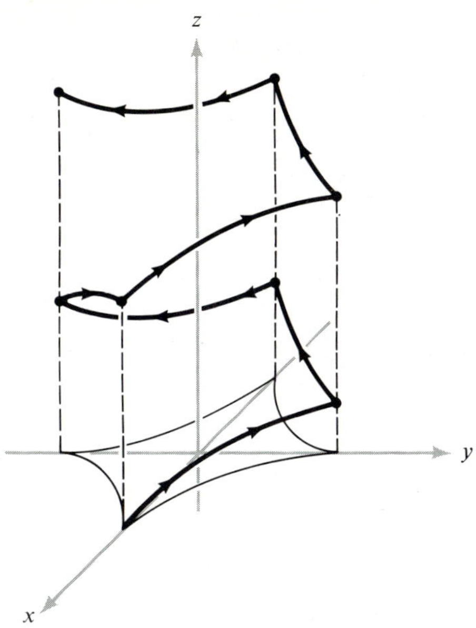

FIGURE 6.2.2
The image of the path $x = \cos^3 \theta$, $y = \sin^3 \theta$, $z = \theta$; $0 \le \theta \le 7\pi/2$.

EXAMPLE 5. Suppose $\mathbf{F}$ is the vector force field $\mathbf{F}(x, y, z) = x^3\mathbf{i} + y\mathbf{j} + z\mathbf{k}$. We may parametrize a circle of radius a in the yz-plane by setting

$$x = 0, \qquad y = a \cos \theta, \qquad z = a \sin \theta, \qquad 0 \le \theta \le 2\pi$$

Since the force field $\mathbf{F}$ is normal to the circle at every point on the circle, $\mathbf{F}$ will not do any work on a particle moving along the circle (Figure 6.2.3). The work done by $\mathbf{F}$ must therefore be 0. We can verify this by a direct computation:

$$W = \int_{\boldsymbol{\sigma}} \mathbf{F} = \int_{\boldsymbol{\sigma}} x^3\, dx + y\, dy + z\, dz$$

$$= \int_0^{2\pi} (0 - a^2 \cos \theta \sin \theta + a^2 \cos \theta \sin \theta)\, d\theta = 0$$

as inferred.

EXAMPLE 6. If we consider the field and curve of Example 4, we see that the work done by the field is $-\frac{1}{2}$, a negative quantity. This means that the field impedes movement along that path.

We have seen that the line integral $\int_{\boldsymbol{\sigma}} \mathbf{F}$ depends not only on the field $\mathbf{F}$ but also on the path $\boldsymbol{\sigma}: [a, b] \to \mathbb{R}^3$. In general, if $\boldsymbol{\sigma}$ and $\boldsymbol{\rho}$ are two

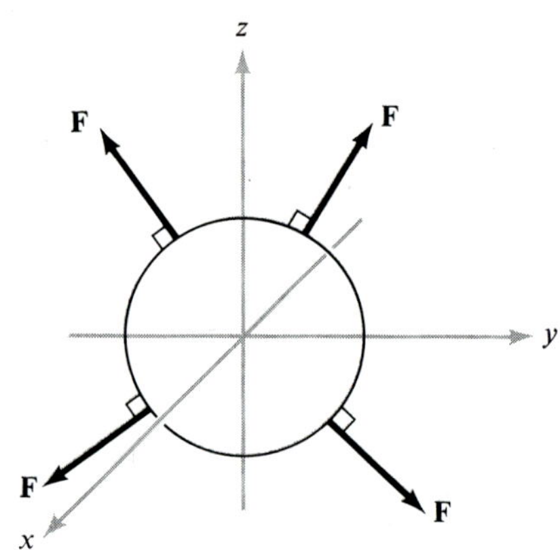

FIGURE 6.2.3
A vector field **F** *normal to a circle in the yz-plane.*

different paths in $\mathbb{R}^3$, $\int_{\boldsymbol{\sigma}} \mathbf{F} \neq \int_{\boldsymbol{\rho}} \mathbf{F}$. On the other hand, we shall see that it is true that $\int_{\boldsymbol{\sigma}} \mathbf{F} = \pm\int_{\boldsymbol{\rho}} \mathbf{F}$ for every vector field $\mathbf{F}$ if $\boldsymbol{\rho}$ is what we call a *reparametrization of* $\boldsymbol{\sigma}$.

Definition. *Let* $h: I \to I_1$ *be a* C^1 *real-valued function that is a one-to-one map of an interval* $I = [a, b]$ *onto another interval* $I_1 = [a_1, b_1]$. *Let* $\boldsymbol{\sigma}: I_1 \to \mathbb{R}^3$ *be a piecewise* C^1 *path. Then we call the composition*

$$\boldsymbol{\rho} = \boldsymbol{\sigma} \circ h: I \to \mathbb{R}^3$$

*a **reparametrization** of* $\boldsymbol{\sigma}$.

This means that $\boldsymbol{\rho}(t) = \boldsymbol{\sigma}(h(t))$, so h changes the variable; alternatively, one can think of h as changing the speed at which a point moves along the path. Indeed, observe that $\boldsymbol{\rho}'(t) = \boldsymbol{\sigma}'(h(t))h'(t)$, so the length of the velocity vector for $\boldsymbol{\sigma}$ is multiplied by the scalar factor $|h'(t)|$.

It is implicit in the definition that h must carry endpoints to endpoints; that is, either $h(a) = a_1$ and $h(b) = b_1$, or $h(a) = b_1$ and $h(b) = a_1$. We thus distinguish two types of reparametrizations. If $\boldsymbol{\sigma} \circ h$ is a reparametrization of $\boldsymbol{\sigma}$ then either

$$\boldsymbol{\sigma} \circ h(a) = \boldsymbol{\sigma}(a_1) \quad \text{and} \quad \boldsymbol{\sigma} \circ h(b) = \boldsymbol{\sigma}(b_1)$$

or

$$\boldsymbol{\sigma} \circ h(a) = \boldsymbol{\sigma}(b_1) \quad \text{and} \quad \boldsymbol{\sigma} \circ h(b) = \boldsymbol{\sigma}(a_1)$$

In the first case, the reparametrization is called *orientation preserving*, and a particle tracing the path $\boldsymbol{\sigma} \circ h$ moves in the same direction as a particle tracing $\boldsymbol{\sigma}$. In the second case, the reparametrization is called *orientation reversing*, and a particle tracing the path $\boldsymbol{\sigma} \circ h$ moves in the opposite direction to that of a particle tracing $\boldsymbol{\sigma}$ (Figure 6.2.4).

For example, if C is the image of a path $\boldsymbol{\sigma}$ as shown in Figure 6.2.5,

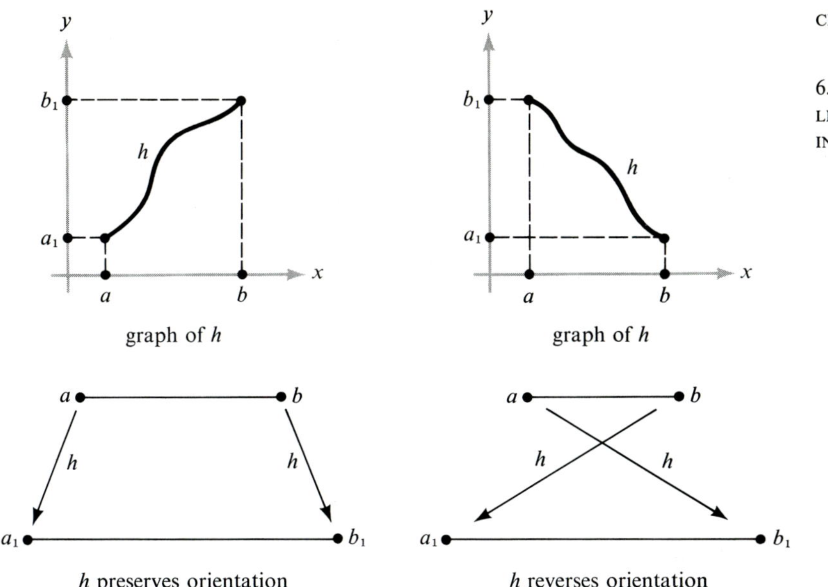

FIGURE 6.2.4
Illustrating an orientation-preserving reparametrization (left) and an orientation-reversing reparametrization (right).

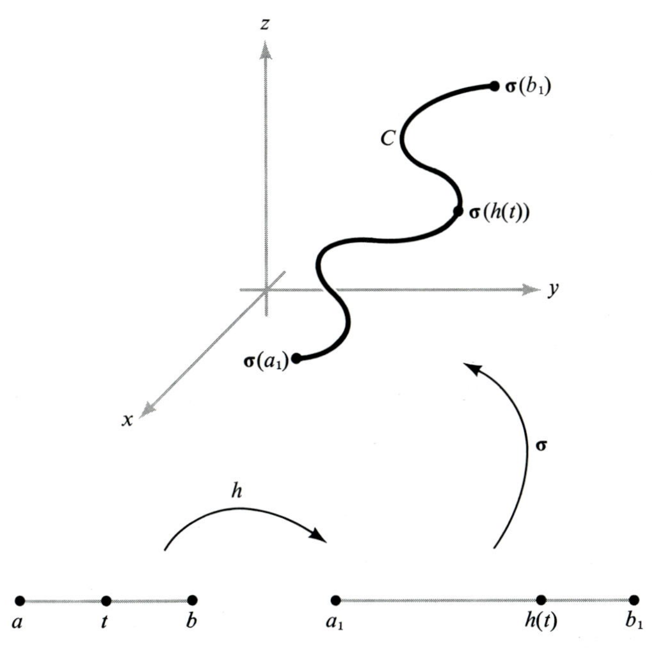

FIGURE 6.2.5
The path $\boldsymbol{\rho} = \boldsymbol{\sigma} \circ h$ is a reparametrization of $\boldsymbol{\sigma}$.

that is, $C = \boldsymbol{\sigma}([a_1, b_1])$, and if h is orientation preserving, then $\boldsymbol{\sigma} \circ h(t)$ will go from $\boldsymbol{\sigma}(a_1)$ to $\boldsymbol{\sigma}(b_1)$ as t goes from a to b; and if h is orientation reversing, $\boldsymbol{\sigma} \circ h(t)$ will go from $\boldsymbol{\sigma}(b_1)$ to $\boldsymbol{\sigma}(a_1)$ as t goes from a to b.

EXAMPLE 7. Let $\boldsymbol{\sigma}: [a, b] \to \mathbb{R}^3$ be any piecewise C^1 path. Then:

(a) The path $\boldsymbol{\sigma}_{\mathrm{op}}: [a, b] \to \mathbb{R}^3$, $t \mapsto \boldsymbol{\sigma}(a + b - t)$, is a reparametrization of $\boldsymbol{\sigma}$ corresponding to the map $h: [a, b] \to [a, b]$, $t \mapsto a + b - t$; we call $\boldsymbol{\sigma}_{\mathrm{op}}$ the *opposite path* to $\boldsymbol{\sigma}$. This reparametrization is orientation reversing.

(b) The path $\boldsymbol{\rho}: [0, 1] \to \mathbb{R}^3$, $t \mapsto \boldsymbol{\sigma}(a + (b - a)t)$, is an orientation-preserving reparametrization of $\boldsymbol{\sigma}$ corresponding to a change of coordinates $h: [0, 1] \to [a, b]$, $t \mapsto a + (b - a)t$.

Theorem 1. *Let* $\mathbf{F}$ *be a vector field continuous on the* C^1 *path* $\boldsymbol{\sigma}: [a_1, b_1] \to \mathbb{R}^3$, *and let* $\boldsymbol{\rho}: [a, b] \to \mathbb{R}^3$ *be a reparametrization of* $\boldsymbol{\sigma}$. *If* $\boldsymbol{\rho}$ *is orientation preserving, then*

$$\int_{\boldsymbol{\rho}} \mathbf{F} = \int_{\boldsymbol{\sigma}} \mathbf{F}$$

while if $\boldsymbol{\rho}$ *is orientation reversing, then*

$$\int_{\boldsymbol{\rho}} \mathbf{F} = -\int_{\boldsymbol{\sigma}} \mathbf{F}$$

Proof. By hypothesis, we have a map h such that $\boldsymbol{\rho} = \boldsymbol{\sigma} \circ h$. By the Chain Rule

$$\boldsymbol{\rho}'(t) = \boldsymbol{\sigma}'(h(t))h'(t)$$

so we have

$$\int_{\boldsymbol{\rho}} \mathbf{F} = \int_a^b [\mathbf{F}(\boldsymbol{\sigma}(h(t))) \cdot \boldsymbol{\sigma}'(h(t))]h'(t)\, dt$$

By Change of Variables (p. 263) this becomes, letting $s = h(t)$,

$$\int_{h(a)}^{h(b)} \mathbf{F}(\boldsymbol{\sigma}(s)) \cdot \boldsymbol{\sigma}'(s)\, ds$$

$$= \begin{cases} \displaystyle\int_{a_1}^{b_1} \mathbf{F}(\boldsymbol{\sigma}(s)) \cdot \boldsymbol{\sigma}'(s)\, ds = \int_{\boldsymbol{\sigma}} \mathbf{F} & \text{if } \boldsymbol{\rho} \text{ is orientation preserving} \\ \displaystyle\int_{b_1}^{a_1} \mathbf{F}(\boldsymbol{\sigma}(s)) \cdot \boldsymbol{\sigma}'(s)\, ds = -\int_{\boldsymbol{\sigma}} \mathbf{F} & \text{if } \boldsymbol{\rho} \text{ is orientation reversing} \end{cases}$$ ∎

Theorem 1 also holds for piecewise C^1 paths, as may be seen by breaking up the intervals into segments on which the paths are C^1 and summing the integrals over the several intervals.

Thus, if it is convenient to reparametrize a path when evaluating an integral, Theorem 1 assures us that the value of the integral will not be affected, except possibly for the sign, depending on the reparametrization.

EXAMPLE 8. Let $\mathbf{F}(x, y, z) = yz\mathbf{i} + xz\mathbf{j} + xy\mathbf{k}$ and $\boldsymbol{\sigma}\colon [-5, 10] \to \mathbb{R}^3$, $t \mapsto (t, t^2, t^3)$. Evaluate $\int_{\boldsymbol{\sigma}} \mathbf{F}$ and $\int_{\boldsymbol{\sigma}_{\text{op}}} \mathbf{F}$.

For $\boldsymbol{\sigma}$ we have $dx/dt = 1$, $dy/dt = 2t$, $dz/dt = 3t^2$, and $\mathbf{F}(\boldsymbol{\sigma}(t)) = t^5\mathbf{i} + t^4\mathbf{j} + t^3\mathbf{k}$. Therefore

$$\int_{\boldsymbol{\sigma}} \mathbf{F} = \int_{-5}^{10} \left(F_1 \frac{dx}{dt} + F_2 \frac{dy}{dt} + F_3 \frac{dz}{dt}\right) dt$$

$$= \int_{-5}^{10} (t^5 + 2t^5 + 3t^5)\, dt = [t^6]_{-5}^{10} = 984{,}375$$

On the other hand, for

$$\boldsymbol{\sigma}_{\text{op}}\colon [-5, 10] \to \mathbb{R}^3,\ t \mapsto \boldsymbol{\sigma}(5 - t) = (5 - t, (5 - t)^2, (5 - t)^3)$$

we have $dx/dt = -1$, $dy/dt = -10 + 2t = -2(5 - t)$, $dz/dt = -75 + 30t - 3t^2 = -3(5 - t)^2$, and $\mathbf{F}(\boldsymbol{\sigma}_{\text{op}}(t)) = (5 - t)^5\mathbf{i} + (5 - t)^4\mathbf{j} + (5 - t)^3\mathbf{k}$. Therefore

$$\int_{\boldsymbol{\sigma}_{\text{op}}} \mathbf{F} = \int_{-5}^{10} (-(5 - t)^5 - 2(5 - t)^5 - 3(5 - t)^5)\, dt$$

$$= [(5 - t)^6]_{-5}^{10} = -984{,}375$$

We are interested in reparametrizations because if the image of a particular $\boldsymbol{\sigma}$ can be represented in many ways, we want to be sure that integrals over this image do not depend on the particular parametrization. For example, for some problems the unit circle may be conveniently represented by the map $\boldsymbol{\rho}$ given by

$$x(t) = \cos 2t, \qquad y(t) = \sin 2t, \qquad 0 \le t \le \pi$$

Theorem 1 guarantees that any integral computed for this representation will be the same as when we represent the circle by the map $\boldsymbol{\sigma}$ given by

$$x(t) = \cos t, \qquad y(t) = \sin t, \qquad 0 \le t \le 2\pi$$

since $\boldsymbol{\rho} = \boldsymbol{\sigma} \circ h$, where $h(t) = 2t$, and thus $\boldsymbol{\rho}$ is a reparametrization of $\boldsymbol{\sigma}$. However, notice that the map γ given by

$$x(t) = \cos t, \qquad y(t) = \sin t, \qquad 0 \le t \le 4\pi$$

is not a reparametrization of $\boldsymbol{\sigma}$. Although it traces out the same image (the circle), it does so twice. (Why does this imply that γ is not a reparametrization of $\boldsymbol{\sigma}$?)

The line integral $\int_{\boldsymbol{\sigma}} \mathbf{F}$ differs from the path integral $\int_{\boldsymbol{\sigma}} f$. We have seen that

$$\int_{\boldsymbol{\sigma}} \mathbf{F} = \int_{\boldsymbol{\sigma}} f \tag{1}$$

where

$$f(\boldsymbol{\sigma}(t)) = \mathbf{F}(\boldsymbol{\sigma}(t)) \cdot \frac{\boldsymbol{\sigma}'(t)}{\|\boldsymbol{\sigma}'(t)\|}$$

Although equation (1) establishes a relationship between line integrals and path integrals, there is a distinct difference between them, other than the obvious fact that the former is an integral of a vector field and the latter of a real-valued function.

The line integral is an *oriented integral*, in that a change of sign occurs (as we have seen in Theorem 1) if the orientation of the curve is reversed. The path integral does not have this property. This follows from the fact that changing t to $-t$ (reversing orientation) just changes the sign of $\boldsymbol{\sigma}'(t)$, not its length.

The technique used in Theorem 1 can be used to prove the next result.

Theorem 2. *Let $\boldsymbol{\sigma}$ be piecewise C^1, f a continuous (real-valued) function on the image of $\boldsymbol{\sigma}$, and let $\boldsymbol{\rho}$ be any reparametrization of $\boldsymbol{\sigma}$. Then*

$$\int_{\boldsymbol{\sigma}} f(x, y, z)\, ds = \int_{\boldsymbol{\rho}} f(x, y, z)\, ds \tag{2}$$

Equation (1) together with Theorem 1 seems, at first sight, to contradict equation (2), since if $\boldsymbol{\rho}$ is an orientation-reversing reparametrization of $\boldsymbol{\sigma}$ we have by Theorem 1

$$\int_{\boldsymbol{\sigma}} \mathbf{F} = -\int_{\boldsymbol{\rho}} \mathbf{F}$$

But by (1) we can conclude (or can we?) that

$$\int_{\boldsymbol{\sigma}} f = \int_{\boldsymbol{\sigma}} \mathbf{F} = -\int_{\boldsymbol{\rho}} \mathbf{F} = -\int_{\boldsymbol{\rho}} f \tag{3}$$

which contradicts (2). The fallacy lies in the fact that the function f in equality (1) depends on $\boldsymbol{\sigma}$, whereas the function f in (2) does not depend on $\boldsymbol{\sigma}$ or $\boldsymbol{\rho}$. Thus equation (3) should read

$$\int_{\boldsymbol{\sigma}} f_{\boldsymbol{\sigma}} = \int_{\boldsymbol{\sigma}} \mathbf{F} = -\int_{\boldsymbol{\rho}} \mathbf{F} = -\int_{\boldsymbol{\rho}} f_{\boldsymbol{\rho}} \tag{3'}$$

where f_σ and f_ρ depend on $\boldsymbol{\sigma}$ and $\boldsymbol{\rho}$ according to the equalities

$$f_\sigma(\boldsymbol{\sigma}(t)) = \mathbf{F}(\boldsymbol{\sigma}(t)) \cdot \frac{\boldsymbol{\sigma}'(t)}{\|\boldsymbol{\sigma}'(t)\|}$$

$$f_\rho(\boldsymbol{\rho}(t)) = \mathbf{F}(\boldsymbol{\rho}(t)) \cdot \frac{\boldsymbol{\rho}'(t)}{\|\boldsymbol{\rho}'(t)\|}$$

Thus there is no contradiction, and theorems 1 and 2 are perfectly consistent.

We next consider a simple but often very useful technique for evaluating line integrals. Recall that a vector field $\mathbf{F}$ is a *gradient vector field* if $\mathbf{F} = \nabla f$ for some real-valued function f. Thus

$$\mathbf{F} = \frac{\partial f}{\partial x}\mathbf{i} + \frac{\partial f}{\partial y}\mathbf{j} + \frac{\partial f}{\partial z}\mathbf{k}$$

Suppose $G, g\colon [a, b] \to \mathbb{R}$ are real-valued continuous functions with $G' = g$. Then by the Fundamental Theorem of Calculus $\int_a^b g(x)\, dx = G(b) - G(a)$. Thus the value of the integral of g depends only on the value of G at the endpoints of the interval $[a, b]$. Since ∇f represents the derivative of f, one can ask whether $\int_{\boldsymbol{\sigma}} \nabla f$ is completely determined by the value of f at the endpoints $\boldsymbol{\sigma}(a)$ and $\boldsymbol{\sigma}(b)$. The answer is contained in the following generalization of the Fundamental Theorem of Calculus.

Theorem 3. *Suppose that $f\colon \mathbb{R}^3 \to \mathbb{R}$ is C^1 and that $\boldsymbol{\sigma}\colon [a, b] \to \mathbb{R}^3$ is a piecewise C^1 path. Then*

$$\int_{\boldsymbol{\sigma}} \nabla f = f(\boldsymbol{\sigma}(b)) - f(\boldsymbol{\sigma}(a))$$

Proof. We apply the Chain Rule to the composite function

$$F\colon t \mapsto f(\boldsymbol{\sigma}(t))$$

to obtain

$$F'(t) = (f \circ \boldsymbol{\sigma})'(t) = \nabla f(\boldsymbol{\sigma}(t)) \cdot \boldsymbol{\sigma}'(t)$$

The function F is a real function of the variable t, and so by the Fundamental Theorem of Calculus, we have

$$\int_a^b F'(t)\, dt = F(b) - F(a) = f(\boldsymbol{\sigma}(b)) - f(\boldsymbol{\sigma}(a))$$

Therefore,

$$\int_{\boldsymbol{\sigma}} \nabla f = \int_a^b \nabla f(\boldsymbol{\sigma}(t)) \cdot \boldsymbol{\sigma}'(t)\, dt = \int_a^b F'(t)\, dt = F(b) - F(a)$$
$$= f(\boldsymbol{\sigma}(b)) - f(\boldsymbol{\sigma}(a)) \quad \blacksquare$$

EXAMPLE 9. Let $\boldsymbol{\sigma}$ be the path $\boldsymbol{\sigma}(t) = (t^4/4, \sin^3(t\pi/2), 0)$, $t \in [0, 1]$. Evaluate

$$\int_{\boldsymbol{\sigma}} y\,dx + x\,dy$$

(which means $\int_{\boldsymbol{\sigma}} y\,dx + x\,dy + 0\,dz$).

We recognize $y\,dx + x\,dy$, or equivalently, the vector field $y\mathbf{i} + x\mathbf{j} + 0\mathbf{k}$, as the gradient of the function $f(x, y, z) = xy$. Thus

$$\begin{aligned}\int_{\boldsymbol{\sigma}} y\,dx + x\,dy &= f(\boldsymbol{\sigma}(1)) - f(\boldsymbol{\sigma}(0)) \\ &= \tfrac{1}{4} \cdot 1 - 0 = \tfrac{1}{4}\end{aligned}$$

Obviously, if one can recognize the integrand as a gradient, then evaluation of the integral becomes much easier. For example, the reader should try to work out the above integral directly. In one-variable calculus, every integral is, in principle, obtainable by finding an antiderivative. For vector fields, however, this is not always true, because a vector field need not always be a gradient. This point will be examined in detail in Section 7.3.

We have seen how to define path integrals (integrals of scalar functions) and line integrals (integrals of vector functions) over parametrized curves. We have also seen that our work is simplified if we make a judicious choice of parametrization.

Since these integrals are independent of the parametrization (except possibly for the sign), it seems natural to try to write out the theory in a way that is independent of the parametrization, and that is thereby more "geometrical." We do this briefly and somewhat informally in the following discussion.

Definition. *We define a* ***simple curve*** *to be the image of a piecewise C^1 map $\boldsymbol{\sigma}: I \to \mathbb{R}^3$ that is one-to-one on an interval I. Thus a simple curve is one that does not intersect itself (Figure 6.2.6). If $I = [a, b]$, we call $\boldsymbol{\sigma}(a)$*

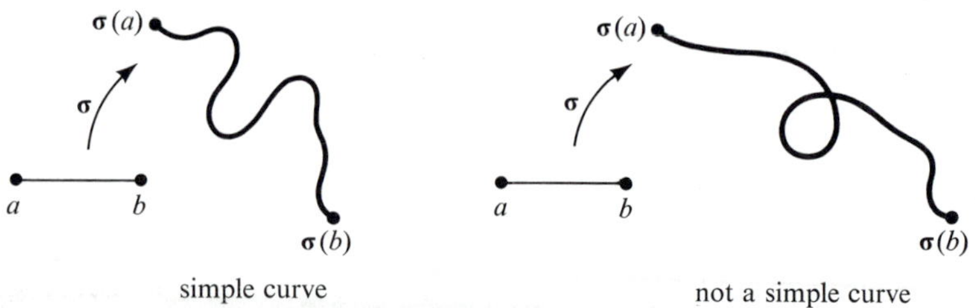

FIGURE 6.2.6
A simple curve has no self intersections.

and $\boldsymbol{\sigma}(b)$ *the* ***endpoints*** *of the curve. Each simple curve* C *has two orientations or directions associated with it. If* P *and* Q *are the endpoints of the curve, then we can consider* C *either as directed from* P *to* Q *or from* Q *to* P*. The simple curve* C *together with a sense of direction is called an* ***oriented simple curve*** *or* ***directed simple curve*** *(Figure 6.2.7).*

FIGURE 6.2.7
There are two possible senses of direction on a curve from P to Q (left).

EXAMPLE 10. If $I = [a, b]$ is a closed interval on the x-axis, then I, as a curve, has two orientations: one corresponding to motion from a to b (left to right) and the other corresponding to motion from b to a (right to left). If f is a real function continuous on I, then denoting I with the first orientation by I^+, and I with the second orientation by I^-, we have

$$\int_{I^+} f(x)\,dx = \int_a^b f(x)\,dx = -\int_b^a f(x)\,dx = -\int_{I^-} f(x)\,dx$$

Definition. *By a* ***simple closed curve*** *we mean the image of a piecewise* C^1 *map* $\boldsymbol{\sigma}\colon [a, b] \to \mathbb{R}^3$ *that is one-to-one on* $[a, b[$ *and satisfies* $\boldsymbol{\sigma}(a) = \boldsymbol{\sigma}(b)$ *(Figure 6.2.8). If* $\boldsymbol{\sigma}$ *satisfies the condition* $\boldsymbol{\sigma}(a) = \boldsymbol{\sigma}(b)$ *but is not*

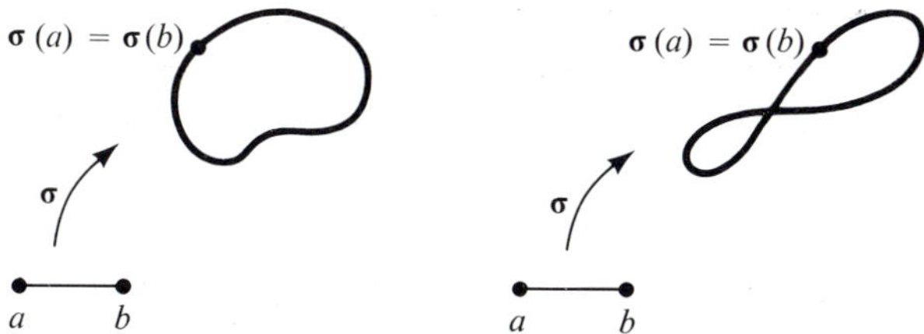

FIGURE 6.2.8
A simple closed curve (left) and a closed curve that is not simple (right).

necessarily one-to-one on $[a, b[$*, we call its image a* ***closed curve****. Simple closed curves have two orientations, corresponding to the two possible directions of motion along the curve (Figure 6.2.9).*

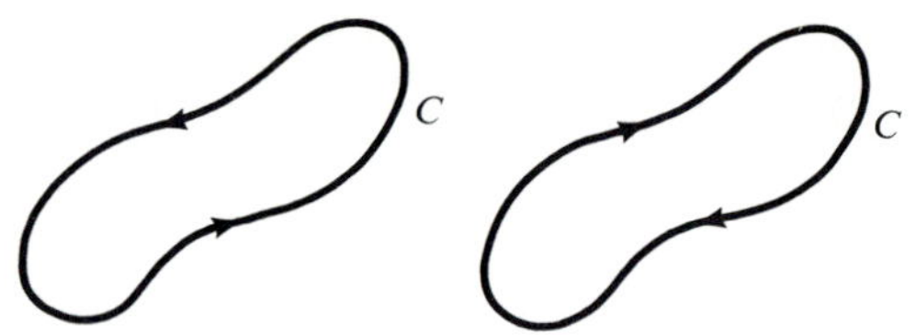

FIGURE 6.2.9
Two possible orientations for a simple closed curve C.

Given a simple curve or a simple closed curve C, we can write $\int_C f$ and (*if C is oriented*) $\int_C \mathbf{F}$ unambiguously, by virtue of theorems 1 and 2. (We have not proved that any two one-to-one paths $\boldsymbol{\sigma}$ and $\boldsymbol{\eta}$ with the same image must be reparametrizations of each other, but this technical point will be omitted here.) The point we want to make here is that while a curve must be parametrized to make integration along it tractable, it is not necessary to include the parametrization in our notation for the integral.

A given simple closed curve can be parametrized in many different ways. Figure 6.2.10 shows C represented as the image of a map $\boldsymbol{\rho}$, with

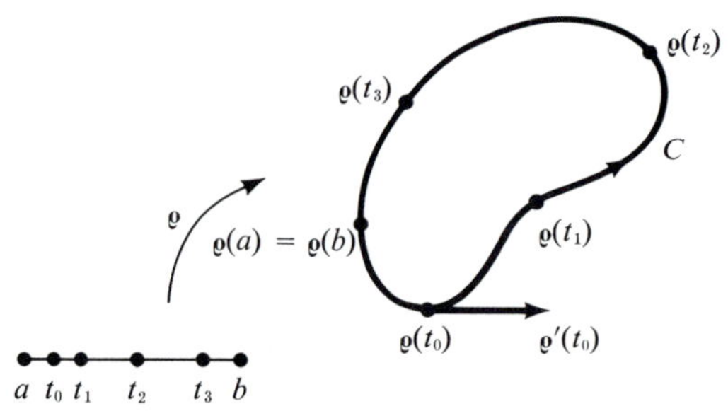

FIGURE 6.2.10
As t goes from a to b $\boldsymbol{\rho}(t)$ moves around the curve C in some fixed direction.

$\boldsymbol{\rho}(t)$ progressing in the correct direction around an oriented curve C as t ranges from a to b. Note that $\boldsymbol{\rho}'(t_0)$ points in this direction also. The speed with which we traverse C may vary from parametrization to parametrization, but the integral will not, according to Theorems 1 and 2.

The following precaution should be noted in regard to these remarks. It is possible to have two mappings $\boldsymbol{\sigma}$ and $\boldsymbol{\eta}$ with the same image and inducing the same orientation on the image, such that

$$\int_{\boldsymbol{\sigma}} \mathbf{F} \neq \int_{\boldsymbol{\eta}} \mathbf{F}$$

For an example, let $\boldsymbol{\sigma}(t) = (\cos t, \sin t, 0)$, $\boldsymbol{\eta}(t) = (\cos 2t, \sin 2t, 0)$, $0 \leq t \leq 2\pi$ with $\mathbf{F}(x, y, z) = (y, 0, 0)$. Then

$$\int_{\boldsymbol{\sigma}} \mathbf{F} = \int_0^{2\pi} F_1(\boldsymbol{\sigma}(t)) \frac{dx}{dt}\, dt$$

(the terms containing F_2 and F_3 are zero)

$$= -\int_0^{2\pi} \sin^2 t\, dt = -\pi$$

But $\int_{\boldsymbol{\eta}} \mathbf{F} = -2 \int_0^{2\pi} \sin^2 2t\, dt = -2\pi$. Clearly $\boldsymbol{\sigma}$ and $\boldsymbol{\eta}$ have the same image, namely, the unit circle in the xy-plane, and moreover, they traverse the unit circle in the same direction; yet $\int_{\boldsymbol{\sigma}} \mathbf{F} \neq \int_{\boldsymbol{\eta}} \mathbf{F}$. The reason for this is that $\boldsymbol{\sigma}$ is one-to-one but $\boldsymbol{\eta}$ is not ($\boldsymbol{\eta}$ traverses the unit circle twice in a counterclockwise direction); therefore $\boldsymbol{\eta}$ is not a parametrization of the unit circle as a simple closed curve.

If C is an oriented simple curve or an oriented simple closed curve we therefore may define

$$\int_C \mathbf{F} = \int_{\boldsymbol{\sigma}} \mathbf{F} \quad \text{and} \quad \int_C f = \int_{\boldsymbol{\sigma}} f \tag{4}$$

where $\boldsymbol{\sigma}$ is any orientation-preserving parametrization of C. As we have mentioned, these integrals do not depend on the choice of $\boldsymbol{\sigma}$ as long as $\boldsymbol{\sigma}$ is one-to-one (except possibly at the endpoints). If $\mathbf{F} = P\mathbf{i} + Q\mathbf{j} + R\mathbf{k}$ is a vector field, then in differential-form notation we write

$$\int_C \mathbf{F} = \int_C P\, dx + Q\, dy + R\, dz$$

If C^- is the same curve as C, but with the opposite orientation, then

$$\int_C \mathbf{F} = -\int_{C^-} \mathbf{F}$$

If C is a (oriented) curve that is made up of several (oriented) component curves C_i, $i = 1, \ldots, k$, as in Figure 6.2.11, then we shall write

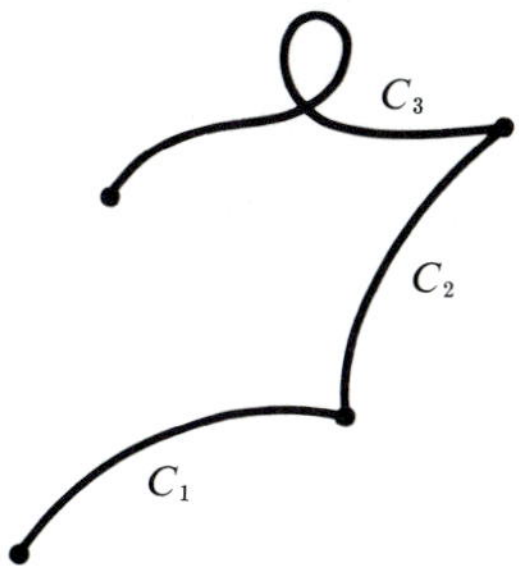

FIGURE 6.2.11
A curve can be made up of several components.

$C = C_1 + C_2 + \cdots + C_k$. Since we can parametrize C by parametrizing the pieces $C_1, \ldots, C_k$ separately, we get

$$\int_C \mathbf{F} = \int_{C_1} \mathbf{F} + \int_{C_2} \mathbf{F} + \cdots + \int_{C_k} \mathbf{F} \tag{5}$$

One reason for writing a curve as a sum of components is that it may be easier to parametrize the components C_i individually than it is to parametrize C as a whole. If that is the case, formula (5) provides a convenient way of evaluating $\int_C \mathbf{F}$.

EXAMPLE 11. Consider C, the perimeter of the unit square in $\mathbb{R}^2$, oriented in the counterclockwise sense (see Figure 6.2.12). Evaluate the line integral $\int_C x^2\,dx + xy\,dy$.

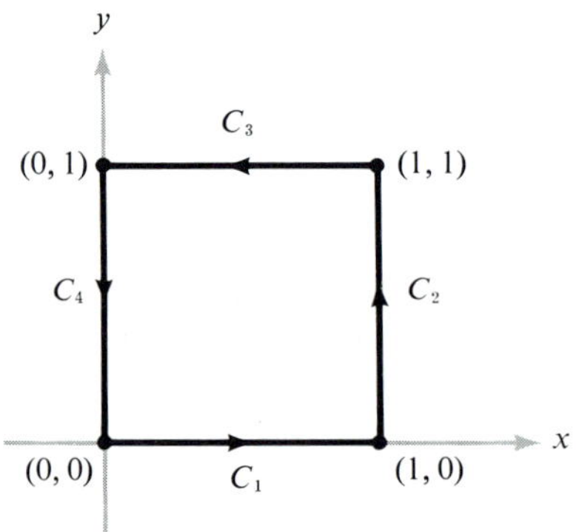

FIGURE 6.2.12
The perimeter of the unit square, parametrized in four pieces.

We may evaluate the integral by using any parametrization of C that induces the given orientation. For example:

$$\boldsymbol{\sigma}\colon [0, 4] \to \mathbb{R}^2,\ t \mapsto \begin{cases} (t, 0) & 0 \le t \le 1 \\ (1, t-1) & 1 \le t \le 2 \\ (3-t, 1) & 2 \le t \le 3 \\ (0, 4-t) & 3 \le t \le 4 \end{cases}$$

Then

$$\int_C x^2\,dx + xy\,dy = \int_0^1 (t^2 + 0)\,dt + \int_1^2 (0 + (t-1))\,dt$$

$$+ \int_2^3 (-(3-t)^2 + 0)\,dt + \int_3^4 (0 + 0)\,dt = \tfrac{1}{3} + \tfrac{1}{2} + (-\tfrac{1}{3}) + 0 = \tfrac{1}{2}$$

Now let us re-evaluate this line integral, using formula (5) and parametrizing the C_i separately. Notice that the curve $C = C_1 + C_2 + C_3 + C_4$, where C_i are the oriented curves pictured in Figure 6.2.12. These can be parametrized as follows:

$$C_1\colon \boldsymbol{\sigma}_1(t) = (t, 0), 0 \le t \le 1$$

$$C_2\colon \boldsymbol{\sigma}_2(t) = (1, t), 0 \le t \le 1$$

$$C_3\colon \boldsymbol{\sigma}_3(t) = (1 - t, 1), 0 \le t \le 1$$

$$C_4\colon \boldsymbol{\sigma}_4(t) = (0, 1 - t), 0 \le t \le 1$$

So

$$\int_{C_1} x^2\,dx + xy\,dy = \int_0^1 t^2\,dt = \tfrac{1}{3}$$

$$\int_{C_2} x^2\,dx + xy\,dy = \int_0^1 t\,dt = \tfrac{1}{2}$$

$$\int_{C_3} x^2\,dx + xy\,dy = \int_0^1 -(1 - t)^2\,dt = -\tfrac{1}{3}$$

$$\int_{C_4} x^2\,dx + xy\,dy = \int_0^1 0\,dt = 0$$

Thus

$$\int_C x^2\,dx + xy\,dy = \tfrac{1}{3} + \tfrac{1}{2} - \tfrac{1}{3} + 0 = \tfrac{1}{2}$$

as before.

Example 12. An interesting application of the line integral is the mathematical formulation of Ampère's law, which relates electric currents to their magnetic effects.* Suppose **H** denotes a magnetic field in $\mathbb{R}^3$, and let C be a closed oriented curve in $\mathbb{R}^3$. Ampère's law states that (in appropriate physical units)

$$\int_C \mathbf{H} = I$$

where I is the net current that passes through any surface bounded by C (see Figure 6.2.13).

Finally, let us mention that the path integral has another important physical meaning, specifically, the interpretation of $\int_C \mathbf{V}$, where **V** is the

* The discovery that electric currents produce magnetic effects was made by Oersted about 1820. See any elementary physics text for discussions of the physical basis of these ideas.

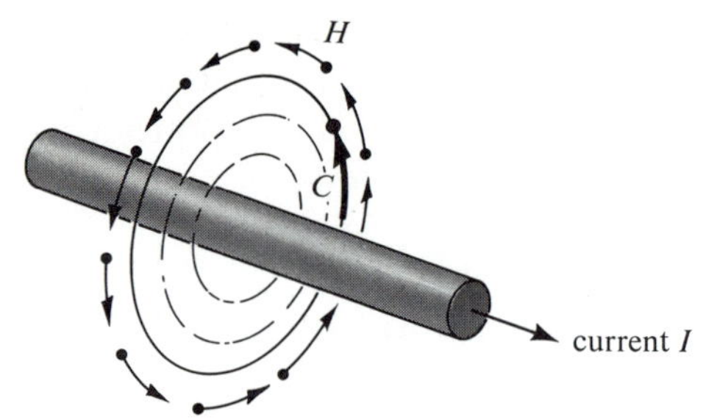

FIGURE 6.2.13
The magnetic field **H** *surrounding a wire carrying a current I satisfies Ampere's law:* $\int_C \mathbf{H} = I$.

velocity field of a fluid. We shall discuss this interpretation in Section 7.2. Thus, a wide variety of physical concepts, from the notion of work to electromagnetic fields and the motions of fluids, can be analyzed with the help of path integrals.

EXERCISES

1. Let $\mathbf{F}(x, y, z) = x\mathbf{i} + y\mathbf{j} + z\mathbf{k}$. Evaluate the integral of $\mathbf{F}$ along each of the following paths.
   (a) $\boldsymbol{\sigma}(t) = (t, t, t), \quad 0 \le t \le 1$
   (b) $\boldsymbol{\sigma}(t) = (\cos t, \sin t, 0), \quad 0 \le t \le 2\pi$
2. Evaluate each of the following integrals.
   (a) $\int_{\boldsymbol{\sigma}} x\,dy - y\,dx, \quad \boldsymbol{\sigma}(t) = (\cos t, \sin t), \quad 0 \le t \le 2\pi$
   (b) $\int_{\boldsymbol{\sigma}} x\,dx + y\,dy, \quad \boldsymbol{\sigma}(t) = (\cos \pi t, \sin \pi t), \quad 0 \le t \le 2$
   (c) $\int_{\boldsymbol{\sigma}} yz\,dx + xz\,dy + xy\,dz$, where $\boldsymbol{\sigma}$ consists of straight line segments joining (1, 0, 0) to (0, 1, 0) to (0, 0, 1)
3. Consider the force $\mathbf{F}(x, y, z) = x\mathbf{i} + y\mathbf{j} + z\mathbf{k}$. Compute the work done in moving a particle along the parabola $y = x^2$, $z = 0$, from $x = -1$ to $x = 2$.
4. Let $\boldsymbol{\sigma}$ be a smooth path.
   (a) Suppose $\mathbf{F}$ is perpendicular to $\boldsymbol{\sigma}'(t)$ at $\boldsymbol{\sigma}(t)$. Show that
   $$\int_{\boldsymbol{\sigma}} \mathbf{F} = 0$$
   (b) If $\mathbf{F}$ is parallel to $\boldsymbol{\sigma}'(t)$ at $\boldsymbol{\sigma}(t)$, show that
   $$\int_{\boldsymbol{\sigma}} \mathbf{F} = \int_{\boldsymbol{\sigma}} \|\mathbf{F}\|$$
   (By parallel to $\boldsymbol{\sigma}'(t)$ we mean that $\mathbf{F}(\boldsymbol{\sigma}(t)) = \lambda \boldsymbol{\sigma}'(t)$ where $\lambda > 0$.)
5. Suppose $\boldsymbol{\sigma}$ has length l, and $\|\mathbf{F}\| \le M$. Then prove
   $$\left| \int_{\boldsymbol{\sigma}} \mathbf{F} \right| \le Ml$$

6. Evaluate $\int_{\sigma} \mathbf{F}$ where $\mathbf{F}(x, y, z) = y\mathbf{i} + 2x\mathbf{j} + y\mathbf{k}$ and $\boldsymbol{\sigma}(t) = t\mathbf{i} + t^2\mathbf{j} + t^3\mathbf{k}$, $0 \leq t \leq 1$.
7. Evaluate $\int_{\sigma} y\,dx + (3y^3 - x)\,dy + z\,dz$ for each of the paths $\boldsymbol{\sigma}(t) = (t, t^n, 0)$, $0 \leq t \leq 1$, $n = 1, 2, 3, \ldots$
8. The image of $t \mapsto (\cos^3 t, \sin^3 t)$, $0 \leq t \leq 2\pi$, in the plane is shown in Figure 6.2.14. Evaluate the integral of the vector field $\mathbf{F}(x, y) = x\mathbf{i} + y\mathbf{j}$ around this curve.

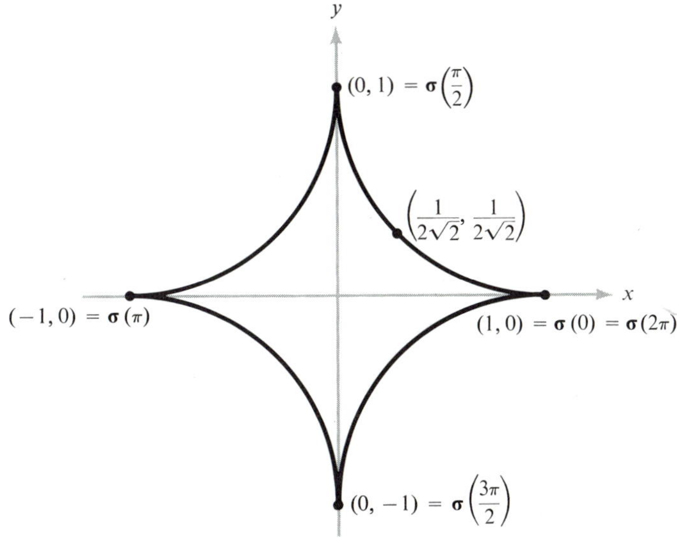

FIGURE 6.2.14
The hypocycloid $\boldsymbol{\sigma}(t) = (\cos^3 t, \sin^3 t)$ *(Exercise 8).*

9. This exercise refers to Example 12. Let L be a very long wire, a planar section of which (with the plane perpendicular to the wire) is shown in Figure 6.2.15. Suppose this plane is the xy-plane. Experiments show that $\mathbf{H}$ is tangent to every circle in the xy-plane whose center is the axis of L, and that the magnitude of $\mathbf{H}$ is constant on every such circle C. Thus $\mathbf{H} = H\mathbf{T}$, where $\mathbf{T}$ is a unit tangent vector to C and H is some scalar. Using this information, show that $H = I/2\pi r$, where r is the radius of circle C, and I is the current flowing in the wire.
10. Suppose $\boldsymbol{\sigma}$, $\boldsymbol{\psi}$ are two paths with the same endpoints, and $\mathbf{F}$ is a vector field. Show that $\int_{\sigma} \mathbf{F} = \int_{\psi} \mathbf{F}$ is equivalent to $\int_C \mathbf{F} = 0$, where C is the closed curve obtained by first moving around $\boldsymbol{\sigma}$ and then moving around $\boldsymbol{\psi}$ in the opposite direction.
11. Let $\boldsymbol{\sigma}(t)$ be a path and $\mathbf{T}$ the unit tangent vector. What is $\int_{\sigma} \mathbf{T}$?
12. Let $\mathbf{F} = (z^3 + 2xy)\mathbf{i} + x^2\mathbf{j} + 3xz^2\mathbf{k}$. Show that the integral of $\mathbf{F}$ around the circumference of the unit square is zero.
13. Using the path in Exercise 8, argue that a C^1 map $\boldsymbol{\sigma}\colon [a, b] \to \mathbb{R}^3$ can have an image that does not "look smooth." Do you think this could happen if $\boldsymbol{\sigma}'(t)$ were always nonzero?

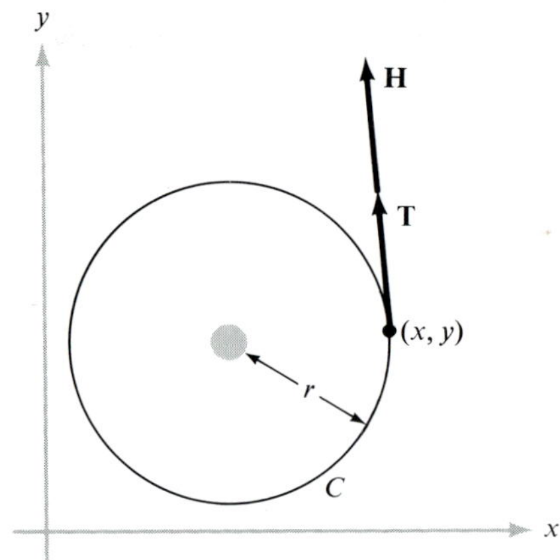

FIGURE 6.2.15
A planar section of a long wire and a curve C about the wire.

14. What is the value of the integral of any gradient field around a closed curve C?
15. Evaluate $\int_C 2xyz\,dx + x^2z\,dy + x^2y\,dz$, where C is an oriented simple curve connecting (1, 1, 1) to (1, 2, 4).
16. Prove Theorem 2.

6.3 PARAMETRIZED SURFACES

In Sections 6.1 and 6.2 we studied integrals of scalar and vector functions along curves. Now we shall turn to integrals over surfaces. Let us begin by studying the geometry of surfaces themselves.

We are already used to one kind of surface, namely, the graph of a function $f(x, y)$. Graphs were extensively studied in Chapter 2, and we know how to compute their tangent planes. However, it would be unduly limiting to restrict ourselves to this case. For example, many surfaces arise as level surfaces of functions. Suppose our surface S is the set of points (x, y, z) where $x - z + z^3 = 0$. Here S is a sheet that (relative to the xy-plane) doubles back on itself (see Figure 6.3.1). Obviously, we want to call S a surface, since it is just a plane with a wrinkle. However, S is *not* the graph of some function $z = f(x, y)$, because this means that for each $(x_0, y_0) \in \mathbb{R}^2$ there must be one z_0 with $(x_0, y_0, z_0) \in S$. As Figure 6.3.1 illustrates, this condition is violated.

Another example is the torus or surface of a doughnut, which is depicted in Figure 6.3.2. Anyone would call a torus a surface; yet, by the same reasoning as above, a torus cannot be the graph of a differen-

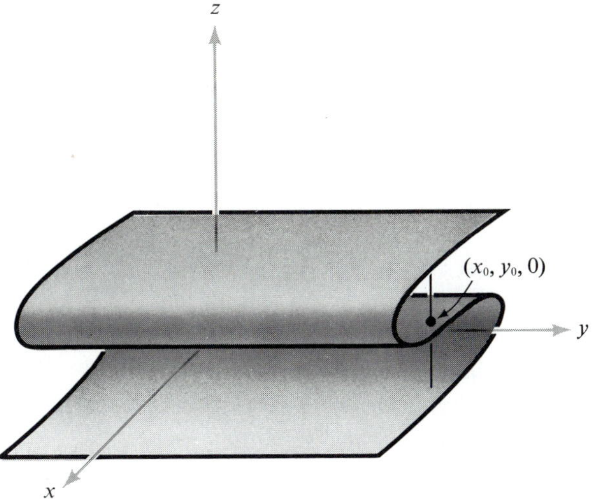

FIGURE 6.3.1
A surface that is not the graph of a function $z = f(x, y)$.

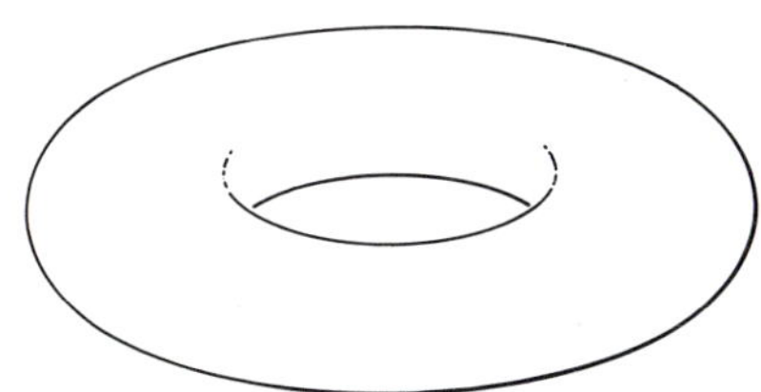

FIGURE 6.3.2
The torus is not the graph of a function of the form $z = f(x, y)$.

tiable function of two variables. These observations encourage us to extend our definition of a surface. The motivation for the definition that follows is partly that a surface can be thought of as being obtained from the plane by "rolling," "bending," and "pushing." For example, to get a torus we take a portion of the plane and roll it (like a cigarette—see Figure 6.3.3), then take the two "ends" and bring them together until they meet (Figure 6.3.4).

FIGURE 6.3.3
The first step in obtaining a torus from a rectangle: making a cylinder.

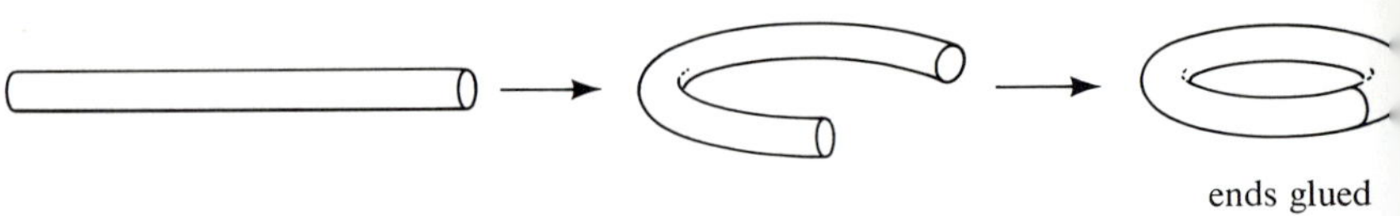

FIGURE 6.3.4
Bending the cylinder and gluing the ends to get a doughnut.

With surfaces, just as with curves, we want to distinguish a map (a parametrization) from its image (a geometrical object).

Definition. *A **parametrized surface** is a function $\mathbf{\Phi}: D \subset \mathbb{R}^2 \to \mathbb{R}^3$, where D is some domain in $\mathbb{R}^2$. The **surface** S corresponding to the function $\mathbf{\Phi}$ is its image: $S = \mathbf{\Phi}(D)$. We can write*

$$\mathbf{\Phi}(u, v) = (x(u, v), y(u, v), z(u, v))$$

If $\mathbf{\Phi}$ is differentiable or C^1 (which is the same as saying that $x(u, v)$, $y(u, v)$, and $z(u, v)$ are differentiable or C^1 functions of (u, v)—see Chapter 2), we call S a differentiable or C^1 surface.

We can think of $\mathbf{\Phi}$ as twisting or bending the region D in the plane to yield the surface S (see Figure 6.3.5). Thus each point (u, v) in D becomes a label for a point $(x(u, v), y(u, v), z(u, v))$ on S.

Suppose that $\mathbf{\Phi}$ is differentiable at $(u_0, v_0) \in \mathbb{R}^2$. Fixing u at u_0, we get a map $\mathbb{R} \to \mathbb{R}^3$ given by $t \mapsto \mathbf{\Phi}(u_0, t)$, whose image is a curve on the surface (Figure 6.3.6). From Chapters 2 and 3 we know that a vector tangent to this curve at the point $\mathbf{\Phi}(u_0, v_0)$ is given by

$$\mathbf{T}_v = \frac{\partial x}{\partial v}(u_0, v_0)\mathbf{i} + \frac{\partial y}{\partial v}(u_0, v_0)\mathbf{j} + \frac{\partial z}{\partial v}(u_0, v_0)\mathbf{k}$$

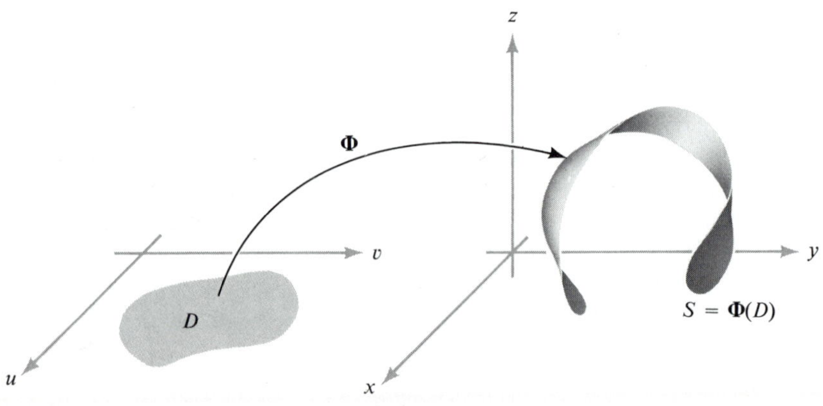

FIGURE 6.3.5
$\mathbf{\Phi}$ *"twists" and "bends" D onto the surface $S = \mathbf{\Phi}(D)$.*

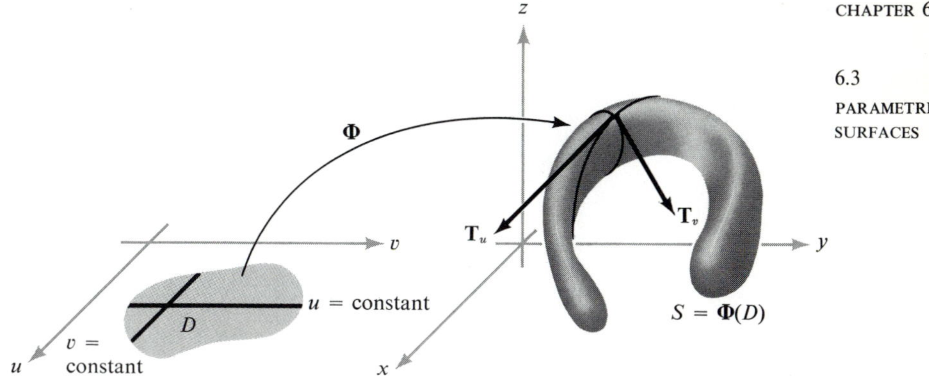

FIGURE 6.3.6
The tangent vectors $\mathbf{T}_u$ and $\mathbf{T}_v$ that are tangent to curves on a surface S, and hence tangent to S.

Similarly, if we fix v and consider the curve $t \mapsto \mathbf{\Phi}(t, v_0)$ we obtain a vector tangent to this curve at $\mathbf{\Phi}(u_0, v_0)$, given by

$$\mathbf{T}_u = \frac{\partial x}{\partial u}(u_0, v_0)\mathbf{i} + \frac{\partial y}{\partial u}(u_0, v_0)\mathbf{j} + \frac{\partial z}{\partial u}(u_0, v_0)\mathbf{k}$$

Since the vectors $\mathbf{T}_u$ and $\mathbf{T}_v$ are tangent to two curves on the surface at $\mathbf{\Phi}(u_0, v_0)$, they ought to determine the plane tangent to the surface at this point; that is, $\mathbf{T}_u \times \mathbf{T}_v$ ought to be normal to the surface.

We say that the surface S is *smooth** at $\mathbf{\Phi}(u_0, v_0)$ if $\mathbf{T}_u \times \mathbf{T}_v \neq \mathbf{0}$ at (u_0, v_0); the surface is smooth if it is smooth at all points $\mathbf{\Phi}(u_0, v_0) \in S$. The nonzero vector $\mathbf{T}_u \times \mathbf{T}_v$ is *normal* to S (recall that the vector product of $\mathbf{T}_u$ and $\mathbf{T}_v$ is perpendicular to the plane spanned by $\mathbf{T}_u$ and $\mathbf{T}_v$); the fact that it is nonzero ensures that there will be a tangent plane. Intuitively, a smooth surface has no "corners."

As an example, consider the surface given by the equations

$$x = u \cos v, \qquad y = u \sin v, \qquad z = u, \qquad u \geq 0$$

These equations describe the surface $z = \sqrt{x^2 + y^2}$ (just square the equations for x, y, and z to check this), which is shown in Figure 6.3.7. This surface is a cone with a "point" at (0, 0, 0); it is a differentiable surface because each component function is differentiable. However, the surface is not smooth at (0, 0, 0). To see this, compute $\mathbf{T}_u$ and $\mathbf{T}_v$ at $(0, 0) \in \mathbb{R}^2$:

$$\mathbf{T}_u = \frac{\partial x}{\partial u}(0, 0)\mathbf{i} + \frac{\partial y}{\partial u}(0, 0)\mathbf{j} + \frac{\partial z}{\partial u}(0, 0)\mathbf{k}$$

$$= (\cos 0)\mathbf{i} + (\sin 0)\mathbf{j} + \mathbf{k} = \mathbf{i} + \mathbf{k}$$

* Strictly speaking, smoothness depends on the parametrization $\mathbf{\Phi}$ and not just on its image S. Therefore this terminology is somewhat imprecise; however, it is descriptive and shouldn't cause confusion. See Exercise 9.

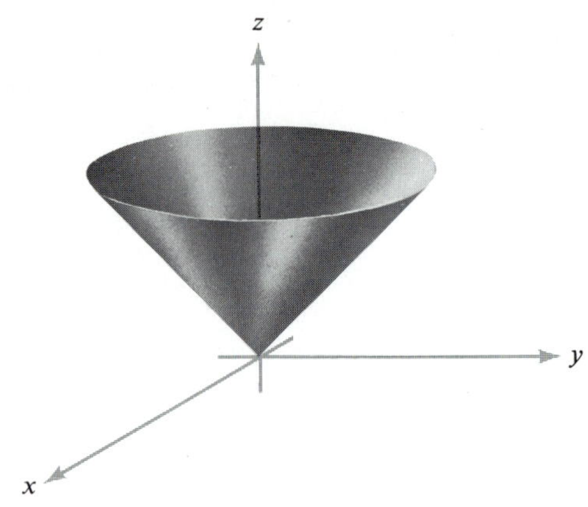

FIGURE 6.3.7
The surface $z = \sqrt{x^2 + y^2}$ *is a cone.*

and similarly

$$\mathbf{T}_v = 0(-\sin 0)\mathbf{i} + 0(\cos 0)\mathbf{j} + 0\mathbf{k} = \mathbf{0}$$

Thus $\mathbf{T}_u \times \mathbf{T}_v = \mathbf{0}$ and so, by definition, the surface is not smooth at (0, 0, 0). Let us summarize our conclusions in a formal definition:

Definition. *If a parametrized surface* $\mathbf{\Phi}: D \subset \mathbb{R}^2 \to \mathbb{R}^3$ *is* ***smooth*** *at* $\mathbf{\Phi}(u_0, v_0)$, *that is, if* $\mathbf{T}_u \times \mathbf{T}_v \neq \mathbf{0}$ *at* (u_0, v_0), *we define the* ***tangent plane*** *of the surface at* $\mathbf{\Phi}(u_0, v_0)$ *to be the plane determined by the vectors* $\mathbf{T}_u$ *and* $\mathbf{T}_v$. *Thus* $\mathbf{n} = \mathbf{T}_u \times \mathbf{T}_v$ *is a normal vector, and an equation of the tangent plane at* (x_0, y_0, z_0) *on the surface is given by*

$$\langle (x - x_0, y - y_0, z - z_0), \mathbf{n} \rangle = 0 \tag{1}$$

where $\mathbf{n}$ *is evaluated at* (u_0, v_0). *If* $\mathbf{n} = (n_1, n_2, n_3) = n_1\mathbf{i} + n_2\mathbf{j} + n_3\mathbf{k}$ *then (1) becomes*

$$n_1(x - x_0) + n_2(y - y_0) + n_3(z - z_0) = 0 \tag{1'}$$

EXAMPLE 1. Let $\mathbf{\Phi}: \mathbb{R}^2 \to \mathbb{R}^3$ be given by

$$x = u \cos v, \qquad y = u \sin v, \qquad z = u^2 + v^2$$

Then

$$\mathbf{T}_u = (\cos v)\mathbf{i} + (\sin v)\mathbf{j} + 2u\mathbf{k}$$

$$\mathbf{T}_v = -u(\sin v)\mathbf{i} + u(\cos v)\mathbf{j} + 2v\mathbf{k}$$

and the tangent plane at (u, v) is the set of vectors through (u, v) perpendicular to

$$\mathbf{T}_u \times \mathbf{T}_v = (-2u^2 \cos v + 2v \sin v, -2u^2 \sin v - 2v \cos v, u)$$

if this vector is nonzero. Since $\mathbf{T}_u \times \mathbf{T}_v$ is equal to $\mathbf{0}$ at $(u, v) = (0, 0)$, there is no tangent plane at $\mathbf{\Phi}(0, 0) = (0, 0, 0)$. However, we can find an equation of the tangent plane at all the other points, where $\mathbf{T}_u \times \mathbf{T}_v \neq \mathbf{0}$. For instance, let us find the equation of the plane tangent to the surface under consideration at the point $\mathbf{\Phi}(1, 0) = (1, 0, 1)$. At this point

$$\mathbf{n} = \mathbf{T}_u \times \mathbf{T}_v = (-2, 0, 1) = -2\mathbf{i} + \mathbf{k}$$

Since we have the vector $\mathbf{n}$ normal to the surface and a point $(1, 0, 1)$ on the surface, we can use formula (1′) to obtain an equation of the tangent plane:

$$-2(x - 1) + (z - 1) = 0$$

that is

$$z = 2x - 1$$

EXAMPLE 2. Suppose a surface S is the graph of a differentiable function $g\colon \mathbb{R}^2 \to \mathbb{R}$. Then the surface is smooth at all points $(u_0, v_0, g(u_0, v_0)) \in \mathbb{R}^3$.

To show this, we write S in parametric form as follows:

$$x = u, \qquad y = v, \qquad z = g(u, v)$$

which is the same as $z = g(x, y)$. Then

$$\mathbf{T}_u = \mathbf{i} + \frac{\partial g}{\partial u}(u_0, v_0)\mathbf{k}$$

$$\mathbf{T}_v = \mathbf{j} + \frac{\partial g}{\partial v}(u_0, v_0)\mathbf{k}$$

and for $(u_0, v_0) \in \mathbb{R}^2$

$$\mathbf{n} = \mathbf{T}_u \times \mathbf{T}_v = -\frac{\partial g}{\partial u}(u_0, v_0)\mathbf{i} - \frac{\partial g}{\partial v}(u_0, v_0)\mathbf{j} + \mathbf{k} \neq \mathbf{0} \tag{2}$$

This is nonzero because the coefficient of $\mathbf{k}$ is 1; consequently the parametrization $(u, v) \mapsto (u, v, g(u, v))$ is smooth at all points. Moreover the tangent plane at $(x_0, y_0, z_0) = (u_0, v_0, g(u_0, v_0))$ is given, by formula (1), as

$$\left\langle (x - x_0, y - y_0, z - z_0), \left(-\frac{\partial g}{\partial u}, -\frac{\partial g}{\partial v}, 1\right) \right\rangle = 0$$

where the partial derivatives are evaluated at (u_0, v_0). Remembering that $x = u$ and $y = v$, we can write this as

$$z - z_0 = \left(\frac{\partial g}{\partial x}\right) \cdot (x - x_0) + \left(\frac{\partial g}{\partial y}\right) \cdot (y - y_0) \tag{3}$$

where $\partial g/\partial x$ and $\partial g/\partial y$ are evaluated at (x_0, y_0).

This example also shows that the definition of the tangent plane for parametrized surfaces agrees with the one for surfaces obtained as graphs, since (3) is the same formula we derived (in Chapter 2) for the plane tangent to S at $(x_0, y_0, z_0) \in S$ (see p. 93).

It is also useful to consider piecewise smooth surfaces, that is, surfaces composed of a certain number of images of smooth parametrized surfaces. For example, the surface of a cube in $\mathbb{R}^3$ is such a surface. These surfaces are considered in Section 6.4.

EXERCISES

In exercises 1 to 3, find an equation for the plane tangent to the given surface at the specified point.

1. $x = 2u, \quad y = u^2 + v, \quad z = v^2$, at $(0, 1, 1)$
2. $x = u^2 - v^2, \quad y = u + v, \quad z = u^2 + 4v$, at $(-\frac{1}{4}, \frac{1}{2}, 2)$
3. $x = u^2, \quad y = u \sin e^v, \quad z = \frac{1}{3} u \cos e^v$, at $(13, -2, 1)$
4. Are the surfaces in exercises 1 and 2 smooth?
5. Find an expression for a unit vector normal to the surface

$$x = \cos v \sin u, \qquad y = \sin v \sin u, \qquad z = \cos u$$

for $u \in [0, \pi]$ and $v \in [0, 2\pi]$. Identify this surface.

*6. Repeat Exercise 5 for the surface

$$x = (2 - \cos v)\cos u, \qquad y = (2 - \cos v)\sin u, \qquad z = \sin v$$

for $-\pi \le u \le \pi$, $-\pi \le v \le \pi$. Is this surface smooth?

7. Develop a formula for the planes tangent to the surfaces $x = h(y, z)$ and $y = f(x, z)$.
8. Find the equation of the plane tangent to each surface at the indicated point.
   (a) $x = u^2$, $y = v^2$, $z = u^2 + v^2$, $u = 1$, $v = 1$
   (b) $z = 3x^2 + 8xy$, $x = 1$, $y = 0$
   (c) $x^3 + 3xy + z^2 = 2$, $x = 1$, $y = \frac{1}{3}$, $z = 0$

*9. Consider the surfaces $\mathbf{\Phi}_1(u, v) = (u, v, 0)$ and $\mathbf{\Phi}_2(u, v) = (u^3, v^3, 0)$.
   (a) Show that the image of $\mathbf{\Phi}_1$ and $\mathbf{\Phi}_2$ is the xy-plane.
   (b) Show that $\mathbf{\Phi}_1$ describes a smooth surface, yet $\mathbf{\Phi}_2$ does not.
   (c) Conclude that the notion of smoothness of a surface S depends on the existence of at least one smooth parametrization for S. Is the tangent plane of S well defined?
   (d) After these remarks, do you think you can find a smooth parametrization of the cone in Figure 6.3.7?

6.4 AREA OF A SURFACE

Before proceeding to general surface integrals, let us first consider the problem of computing the area of a surface, just as we considered the problem of finding the arc length of a curve before discussing path integrals.

Our object here is to derive a formula for the area of a surface. There are various ways of obtaining such formulas. In order to avoid certain difficulties involved in deriving surface area from a limiting process involving Riemann sums, we shall take a simpler route and *define* surface area as a double integral. Then we shall present an argument for the plausibility of this definition.

In Section 6.3 we defined a surface S as the *image* of a function $\mathbf{\Phi}: D \subset \mathbb{R}^2 \to \mathbb{R}^3$, written as $\mathbf{\Phi}(u, v) = (x(u, v), y(u, v), z(u, v))$. The map $\mathbf{\Phi}$ was called the parametrization of S. Then S was said to be smooth at $\mathbf{\Phi}(u, v) \in S$ if $\mathbf{T}_u \times \mathbf{T}_v \neq \mathbf{0}$, where

$$\mathbf{T}_u = \frac{\partial x}{\partial u}(u, v)\mathbf{i} + \frac{\partial y}{\partial u}(u, v)\mathbf{j} + \frac{\partial z}{\partial u}(u, v)\mathbf{k}$$

and

$$\mathbf{T}_v = \frac{\partial x}{\partial v}(u, v)\mathbf{i} + \frac{\partial y}{\partial v}(u, v)\mathbf{j} + \frac{\partial z}{\partial v}(u, v)\mathbf{k}$$

Recall that a smooth surface (loosely speaking) is one that has no corners or breaks.

In the rest of this chapter and in the next one we shall consider only piecewise smooth surfaces that are unions of images of parametrized surfaces $\mathbf{\Phi}_i: D_i \to \mathbb{R}^3$ for which:

(*i*) D_i is an elementary region in the plane;

(*ii*) $\mathbf{\Phi}_i$ is C^1 and one-to-one, except possibly on the boundary of D_i; and

(*iii*) S_i, the image of Φ_i is smooth, except possibly at a finite number of points.

Definition. *We define the* ***surface area**** $A(S)$ *of a parametrized surface by*

$$A(S) = \int_D \|\mathbf{T}_u \times \mathbf{T}_v\| \, du \, dv \tag{1}$$

where $\|\mathbf{T}_u \times \mathbf{T}_v\|$ *is the norm of* $\mathbf{T}_u \times \mathbf{T}_v$. *If* S *is a union of surfaces* S_i, *its area is the sum of the areas of the* S_i.

* As we have not yet discussed the independence of parametrization, it may seem that $A(S)$ depends on the parametrization $\mathbf{\Phi}$. We shall discuss independence of parametrization in Section 6.6; the use of this notation here should not cause confusion.

As the reader can easily verify, we have

$$\|\mathbf{T}_u \times \mathbf{T}_v\| = \sqrt{\left[\frac{\partial(x, y)}{\partial(u, v)}\right]^2 + \left[\frac{\partial(y, z)}{\partial(u, v)}\right]^2 + \left[\frac{\partial(x, z)}{\partial(u, v)}\right]^2} \tag{2}$$

where

$$\frac{\partial(x, y)}{\partial(u, v)} = \begin{vmatrix} \dfrac{\partial x}{\partial u} & \dfrac{\partial x}{\partial v} \\ \dfrac{\partial y}{\partial u} & \dfrac{\partial y}{\partial v} \end{vmatrix}$$

and so on. Thus formula (1) becomes

$$A(S) = \int_D \sqrt{\left[\frac{\partial(x, y)}{\partial(u, v)}\right]^2 + \left[\frac{\partial(y, z)}{\partial(u, v)}\right]^2 + \left[\frac{\partial(x, z)}{\partial(u, v)}\right]^2}\, du\, dv \tag{3}$$

We can justify this definition by analyzing the integral $\int_D \|\mathbf{T}_u \times \mathbf{T}_v\|\, du\, dv$ in terms of Riemann sums. For simplicity, suppose D is a rectangle; consider the nth regular partition of D, and let R_{ij} be the ijth rectangle in the partition, with vertices (u_i, v_j), (u_{i+1}, v_j), (u_i, v_{j+1}), and (u_{i+1}, v_{j+1}), $0 \le i \le n-1$, $0 \le j \le n-1$. Denote the values of $\mathbf{T}_u$ and $\mathbf{T}_v$ at (u_i, v_j) by $\mathbf{T}_{u_i}$ and $\mathbf{T}_{v_j}$. We can think of the vectors $\Delta u \mathbf{T}_{u_i}$ and $\Delta v \mathbf{T}_{v_j}$ as tangent to the surface at $\mathbf{\Phi}(u_i, v_j) = (x_{ij}, y_{ij}, z_{ij})$ where $\Delta u = u_{i+1} - u_i$, $\Delta v = v_{j+1} - v_j$. Then these vectors form a parallelogram P_{ij} that lies in the plane tangent to the surface at (x_{ij}, y_{ij}, z_{ij}) (see Figure 6.4.1). We thus have a "patchwork cover" of the surface by the P_{ij}.

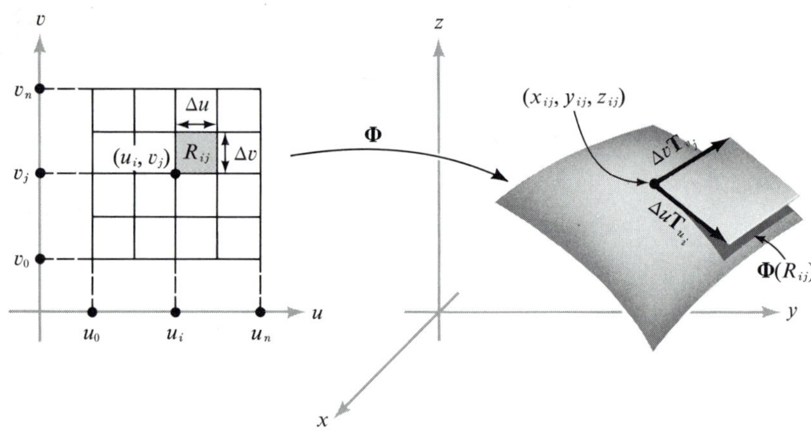

FIGURE 6.4.1

$\|\mathbf{T}_{u_i} \times \mathbf{T}_{v_j}\|\, \Delta u\, \Delta v$ *is equal to the area of a parallelogram that approximates the area of a patch on a surface* $S = \mathbf{\Phi}(D)$.

For n large, the area of P_{ij} is a good approximation to the area of $\mathbf{\Phi}(R_{ij})$. Since the area of the parallelogram spanned by two vectors $\mathbf{v}_1$ and $\mathbf{v}_2$ is $\|\mathbf{v}_1 \times \mathbf{v}_2\|$ (see Chapter 1), we see that

$$A(P_{ij}) = \|\Delta u \mathbf{T}_{u_i} \times \Delta v \mathbf{T}_{v_j}\| = \|\mathbf{T}_{u_i} \times \mathbf{T}_{v_j}\|\, \Delta u\, \Delta v$$

Therefore, the area of the cover is

$$A_n = \sum_{i=0}^{n-1} \sum_{j=0}^{n-1} A(P_{ij}) = \sum_{i=0}^{n-1} \sum_{j=0}^{n-1} \|\mathbf{T}_{u_i} \times \mathbf{T}_{v_j}\| \,\Delta u\, \Delta v$$

As $n \to \infty$, the sums A_n converge to the integral

$$\int_D \|\mathbf{T}_u \times \mathbf{T}_v\| \, du\, dv$$

Since A_n should approximate the surface area better and better as $n \to \infty$, we are led to formula (1) for a reasonable definition of $A(S)$.

Example 1. Let D be the region determined by $0 \le \theta \le 2\pi, 0 \le r \le 1$ and let the function $\mathbf{\Phi}\colon D \to \mathbb{R}^3$, where

$$x = r\cos\theta, \qquad y = r\sin\theta, \qquad z = r$$

be a parametrization of a cone S. We compute

$$\frac{\partial(x, y)}{\partial(r, \theta)} = \begin{vmatrix} \cos\theta & -r\sin\theta \\ \sin\theta & r\cos\theta \end{vmatrix} = r$$

$$\frac{\partial(y, z)}{\partial(r, \theta)} = \begin{vmatrix} \sin\theta & r\cos\theta \\ 1 & 0 \end{vmatrix} = -r\cos\theta$$

and

$$\frac{\partial(x, z)}{\partial(r, \theta)} = \begin{vmatrix} \cos\theta & -r\sin\theta \\ 1 & 0 \end{vmatrix} = r\sin\theta$$

so the area integrand is

$$\begin{aligned} \|\mathbf{T}_r \times \mathbf{T}_\theta\| &= \sqrt{r^2 + r^2\cos^2\theta + r^2\sin^2\theta} \\ &= r\sqrt{2} \end{aligned}$$

Clearly, $\|\mathbf{T}_r \times \mathbf{T}_\theta\|$ vanishes for $r = 0$, but $\mathbf{\Phi}(0, \theta) = (0, 0, 0)$ for any θ. Thus $(0, 0, 0)$ is the only point where the surface is not smooth. We have

$$\begin{aligned} \int_D \|\mathbf{T}_r \times \mathbf{T}_\theta\| \, dr\, d\theta &= \int_0^{2\pi} \int_0^1 \sqrt{2}\, r\, dr\, d\theta \\ &= \int_0^{2\pi} \tfrac{1}{2}\sqrt{2}\, d\theta \\ &= \sqrt{2}\,\pi \end{aligned}$$

To confirm that this is the area of $\mathbf{\Phi}(D)$ we must verify that $\mathbf{\Phi}$ is one-to-one (for points not on the boundary of D). Let D^0 be the set of (r, θ) with $0 < r < 1$ and $0 < \theta < 2\pi$. Hence, D^0 is D without its boundary. To see that $\mathbf{\Phi}\colon D^0 \to \mathbb{R}^3$ is one-to-one, assume that $\mathbf{\Phi}(r, \theta) = \mathbf{\Phi}(r', \theta')$ for (r, θ) and $(r', \theta') \in D^0$. Then

$$r\cos\theta = r'\cos\theta', \qquad r\sin\theta = r'\sin\theta', \qquad r = r'$$

From these equations it follows that $\cos\theta = \cos\theta'$ and $\sin\theta = \sin\theta'$. Thus either $\theta = \theta'$ or $\theta = \theta' + 2\pi n$. But the second case is impossible, since both θ and θ' belong to the open interval $]0, 2\pi[$ and thus cannot be 2π radians apart. This proves that off the boundary $\mathbf{\Phi}$ is one-to-one. (Is $\mathbf{\Phi}: D \to \mathbb{R}^3$ one-to-one?)

In future examples we shall not usually verify that the parametrization is one-to-one.

EXAMPLE 2. A helicoid is defined by $\mathbf{\Phi}: D \to \mathbb{R}^3$ where

$$x = r\cos\theta, \qquad y = r\sin\theta, \qquad z = \theta$$

and D is the region where $0 \le \theta \le 2\pi$ and $0 \le r \le 1$ (Figure 6.4.2). We compute $\partial(x, y)/\partial(r, \theta) = r$ as before, and

$$\frac{\partial(y, z)}{\partial(r, \theta)} = \begin{vmatrix} \sin\theta & r\cos\theta \\ 0 & 1 \end{vmatrix} = \sin\theta$$

$$\frac{\partial(x, z)}{\partial(r, \theta)} = \begin{vmatrix} \cos\theta & -r\sin\theta \\ 0 & 1 \end{vmatrix} = \cos\theta$$

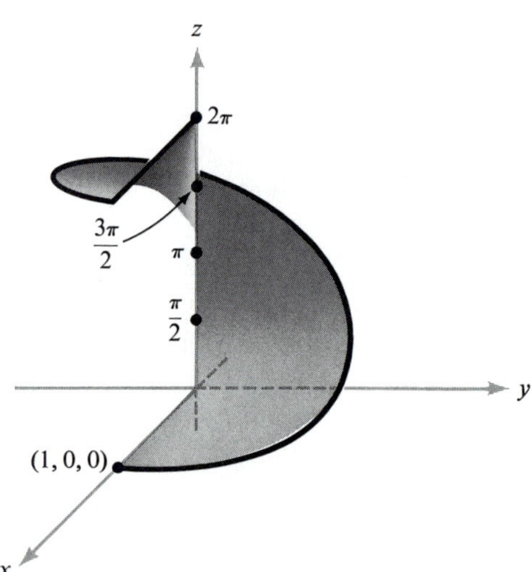

FIGURE 6.4.2
The helicoid $x = r\cos\theta$, $y = r\sin\theta$, $z = \theta$.

The area integrand is therefore $\sqrt{r^2 + 1}$, which never vanishes, so the surface is smooth. The area of the helicoid is

$$\int_D \|\mathbf{T}_r \times \mathbf{T}_\theta\|\, dr\, d\theta = \int_0^{2\pi}\int_0^1 \sqrt{r^2 + 1}\, dr\, d\theta = 2\pi \int_0^1 \sqrt{r^2 + 1}\, dr$$

After a little computation (using the integral tables in Appendix A), we find that this integral is equal to

$$\pi(\sqrt{2} + \log(1 + \sqrt{2}))$$

A surface S given in the form $z = f(x, y)$, where $(x, y) \in D$, admits the parametrization

$$x = u, \qquad y = v, \qquad z = f(u, v)$$

for $(u, v) \in D$. When f is C^1, this parametrization is smooth, and the formula for surface area reduces to

$$A(S) = \int_D \left(\sqrt{\left(\frac{\partial f}{\partial x}\right)^2 + \left(\frac{\partial f}{\partial y}\right)^2 + 1}\right) dA \tag{4}$$

after applying the formulas

$$\mathbf{T}_u = \mathbf{i} + \frac{\partial f}{\partial u}\mathbf{k}$$

$$\mathbf{T}_v = \mathbf{j} + \frac{\partial f}{\partial v}\mathbf{k}$$

and

$$\mathbf{T}_u \times \mathbf{T}_v = -\frac{\partial f}{\partial u}\mathbf{i} - \frac{\partial f}{\partial v}\mathbf{j} + \mathbf{k} = -\frac{\partial f}{\partial x}\mathbf{i} - \frac{\partial f}{\partial y}\mathbf{j} + \mathbf{k}$$

noted in Example 2 of Section 6.3.

In formula (4) we have assumed that $\partial f/\partial x$ and $\partial f/\partial y$ are continuous (and hence, bounded) functions on D. However, it is important to consider areas of surfaces for which either $(\partial f/\partial x)(x_0, y_0)$ or $(\partial f/\partial y)(x_0, y_0)$ gets arbitrarily large as (x_0, y_0) approaches the boundary of D. For example, consider the hemisphere

$$z = \sqrt{1 - x^2 - y^2}$$

where D is the region $x^2 + y^2 \leq 1$ (see Figure 6.4.3). We have

$$\frac{\partial f}{\partial x} = \frac{-x}{\sqrt{1 - x^2 - y^2}}, \qquad \frac{\partial f}{\partial y} = \frac{-y}{\sqrt{1 - x^2 - y^2}} \tag{5}$$

The boundary of D is the unit circle $x^2 + y^2 = 1$, so as (x, y) gets close to ∂D, the value of $x^2 + y^2$ approaches 1. Hence, the denominators in (5) go to zero.

We want to be able to deal with cases such as these, so we define the area $A(S)$ of a surface S described by $z = f(x, y)$ over a region D, where f is differentiable with possible discontinuities of $\partial f/\partial x$ and $\partial f/\partial y$ on ∂D, as

$$A(S) = \int_D \sqrt{\left(\frac{\partial f}{\partial x}\right)^2 + \left(\frac{\partial f}{\partial y}\right)^2 + 1}\, dA$$

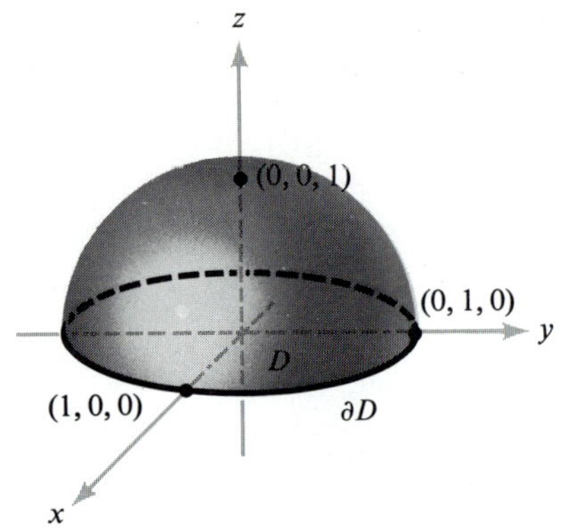

FIGURE 6.4.3
The hemisphere $z = \sqrt{1 - x^2 - y^2}$.

whenever $\sqrt{(\partial f/\partial x)^2 + (\partial f/\partial y)^2 + 1}$ is integrable over D, *even if the integral is improper*; in fact, this is one of the reasons we introduced the notion of improper integral, in Chapter 5.

EXAMPLE 3. Compute the area of the surface of the sphere S described by $x^2 + y^2 + z^2 = 1$.

We shall compute the area of the upper hemisphere S^+, where

$$x^2 + y^2 + z^2 = 1, \qquad z \geq 0$$

and then multiply our result by two. Hence we have

$$f(x, y) = \sqrt{1 - x^2 - y^2}, \qquad x^2 + y^2 \leq 1$$

Let D be the region $x^2 + y^2 \leq 1$. Then applying formula (4) and calculations (5) above we get

$$\begin{aligned} A(S^+) &= \int_D \sqrt{\left(\frac{\partial f}{\partial x}\right)^2 + \left(\frac{\partial f}{\partial y}\right)^2 + 1}\, dA \\ &= \int_D \sqrt{\frac{x^2}{1 - x^2 - y^2} + \frac{y^2}{1 - x^2 - y^2} + 1}\, dA \\ &= \int_D \frac{1}{\sqrt{1 - x^2 - y^2}}\, dy\, dx \end{aligned}$$

which is an improper integral. However we may apply Fubini's Theorem in this case to obtain the iterated improper integral

$$\begin{aligned} \int_{-1}^{1} \int_{-(1-x^2)^{1/2}}^{(1-x^2)^{1/2}} \frac{1}{\sqrt{1 - x^2 - y^2}}\, dy\, dx &= \int_{-1}^{1} \left[\sin^{-1} \frac{y}{(1 - x^2)^{1/2}}\right]_{-(1-x^2)^{1/2}}^{(1-x^2)^{1/2}} dx \\ &= \int_{-1}^{1} \left[\frac{\pi}{2} + \frac{\pi}{2}\right] dx = \int_{-1}^{1} \pi\, dx = 2\pi \end{aligned}$$

Thus, the area of the entire sphere is 4π. For another way of computing this area see Exercise 1.

In most books on one-variable calculus, it is shown that the lateral surface area generated by revolving the graph of a function $y = f(x)$ about the x-axis is given by

$$A_1 = 2\pi \int_a^b |f(x)|\sqrt{1 + [f'(x)]^2}\, dx \tag{6}$$

If the graph is revolved about the y-axis, we have

$$A_2 = 2\pi \int_a^b |x|\sqrt{1 + [f'(x)]^2}\, dx \tag{7}$$

We shall derive (6) by using the methods developed above; one can obtain (7) in a similar fashion (Exercise 10).

To derive formula (6) from formula (3), we must give a parametrization of S. Define the parametrization by

$$x = u, \qquad y = f(u)\cos v, \qquad z = f(u)\sin v$$

over the region D given by

$$a \leq u \leq b, \qquad 0 \leq v \leq 2\pi$$

This is indeed a parametrization of S, because for fixed u

$$(u, f(u)\cos v, f(u)\sin v)$$

traces out a circle of radius $|f(u)|$ with the center $(u, 0, 0)$. Now

$$\frac{\partial(x, y)}{\partial(u, v)} = -f(u)\sin v$$

$$\frac{\partial(y, z)}{\partial(u, v)} = f(u)f'(u)$$

$$\frac{\partial(x, z)}{\partial(u, v)} = f(u)\cos v$$

So by formula (3)

$$\begin{aligned}
A(S) &= \int_D \sqrt{\left[\frac{\partial(x, y)}{\partial(u, v)}\right]^2 + \left[\frac{\partial(y, z)}{\partial(u, v)}\right]^2 + \left[\frac{\partial(x, z)}{\partial(u, v)}\right]^2}\, du\, dv \\
&= \int_D \sqrt{[f(u)]^2 \sin^2 v + [f(u)]^2[f'(u)]^2 + [f(u)]^2 \cos^2 v}\, du\, dv \\
&= \int_D |f(u)|\sqrt{1 + [f'(u)]^2}\, du\, dv \\
&= \int_a^b \int_0^{2\pi} |f(u)|\sqrt{1 + [f'(u)]^2}\, dv\, du \\
&= 2\pi \int_a^b |f(u)|\sqrt{1 + [f'(u)]^2}\, du
\end{aligned}$$

which is formula (6).

If S is the surface of revolution then $2\pi|f(x)|$ is the circumference of the vertical cross section to S at the point x (Figure 6.4.4). Observe that we can write

$$2\pi \int_a^b |f(x)|\sqrt{1+[f'(x)]^2}\,dx = \int_{\boldsymbol{\sigma}} 2\pi|f(x)|\,ds$$

where the expression on the right is the integral of $2\pi|f(x)|$ along the path $\boldsymbol{\sigma}\colon [a, b] \to \mathbb{R}^2$, $t \mapsto (t, f(t))$. Therefore, the lateral surface area of a solid of revolution is obtained by integrating the cross-sectional circumference along the path determined by the given function.

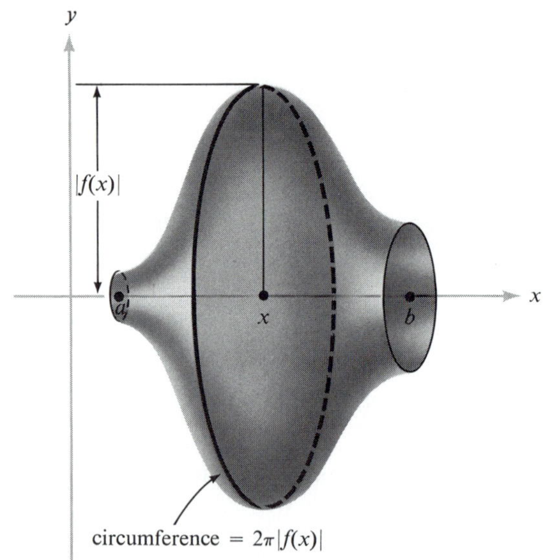

FIGURE 6.4.4
The curve $y = f(x)$ rotated about the x-axis.

EXERCISES

1. Find the surface area of the unit sphere S represented parametrically by $\boldsymbol{\Phi}\colon D \to S \subset \mathbb{R}^3$ where D is the rectangle $0 \le \theta \le 2\pi$, $0 \le \phi \le \pi$ and $\boldsymbol{\Phi}$ is given by the equations

$$x = \cos\theta \sin\phi, \qquad y = \sin\theta\sin\phi, \qquad z = \cos\phi$$

Note that we can represent the entire sphere parametrically, but we cannot represent it in the form $z = f(x, y)$. Compare with Example 3.

2. In Exercise 1, what happens if we allow ϕ to vary from $-\pi/2$ to $\pi/2$; from 0 to 2π? Why do we obtain different answers?

3. The torus T can be represented parametrically by the function $\boldsymbol{\Phi}\colon D \to \mathbb{R}^3$, where $\boldsymbol{\Phi}$ is given by the coordinate functions $x =$

$(R + \cos\phi)\cos\theta$, $y = (R + \cos\phi)\sin\theta$, $z = \sin\phi$; D is the rectangle $[0, 2\pi] \times [0, 2\pi]$, that is, $0 \le \theta$, $\phi \le 2\pi$; and $R > 1$ is fixed (see Figure 6.4.5). Show that $A(T) = (2\pi)^2 R$, first by using formula (3) and then by using formula (7).

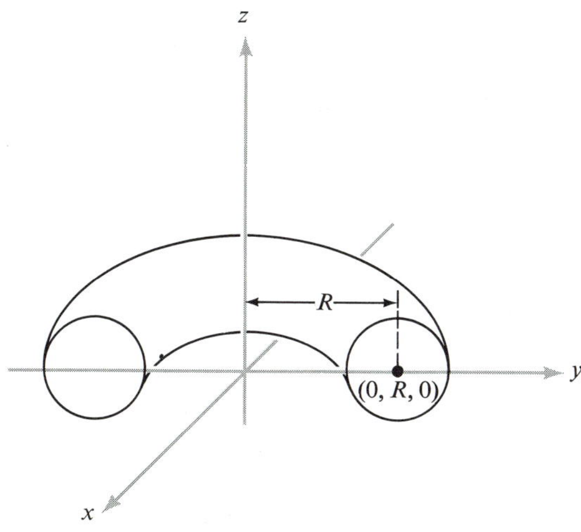

FIGURE 6.4.5
A cross section of a torus.

4. Find the area of the helicoid in Example 2 if the domain D is $0 \le r \le 1$ and $0 \le \theta \le 3\pi$.
5. Let $\mathbf{\Phi}(u, v) = (u - v, u + v, uv)$ and let D be the unit disc in the uv-plane. Find the area of $\mathbf{\Phi}(D)$.
6. Find the area of the portion of the unit sphere that is cut out by the cone $x^2 + y^2 = z^2$, $z \ge 0$ (see Exercise 1).
*7. The cylinder $x^2 + y^2 = x$ divides the unit sphere S into two regions S_1 and S_2, where S_1 is inside the cylinder and S_2 outside. Find the ratio of areas $A(S_2)/A(S_1)$.
*8. Suppose a surface S that is the graph of a function $z = f(x, y)$, $(x, y) \in D \subset \mathbb{R}^2$, can also be described as the set of $(x, y, z) \in \mathbb{R}^3$ with $F(x, y, z) = 0$ (a level surface). Derive a formula for $A(S)$ that involves only F.
9. Represent the ellipsoid

$$E\colon \frac{x^2}{a^2} + \frac{y^2}{b^2} + \frac{z^2}{c^2} = 1$$

parametrically and write out the integral for its surface area $A(E)$ (do not evaluate the integral).
10. Let the curve $y = f(x)$, $a \le x \le b$ be rotated about the y-axis. Show that the area of the surface swept out is

$$A = 2\pi \int_a^b \sqrt{1 + [f'(x)]^2}\,|x|\,dx$$

11. Use formula (4) to compute the surface area of the cone in Example 1.

12. Find the area of the surface obtained by rotating the curve $y = x^2$, $0 \le x \le 1$, about the y-axis.
13. Show that for the vectors $\mathbf{T}_u$ and $\mathbf{T}_v$ we have the formula

$$\|\mathbf{T}_u \times \mathbf{T}_v\| = \sqrt{\left[\frac{\partial(x, y)}{\partial(u, v)}\right]^2 + \left[\frac{\partial(y, z)}{\partial(u, v)}\right]^2 + \left[\frac{\partial(x, z)}{\partial(u, v)}\right]^2}$$

14. Find the area of the surface defined by $x + y + z = 1$, $x^2 + 2y^2 \le 1$.
15. Compute the area of the surface given by

$$x = r\cos\theta, \qquad y = 2r\cos\theta, \qquad z = \theta, \qquad 0 \le r \le 1, \qquad 0 \le \theta \le 2\pi$$

Sketch.

16. Prove *Pappus' Theorem*: Let $\boldsymbol{\sigma}\colon [a, b] \to \mathbb{R}^2$ be a C^1 path. The area of the lateral surface generated by rotating the graph of $\boldsymbol{\sigma}$ about the y-axis is equal to $2\pi\bar{x}l(\boldsymbol{\sigma})$ where $\bar{x}$ is the average value of x coordinates of points on $\boldsymbol{\sigma}$, and $l(\boldsymbol{\sigma})$ is the length of $\boldsymbol{\sigma}$.

6.5 INTEGRALS OF SCALAR FUNCTIONS OVER SURFACES

Now we are ready to define the integral of a *scalar* function f over a surface S. This concept is a natural generalization of the area of a surface, which corresponds to the integral over S of the scalar function $f(x, y, z) = 1$. This is quite analogous to considering the path integral as a generalization of arc length. In the next section we shall deal with the integral of a *vector* function $\mathbf{F}$ over a surface. These concepts will play a crucial role in the vector analysis treated in the final chapter.

Let us start with a surface S parametrized by $\mathbf{\Phi}\colon D \to S \subset \mathbb{R}^3$, $\mathbf{\Phi}(u, v) = (x(u, v), y(u, v), z(u, v))$.

Definition. *If $f(x, y, z)$ is a real-valued continuous function defined on S, we define the **integral of f over S** to be*

$$\int_S f(x, y, z)\, dS = \int_S f\, dS = \int_D f(\mathbf{\Phi}(u, v))\|\mathbf{T}_u \times \mathbf{T}_v\|\, du\, dv \qquad (1)$$

where $\mathbf{T}_u$ and $\mathbf{T}_v$ have the same meaning as in Section 6.3. Written out, equation (1) becomes

$$\int_S f\, dS = \int_D f(x(u, v), y(u, v), z(u, v)) \times \sqrt{\left[\frac{\partial(x, y)}{\partial(u, v)}\right]^2 + \left[\frac{\partial(y, z)}{\partial(u, v)}\right]^2 + \left[\frac{\partial(x, z)}{\partial(u, v)}\right]^2}\, du\, dv \qquad (2)$$

Thus if f is identically 1, we recover the area formula (3) of Section

6.4. Like surface area, the surface integral is independent of the particular parametrization used. This will be discussed in Section 6.6.

We can gain some intuition of this integral by considering it as a limit of sums. Let D be a rectangle partitioned into n^2 rectangles R_{ij}. Let $S_{ij} = \mathbf{\Phi}(R_{ij})$ be the portion of the surface $\mathbf{\Phi}(D)$ corresponding to R_{ij} (see Figure 6.5.1), and let $A(S_{ij})$ be the area of this portion of the surface.

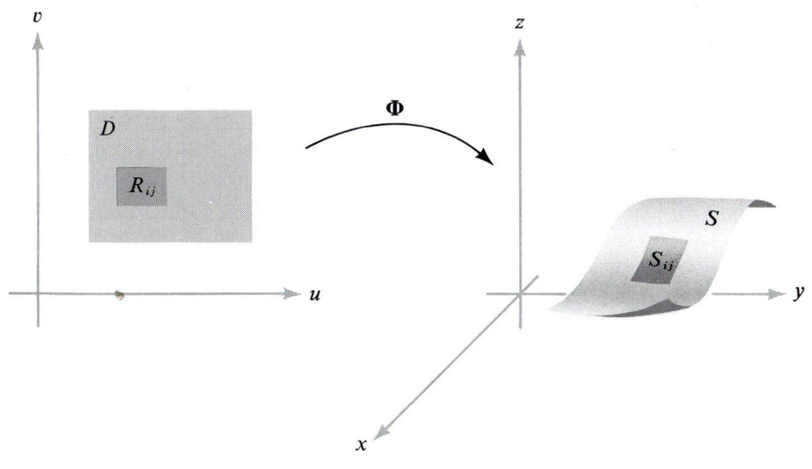

FIGURE 6.5.1
$\mathbf{\Phi}$ takes a portion R_{ij} of D to a portion of S.

For large n, f will be approximately constant on S_{ij} and we form the sum

$$S_n = \sum_{i=0}^{n-1} \sum_{j=0}^{n-1} f(\mathbf{\Phi}(u_i, v_j))A(S_{ij}) \tag{3}$$

where $(u_i, v_j) \in R_{ij}$. But from Section 6.4 we have a formula for $A(S_{ij})$

$$A(S_{ij}) = \int_{R_{ij}} \|\mathbf{T}_u \times \mathbf{T}_v\| \, du \, dv$$

which, by the Mean Value Theorem for Integrals (see p. 222), equals $\|\mathbf{T}_{u_i*} \times \mathbf{T}_{v_j*}\| \, \Delta u \, \Delta v$ for some point (u_i^*, v_j^*) in R_{ij}. Hence our sum becomes

$$S_n = \sum_{i=0}^{n-1} \sum_{j=0}^{n-1} f(\mathbf{\Phi}(u_i, v_j))\|\mathbf{T}_{u_i*} \times \mathbf{T}_{v_j*}\| \, \Delta u \, \Delta v$$

which is an approximating sum for the last integral in (1). Therefore, $\lim_{n\to\infty} S_n = \int_S f \, dS$. Note that each term in the sum (3) is the value of f at some point $\mathbf{\Phi}(u_i, v_j)$ times the area of S_{ij}. Compare this with the Riemann-sum interpretation of the path integral in Section 6.1.

If S is a union of parametrized surfaces S_i, $i = 1, \ldots, N$, that do not intersect except possibly along curves defining their boundaries, then the integral of f over S is defined by

$$\int_S f = \sum_{i=1}^{N} \int_{S_i} f$$

as we should expect. For example, the integral over the surface of a cube may be expressed as the sum of the integrals over the six sides.

EXAMPLE 1. Suppose a helicoid is described as in Example 2, Section 6.4, and let f be given by $f(x, y, z) = \sqrt{x^2 + y^2 + 1}$. As before

$$\frac{\partial(x, y)}{\partial(r, \theta)} = r, \qquad \frac{\partial(y, z)}{\partial(r, \theta)} = \sin\theta, \qquad \frac{\partial(x, z)}{\partial(r, \theta)} = \cos\theta$$

Also, $f(r\cos\theta, r\sin\theta, \theta) = \sqrt{r^2 + 1}$. Therefore

$$\begin{aligned}
\int_S f(x, y, z)\, dS &= \int_D f(\mathbf{\Phi}(r, \theta)) \|\mathbf{T}_r \times \mathbf{T}_\theta\|\, dr\, d\theta \\
&= \int_0^{2\pi} \int_0^1 \sqrt{r^2 + 1}\sqrt{r^2 + 1}\, dr\, d\theta \\
&= \int_0^{2\pi} \tfrac{4}{3}\, d\theta \\
&= \tfrac{8}{3}\pi
\end{aligned}$$

Suppose S is the graph of a C^1 function $z = g(x, y)$. Then we can parametrize S by

$$x = u, \qquad y = v, \qquad z = g(u, v)$$

In this case

$$\|\mathbf{T}_u \times \mathbf{T}_v\| = \sqrt{1 + \left(\frac{\partial g}{\partial u}\right)^2 + \left(\frac{\partial g}{\partial v}\right)^2}$$

and so

$$\int_S f(x, y, z)\, dS = \int_D f(x, y, g(x, y)) \sqrt{1 + \left(\frac{\partial g}{\partial x}\right)^2 + \left(\frac{\partial g}{\partial y}\right)^2}\, dx\, dy \qquad (4)$$

EXAMPLE 2. Let S be the surface defined by $z = x^2 + y$, where D is the region $0 \le x \le 1$, $-1 \le y \le 1$. Evaluate $\int_S x\, dS$.

If we let $z = g(x, y) = x^2 + y$, formula (4) gives

$$\int_S x\, dS = \int_D x \sqrt{1 + \left(\frac{\partial g}{\partial x}\right)^2 + \left(\frac{\partial g}{\partial y}\right)^2}\, dx\, dy$$

$$= \int_{-1}^{1} \int_{0}^{1} x\sqrt{1 + 4x^2 + 1}\, dx\, dy$$

$$= \tfrac{1}{8}\int_{-1}^{1} \left[\int_0^1 [2 + 4x^2]^{1/2}(8x\, dx)\right] dy$$

$$= \tfrac{2}{3}\cdot\tfrac{1}{8}\int_{-1}^{1} [(2 + 4x^2)^{3/2}]_0^1\, dy$$

$$= \tfrac{1}{12}\int_{-1}^{1} [6^{3/2} - 2^{3/2}]\, dy = \tfrac{1}{6}[6^{3/2} - 2^{3/2}]$$

$$= \sqrt{6} - \frac{\sqrt{2}}{3} = \sqrt{2}\left(\sqrt{3} - \frac{1}{3}\right)$$

EXAMPLE 3. Evaluate $\int_S z^2\, dS$ where S is the unit sphere $x^2 + y^2 + z^2 = 1$.

For this problem it is convenient to represent the sphere parametrically by the equations $x = \cos\theta \sin\psi$, $y = \sin\theta \sin\psi$, $z = \cos\psi$, over the region D in the $\theta\psi$-plane given by $0 \le \psi \le \pi$, $0 \le \theta \le 2\pi$. From equation (1) we get

$$\int_S z^2\, dS = \int_D (\cos\psi)^2 \|\mathbf{T}_\theta \times \mathbf{T}_\psi\|\, d\theta\, d\psi$$

Now a little computation [use formula (2) of Section 6.4] shows that

$$\|\mathbf{T}_\theta \times \mathbf{T}_\psi\| = |\sin\psi|$$

so

$$\int_S z^2\, dS = \int_0^{2\pi}\int_0^{\pi} \cos^2\psi\, |\sin\psi|\, d\psi\, d\theta$$

$$= \int_0^{2\pi}\int_0^{\pi} \cos^2\psi \sin\psi\, d\psi\, d\theta = \tfrac{1}{3}\int_0^{2\pi} [-\cos^3\psi]_0^{\pi}\, d\theta$$

$$= \frac{2}{3}\int_0^{2\pi} d\theta = \frac{4\pi}{3}$$

Once again, let S be the graph of $z = g(x, y)$ and consider formula (4); we wish to interpret this result geometrically. We claim that

$$\int_S f(x, y, z)\, dS = \int_D \frac{f(x, y, g(x, y))}{\cos\theta}\, dx\, dy \tag{5}$$

where θ is the angle the normal to the surface makes with the unit vector $\mathbf{k}$ at the point $(x, y, g(x, y))$ (see Figure 6.5.2).

Since $\phi(x, y, z) = z - g(x, y) = 0$, a normal vector is $\nabla\phi$; that is,

$$\mathbf{n} = -(\partial g/\partial x)\mathbf{i} - (\partial g/\partial y)\mathbf{j} + \mathbf{k}$$

(see Example 2 of Section 6.3, or recall that the normal to a surface $g(x, y, z) = \text{constant}$ is ∇g.) Thus

$$\cos\theta = \frac{\mathbf{n}\cdot\mathbf{k}}{\|\mathbf{n}\|} = 1\bigg/\sqrt{\left(\frac{\partial g}{\partial x}\right)^2 + \left(\frac{\partial g}{\partial y}\right)^2 + 1}$$

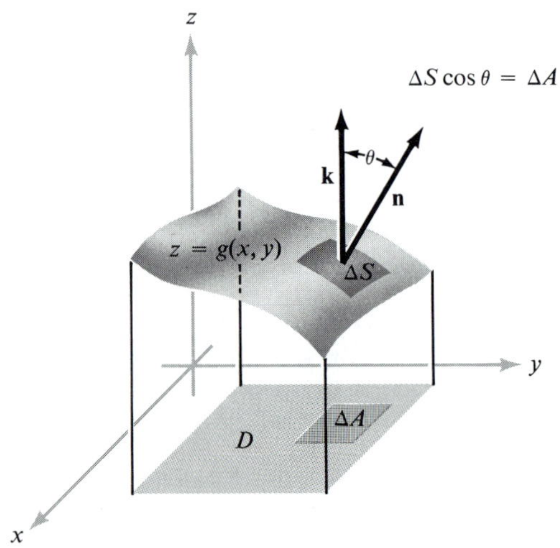

FIGURE 6.5.2
The area of a patch of area ΔS over a patch ΔA is $\Delta S = \Delta A/\cos\theta$ where θ is the angle the normal **n** *makes with* **k**.

The result is, in fact, obvious geometrically, for if a small rectangle in the xy-plane has area ΔA then the area of the portion above it on the surface is $\Delta S = \Delta A/\cos\theta$ (Figure 6.5.2). This approach can help us to remember formula (5) and to apply it in problems.

EXAMPLE 4. Compute $\int_S x\,dS$ where S is the triangle with vertices $(1, 0, 0)$, $(0, 1, 0)$, $(0, 0, 1)$ (see Figure 6.5.3).

This surface is the plane described by the equation $x + y + z = 1$. Since the surface is a plane, the angle θ is constant and a unit normal vector is $\mathbf{n} = (1/\sqrt{3}, 1/\sqrt{3}, 1/\sqrt{3})$. Thus, $\cos\theta = 1/\sqrt{3}$ and by (5)

$$\int_S x\,dS = \sqrt{3}\int_D x\,dx\,dy$$

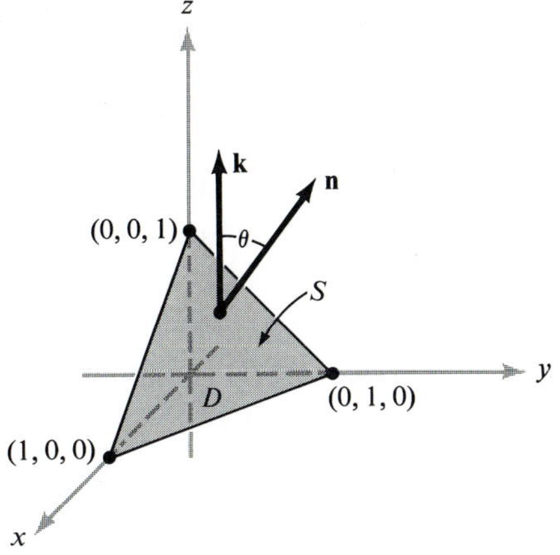

FIGURE 6.5.3
In computing a specific surface integral, one finds a formula for the normal **n** *and computes the angle* θ *in preparation for formula* (5).

where D is the domain in the xy-plane. But

$$\sqrt{3}\int_D x\,dx\,dy = \sqrt{3}\int_0^1\int_0^{1-x} x\,dy\,dx = \sqrt{3}\int_0^1 x(1-x)\,dx = \frac{\sqrt{3}}{6}$$

Definition. *The* ***average value*** *of* $f(x, y, z)$ *over* S *is*

$$\frac{\int_S f(x, y, z)\,dS}{A(S)} = \frac{\int_S f(x, y, z)\,dS}{\int_S dS} \tag{6}$$

This definition is perfectly analogous to the definition of the average value of $f(x, y, z)$ along a path $\boldsymbol{\sigma}$ as $\int_{\boldsymbol{\sigma}} f(x, y, z)\,ds/\int_{\boldsymbol{\sigma}} ds$. We can justify equation (6) in the terms of Riemann sums as follows. The ratio

$$R_n = \frac{\sum_{i=0}^{n-1}\sum_{j=0}^{n-1} f(x_{ij}, y_{ij}, z_{ij})A(S_{ij})}{\sum_{i=0}^{n-1}\sum_{j=0}^{n-1} A(S_{ij})} = \frac{\sum_{i=0}^{n-1}\sum_{j=0}^{n-1} f(x_{ij}, y_{ij}, z_{ij})A(S_{ij})}{A(S)}$$

is the approximate average value of f on S obtained by considering f as constant on S_{ij}. The limit of the ratios R_n as $n \to \infty$ is the average value as given in equation (6).

In Example 3 above, if $f(x, y, z) = z^2$ represents the temperature at (x, y, z), then the average temperature on the surface of the sphere is $\frac{1}{3}(4\pi)/4\pi = 1/3$ degree. In Example 4, we computed $\int_S x\,dS$ for the triangle whose vertices are the tips of the vectors **i**, **j**, and **k**. The area of this triangle is

$$\int_S dS = \int_D \frac{1}{\cos\theta}\,dx\,dy = \int_0^1\int_0^{1-x} \sqrt{3}\,dy\,dx = \frac{\sqrt{3}}{2}$$

Hence, the average x-coordinate of points on the triangle is the ratio $(\sqrt{3}/6)/(\sqrt{3}/2) = 1/3$. This by definition is the x-coordinate of the "center of gravity" of the triangle. From symmetry we see that the average z- and y-coordinates are also 1/3. Consequently the center of gravity of the triangle is at (1/3, 1/3, 1/3).

To paraphrase these ideas, we define the (x, y, z) coordinates of the *center of gravity* $(\bar{x}, \bar{y}, \bar{z})$ of a surface S to be the average x-, y-, and z-coordinates. They are given by the formulas

$$\begin{aligned} \bar{x} &= \frac{1}{A(S)} \int_S x \, dS \\ \bar{y} &= \frac{1}{A(S)} \int_S y \, dS \\ \bar{z} &= \frac{1}{A(S)} \int_S z \, dS \end{aligned} \tag{7}$$

The integrals of functions over surfaces are also useful for computing the mass of a surface when the mass density function m is known. The total mass of a surface with mass density m is given by

$$M(S) = \int_S m(x, y, z) \, dS \tag{8}$$

EXAMPLE 5. Let $\mathbf{\Phi}: D \to \mathbb{R}^3$ be the parametrization of the helicoid $S = \mathbf{\Phi}(D)$ of Example 2 of Section 6.4. Recall that $\mathbf{\Phi}(r, \theta) = (r \cos \theta, r \sin \theta, \theta)$, $0 \le \theta \le 2\pi$, $0 \le r \le 1$. Suppose S has a mass density at $(x, y, z) \in S$ equal to twice the distance of (x, y, z) to the central axis (see Figure 6.4.2), namely $m(x, y, z) = 2\sqrt{x^2 + y^2} = 2r$, in the cylindrical coordinate system. Find the total mass of the surface.

Applying formula (8) we have that

$$M(S) = \int_S 2\sqrt{x^2 + y^2} \, dS = \int_S 2r \, dS = \int_D 2r \|\mathbf{T}_r \times \mathbf{T}_\theta\| \, dr \, d\theta$$

From Example 2 of Section 6.4 we see that $\|\mathbf{T}_r \times \mathbf{T}_\theta\| = \sqrt{1 + r^2}$. Thus

$$\begin{aligned} M(S) &= \int_D 2r\sqrt{1 + r^2} \, dr \, d\theta = \int_0^{2\pi} \int_0^1 2r\sqrt{1 + r^2} \, dr \, d\theta \\ &= \int_0^{2\pi} [\tfrac{2}{3}(1 + r^2)^{3/2}]_0^1 \, d\theta = \int_0^{2\pi} \tfrac{2}{3}[2^{3/2} - 1] \, d\theta \\ &= \frac{4\pi}{3} [2^{3/2} - 1] \end{aligned}$$

EXERCISES

1. Compute $\int_S xy\, dS$ where S is the surface of the tetrahedron with sides $z = 0$, $y = 0$, $x + z = 1$, and $x = y$.
2. Let $\mathbf{\Phi}: D \subset \mathbb{R}^2 \to \mathbb{R}^3$ be a parametrization of a surface S defined by

$$x = x(u, v), \qquad y = y(u, v), \qquad z = z(u, v)$$

(a) Let

$$\frac{\partial \mathbf{\Phi}}{\partial u} = \left(\frac{\partial x}{\partial u}, \frac{\partial y}{\partial u}, \frac{\partial z}{\partial u}\right), \qquad \frac{\partial \mathbf{\Phi}}{\partial v} = \left(\frac{\partial x}{\partial v}, \frac{\partial y}{\partial v}, \frac{\partial z}{\partial v}\right)$$

that is, $\partial \mathbf{\Phi}/\partial u = \mathbf{T}_u$ and $\partial \mathbf{\Phi}/\partial v = \mathbf{T}_v$, and set

$$E = \left|\frac{\partial \mathbf{\Phi}}{\partial u}\right|^2, \qquad F = \frac{\partial \mathbf{\Phi}}{\partial u} \cdot \frac{\partial \mathbf{\Phi}}{\partial v}, \qquad G = \left|\frac{\partial \mathbf{\Phi}}{\partial v}\right|^2$$

Show that the surface area of S is $\int_D \sqrt{EG - F^2}\, du\, dv$. In this notation, how can we express $\int_S f\, dS$ for a general function f?

(b) What does the formula become if the vectors $\partial \mathbf{\Phi}/\partial u$ and $\partial \mathbf{\Phi}/\partial v$ are orthogonal?

(c) Use parts (a) and (b) to compute the surface area of a sphere of radius a.

3. Evaluate $\int_S z\, dS$, where S is the upper hemisphere of radius a, that is, the set of (x, y, z) with $z = \sqrt{a^2 - x^2 - y^2}$.
4. Evaluate $\int_S (x + y + z)\, dS$, where S is the boundary of the unit ball B; that is, S is the set of (x, y, z) with $x^2 + y^2 + z^2 = 1$. (HINT: Use the symmetry of the problem.)
5. Evaluate $\int_S xyz\, dS$, where S is the triangle with vertices $(1, 0, 0)$, $(0, 2, 0)$, and $(0, 1, 1)$.
6. Let a surface S be defined implicitly by $F(x, y, z) = 0$ for (x, y) in a domain D of $\mathbb{R}^2$. Show that

$$\int_S \left|\frac{\partial F}{\partial z}\right| dS = \int_D \sqrt{\left[\frac{\partial F}{\partial x}\right]^2 + \left[\frac{\partial F}{\partial y}\right]^2 + \left[\frac{\partial F}{\partial z}\right]^2}\, dx\, dy$$

Compare with Exercise 8 of Section 6.4.

7. Evaluate $\int_S z\, dS$, where S is the surface $z = x^2 + y^2$, $x^2 + y^2 \leq 1$.
8. Evaluate $\int_S z^2\, dS$, where S is the boundary of the cube $C = [-1, 1] \times [-1, 1] \times [-1, 1]$. (HINT: Do each face separately and add the results.)
9. Find the mass of a spherical surface S of radius R such that at each point $(x, y, z) \in S$ the mass density is equal to the distance of (x, y, z) to some fixed point $(x_0, y_0, z_0) \in S$.
10. Find the x-, y-, and z-coordinates of the center of gravity of the octant of the sphere of radius R determined by $x \geq 0$, $y \geq 0$, $z \geq 0$. (HINT: Write this octant as a parametrized surface—see Example 3 of this section.)
11. A metallic surface S is in the shape of a hemisphere $z = \sqrt{R^2 - x^2 - y^2}$, $0 \leq x^2 + y^2 \leq R^2$. The mass density at $(x, y, z) \in S$ is given by $m(x, y, z) = x^2 + y^2$. Find the total mass of S.

12. Let S be the sphere of radius R.
    (a) Argue by symmetry that

$$\int_S x^2\, dS = \int_S y^2\, dS = \int_S z^2\, dS$$

    (b) Use this fact and some clever thinking to evaluate, with very little computation, the integral

$$\int_S x^2\, dS$$

    (c) Does this help in Exercise 11?
13. Find the z-coordinate of the center of gravity (the average z-coordinate) of the surface of a hemisphere ($z \geq 0$) of radius R. Argue by symmetry that the average x- and y-coordinates are both zero.

6.6 SURFACE INTEGRALS OF VECTOR FUNCTIONS

In this section we shall turn our attention to integrals of *vector* functions over surfaces. The definition we give here is a natural extension of that for scalar functions discussed in Section 6.5.

Definition. *Let* **F** *be a vector field defined on S, the image of a parametrized surface* **Φ**. *The* ***surface integral*** *of* **F** *over* **Φ**, *denoted by*

$$\int_{\mathbf{\Phi}} \mathbf{F}, \quad \int_{\mathbf{\Phi}} \mathbf{F} \cdot d\mathbf{S}, \quad \textit{or} \quad \iint_{\mathbf{\Phi}} \mathbf{F} \cdot d\mathbf{S}$$

is defined by

$$\int_{\mathbf{\Phi}} \mathbf{F} = \int_D \mathbf{F} \cdot (\mathbf{T}_u \times \mathbf{T}_v)\, du\, dv$$

where $\mathbf{T}_u$ *and* $\mathbf{T}_v$ *are defined as on p. 308 (see Figure 6.6.1).*

EXAMPLE 1. Let D be the rectangle in the $\theta\phi$-plane defined by

$$0 \leq \theta \leq 2\pi, \qquad 0 \leq \phi \leq \pi$$

and let the surface S be defined by the parametrization $\mathbf{\Phi}: D \to \mathbb{R}^3$ given by

$$x = \cos\theta \sin\phi, \qquad y = \sin\theta\sin\phi, \qquad z = \cos\phi$$

Thus θ and ϕ are the angles of spherical coordinates (see p. 274), and S is the unit sphere parametrized by **Φ**. Let **r** be the position vector $\mathbf{r}(x, y, z) = x\mathbf{i} + y\mathbf{j} + z\mathbf{k}$. We compute $\int_{\mathbf{\Phi}} \mathbf{r}$ as follows. First we find

$$\mathbf{T}_\theta = (-\sin\phi\sin\theta)\mathbf{i} + (\sin\phi\cos\theta)\mathbf{j}$$

$$\mathbf{T}_\phi = (\cos\theta\cos\phi)\mathbf{i} + (\sin\theta\cos\phi)\mathbf{j} - (\sin\phi)\mathbf{k}$$

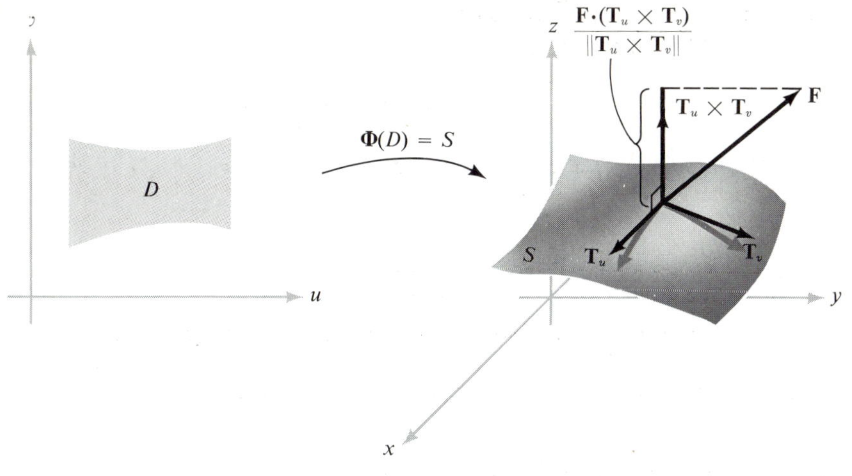

FIGURE 6.6.1
The geometrical significance of $\mathbf{F}\cdot(\mathbf{T}_u \times \mathbf{T}_v)$.

and hence

$$\mathbf{T}_\theta \times \mathbf{T}_\phi = (-\sin^2 \phi \cos \theta)\mathbf{i} - (\sin^2 \phi \sin \theta)\mathbf{j} - (\sin \phi \cos \phi)\mathbf{k}$$

Then we evaluate

$$\begin{aligned}
\mathbf{r}\cdot(\mathbf{T}_\theta \times \mathbf{T}_\phi) &= (x\mathbf{i} + y\mathbf{j} + z\mathbf{k})\cdot(\mathbf{T}_\theta \times \mathbf{T}_\phi) \\
&= [(\cos\theta \sin\phi)\mathbf{i} + (\sin\theta\sin\phi)\mathbf{j} + (\cos\phi)\mathbf{k}] \\
&\quad \cdot (-\sin\phi)[(\sin\phi\cos\theta)\mathbf{i} + (\sin\phi\sin\theta)\mathbf{j} + (\cos\phi)\mathbf{k}] \\
&= (-\sin\phi)(\sin^2\phi\cos^2\theta + \sin^2\phi\sin^2\theta + \cos^2\phi) \\
&= -\sin\phi
\end{aligned}$$

Then

$$\int_{\Phi} \mathbf{r} = \int_D -\sin\phi \, d\phi \, d\theta = \int_0^{2\pi} (-2)\, d\theta = -4\pi$$

An analogy can be drawn between the surface integral $\int_{\Phi} \mathbf{F}$ and the line integral $\int_{\sigma} \mathbf{F}$. Recall that the line integral is an oriented integral. We needed the notion of orientation of a curve to extend the definition of $\int_{\sigma} \mathbf{F}$ to line integrals $\int_C \mathbf{F}$ over oriented curves. We should like to extend the definition of $\int_{\Phi} \mathbf{F}$ to surfaces in a similar fashion; that is, given a surface S parametrized by a mapping Φ, we want to define $\int_S \mathbf{F} = \int_{\Phi} \mathbf{F}$ and show that it is independent of the parametrization, except possibly for the sign. In order to accomplish this, we need the notion of orientation of a surface.

Definition. *An **oriented surface** is a two-sided surface with one side specified as the **outside** or **positive side**; we call the other side the **inside** or **negative side**.* At each point $(x, y, z) \in S$ there are two unit normal vectors $\mathbf{n}_1$ and $\mathbf{n}_2$, where $\mathbf{n}_1 = -\mathbf{n}_2$ (see Figure 6.6.2). Each of these two normals can be associated with one side of the surface. Thus to specify a side of a surface S, at each point we choose a unit normal vector **n** that points away from the positive side of S at that point.*

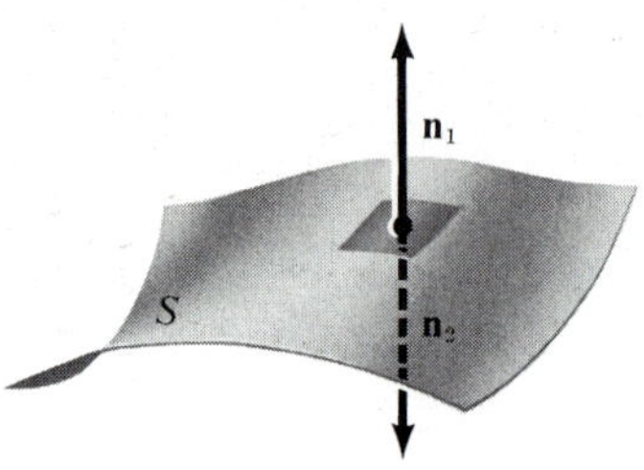

FIGURE 6.6.2
The two possible unit normals to a surface at a point.

This definition assumes that our surface does have two sides. We should give an example of a surface with only one side. The first known example of such a surface was the Möbius strip (named after the Dutch mathematician A. F. Möbius, who, along with the mathematician J. B. Listing, discovered it in 1858). Pictures of such a surface are given in figures 6.6.3 and 6.6.4.

At each point of M there are two unit normals $\mathbf{n}_1$ and $\mathbf{n}_2$. However, $\mathbf{n}_1$ doesn't determine a unique side of M, and neither does $\mathbf{n}_2$. To see this intuitively, we can slide $\mathbf{n}_2$ around the closed curve C (Figure

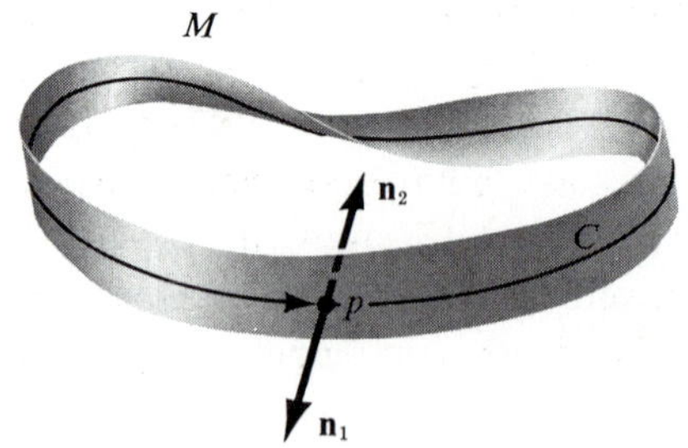

FIGURE 6.6.3
The Möbius strip: slide $\mathbf{n}_2$ around C once; when $\mathbf{n}_2$ returns to its initial point p, $\mathbf{n}_2$ will coincide with $\mathbf{n}_1 = -\mathbf{n}_2$.

* We use the phrase "side" in an intuitive sense. This concept can be developed rigorously.

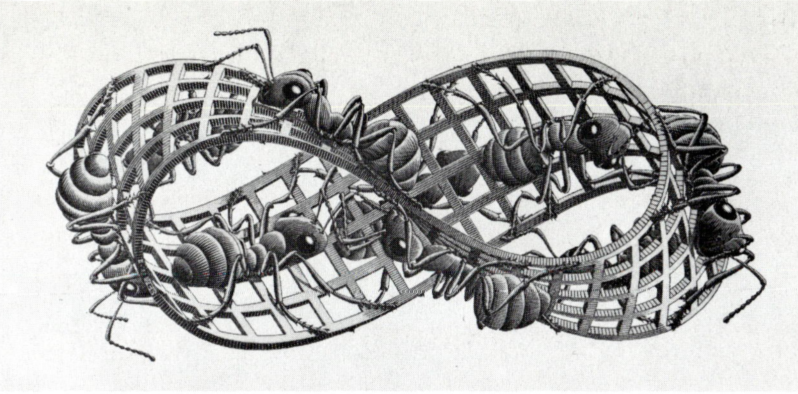

FIGURE 6.6.4
Ants walking on a Möbius strip. (Moebius Strip II 1963, by M. C. Escher. Escher Foundation, Haags Gemeentemuseum, The Hague.)

6.6.3). When $\mathbf{n}_2$ returns to a fixed point p on C it will coincide with $\mathbf{n}_1$, showing that both $\mathbf{n}_1$ and $\mathbf{n}_2$ point away from the same side of M and, consequently, that M has only one side.

Figure 6.6.4 is a Möbius strip as drawn by the well-known twentieth-century mathematician and artist M. C. Escher. It depicts ants crawling along the Möbius band. They walk in a circle, and after one complete trip around (without crossing an edge) they end up on the "opposite side" of the surface.

Let $\mathbf{\Phi}\colon D \to \mathbb{R}^3$ be a parametrization of an oriented surface S and suppose S is smooth at $\mathbf{\Phi}(u_0, v_0)$, $(u_0, v_0) \in D$; that is, the unit normal vector $(\mathbf{T}_{u_0} \times \mathbf{T}_{v_0})/\|\mathbf{T}_{u_0} \times \mathbf{T}_{v_0}\|$ is defined. If $\mathbf{n}(\mathbf{\Phi}(u_0, v_0))$ denotes the unit normal to S at $\mathbf{\Phi}(u_0, v_0)$ pointing to the positive side of S at that point, it follows that $(\mathbf{T}_{u_0} \times \mathbf{T}_{v_0})/\|\mathbf{T}_{u_0} \times \mathbf{T}_{v_0}\| = \pm\mathbf{n}(\mathbf{\Phi}(u_0, v_0))$. The parametrization $\mathbf{\Phi}$ is said to be *orientation preserving* if $(\mathbf{T}_u \times \mathbf{T}_v)/\|\mathbf{T}_u \times \mathbf{T}_v\| = \mathbf{n}(\mathbf{\Phi}(u, v))$ at all $(u, v) \in D$ for which S is smooth at $\mathbf{\Phi}(u, v)$. In other words, $\mathbf{\Phi}$ is orientation preserving if the vector $\mathbf{T}_u \times \mathbf{T}_v$ points away from the outside of the surface. If $\mathbf{T}_u \times \mathbf{T}_v$ points away from the inside of the surface at all $(u, v) \in D$ for which S is smooth at $\mathbf{\Phi}(u, v)$, then $\mathbf{\Phi}$ is said to be *orientation reversing*. Using the above notation, this condition corresponds to $(\mathbf{T}_u \times \mathbf{T}_v)/\|\mathbf{T}_u \times \mathbf{T}_v\| = -\mathbf{n}(\mathbf{\Phi}(u, v))$.

EXAMPLE 2. We can give the unit sphere $x^2 + y^2 + z^2 = 1$ in $\mathbb{R}^3$ (Figure 6.6.5) an orientation by selecting the unit vector $\mathbf{n}(x, y, z) = \mathbf{r}$, where $\mathbf{r} = x\mathbf{i} + y\mathbf{j} + z\mathbf{k}$, which points away from the outside of the surface. This choice corresponds to our intuitive notion of outside for the sphere.

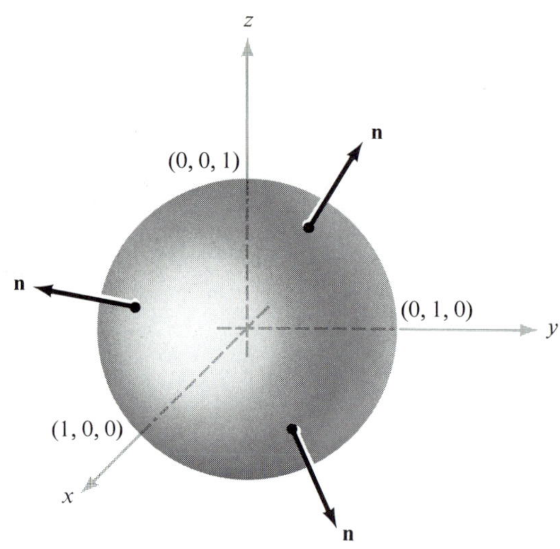

FIGURE 6.6.5
The unit sphere oriented by its outward normal **n**.

Now that the sphere S is an oriented surface, consider the parametrization $\mathbf{\Phi}$ of S given in Example 1. The cross product of the tangent vectors $\mathbf{T}_\theta$ and $\mathbf{T}_\phi$—that is, a normal to S—is given by

$$(-\sin\phi)[(\cos\theta\sin\phi)\mathbf{i} + (\sin\theta\sin\phi)\mathbf{j} + (\cos\phi)\mathbf{k}] = -\mathbf{r}\sin\phi$$

Since $-\sin\phi \le 0$ for $0 \le \phi \le \pi$, this normal vector points inward from the sphere. Thus the given parametrization $\mathbf{\Phi}$ is orientation reversing.

EXAMPLE 3. Let S be a surface described by $z = f(x, y)$. There are two unit normal vectors to S at $(x_0, y_0, f(x_0, y_0))$, namely $\pm\mathbf{n}$, where

$$\mathbf{n} = \frac{-\dfrac{\partial f}{\partial x}(x_0, y_0)\mathbf{i} - \dfrac{\partial f}{\partial y}(x_0, y_0)\mathbf{j} + \mathbf{k}}{\sqrt{\left(\dfrac{\partial f}{\partial x}(x_0, y_0)\right)^2 + \left(\dfrac{\partial f}{\partial y}(x_0, y_0)\right)^2 + 1}}$$

We can orient all such surfaces* by taking the positive side of S to be the side away from which **n** points (Figure 6.6.6). Thus the positive side of such a surface is determined by the unit normal **n** with positive **k** component. If we parametrize this surface by $\mathbf{\Phi}(u, v) = (u, v, f(u, v))$, then $\mathbf{\Phi}$ will be orientation preserving.

* If we had given a rigorous definition of orientation we could use this argument to show all surfaces $z = f(x, y)$ are in fact orientable; i.e., "have two sides."

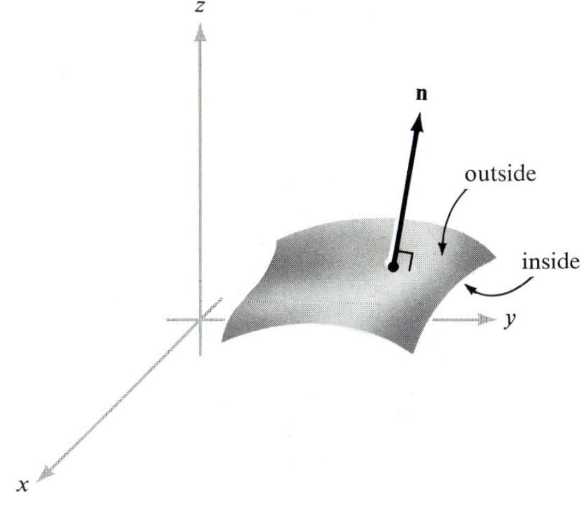

FIGURE 6.6.6
n *points away from the outside of the surface.*

We now state without proof a theorem showing that the integral over an oriented surface is independent of the parametrization.

We should mention that the proof of Theorem 4 is analogous to that of Theorem 1 (Section 6.2); the heart of the proof is in the Change of Variables formula—this time applied to double integrals.

Theorem 4. *Let S be an oriented surface and let* $\mathbf{\Phi}_1$ *and* $\mathbf{\Phi}_2$ *be two smooth orientation-preserving parametrizations, with* $\mathbf{F}$ *a continuous vector field defined on S. Then*

$$\int_{\mathbf{\Phi}_1} \mathbf{F} = \int_{\mathbf{\Phi}_2} \mathbf{F}$$

If $\mathbf{\Phi}_1$ *is orientation preserving and* $\mathbf{\Phi}_2$ *orientation reversing*

$$\int_{\mathbf{\Phi}_1} \mathbf{F} = -\int_{\mathbf{\Phi}_2} \mathbf{F}$$

Moreover, if f is a real-valued continuous function defined on S, and if $\mathbf{\Phi}_1$ *and* $\mathbf{\Phi}_2$ *are parametrizations of S, then*

$$\int_{\mathbf{\Phi}_1} f\, dS = \int_{\mathbf{\Phi}_2} f\, dS$$

If $f = 1$ *we obtain*

$$A(S) = \int_{\mathbf{\Phi}_1} dS = \int_{\mathbf{\Phi}_2} dS$$

thus showing that area is independent of parametrization.

We can thus unambiguously use the notation

$$\int_S \mathbf{F} = \int_{\mathbf{\Phi}} \mathbf{F} \quad \text{or} \quad \int_S \mathbf{F} \cdot d\mathbf{S} = \int_{\mathbf{\Phi}} \mathbf{F} \cdot d\mathbf{S}$$

(or a sum of such integrals if S is a union of parametrized surfaces that intersect only along their boundary curves) where $\mathbf{\Phi}$ is an orientation-preserving parametrization. Theorem 4 guarantees that the value of the integral does not depend on the selection of $\mathbf{\Phi}$.

We can also express $\int_S \mathbf{F}$ as an integral of a real-valued function f over the surface S (see Section 6.5). This is analogous to expressing $\int_{\boldsymbol{\sigma}} \mathbf{F}$ as a path integral $\int_{\boldsymbol{\sigma}} f$ for a particular real-valued function f. Let S be an oriented smooth surface and let $\mathbf{\Phi}\colon D \to \mathbb{R}^3$ be an orientation-preserving parametrization. Then $\mathbf{n} = (\mathbf{T}_u \times \mathbf{T}_v)/\|\mathbf{T}_u \times \mathbf{T}_v\|$ is a unit normal pointing to the outside of S, and

$$\begin{aligned}\int_S \mathbf{F} = \int_{\mathbf{\Phi}} \mathbf{F} &= \int_D \mathbf{F} \cdot (\mathbf{T}_u \times \mathbf{T}_v)\, du\, dv \\ &= \int_D \mathbf{F} \cdot \left(\frac{\mathbf{T}_u \times \mathbf{T}_v}{\|\mathbf{T}_u \times \mathbf{T}_v\|}\right) \|\mathbf{T}_u \times \mathbf{T}_v\|\, du\, dv \\ &= \int_D (\mathbf{F} \cdot \mathbf{n})\|\mathbf{T}_u \times \mathbf{T}_v\|\, du\, dv = \int_S (\mathbf{F} \cdot \mathbf{n})\, dS = \int_S f\, dS\end{aligned}$$

where $f = \mathbf{F} \cdot \mathbf{n}$.

Thus $\int_S \mathbf{F}$, *the surface integral of* $\mathbf{F}$ *over* S, *is equal to the integral of the normal component of* $\mathbf{F}$ *over the surface*. This observation can often save a great deal of computational effort, as Example 4 below demonstrates.

The geometric and physical significance of the surface integral can be understood by expressing it as a limit of Riemann sums. For simplicity, we assume D is a rectangle. Fix a parametrization $\mathbf{\Phi}$ of S that preserves orientation and partition the region D into n^2 pieces D_{ij}, $0 \le i \le n-1$, $0 \le j \le n-1$. We let Δu denote the length of the horizontal side of D_{ij} and Δv denote the length of the vertical side of D_{ij}. Let (u, v) be a point in D_{ij} and $(x, y, z) = \mathbf{\Phi}(u, v)$, the corresponding point on the surface. We consider the parallelogram with sides $\Delta u \mathbf{T}_u$ and $\Delta v \mathbf{T}_v$ lying in the plane tangent to S at (x, y, z) and the parallelepiped formed by $\mathbf{F}$, $\Delta u \mathbf{T}_u$, and $\Delta v \mathbf{T}_v$. The volume of the parallelepiped is the absolute value of the triple product

$$\mathbf{F} \cdot (\Delta u \mathbf{T}_u \times \Delta v \mathbf{T}_v) = \mathbf{F} \cdot (\mathbf{T}_u \times \mathbf{T}_v)\, \Delta u\, \Delta v$$

The vector $\mathbf{T}_u \times \mathbf{T}_v$ is normal to the surface at (x, y, z) and points away from the outside of the surface. Thus the number $\mathbf{F} \cdot (\mathbf{T}_u \times \mathbf{T}_v)$ is positive when the parallelepiped lies on the outside of the surface (Figure 6.6.7).

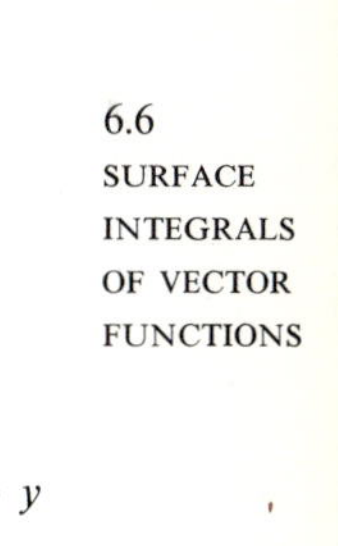

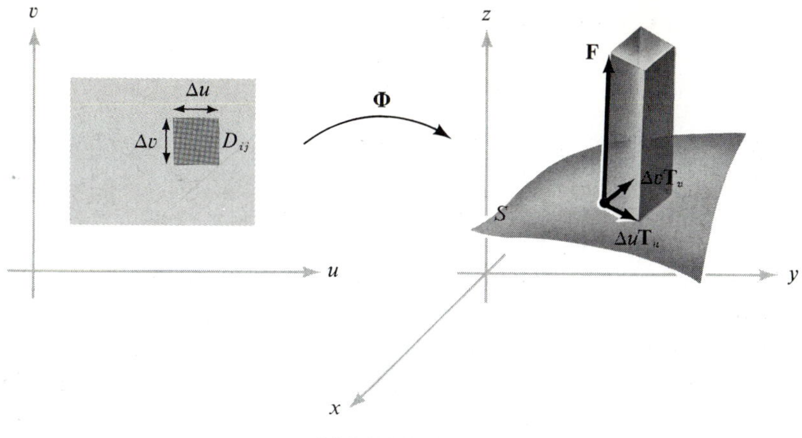

FIGURE 6.6.7
$\mathbf{F} \cdot (\mathbf{T}_u \times \mathbf{T}_v) > 0$ *when the parallelepiped formed by* $\Delta v\mathbf{T}_v$, $\Delta u\mathbf{T}_u$, *and* **F** *lies to the "outside" of the surface S.*

In general, the parallelepiped lies on that side of the surface away from which **F** is pointing. If we think of **F** as the velocity field of a fluid, $\mathbf{F}(x, y, z)$ is pointing in the direction in which fluid is moving across the surface near (x, y, z). Moreover, the number $|\mathbf{F} \cdot (\mathbf{T}_u\,\Delta u \times \mathbf{T}_v\,\Delta v)|$ measures the amount of fluid that passes through the tangent parallelogram per unit time. Since the sign of $\mathbf{F} \cdot (\Delta u\mathbf{T}_u \times \Delta v\mathbf{T}_v)$ is positive if the vector **F** is pointing outward at (x, y, z) and negative if **F** is pointing inward, the sum $\sum_{i,j} \mathbf{F} \cdot (\mathbf{T}_u \times \mathbf{T}_v)\,\Delta u\,\Delta v$ is an approximate measure of the net quantity of fluid to flow outward across the surface per unit time. (Remember that "outward" or "inward" depends on our choice of parametrization. Figure 6.6.8 illustrates **F**

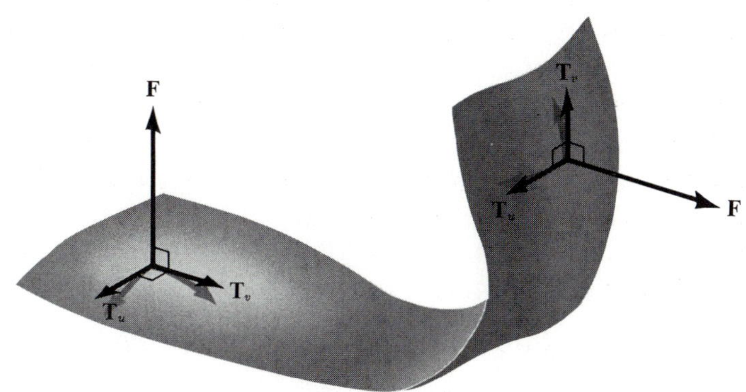

FIGURE 6.6.8
(*left*) *When* $\mathbf{F} \cdot \mathbf{T}_u \times \mathbf{T}_v > 0$, **F** *points outward;* (*right*) *when* $\mathbf{F} \cdot \mathbf{T}_u \times \mathbf{T}_v < 0$, **F** *points inward.*

directed outward and inward, given $\mathbf{T}_u$ and $\mathbf{T}_v$.) Hence, the integral $\int_S \mathbf{F}$ is the net quantity of fluid to flow across the surface per unit time, that is, the rate of fluid flow. Therefore this integral is also called the *flux* of $\mathbf{F}$ across the surface.

In case $\mathbf{F}$ represents electric or magnetic fields, $\int_S \mathbf{F}$ is also commonly known as the flux. The reader may be familiar with physical laws (such as Faraday's law) that relate flux of a vector field to a circulation (or current) in a bounding loop. This is the historical and physical basis of Stokes' Theorem, which we shall meet in Section 7.2. The corresponding principle in fluid mechanics is called Kelvin's Circulation Theorem.

Surface integrals also apply to the study of heat flow. Let $T(x, y, z)$ be the temperature at a point $(x, y, z) \in W \subset \mathbb{R}^3$, where W is some region and T is a C^1 function. Then

$$\nabla T = \frac{\partial T}{\partial x}\mathbf{i} + \frac{\partial T}{\partial y}\mathbf{j} + \frac{\partial T}{\partial z}\mathbf{k}$$

represents the temperature gradient, and heat "flows" with the vector field $-k\nabla T = \mathbf{F}$, where k is a positive constant (see Section 7.5). Therefore $\int_S \mathbf{F} \cdot d\mathbf{S}$ is the total rate of heat flow or flux across the surface S.

EXAMPLE 4. Suppose a temperature function is given as $T(x, y, z) = x^2 + y^2 + z^2$, and let S be the unit sphere $x^2 + y^2 + z^2 = 1$ oriented with the outward normal (see Example 2). Find the heat flux across the surface S if $k = 1$.

We have

$$\mathbf{F} = -\nabla T(x, y, z) = -2x\mathbf{i} - 2y\mathbf{j} - 2z\mathbf{k}$$

On S, $\mathbf{n}(x, y, z) = x\mathbf{i} + y\mathbf{j} + z\mathbf{k}$ is the unit "outward" normal to S at (x, y, z), and $f(x, y, z) = \mathbf{F} \cdot \mathbf{n} = -2x^2 - 2y^2 - 2z^2 = -2$ is the normal component of $\mathbf{F}$. From the remarks following Theorem 4 we can see that the surface integral of $\mathbf{F}$ is equal to the integral of its normal component $f = \mathbf{F} \cdot \mathbf{n}$ over S. Thus $\int_S \mathbf{F} \cdot d\mathbf{S} = \int_S f\, dS = -2\int_S dS = -2A(S) = -2(4\pi) = -8\pi$. The flux of heat is directed towards the center of the sphere (Why towards?). Consequently our observation that $\int_S \mathbf{F} \cdot d\mathbf{S} = \int_S f\, dS$ has saved us considerable computational time.

In this example $\mathbf{F}(x, y, z) = -2x\mathbf{i} - 2y\mathbf{j} - 2z\mathbf{k}$ could also represent a magnetic field, in which case $\int_S \mathbf{F} = -8\pi$ would be the magnetic flux across S.

EXAMPLE 5. There is an important physical law, due to the great mathematician and physicist K. F. Gauss, that relates the flux of an electric field $\mathbf{E}$ over a "closed" surface S (for example, a sphere or an ellipsoid) to the net charge Q enclosed by the surface, namely

$$\int_S \mathbf{E} = Q \tag{1}$$

(see Figure 6.6.9). Gauss' law will be discussed in detail in Chapter 7. This law is analogous to Ampère's law (see Example 12, Section 6.2).

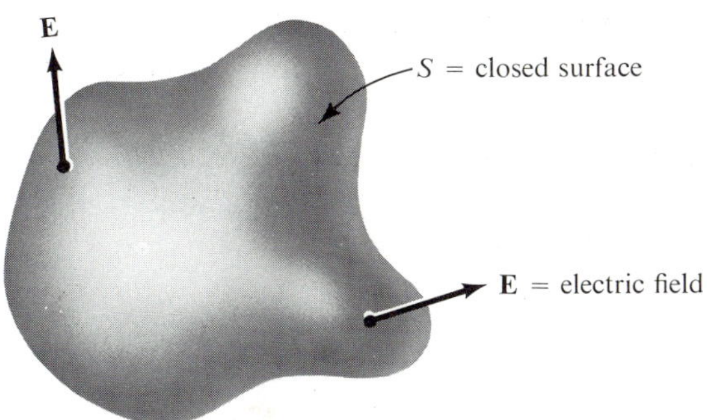

FIGURE 6.6.9
Gauss' Law: $\int_S \mathbf{E} = Q$, *where* Q *is the net charge inside* S.

Suppose that $\mathbf{E} = E\mathbf{n}$; that is, $\mathbf{E}$ is a constant scalar multiple of the unit normal to S. Then Gauss' law (1) becomes

$$\int_S \mathbf{E} = \int_S E\, dS = E \int_S dS = Q$$

Hence

$$E = \frac{Q}{A(S)} \tag{2}$$

In case S is the sphere of radius R, (2) becomes

$$E = \frac{Q}{4\pi R^2} \tag{3}$$

(see Figure 6.6.10).

Now suppose that $\mathbf{E}$ arises from an isolated point charge Q. From symmetry it follows that $\mathbf{E} = E\mathbf{n}$, where $\mathbf{n}$ is the unit normal to any sphere centered at Q. Hence (3) holds. Consider a second point charge Q_0 located at a distance R from Q. The force $\mathbf{F}$ that acts on this second charge Q_0 is given by

$$\mathbf{F} = \mathbf{E}Q_0 = EQ_0\,\mathbf{n} = \frac{QQ_0}{4\pi R^2}\,\mathbf{n}$$

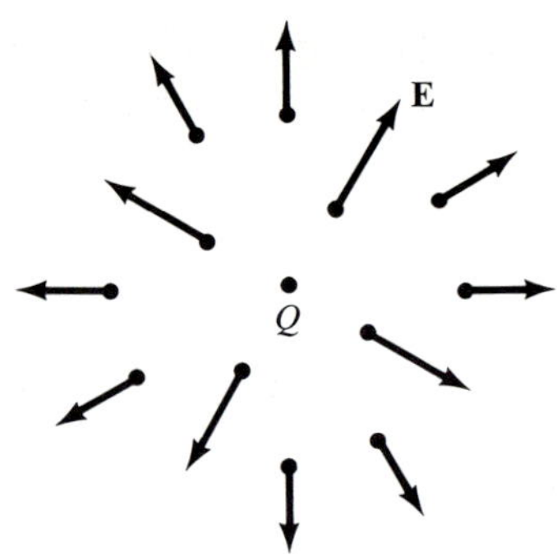

FIGURE 6.6.10
The field $\mathbf{E}$ *due to a point charge* Q *is* $\mathbf{E} = Q\mathbf{n}/4\pi R^2$.

If F is the magnitude of $\mathbf{F}$ we have

$$F = \frac{QQ_0}{4\pi R^2}$$

which is the well-known *Coulomb's law* for the force between two point charges.*

An important kind of surface, as we know from the preceding section, is the graph of a function. Let us work out the surface-integral formulas for this case. Consider the surface S described by $z = f(x, y)$, $(x, y) \in D$, where S is oriented so that

$$\mathbf{n} = \frac{-\dfrac{\partial f}{\partial x}\mathbf{i} - \dfrac{\partial f}{\partial y}\mathbf{j} + \mathbf{k}}{\sqrt{\left(\dfrac{\partial f}{\partial x}\right)^2 + \left(\dfrac{\partial f}{\partial y}\right)^2 + 1}}$$

points outward. We have seen that we can parametrize S by $\mathbf{\Phi}: D \to \mathbb{R}^3$ given by $\mathbf{\Phi}(x, y) = (x, y, f(x, y))$. In this case, $\int_S \mathbf{F}$ can be written in a particularly simple form. We have

$$\mathbf{T}_x = \mathbf{i} + \frac{\partial f}{\partial x}\mathbf{k}$$

$$\mathbf{T}_y = \mathbf{j} + \frac{\partial f}{\partial y}\mathbf{k}$$

* Sometimes one sees the formula $F = (1/4\pi\varepsilon_0)QQ_0/R^2$. The extra constant ε_0 appears when the MKS units are used for measuring charge. We are using CGS, or Gaussian units.

Thus $\mathbf{T}_x \times \mathbf{T}_y = -(\partial f/\partial x)\mathbf{i} - (\partial f/\partial y)\mathbf{j} + \mathbf{k}$. If $\mathbf{F} = F_1\mathbf{i} + F_2\mathbf{j} + F_3\mathbf{k}$ is a continuous vector field then we get the formula

$$\begin{aligned}\int_S \mathbf{F} &= \int_D \mathbf{F} \cdot (\mathbf{T}_x \times \mathbf{T}_y)\, dx\, dy \\ &= \int_D \left[F_1\left(-\frac{\partial f}{\partial x}\right) + F_2\left(-\frac{\partial f}{\partial y}\right) + F_3\right] dx\, dy \qquad (4)\end{aligned}$$

EXAMPLE 6. The equations

$$z = 12, \qquad x^2 + y^2 \leq 25$$

describe a disc of radius 5 lying in the plane $z = 12$. Suppose $\mathbf{r}$ is the field

$$\mathbf{r}(x, y, z) = x\mathbf{i} + y\mathbf{j} + z\mathbf{k}$$

Then $\int_S \mathbf{r}$ is easily computed; we have $\partial z/\partial x = \partial z/\partial y = 0$ since $z = 12$ is constant on the disc, so

$$\begin{aligned}\mathbf{r}(x, y, z) \cdot \mathbf{T}_x \times \mathbf{T}_y &= \mathbf{r}(x, y, z) \cdot (\mathbf{i} \times \mathbf{j}) \\ &= \mathbf{r}(x, y, z) \cdot \mathbf{k} = z\end{aligned}$$

and the integral becomes

$$\int_S \mathbf{r} = \int_D z\, dx\, dy = \int_D 12\, dx\, dy = 12(\text{area of } D) = 300\pi$$

On the other hand, since the disc is parallel to the xy-plane, the outward unit normal is $\mathbf{k}$. So $\mathbf{n}(x, y, z) = \mathbf{k}$ and $\mathbf{r} \cdot \mathbf{n} = z$. However, $\|\mathbf{T}_x \times \mathbf{T}_y\| = \|\mathbf{k}\| = 1$, so we know from the discussion following Theorem 4 (p. 336) that

$$\int_S \mathbf{r} = \int_S \mathbf{r} \cdot \mathbf{n}\, dS = \int_S z\, dS = \int_D 12\, dx\, dy = 300\pi$$

Alternatively, we may solve this problem by using formula (4) directly, with $f(x, y) = 12$ and D the disc $x^2 + y^2 \leq 25$:

$$\int_S \mathbf{r} = \int_D [x \cdot 0 + y \cdot 0 + 12]\, dx\, dy = 12(\text{Area of } D) = 300\pi$$

EXERCISES

1. Let the temperature of a point in $\mathbb{R}^3$ be given by $T(x, y, z) = 3x^2 + 3z^2$. Compute the heat flux across the surface $x^2 + z^2 = 2, 0 \leq y \leq 2$, if $k = 1$.
2. Compute the heat flux across the unit sphere S if $T(x, y, z) = x$ (see Example 4). Can you interpret your answer physically?

3. Let S be the closed surface that consists of the hemisphere $x^2 + y^2 + z^2 = 1$, $z \geq 0$, and its base $x^2 + y^2 \leq 1$, $z = 0$. Let $\mathbf{E}$ be the electric field defined by $\mathbf{E}(x, y, z) = 2x\mathbf{i} + 2y\mathbf{j} + 2z\mathbf{k}$. Find the electric flux across S. (HINT: Break S into two pieces S_1 and S_2 and evaluate $\int_{S_1} \mathbf{E}$ and $\int_{S_2} \mathbf{E}$ separately.)
4. Let the velocity field of a fluid be described by $\mathbf{F} = \sqrt{y}\,\mathbf{j}$ (measured in meters/second). Compute how many cubic meters of fluid per second is crossing the surface $x^2 + z^2 = y$, $0 \leq y \leq 1$.
5. Evaluate $\int_S \nabla \times \mathbf{F}$, where S is the surface $x^2 + y^2 + 3z^2 = 1$, $z \leq 0$, and $\mathbf{F} = y\mathbf{i} - x\mathbf{j} + zx^3y^2\mathbf{k}$.
6. Evaluate $\int_S \nabla \times \mathbf{F}$ where $\mathbf{F} = (x^2 + y - 4)\mathbf{i} + 3xy\mathbf{j} + (2xz + z^2)\mathbf{k}$ and S is the surface $x^2 + y^2 + z^2 = 16$, $z \geq 0$.
7. Let S be the surface of the unit sphere. Let $\mathbf{F}$ be a vector field and F_r its radial component. Prove

$$\int_S \mathbf{F} = \int_{\theta=0}^{2\pi} \int_{\phi=0}^{\pi} F_r \sin \phi \, d\phi \, d\theta$$

What is the formula for real-valued functions f?
8. Work out a formula like that in Exercise 7 for integration over the surface of a cylinder.
*9. Prove the Mean Value Theorem for surface integrals:

$$\int_S \mathbf{F} \cdot \mathbf{n} \, dS = [\mathbf{F}(Q) \cdot \mathbf{n}(Q)]A(S)$$

for some $Q \in S$, where $A(S)$ is the area of S. (HINT: Prove it for real functions first, by reducing the problem to one of a double integral: show that if $g \geq 0$, then

$$\int_D fg = f(Q) \int_D g$$

for some $Q \in D$ (do it by considering $\int_D fg/\int_D g$ and using the Intermediate Value Theorem).)
10. Let the velocity field of a fluid be described by $\mathbf{F} = \mathbf{i} + x\mathbf{j} + z\mathbf{k}$ (measured in meters/second). Compute how many cubic meters of fluid per second is crossing the surface described by $x^2 + y^2 + z^2 = 1$, $z \geq 0$.
11. Let S be a surface in $\mathbb{R}^3$ that is actually a subset D of the xy-plane. Show that the integral of a scalar function $f(x, y, z)$ over S reduces to the double integral of $f(x, y, z)$ over D. What does the surface integral of a vector field over S become?

REVIEW EXERCISES FOR CHAPTER 6

1. Integrate $f(x, y, z) = xyz$ along the following paths.
   (a) $\boldsymbol{\sigma}(t) = (e^t \cos t, e^t \sin t, 3)$, $0 \leq t \leq 2\pi$
   (b) $\boldsymbol{\sigma}(t) = (\cos t, \sin t, t)$, $0 \leq t \leq 2\pi$
   (c) $\boldsymbol{\sigma}(t) = \frac{3}{2}t^2\mathbf{i} + 2t^2\mathbf{j} + t\mathbf{k}$, $0 \leq t \leq 1$
   (d) $\boldsymbol{\sigma}(t) = t\mathbf{i} + (1/\sqrt{2})t^2\mathbf{j} + (1/3)t^3\mathbf{k}$, $0 \leq t \leq 1$

2. Find the area of the surface defined by $\mathbf{\Phi}$: $(u, v) \mapsto (x, y, z)$ where

$$x = h(u, v) = u + v, \qquad y = g(u, v) = u, \qquad z = f(u, v) = v$$

$0 \le u \le 1$, $0 \le v \le 1$. Sketch.

3. Write a formula for the surface area of $\mathbf{\Phi}$: $(r, \theta) \mapsto (x, y, z)$ where

$$x = r \cos \theta, \qquad y = 2r \sin \theta, \qquad z = r$$

$0 \le r \le 1$, $0 \le \theta \le 2\pi$. Sketch.

4. Compute the integral of $f(x, y, z) = x^2 + y^2 + z^2$ over the surface in Exercise 2.

5. Compute the integral of $f(x, y, z) = xyz$ over the rectangle with vertices $(1, 0, 1)$, $(2, 0, 0)$, $(1, 1, 1)$, and $(2, 1, 0)$.

6. Compute the integral of $x + y$ over the surface of the unit sphere.

7. Compute the integral of x over the triangle with vertices $(1, 1, 1)$, $(2, 1, 1)$, and $(2, 0, 3)$.

8. Let $f(x, y, z) = xe^y \cos \pi z$.
   (a) Compute $\mathbf{F} = \nabla f$.
   (b) Evaluate $\int_{\mathbf{c}} \mathbf{F}$ where $\mathbf{c}(t) = (3 \cos^4 t, 5 \sin^7 t, 0)$, $0 \le t \le \pi$.

9. Let $\mathbf{F}(x, y, z) = x\mathbf{i} + y\mathbf{j} + z\mathbf{k}$. Evaluate $\int_S \mathbf{F} \cdot d\mathbf{S}$ where S is the upper hemisphere of the unit sphere $x^2 + y^2 + z^2 = 1$.

10. Let $\mathbf{F}(x, y, z) = x\mathbf{i} + y\mathbf{j} + z\mathbf{k}$. Evaluate $\int_{\mathbf{c}} \mathbf{F} \cdot d\mathbf{s}$ where $\mathbf{c}(t) = (e^t, t, t^2)$, $0 \le t \le 1$.

11. Let $\mathbf{F} = \nabla f$ for a given scalar function. Let $\mathbf{c}(t)$ be a closed curve, that is, $\mathbf{c}(b) = \mathbf{c}(a)$. Show $\int_{\mathbf{c}} \mathbf{F} = 0$.

12. Consider the surface $\mathbf{\Phi}(u, v) = (u^2 \cos v, u^2 \sin v, u)$. Compute the unit normal at $u = 1$, $v = 0$. Compute the equation of the tangent plane at this point.

13. Let $\mathbf{F} = x\mathbf{i} + x^2\mathbf{j} + yz\mathbf{k}$ represent the velocity field of a fluid (velocity measured in meters/second). Compute how many cubic meters of fluid per second is crossing the xy-plane through the square $0 \le x \le 1$, $0 \le y \le 1$.

14. Let $\mathbf{F} = x^3\mathbf{i} + y^3\mathbf{j} + z^3\mathbf{k}$. Compute (a) div $\mathbf{F} = \nabla \cdot \mathbf{F}$ and (b) $\nabla \times \mathbf{F}$.

15. Let S be a surface and $\mathbf{c}$ a closed curve bounding S. Verify the equality

$$\int_S \nabla \times \mathbf{F} \cdot d\mathbf{S} = \int_{\mathbf{c}} \mathbf{F} \cdot d\mathbf{s}$$

if $\mathbf{F}$ is a gradient field (use Exercise 11).

CHAPTER 7

VECTOR ANALYSIS

We are now prepared to tie together the vector differential calculus (see Chapter 3) and the vector integral calculus (see Chapter 6). This will be done by means of the important theorems of Green, Gauss, and Stokes. We shall also point out some of the physical applications of these theorems to the study of electricity and magnetism, hydrodynamics, heat conduction, and differential equations (the last through a brief introduction to potential theory).

Many of these basic theorems had their origins in physics. For example, Green's Theorem, discovered in the early part of the 19th century, arose in connection with potential theory (this includes gravitational and electrical potentials). Gauss' Theorem—the Divergence Theorem—arose in connection with electrostatics (there seems to be valid historical evidence that this theorem should be jointly credited to the Russian mathematician Ostrogradsky). Stokes' Theorem first appeared in a letter to Stokes from the physicist Lord Kelvin, and was used by Stokes on the examination for the Smith Prize.*

* The history of vector analysis is complicated and fascinating. For this and the history of other topics, see M. Kline, *Mathematical Thought from Ancient to Modern Times*, Oxford University Press, New York, 1972.

7.1 GREEN'S THEOREM

Green's Theorem relates a line integral along a closed curve C in the plane $\mathbb{R}^2$ to a double integral over the region enclosed by C. This important result will be generalized, in the following sections, to curves and surfaces in $\mathbb{R}^3$. We shall be referring to line integrals around curves that are the boundaries of elementary regions of type 1, 2, or 3 (see Section 5.3). To understand the ideas in this section you may need to refer to Section 6.2.

A simple closed curve C that is the boundary of a region of type 1, 2, or 3 has two orientations—counterclockwise (positive) and clockwise (negative). We denote C with the counterclockwise orientation as C^+, and with the clockwise orientation as C^- (Figure 7.1.1).

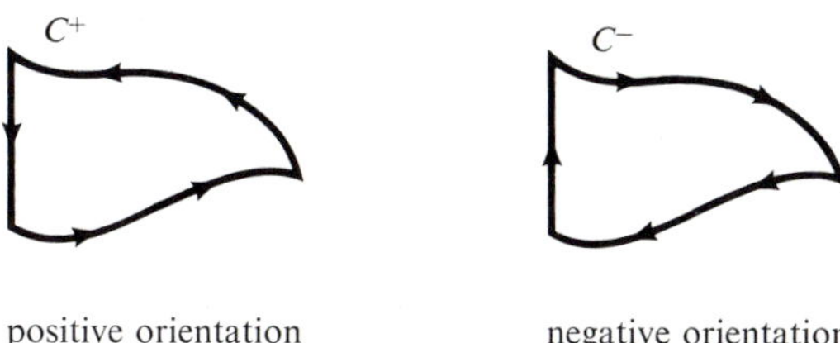

FIGURE 7.1.1
Positive orientation of C (left) and negative orientation of C (right).

The boundary C of a region of type 1 can be decomposed into top and bottom portions, C_1 and C_2, and (if applicable) left and right vertical portions, B_1 and B_2. Then we write, following Figure 7.1.2,

$$C^+ = C_2^+ + B_2^+ + C_1^- + B_1^-$$

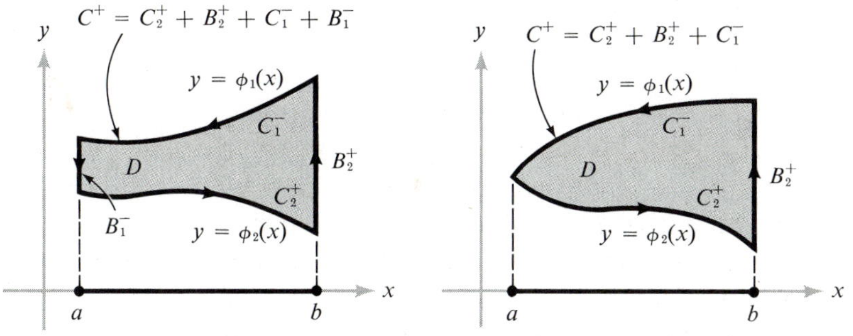

FIGURE 7.1.2
Breaking the positively oriented boundary of a region D of type 1 into oriented components.

where the pluses denote the curves oriented in the direction of left to right or bottom to top, and the minuses denote the curves oriented from right to left or from top to bottom.

We may make a similar decomposition of the boundary of a region of type 2 into left and right portions, and upper and lower horizontal portions (if applicable) (Figure 7.1.3).

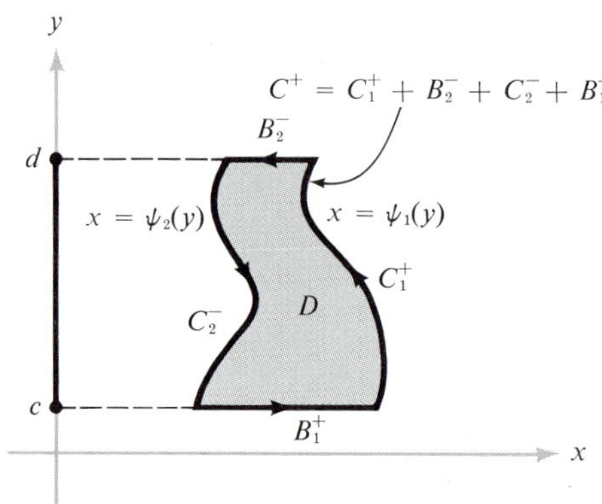

FIGURE 7.1.3
Breaking the positively oriented boundary of a region D of type 2 into oriented components.

The boundary of a region of type 3 has two decompositions—one into upper and lower halves, the other into left and right halves.

We shall now prove two lemmas in preparation for Green's Theorem.

Lemma 1. *Let D be a region of type 1 and let C be its boundary. Suppose $P: D \to \mathbb{R}$ is C^1. Then*

$$\int_{C^+} P\,dx = -\int_D \frac{\partial P}{\partial y}\,dx\,dy$$

(The left-hand side denotes the line integral $\int_{C^+} P\,dx + Q\,dy + R\,dz$ where $Q = 0$ and $R = 0$.)

Proof. Suppose D is described by

$$a \le x \le b, \qquad \phi_2(x) \le y \le \phi_1(x)$$

We decompose C^+ by writing $C^+ = C_2^+ + B_2^+ + C_1^- + B_1^-$ (see Figure 7.1.2). By Fubini's Theorem, we may evaluate the double integral as an iterated integral (see p. 222)

$$\int_D \frac{\partial P}{\partial y}(x, y)\,dx\,dy = \int_a^b \int_{\phi_2(x)}^{\phi_1(x)} \frac{\partial P}{\partial y}(x, y)\,dy\,dx$$

$$= \int_a^b [P(x, \phi_1(x)) - P(x, \phi_2(x))]\,dx$$

However, since C_2^+ can be parametrized by $x \mapsto (x, \phi_2(x))$, $a \le x \le b$, and C_1^+ can be parametrized by $x \mapsto (x, \phi_1(x))$, $a \le x \le b$, we have

$$\int_a^b P(x, \phi_2(x))\, dx = \int_{C_2^+} P(x, y)\, dx$$

and

$$\int_a^b P(x, \phi_1(x))\, dx = \int_{C_1^+} P(x, y)\, dx$$

Thus by reversing orientations

$$-\int_a^b P(x, \phi_1(x))\, dx = \int_{C_1^-} P(x, y)\, dx$$

Hence

$$\int_D \frac{\partial P}{\partial y}\, dx\, dy = -\int_{C_2^+} P\, dx - \int_{C_1^-} P\, dx$$

Since x is constant on B_2^+ and B_1^-, we have

$$\int_{B_2^+} P\, dx = 0 = \int_{B_1^-} P\, dx$$

and so

$$\int_{C^+} P\, dx = \int_{C_2^+} P\, dx + \int_{B_2^+} P\, dx + \int_{C_1^-} P\, dx + \int_{B_1^-} P\, dx$$

$$= \int_{C_2^+} P\, dx + \int_{C_1^-} P\, dx$$

Thus

$$\int_D \frac{\partial P}{\partial y}\, dx\, dy = -\int_{C_2^+} P\, dx - \int_{C_1^-} P\, dx = -\int_{C^+} P\, dx \quad \blacksquare$$

We now prove the analogous lemma with the roles of x and y interchanged.

Lemma 2. *Let D be a region of type 2 with boundary C. Then if $Q: D \to \mathbb{R}$ is C^1,*

$$\int_{C^+} Q\, dy = \int_D \frac{\partial Q}{\partial x}\, dx\, dy$$

The negative sign does not occur here because reversing the role of x and y corresponds to a change of orientation for the plane.

Proof. Suppose D is given by

$$\psi_2(y) \le x \le \psi_1(y), \qquad c \le y \le d$$

Using the notation of Figure 7.1.3 we have

$$\int_{C^+} Q\,dy = \int_{C_2^- + B_1^+ + C_1^+ + B_2^-} Q\,dy = \int_{C_1^+} Q\,dy + \int_{C_2^-} Q\,dy$$

where C_1^+ is the curve parametrized by $y \mapsto (\psi_1(y), y)$, $c \le y \le d$, and C_2^+ is the curve $y \mapsto (\psi_2(y), y)$, $c \le y \le d$. Applying Fubini's Theorem, we obtain

$$\begin{aligned}\int_D \frac{\partial Q}{\partial x}\,dx\,dy &= \int_c^d \int_{\psi_2(y)}^{\psi_1(y)} \frac{\partial Q}{\partial x}\,dx\,dy \\ &= \int_c^d [Q(\psi_1(y), y) - Q(\psi_2(y), y)]\,dy \\ &= \int_{C_1^+} Q\,dy - \int_{C_2^+} Q\,dy = \int_{C_1^+} Q\,dy + \int_{C_2^-} Q\,dy \\ &= \int_{C^+} Q\,dy \quad \blacksquare\end{aligned}$$

Adding the results of Lemmas 1 and 2 proves the following important theorem.

Theorem 1 (Green's Theorem). *Let D be a region of type 3 and let C be its boundary. Suppose $P\colon D \to \mathbb{R}$ and $Q\colon D \to \mathbb{R}$ are C^1. Then*

$$\int_{C^+} P\,dx + Q\,dy = \int_D \left(\frac{\partial Q}{\partial x} - \frac{\partial P}{\partial y}\right) dx\,dy$$

Green's Theorem actually applies to any "decent" region in $\mathbb{R}^2$, but such a result is too complicated to be included here. However, a somewhat more general version is considered in Exercise 8.

Let us use the notation ∂D for the oriented curve C^+, that is, the boundary curve of D oriented in the counterclockwise direction. Then we may write Green's Theorem as

$$\int_{\partial D} P\,dx + Q\,dy = \int_D \left(\frac{\partial Q}{\partial x} - \frac{\partial P}{\partial y}\right) dx\,dy$$

Green's Theorem is very useful because it relates a line integral around the boundary of a region to an area integral over the interior of the region, and in many cases it is easier to evaluate the line integral than the area integral. For example, if we know that P vanishes on the boundary, we can immediately conclude that $\int_D (\partial P/\partial y)\,dx\,dy = 0$ even though $\partial P/\partial y$ need not vanish on the interior. (Can you construct such a P on the unit square?)

EXAMPLE 1. Verify Green's Theorem for $P(x, y) = x$ and $Q(x, y) = xy$ where D is the unit disc $x^2 + y^2 \leq 1$.

We can evaluate both sides in Green's Theorem directly. The boundary of D is the unit circle parametrized by $x = \cos t$, $y = \sin t$, $0 \leq t \leq 2\pi$, so

$$\int_{\partial D} P\,dx + Q\,dy = \int_0^{2\pi} [(\cos t)(-\sin t) + \cos t \sin t \cos t]\,dt$$

$$= \left[\frac{\cos^2 t}{2}\right]_0^{2\pi} + \left[-\frac{\cos^3 t}{3}\right]_0^{2\pi} = 0$$

On the other hand

$$\int_D \left(\frac{\partial Q}{\partial x} - \frac{\partial P}{\partial y}\right) dx\,dy = \int_D y\,dx\,dy$$

which is zero also. Thus Green's Theorem is verified in this case.

We can use Green's Theorem to obtain a formula for the area of a region bounded by a simple closed curve.

Theorem 2. *If C is a simple closed curve that bounds a region to which Green's Theorem applies, then the area of the region D bounded by C is*

$$A = \frac{1}{2}\int_{\partial D} x\,dy - y\,dx$$

Proof. Let $P(x, y) = -y$, $Q(x, y) = x$; then by Green's Theorem we have

$$\frac{1}{2}\int_{\partial D} x\,dy - y\,dx = \frac{1}{2}\int_D \left(\frac{\partial x}{\partial x} - \frac{\partial(-y)}{\partial y}\right) dx\,dy$$

$$= \int_D dx\,dy = A \quad \blacksquare$$

EXAMPLE 2. The area of the region enclosed by the hypocycloid $x^{2/3} + y^{2/3} = a^{2/3}$ can be computed using the parametrization

$$x = a\cos^3\theta, \qquad y = a\sin^3\theta, \qquad 0 \leq \theta \leq 2\pi$$

(see Figure 7.1.4). Thus

$$A = \frac{1}{2}\int_{\partial D} x\,dy - y\,dx = \frac{1}{2}\int_0^{2\pi} [(a\cos^3\theta)(3a\sin^2\theta\cos\theta) - (a\sin^3\theta)(-3a\cos^2\theta\sin\theta)]\,d\theta$$

$$= \frac{3}{2}a^2\int_0^{2\pi} [\sin^2\theta\cos^4\theta + \cos^2\theta\sin^4\theta]\,d\theta$$

$$= \frac{3}{2}a^2\int_0^{2\pi} \sin^2\theta\cos^2\theta(\sin^2\theta + \cos^2\theta)\,d\theta$$

$$= \frac{3}{2}a^2\int_0^{2\pi} \sin^2\theta\cos^2\theta\,d\theta$$

$$= \frac{3}{8}a^2\int_0^{2\pi} \sin^2 2\theta\,d\theta$$

$$= \frac{3}{8}a^2\int_0^{2\pi} \left[\frac{1-\cos 4\theta}{2}\right] d\theta$$

$$= \frac{3}{16}a^2\int_0^{2\pi} d\theta - \frac{3}{16}a^2\int_0^{2\pi}\cos 4\theta\,d\theta$$

$$= \frac{3}{8}\pi a^2$$

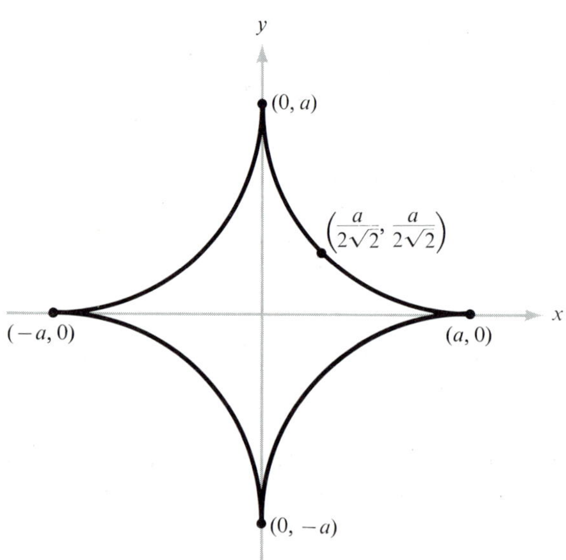

FIGURE 7.1.4
The hypocycloid $x = a\cos^3\theta$, $y = a\sin^3\theta$, $0 \le \theta \le 2\pi$.

The statement of Green's Theorem contained in Theorem 1 is not the form that we shall generalize. We can rewrite the theorem neatly in the language of vector fields.

Theorem 3 (Vector form of Green's Theorem). *Let $D \subset \mathbb{R}^2$ be a region of type 3 and let ∂D be its boundary (oriented counterclockwise). Let* $\mathbf{F} = P\mathbf{i} + Q\mathbf{j}$ *be a C^1 vector field on D. Then*

$$\int_{\partial D} \mathbf{F} = \int_D (\text{curl } \mathbf{F}) \cdot \mathbf{k}\, dA = \int_D (\nabla \times \mathbf{F}) \cdot \mathbf{k}\, dA$$

(see Figure 7.1.5).

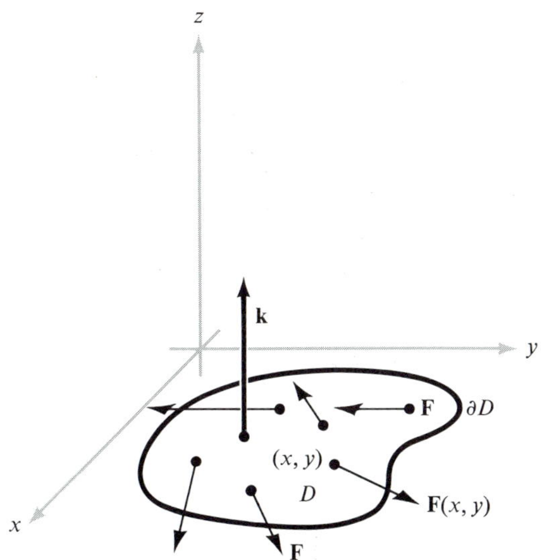

FIGURE 7.1.5
The elements of the vector form of Green's Theorem.

This result follows easily from Theorem 1 after we interpret the various symbols. We shall ask the reader to supply the details in Exercise 15.

There is yet another form of Green's Theorem that is capable of being generalized to $\mathbb{R}^3$.

Theorem 4 (Divergence Theorem in the plane). *Let $D \subset \mathbb{R}^2$ be a region of type 3 and let ∂D be its boundary. Let* **n** *denote the outward unit normal to ∂D, which is given by*

$$\mathbf{n} = (y'(t), \ -x'(t))/\sqrt{(x'(t))^2 + (y'(t))^2}$$

if $\boldsymbol{\sigma}\colon [a, b] \to \mathbb{R}^2$, $t \mapsto \boldsymbol{\sigma}(t) = (x(t), y(t))$ is a parametrization of ∂D (see Figure 7.1.6).

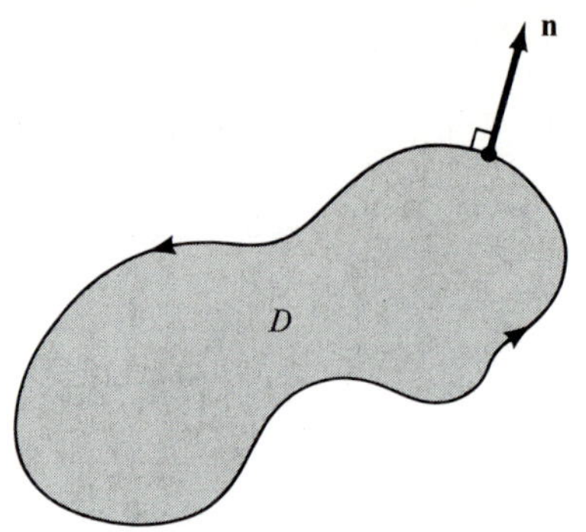

FIGURE 7.1.6
n *is the outward unit normal to* ∂D.

Let $\mathbf{F} = P\mathbf{i} + Q\mathbf{j}$ *be a* C^1 *vector field on D. Then*

$$\int_{\partial D} \mathbf{F} \cdot \mathbf{n} = \int_D \operatorname{div} \mathbf{F}$$

Proof. Since $\boldsymbol{\sigma}'(t) = (x'(t), y'(t))$ is tangent to ∂D it is clear that $\mathbf{n} \cdot \boldsymbol{\sigma}' = 0$, so $\mathbf{n}$ is normal to the boundary. The sign of $\mathbf{n}$ is chosen to make it correspond to the *outward* (rather than the inward) direction. By definition of the path integral (see Section 6.1)

$$\int_{\partial D} \mathbf{F} \cdot \mathbf{n} =$$

$$\int_a^b \left(\frac{P(x(t), y(t))y'(t) - Q(x(t), y(t))x'(t)}{\sqrt{(x'(t))^2 + (y'(t))^2}} \right) \sqrt{(x'(t))^2 + (y'(t))^2}\, dt$$

$$= \int_a^b [P(x(t), y(t))y'(t) - Q(x(t), y(t))x'(t)]\, dt$$

$$= \int_{\partial D} P\, dy - Q\, dx$$

By Green's Theorem, this equals

$$\int_D \left(\frac{\partial P}{\partial x} + \frac{\partial Q}{\partial y} \right) dx\, dy = \int_D \operatorname{div} \mathbf{F} \quad \blacksquare$$

EXAMPLE 3. Let $\mathbf{F} = y^3\mathbf{i} + x^5\mathbf{j}$. Compute the integral of the normal component of $\mathbf{F}$ around the unit square.

This can be done by the Divergence Theorem. Indeed,

$$\int_{\partial D} \mathbf{F} \cdot \mathbf{n} = \int_D \operatorname{div} \mathbf{F}$$

But $\operatorname{div} \mathbf{F} = 0$, so the integral is zero.

EXERCISES

1. Find the area of the disc D of radius R using Green's Theorem.
2. Evaluate $\int_C y\,dx - x\,dy$ where C is the boundary of the square $[-1, 1] \times [-1, 1]$ oriented in the counterclockwise direction (use Green's Theorem).
3. Using the Divergence Theorem show that $\int_{\partial D} \mathbf{F} \cdot \mathbf{n} = 0$ where $\mathbf{F}(x, y) = y\mathbf{i} - x\mathbf{j}$ and D is the unit disc. Verify this directly.
4. Verify Green's Theorem for the disc D with center $(0, 0)$ and radius R and the functions:
   (a) $P(x, y) = xy^2$, $Q(x, y) = -yx^2$
   (b) $P(x, y) = x + y$, $Q(x, y) = y$
   (c) $P(x, y) = xy = Q(x, y)$
5. Find the area bounded by one arc of the cycloid $x = a(\theta - \sin\theta)$, $y = a(1 - \cos\theta)$, $a > 0$, $0 \le \theta \le 2\pi$, and the x-axis (use Green's Theorem).
6. Under the conditions of Green's Theorem, prove that

   (a) $$\int_{\partial D} PQ\,dx + PQ\,dy = \int_D \left[Q\left(\frac{\partial P}{\partial x} - \frac{\partial P}{\partial y}\right) + P\left(\frac{\partial Q}{\partial x} - \frac{\partial Q}{\partial y}\right)\right] dx\,dy$$

   (b) $$\int_{\partial D} \left(Q\frac{\partial P}{\partial x} - P\frac{\partial Q}{\partial x}\right) dx + \left(P\frac{\partial Q}{\partial y} - Q\frac{\partial P}{\partial y}\right) dy$$
   $$= 2\int_D \left(P\frac{\partial^2 Q}{\partial x\,\partial y} - Q\frac{\partial^2 P}{\partial x\,\partial y}\right) dx\,dy$$

7. Evaluate $\int_C (2x^3 - y^3)\,dx + (x^3 + y^3)\,dy$, where C is the unit circle, and verify Green's Theorem for this case.
8. Prove the following generalization of Green's Theorem: Let D be a region in the xy-plane with boundary a finite number of oriented simple closed curves. Suppose that by means of a finite number of line segments parallel to the coordinate axes, D can be decomposed into a finite number of regions D_i of type 3 with the boundary of each D_i oriented counterclockwise (see Figure 7.1.7). Then if P and Q are C^1 on D,

   $$\int_D \left(\frac{\partial Q}{\partial x} - \frac{\partial P}{\partial y}\right) dx\,dy = \int_{C^+} P\,dx + Q\,dy$$

   where C^+ is the oriented boundary of D. (HINT: Apply Green's Theorem to each D_i.)
9. Verify Green's Theorem for the integrand of Exercise 7 ($P = 2x^3 - y^3$, $Q = x^3 + y^3$) and the annular region D described by $a \le x^2 + y^2 \le b$, with boundaries oriented as in Figure 7.1.7.
10. Let D be a region for which Green's Theorem holds. Suppose f is harmonic; that is,

    $$\frac{\partial^2 f}{\partial x^2} + \frac{\partial^2 f}{\partial y^2} = 0$$

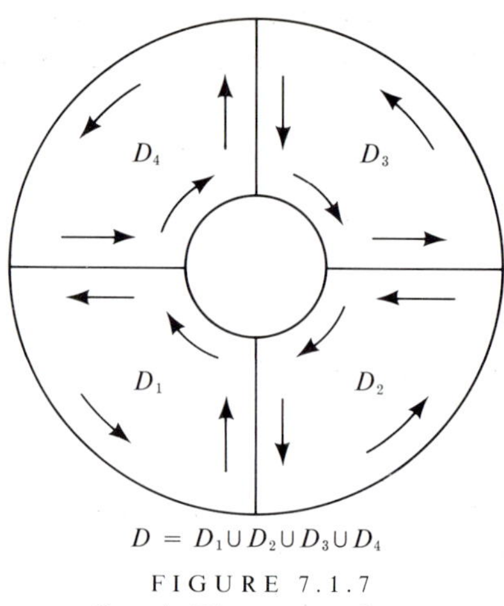

FIGURE 7.1.7
Green's Theorem applies to $D = D_1 \cup D_2 \cup D_3 \cup D_4$.

on D. Prove that

$$\int_{\partial D} \frac{\partial f}{\partial y}\, dx - \frac{\partial f}{\partial x}\, dy = 0$$

11. (a) Verify the Divergence Theorem for $\mathbf{F} = x\mathbf{i} + y\mathbf{j}$ and D the unit disc $x^2 + y^2 \leq 1$.
    (b) Evaluate the integral of the normal component of $2xy\mathbf{i} - y^2\mathbf{j}$ around the ellipse $x^2/a^2 + y^2/b^2 = 1$.
12. Let $P(x, y) = -y/(x^2 + y^2)$, $Q(x, y) = x/(x^2 + y^2)$. Assuming D is the unit disc, investigate why Green's Theorem fails for this P and Q.
13. Use Green's Theorem to evaluate $\int_{C^+} (y^2 + x^3)\, dx + x^4\, dy$, where C^+ is the perimeter of $[0, 1] \times [0, 1]$ in the counterclockwise direction.
14. Use Theorem 2 to compute the area inside the ellipse $x^2/a^2 + y^2/b^2 = 1$.
15. Verify Theorem 3.
16. Evaluate $\int_{\boldsymbol{\sigma}} (x^5 - 2xy^3)\, dx - 3x^2y^2\, dy$, where $\boldsymbol{\sigma}$ is the path $\boldsymbol{\sigma}(t) = (t^8, t^{10})$, $0 \leq t \leq 1$.
*17. Use Green's Theorem to prove the Change of Variables formula in the following special case:

$$\int_D dx\, dy = \int_{D^*} \left| \frac{\partial(x, y)}{\partial(u, v)} \right| du\, dv$$

for a transformation $(u, v) \mapsto (x(u, v), y(u, v))$. Formulate the necessary hypotheses on the functions $x = x(u, v)$ and $y = y(u, v)$ and on $\partial(x, y)/\partial(u, v)$ for your proof.

18. Prove the identity

$$\int_{\partial D} \phi\, \nabla\phi \cdot \mathbf{n} = \int_D (\phi\nabla^2\phi + \nabla\phi \cdot \nabla\phi)$$

19. Use Green's Theorem to find the area of one loop of the four-leafed rose $r = 3 \sin 2\theta$. (HINT: $x\, dy - y\, dx = r^2\, d\theta$).

20. Show that if C is a simple closed curve that bounds a region to which Green's Theorem applies, then the area of the region D bounded by C is

$$A = \int_{\partial D} x\,dy = -\int_{\partial D} y\,dx$$

Show how this implies Theorem 2.

7.2 STOKES' THEOREM

Stokes' Theorem relates the line integral of a vector field around a simple closed curve C in $\mathbb{R}^3$ to an integral over a surface S for which C is the boundary. In this regard it is very much like Green's Theorem.

Let us begin by recalling a few facts from Chapter 6. Consider a surface S that is the graph of a function $f(x, y)$, so S is parametrized by

$$\begin{cases} x = u \\ y = v \\ z = f(u, v) = f(x, y) \end{cases}$$

for (u, v) in some domain D. The integral of a vector function $\mathbf{F}$ over S was developed in Section 6.6 as

$$\int_S \mathbf{F} = \int_D \left[F_1\left(-\frac{\partial z}{\partial x}\right) + F_2\left(-\frac{\partial z}{\partial y}\right) + F_3 \right] dx\,dy \qquad (1)$$

where $\mathbf{F} = F_1\mathbf{i} + F_2\mathbf{j} + F_3\mathbf{k}$.

In Section 7.1 we assumed that the regions D under consideration were of type 3; this was an essential requirement in our proof of Green's Theorem. In fact Green's Theorem generalizes to a much wider class of regions, but unfortunately the proofs of such generalizations are somewhat complicated (for example, see Exercise 8, Section 7.1). In this section, for the sake of generality we shall assume that D is a region with boundary a simple closed curve and to which Green's Theorem applies.

Now suppose that $\boldsymbol{\sigma}\colon [a, b] \to \mathbb{R}^3$, $\boldsymbol{\sigma}(t) = (x(t), y(t))$, is a parametrization of ∂D in the counterclockwise direction. Then we define the *boundary curve* ∂S to be the oriented simple closed curve that is the image of the mapping $\boldsymbol{\eta}\colon t \mapsto (x(t), y(t), f(x(t), y(t)))$ with the *orientation* induced by $\boldsymbol{\eta}$ (Figure 7.2.1).

To remember this orientation (that is, the positive direction) on ∂S, think of an "observer" walking along the boundary of the surface with the normal pointing to the outside (p. 332) being his upright direction; he is moving in the positive direction if the surface is on his left. This

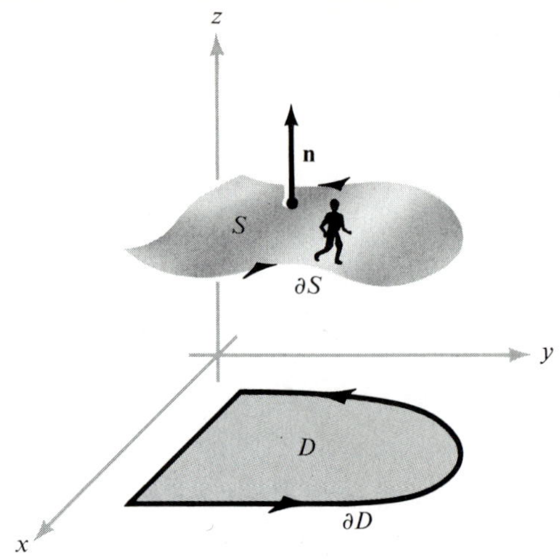

FIGURE 7.2.1
The induced orientation on ∂S; as you walk around the boundary, the surface should be on your left.

orientation on ∂S is often called the *orientation induced by an upward normal* **n**.

We are now ready to state and prove one of the primary results of this section.

Theorem 5 **(*Stokes' Theorem for graphs*).** *Let S be the oriented surface defined by a C^2 function $z = f(x, y)$, $(x, y) \in D$, and let* **F** *be a C^1 vector field on S. Then if ∂S denotes the oriented boundary curve of S as defined above we have*

$$\int_S \operatorname{curl} \mathbf{F} = \int_S \nabla \times \mathbf{F} = \int_{\partial S} \mathbf{F}$$

Remember that $\int_{\partial S} \mathbf{F}$ is the integral around ∂S of the tangential component of **F**, while $\int_S \mathbf{G}$ is the integral over S of $\mathbf{G} \cdot \mathbf{n}$, the normal component of **G** (see sections 6.2 and 6.6). Thus Stokes' Theorem says that the integral of the normal component of the curl of a vector field **F** over a surface S is equal to the integral of the tangential component of **F** around the boundary ∂S.

Proof. If $\mathbf{F} = F_1\mathbf{i} + F_2\mathbf{j} + F_3\mathbf{k}$ then

$$\operatorname{curl} \mathbf{F} = \left(\frac{\partial F_3}{\partial y} - \frac{\partial F_2}{\partial z}\right)\mathbf{i} + \left(\frac{\partial F_1}{\partial z} - \frac{\partial F_3}{\partial x}\right)\mathbf{j} + \left(\frac{\partial F_2}{\partial x} - \frac{\partial F_1}{\partial y}\right)\mathbf{k}$$

Therefore we use formula (1) to write

$$\int_S \operatorname{curl} \mathbf{F} = \int_D \left[\left(\frac{\partial F_3}{\partial y} - \frac{\partial F_2}{\partial z}\right)\left(-\frac{\partial z}{\partial x}\right) + \left(\frac{\partial F_1}{\partial z} - \frac{\partial F_3}{\partial x}\right)\left(-\frac{\partial z}{\partial y}\right) + \left(\frac{\partial F_2}{\partial x} - \frac{\partial F_1}{\partial y}\right)\right] dA \tag{2}$$

On the other hand,

$$\int_{\partial S} \mathbf{F} = \int_{\boldsymbol{\eta}} \mathbf{F} = \int_{\boldsymbol{\eta}} F_1\, dx + F_2\, dy + F_3\, dz$$

where $\boldsymbol{\eta}$: $[a, b] \to \mathbb{R}^3$, $\boldsymbol{\eta}(t) = (x(t),\ y(t),\ f(x(t),\ y(t)))$ is the orientation-preserving parametrization of the oriented simple closed curve ∂S discussed above. Thus

$$\int_{\partial S} \mathbf{F} = \int_a^b \left(F_1 \frac{dx}{dt} + F_2 \frac{dy}{dt} + F_3 \frac{dz}{dt}\right) dt \tag{3}$$

But by the Chain Rule

$$\frac{dz}{dt} = \frac{\partial z}{\partial x}\frac{dx}{dt} + \frac{\partial z}{\partial y}\frac{dy}{dt}$$

Substituting this expression into (3) we obtain

$$\begin{aligned}\int_{\partial S} \mathbf{F} &= \int_a^b \left[\left(F_1 + F_3 \frac{\partial z}{\partial x}\right)\frac{dx}{dt} + \left(F_2 + F_3 \frac{\partial z}{\partial y}\right)\frac{dy}{dt}\right] dt \\ &= \int_{\boldsymbol{\sigma}} \left(F_1 + F_3 \frac{\partial z}{\partial x}\right) dx + \left(F_2 + F_3 \frac{\partial z}{\partial y}\right) dy \\ &= \int_{\partial D} \left(F_1 + F_3 \frac{\partial z}{\partial x}\right) dx + \left(F_2 + F_3 \frac{\partial z}{\partial y}\right) dy \end{aligned} \tag{4}$$

Applying Green's Theorem to (4) yields (we are assuming that Green's Theorem applies to D)

$$\int_D \left[\frac{\partial(F_2 + F_3\ \partial z/\partial y)}{\partial x} - \frac{\partial(F_1 + F_3\ \partial z/\partial x)}{\partial y}\right] dA$$

Now we use the Chain Rule, remembering that F_1, F_2, and F_3 are functions of x, y, and z, and z is a function of x and y, to obtain

$$\int_D \left[\left(\frac{\partial F_2}{\partial x} + \frac{\partial F_2}{\partial z}\cdot\frac{\partial z}{\partial x} + \frac{\partial F_3}{\partial x}\cdot\frac{\partial z}{\partial y} + \frac{\partial F_3}{\partial z}\cdot\frac{\partial z}{\partial x}\cdot\frac{\partial z}{\partial y} + F_3 \cdot \frac{\partial^2 z}{\partial x\, \partial y}\right) - \left(\frac{\partial F_1}{\partial y} + \frac{\partial F_1}{\partial z}\cdot\frac{\partial z}{\partial y} + \frac{\partial F_3}{\partial y}\cdot\frac{\partial z}{\partial x} + \frac{\partial F_3}{\partial z}\cdot\frac{\partial z}{\partial y}\cdot\frac{\partial z}{\partial x} + F_3 \frac{\partial^2 z}{\partial y\, \partial x}\right)\right] dA$$

The last two terms in each parenthesis cancel each other, and we can rearrange terms to obtain the integral (2), which completes the proof. ▮

EXAMPLE 1. Let $\mathbf{F} = ye^z\mathbf{i} + xe^z\mathbf{j} + xye^z\mathbf{k}$. Show that the integral of $\mathbf{F}$ around an oriented simple closed curve C that is the boundary of a surface S is 0. (Assume S is the graph of a function, as in Theorem 5.)

Indeed, $\int_C \mathbf{F} = \int_S \nabla \times \mathbf{F}$ by Stokes' Theorem. But we compute

$$\nabla \times \mathbf{F} = \begin{vmatrix} \mathbf{i} & \mathbf{j} & \mathbf{k} \\ \dfrac{\partial}{\partial x} & \dfrac{\partial}{\partial y} & \dfrac{\partial}{\partial z} \\ ye^z & xe^z & xye^z \end{vmatrix} = \mathbf{0}$$

so $\int_C \mathbf{F} = 0$.

EXAMPLE 2. Use Stokes' Theorem to evaluate the line integral

$$\int_C -y^3\,dx + x^3\,dy - z^3\,dz$$

where C is the intersection of the cylinder $x^2 + y^2 = 1$ and the plane $x + y + z = 1$, and the orientation on C corresponds to counterclockwise motion in the xy-plane.

The curve C bounds the surface S defined by $z = 1 - x - y = f(x, y)$ for (x, y) in $D = \{(x, y) \mid x^2 + y^2 \le 1\}$ (Figure 7.2.2). We set

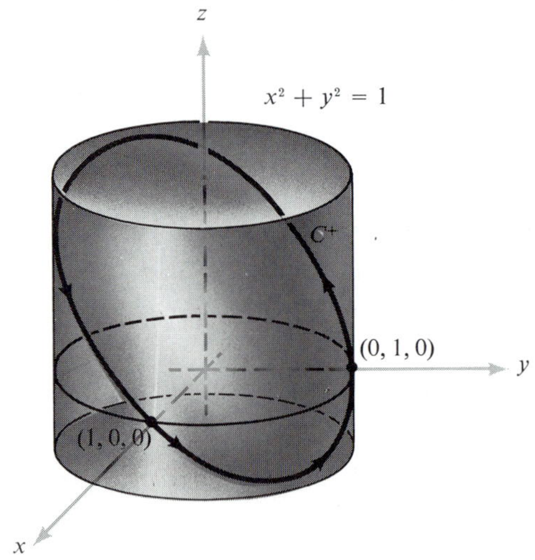

FIGURE 7.2.2
The curve C is the intersection of the cylinder $x^2 + y^2 = 1$ and the plane $x + y + z = 1$.

$\mathbf{F} = -y^3\mathbf{i} + x^3\mathbf{j} - z^3\mathbf{k}$, which has curl $\nabla \times \mathbf{F} = (3x^2 + 3y^2)\mathbf{k}$. Then by Stokes' Theorem the line integral is equal to the surface integral

$$\int_S \nabla \times \mathbf{F}$$

But $\nabla \times \mathbf{F}$ has only a $\mathbf{k}$ component. Thus by formula (1) we have

$$\int_S \nabla \times \mathbf{F} = \int_D (3x^2 + 3y^2)\, dx\, dy$$

This integral can be evaluated by changing to polar coordinates. Having done this, we get

$$3 \int_D (x^2 + y^2)\, dx\, dy = 3 \int_0^1 \int_0^{2\pi} r^2 \cdot r\, d\theta\, dr = 6\pi \int_0^1 r^3\, dr = \frac{6\pi}{4} = \frac{3\pi}{2}$$

Let us verify this result by directly evaluating the line integral

$$\int_C -y^3\, dx + x^3\, dy - z^3\, dz$$

We can parametrize the curve ∂D by the equations

$$x = \cos t, \quad y = \sin t, \quad z = 0, \quad 0 \le t \le 2\pi$$

The curve C is therefore parametrized by the equations

$$x = \cos t, \quad y = \sin t, \quad z = 1 - \sin t - \cos t, \quad 0 \le t \le 2\pi$$

Thus

$$\int_C -y^3\, dx + x^3\, dy - z^3\, dz$$

$$= \int_0^{2\pi} [(-\sin^3 t)(-\sin t) + (\cos^3 t)(\cos t)$$

$$- (1 - \sin t - \cos t)^3(-\cos t + \sin t)]\, dt$$

$$= \int_0^{2\pi} (\cos^4 t + \sin^4 t)\, dt - \int_0^{2\pi} (1 - \sin t - \cos t)^3(-\cos t + \sin t)\, dt$$

The second integrand is of the form $u^3\, du$, where $u = 1 - \sin t - \cos t$, and thus the integral is equal to

$$\tfrac{1}{4}[(1 - \sin t - \cos t)^4]_0^{2\pi} = 0$$

Hence we are left with

$$\int_0^{2\pi} (\cos^4 t + \sin^4 t)\, dt$$

Since

$$\sin^2 t = \frac{1 - \cos 2t}{2}, \qquad \cos^2 t = \frac{1 + \cos 2t}{2}$$

we reduce the above integral to

$$\tfrac{1}{2}\int_0^{2\pi}(1+\cos^2 2t)\,dt = \pi + \tfrac{1}{2}\int_0^{2\pi}\cos^2 2t\,dt$$

Again using the fact that

$$\cos^2 2t = \frac{1+\cos 4t}{2}$$

we find

$$\pi + \tfrac{1}{4}\int_0^{2\pi}(1+\cos 4t)\,dt = \pi + \tfrac{1}{4}\int_0^{2\pi} dt + \tfrac{1}{4}\int_0^{2\pi}\cos 4t\,dt$$

$$= \pi + \frac{\pi}{2} + 0 = \frac{3\pi}{2}$$

In order to simplify the proof of Stokes' Theorem above we assumed that the surface S could be described as the graph of a function $z=f(x, y)$, $(x, y)\in D$, where D is some region to which Green's Theorem applies. However, without too much more effort we can obtain a more general theorem for oriented parametrized surfaces S. The main complication is in the definition of ∂S.

Suppose $\mathbf{\Phi}: D\to\mathbb{R}^3$ is a parametrization of a surface S and $\boldsymbol{\sigma}(t)=(u(t), v(t))$ is a parametrization of ∂D. We might be tempted to define ∂S as the curve parametrized by $t\mapsto\boldsymbol{\eta}(t)=\mathbf{\Phi}(u(t), v(t))$. However with this definition ∂S might not be the boundary of S in any reasonable geometric sense.

For example, we would conclude that the boundary of the unit sphere S parametrized by spherical coordinates in $\mathbb{R}^3$ is half of the great circle on S lying in the xz-plane, but clearly in a geometric sense S is a smooth surface (no points or cusps) with no boundary or edge at all (see Figure 7.2.3 and Exercise 14). Thus this great circle is in some sense the "mistaken'— boundary of S.

We can get around this difficulty by assuming that $\mathbf{\Phi}$ is one-to-one on all of D. Then the image of ∂D under $\mathbf{\Phi}$, namely $\mathbf{\Phi}(\partial D)$ will be the geometric boundary of $S=\mathbf{\Phi}(D)$. If $\boldsymbol{\sigma}(t)=(u(t), v(t))$ is a parametrization of ∂D in the counterclockwise direction, we define ∂S to be the oriented simple closed curve that is the image of the mapping $\boldsymbol{\eta}$: $t\mapsto\mathbf{\Phi}(u(t), v(t))$ with the orientation on ∂S induced by $\boldsymbol{\eta}$ (see Figure 7.2.1).

Theorem 6 (Stokes' Theorem for parametrized surfaces). *Suppose S is an oriented surface defined by a one-to-one parametrization*

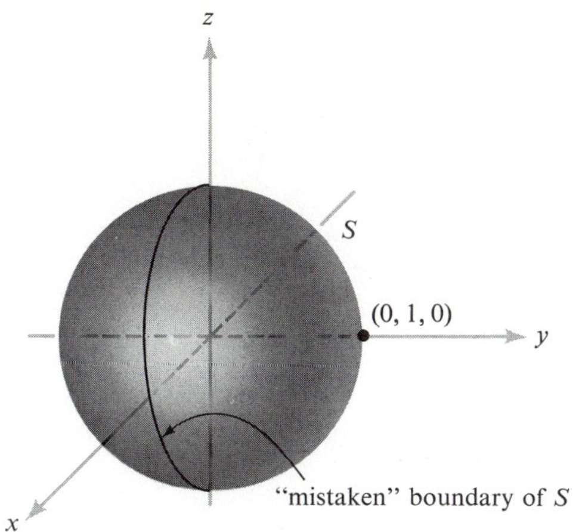

FIGURE 7.2.3
The boundary of a surface S that is parametrized by $\mathbf{\Phi}: D \to \mathbb{R}^3$ *is the image of the boundary of D only if* $\mathbf{\Phi}$ *is one-to-one on D.*

$\mathbf{\Phi}: D \subset \mathbb{R}^2 \to S$. *Let* ∂S *denote the oriented boundary of S and let* $\mathbf{F}$ *be a* C^1 *vector field on S. Then*

$$\int_S \nabla \times \mathbf{F} = \int_{\partial S} \mathbf{F}$$

This is proved in exactly the same way as Theorem 5.

Let us now use Stokes' Theorem to justify the physical interpretation of $\nabla \times \mathbf{F}$ in terms of paddle wheels that was proposed in Chapter 3. Paraphrasing Theorem 6 we have

$$\int_S (\text{curl } \mathbf{F}) \cdot \mathbf{n}\, dS = \int_S \text{curl } \mathbf{F} = \int_{\partial S} \mathbf{F} = \int_{\partial S} F_{\mathrm{T}}$$

where F_{T} is the tangential component of $\mathbf{F}$. This says that the integral of the normal component of the curl of a vector field over an oriented surface S is equal to the line integral of $\mathbf{F}$ along ∂S, which in turn is equal to the path integral of the tangential component of $\mathbf{F}$ over ∂S.

Suppose $\mathbf{V}$ represents the velocity vector field of a fluid. Consider a point P and a unit vector $\mathbf{n}$. Let S_ρ denote the disc of radius ρ and center P, which is perpendicular to $\mathbf{n}$. By Stokes' Theorem

$$\int_{S_\rho} \text{curl } \mathbf{V} = \int_{S_\rho} \text{curl } \mathbf{V} \cdot \mathbf{n}\, dS = \int_{\partial S_\rho} \mathbf{V}$$

where ∂S_ρ has the orientation induced by $\mathbf{n}$ (see Figure 7.2.4). It is not

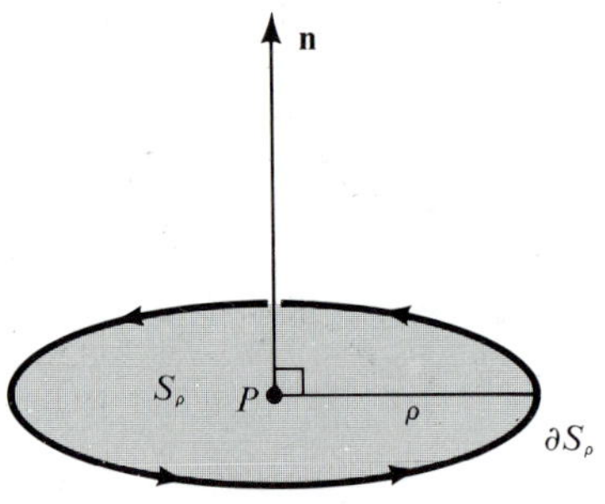

FIGURE 7.2.4
A normal **n** *induces an orientation on the boundary* ∂S_ρ *of the disc* S_ρ.

difficult to show (see Exercise 9, Section 6.6) that there is a point Q in S_ρ such that

$$\int_{S_\rho} \text{curl } \mathbf{V} \cdot \mathbf{n}\, dS = (\text{curl } \mathbf{V}(Q) \cdot \mathbf{n})A(S_\rho)$$

(this is the Mean Value Theorem for Integrals; see footnote, p. 222) where $A(S_\rho) = \pi\rho^2$ is the area of S_ρ, curl $\mathbf{V}(Q)$ is the value of curl $\mathbf{V}$ at Q, and $\mathbf{n}$ is evaluated at Q as well. Thus

$$\begin{aligned}\lim_{\rho\to 0} \frac{1}{A(S_\rho)} \int_{\partial S_\rho} \mathbf{V} &= \lim_{\rho\to 0} \frac{1}{A(S_\rho)} \int_{S_\rho} \text{curl } \mathbf{V} \\ &= \lim_{\rho\to 0} \text{curl } \mathbf{V}(Q) \cdot \mathbf{n}(Q) \\ &= \text{curl } \mathbf{V}(P) \cdot \mathbf{n}(P)\end{aligned}$$

Thus*

$$\text{curl } \mathbf{V}(P) \cdot \mathbf{n}(P) = \lim_{\rho\to 0} \frac{1}{A(S_\rho)} \int_{\partial S_\rho} \mathbf{V} \qquad (5)$$

Let us pause to consider the physical meaning of $\int_C \mathbf{V}$ when $\mathbf{V}$ is the velocity field of a fluid (see p. 155). Suppose, for example, that $\mathbf{V}$ points in the direction tangent to the oriented curve C (Figure 7.2.5). Then clearly $\int_C \mathbf{V} > 0$, and particles on C tend to rotate counterclockwise. If $\mathbf{V}$ is pointing in the opposite direction, $\int_C \mathbf{V} < 0$. If $\mathbf{V}$ is perpendicular to C then particles don't rotate on C at all and $\int_C \mathbf{V} = 0$. In general $\int_C \mathbf{V}$, being the integral of the tangential component of $\mathbf{V}$, represents the net amount of turning of the fluid in a counterclockwise direction

* Some physics texts adopt (5) as the definition of curl, and use it to "prove" Stokes' Theorem easily. However, this is circular reasoning, for to show that (5) really defines a vector "curl $\mathbf{V}(P)$" requires Stokes' Theorem.

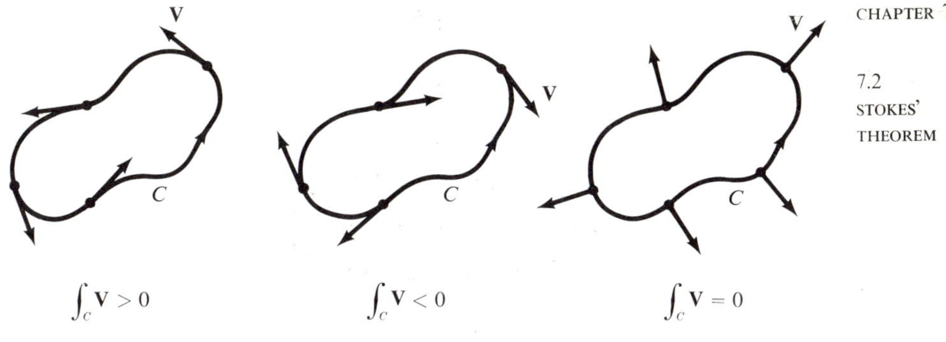

FIGURE 7.2.5
The intuitive meaning of the possible signs of $\int_C \mathbf{V}$.

around C. One therefore refers to $\int_C \mathbf{V}$ as the *circulation* of $\mathbf{V}$ around C (see Figure 7.2.6).

These results allow us to see just what curl $\mathbf{V}$ means for the motion of a fluid. The circulation $\int_{\partial S_\rho} V$ is the net velocity of the fluid around ∂S_ρ, so curl $\mathbf{V} \cdot \mathbf{n}$ represents the turning or rotating effect of the fluid around the axis $\mathbf{n}$. More precisely, formula (5) states that curl $\mathbf{V}(P) \cdot \mathbf{n}$

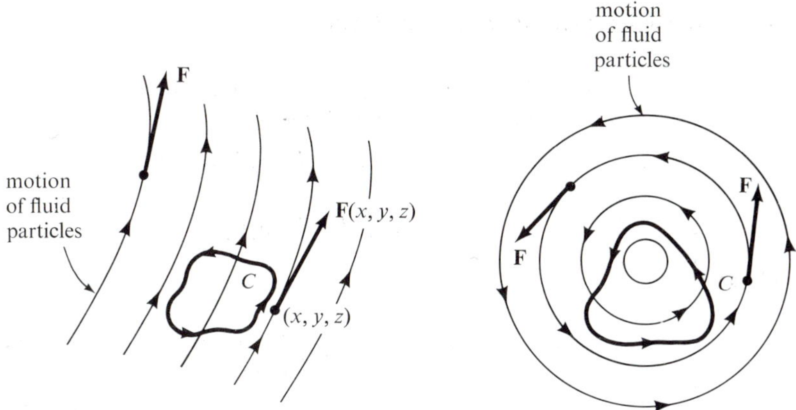

FIGURE 7.2.6
Circulation of a vector field (velocity field of a fluid): (left) circulation about C is zero; (right) non-zero circulation about C ("whirlpool").

is the circulation of $\mathbf{V}$ *per unit area on a surface perpendicular to* $\mathbf{n}$. Observe that the magnitude of curl $\mathbf{V} \cdot \mathbf{n}$ is maximized when $\mathbf{n} =$ curl $\mathbf{V}/\|\text{curl } \mathbf{V}\|$. Therefore, the rotating effect at P is greatest about the axis parallel to curl $\mathbf{V}/\|\text{curl } \mathbf{V}\|$. Thus curl $\mathbf{V}$ is aptly called the *vorticity vector.*

In the next section we shall prove that $\nabla \times \mathbf{F} = \mathbf{0}$ on all of $\mathbb{R}^3$ if and only if $\mathbf{F} = -\nabla V$ for some real-valued function V called the potential.

The negative sign is used to conform to certain physical conventions.*

For electric fields $\mathbf{E}$ we do not generally have $\nabla \times \mathbf{E} = \mathbf{0}$ unless the situation is static. On the other hand, a basic law of electromagnetic theory is that if $\mathbf{E}(t, x, y, z)$ and $\mathbf{H}(t, x, y, z)$ represent the electric and magnetic fields at time t, then $\nabla \times \mathbf{E} = -\partial \mathbf{H}/\partial t$, where $\nabla \times \mathbf{E}$ is computed by holding t fixed, and $\partial \mathbf{H}/\partial t$ is computed by holding x, y, and z constant.

Let us use Stokes' Theorem to determine what this means physically. Assume S is a surface to which Stokes' Theorem applies. Then

$$\int_{\partial S} \mathbf{E} = \int_S \nabla \times \mathbf{E} = -\int_S \frac{\partial \mathbf{H}}{\partial t}$$

$$= -\frac{\partial}{\partial t} \int_S \mathbf{H}$$

(The last equality may be justified if $\mathbf{H}$ is C^1.) Thus we obtain

$$\int_{\partial S} \mathbf{E} = -\frac{\partial}{\partial t} \int_S \mathbf{H}$$

This equality is known as *Faraday's law*. The quantity $\int_{\partial S} \mathbf{E}$ represents the "voltage" around ∂S, and if ∂S were a wire, a current would flow in proportion to this voltage. Also $\int_S \mathbf{H}$ is called the flux of $\mathbf{H}$, or the magnetic flux. Thus, Faraday's law says that the voltage around a loop equals the negative of the rate of change of magnetic flux through the loop.

EXERCISES

1. Redo Exercise 5 of Section 6.6 using Stokes' Theorem.
2. Redo Exercise 6 of Section 6.6 using Stokes' Theorem.
3. Verify Stokes' Theorem for the upper hemisphere $z = \sqrt{1 - x^2 - y^2}$, $z \geq 0$, and the radial vector field $\mathbf{F}(x, y, z) = x\mathbf{i} + y\mathbf{j} + z\mathbf{k}$.
4. Let S be a surface with boundary ∂S, and suppose $\mathbf{E}$ is an electric field that is perpendicular to ∂S. Show that the induced magnetic flux across S is constant in time. (HINT: Use Faraday's law.)
5. Let S be the capped cylindrical surface shown in Figure 7.2.7. S is the union of two surfaces S_1 and S_2, where S_1 is the set of (x, y, z) with $x^2 + y^2 = 1$, $0 \leq z \leq 1$, and S_2 is the set of (x, y, z) with $x^2 + y^2 + (z - 1)^2 = 1$, $z \geq 1$. Set $\mathbf{F}(x, y, z) = (zx + z^2 y + x)\mathbf{i} + (z^3 yx + y)\mathbf{j} + z^4 x^2 \mathbf{k}$. Compute $\int_S \nabla \times \mathbf{F}$. (HINT: Stokes' Theorem holds for this surface.)

* If the minus sign is used, then V is decreasing in the direction $\mathbf{F}$. Thus a particle acted on by $\mathbf{F}$ moves in a direction that decreases the potential.

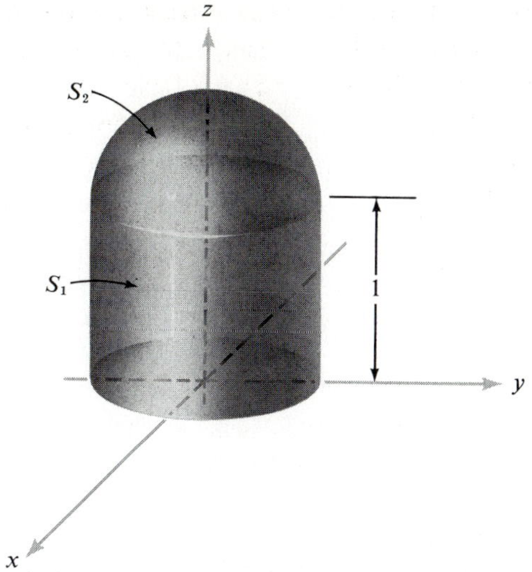

FIGURE 7.2.7
The capped cylinder is the union of S_1 and S_2.

6. Let $\boldsymbol{\sigma}$ consist of straight lines joining (1, 0, 0), (0, 1, 0), and (0, 0, 1) and let S be the triangle with these vertices. Verify Stokes' Theorem directly with $\mathbf{F} = yz\mathbf{i} + xz\mathbf{j} + xy\mathbf{k}$.
7. Prove that Faraday's law implies $\nabla \times \mathbf{E} = -\partial \mathbf{H}/\partial t$.
8. Let S be a surface and let $\mathbf{F}$ be perpendicular to the tangent to the boundary of S. Show that

$$\int_S \nabla \times \mathbf{F} = 0$$

What does this mean physically?

9. Consider two surfaces S_1, S_2 with the same boundary ∂S. Describe with sketches how S_1 and S_2 must be oriented to ensure that

$$\int_{S_1} \nabla \times \mathbf{F} = \int_{S_2} \nabla \times \mathbf{F}$$

10. For a surface S and a fixed vector $\mathbf{v}$ prove

$$2\int_S \mathbf{v} \cdot \mathbf{n}\, dS = \int_{\partial S} \mathbf{v} \times \mathbf{r}$$

where $\mathbf{r}(x, y, z) = (x, y, z)$.

11. Argue informally that if S is a closed surface then

$$\int_S \nabla \times \mathbf{F} = 0$$

(see Exercise 9). (A *closed surface* is one that forms the boundary of a region in space; thus, for example, a sphere is a closed surface.)

12. If C is a closed curve which is the boundary of a surface S, show that
    (a) $\int_C f\nabla g = \int_S \nabla f \times \nabla g$
    (b) $\int_C (f\nabla g + g\nabla f) = 0$
13. If C is a closed curve that is the boundary of a surface S and $\mathbf{v}$ is a constant vector, show that

$$\int_C \mathbf{v} = 0$$

Show that this is true even if C is not the boundary of a surface S.
14. Show that the parametrization $\mathbf{\Phi}: D \to \mathbb{R}^3$, $D = [0, \pi] \times [0, 2\pi]$, $\mathbf{\Phi}(\phi, \theta) = (\cos\theta \sin\phi, \sin\theta \sin\phi, \cos\phi)$, of the unit sphere takes the boundary of D to half of a great circle on S.
15. Verify Theorem 6 for the helicoid $\mathbf{\Phi}(r, \theta) = (r\cos\theta, r\sin\theta, \theta)$, $(r, \theta) \in [0, 1] \times [0, \pi/2]$, and the vector field $\mathbf{F}(x, y, z) = (z, x, y)$.
16. Let $\mathbf{F} = x^2\mathbf{i} + (2xy + x)\mathbf{j} + z\mathbf{k}$. Let C be the circle $x^2 + y^2 = 1$ and S the disc $x^2 + y^2 \le 1$. Determine
    (a) the flux of $\mathbf{F}$ out of S.
    (b) the circulation of $\mathbf{F}$ around C.
    (c) Find the flux of $\nabla \times \mathbf{F}$. Verify Stokes' Theorem directly in this case.

*17. Prove Theorem 6.

7.3 CONSERVATIVE FIELDS

We saw in Section 6.2 that in the case of a gradient force field $\mathbf{F} = \nabla f$, line integrals of $\mathbf{F}$ were evaluated as follows:

$$\int_{\boldsymbol{\sigma}} \mathbf{F} = f(\boldsymbol{\sigma}(b)) - f(\boldsymbol{\sigma}(a))$$

The value of the integral depends only on the endpoints $\boldsymbol{\sigma}(b)$ and $\boldsymbol{\sigma}(a)$ of the path. In other words, if we used another path with the same endpoints, we would still get the same answer. This leads us to say that the integral is *path independent*.

Gradient fields are very important in physical problems. Usually $V = -f$ represents a potential energy (gravitational, electrical, and so on), and $\mathbf{F}$ represents a force. Consider the example of a particle of mass m in the field of the earth; in this case one takes f to be GmM/r or $V = -GmM/r$, where G is the gravitational constant, M is the mass of the earth, and r is the distance from the center of the earth. Then the corresponding force is $\mathbf{F} = (GmM/r^3)\mathbf{r} = (GmM/r^2)\mathbf{n}$, where $\mathbf{n}$ is the unit radial vector. (We shall discuss this case further below.) Note that $\mathbf{F}$ fails to be defined at the one point $r = 0$.

We now wish to characterize those vector fields which can be written as a gradient. Our task is simplified considerably by Stokes' Theorem.

Theorem 7. *Let* **F** *be a* C^1 *vector field defined on* $\mathbb{R}^3$ *except possibly for a finite number of points. The following conditions on* **F** *are all equivalent:*

(*i*) *For any oriented simple closed curve* C, $\int_C \mathbf{F} = 0$.

(*ii*) *For any two oriented simple curves* C_1, C_2 *with the same endpoints,* $\int_{C_1} \mathbf{F} = \int_{C_2} \mathbf{F}$.

(*iii*) **F** *is the gradient of some function* f; *that is,* $\mathbf{F} = \nabla f$ (*and if* **F** *has an exceptional point where it fails to be defined,* f *is also undefined*).

(*iv*) $\nabla \times \mathbf{F} = \mathbf{0}$.

A vector field satisfying one (and hence, all) of these conditions is called a *conservative* vector field.*

Proof. We shall establish the following chain of implications, which will prove the theorem:

$$(i) \Rightarrow (ii) \Rightarrow (iii) \Rightarrow (iv) \Rightarrow (i)$$

First we show that condition (*i*) implies condition (*ii*). Suppose $\boldsymbol{\sigma}_1$ and $\boldsymbol{\sigma}_2$ are parametrizations representing C_1 and C_2, with the same endpoints. Construct the closed curve $\boldsymbol{\sigma}$ obtained by first traversing $\boldsymbol{\sigma}_1$, then $-\boldsymbol{\sigma}_2$ (Figure 7.3.1), or symbolically $\boldsymbol{\sigma} = \boldsymbol{\sigma}_1 - \boldsymbol{\sigma}_2$. Assuming $\boldsymbol{\sigma}$ is simple, (*i*) gives

$$\int_{\boldsymbol{\sigma}} \mathbf{F} = \int_{\boldsymbol{\sigma}_1} \mathbf{F} - \int_{\boldsymbol{\sigma}_2} \mathbf{F} = 0$$

so (*ii*) holds. (If $\boldsymbol{\sigma}$ is not simple, an additional argument, omitted here, is needed.)

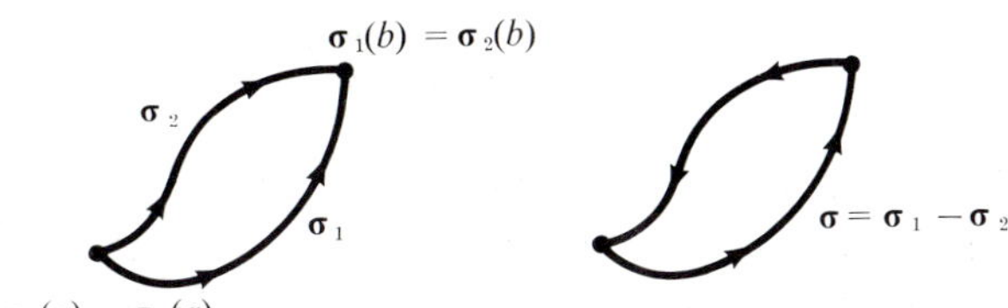

FIGURE 7.3.1
Constructing an oriented simple closed curve $\boldsymbol{\sigma}_1 - \boldsymbol{\sigma}_2$ *from two oriented simple curves.*

Next we prove (*ii*) implies (*iii*). Let C be any oriented simple curve joining a point such as (0, 0, 0) to (x, y, z), and suppose C is represented by the parametrization $\boldsymbol{\sigma}$ (if (0, 0, 0) is the exceptional point of **F**, we can choose a different starting point for $\boldsymbol{\sigma}$ without affecting the argument). Define f to be $\int_{\boldsymbol{\sigma}} \mathbf{F}$. By

* In the plane $\mathbb{R}^2$ exceptional points are not allowed (see Exercise 12). Theorem 7 can be proved in the same way if **F** is defined and C^1 only on an open convex set in $\mathbb{R}^2$ or $\mathbb{R}^3$. (A set D is convex if $P, Q \in D$ implies the line joining P and Q belongs to D.)

hypothesis (*ii*) f is independent of C. We shall show that $\mathbf{F} = \text{grad } f$. Indeed, choose $\boldsymbol{\sigma}$ to be the path shown in Figure 7.3.2, so that

$$f(x, y, z) = \int_0^x F_1(t, 0, 0)\,dt + \int_0^y F_2(x, t, 0)\,dt + \int_0^z F_3(x, y, t)\,dt$$

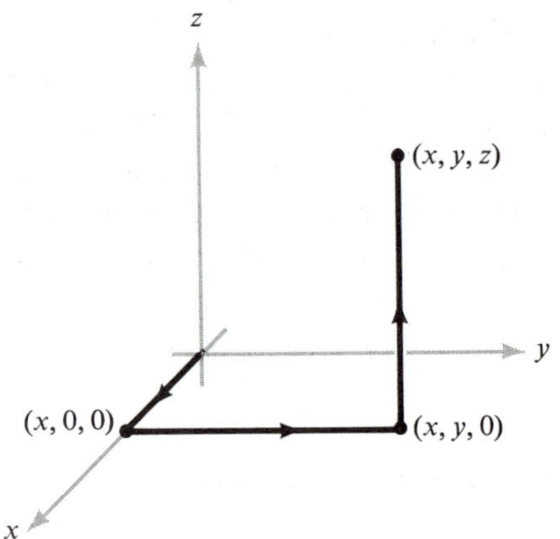

A path joining (0, 0, 0) *to* (*x*, *y*, *z*).

where $\mathbf{F} = (F_1, F_2, F_3)$. It then follows immediately that $\partial f/\partial z = F_3$. Permuting x, y, and z, we can similarly show $\partial f/\partial x = F_1$ and $\partial f/\partial y = F_2$; i.e., $\nabla f = \mathbf{F}$. Third, (*iii*) implies (*iv*) because, as proved in Section 3.4,

$$\nabla \times \nabla f = \mathbf{0}$$

Finally, let $\boldsymbol{\sigma}$ represent a closed curve C and let S be any surface whose boundary is $\boldsymbol{\sigma}$ (if $\mathbf{F}$ has exceptional points, choose S to avoid them). Figure 7.3.3 indicates that we can probably always find such a surface; however, a formal proof of this would require the development of many more sophisticated mathematical ideas than we can present here. By Stokes' Theorem

$$\int_C \mathbf{F} = \int_{\boldsymbol{\sigma}} \mathbf{F} = \int_S (\nabla \times \mathbf{F}) \cdot \mathbf{n}\,dS = \int_S (\text{curl } \mathbf{F}) \cdot \mathbf{n}\,dS$$

Since $\nabla \times \mathbf{F} = \mathbf{0}$, this integral vanishes, so that (*iv*) $\Rightarrow$ (*i*). ▮

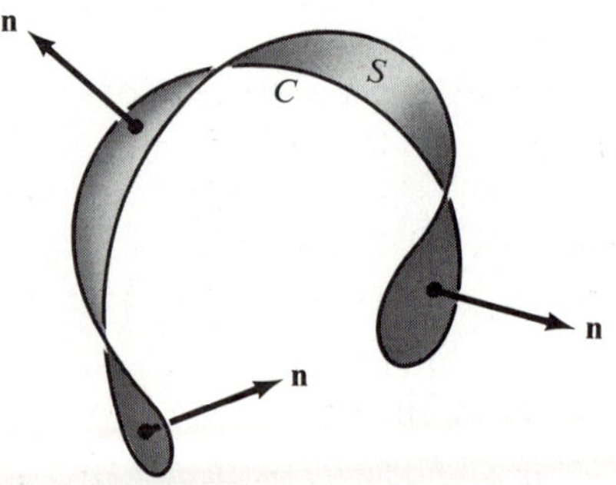

FIGURE 7.3.3
A surface S spanning a curve C.

There are several useful physical interpretations of $\int_C \mathbf{F}$. We have already seen that one is the work done by $\mathbf{F}$ in moving a particle along C. A second interpretation is the notion of circulation, which we encountered at the end of the last section. In this case we think of $\mathbf{F}$ as the velocity field of a fluid; that is, to each point P in space $\mathbf{F}$ assigns the velocity vector of the fluid at P. Take C to be a closed curve, and let $\Delta\mathbf{s}$ be a small directed chord of C. Then $\mathbf{F} \cdot \Delta\mathbf{s}$ is approximately the tangential component of $\mathbf{F}$ times $\|\Delta\mathbf{s}\|$. The integral $\int_C \mathbf{F}$ is the net tangential component around C. This means that a small paddle wheel placed in the fluid would rotate if the circulation of the fluid were nonzero, $\int_C \mathbf{F} \neq 0$ (see Figure 7.3.4). Thus we often speak of the line integral

$$\int_C \mathbf{F}$$

as being the circulation of $\mathbf{F}$ around C.

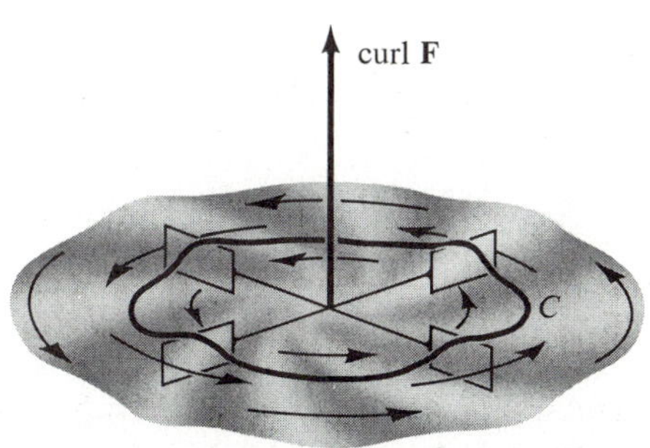

FIGURE 7.3.4
$\int_C \mathbf{F} \neq 0$ *implies that a paddle wheel in a fluid with velocity field* $\mathbf{F}$ *will rotate.*

There is a similar interpretation in electromagnetic theory: if $\mathbf{F}$ represents an electric field, then a current will flow around a loop C if $\int_C \mathbf{F} \neq 0$.

By Theorem 7, a field $\mathbf{F}$ has no circulation if and only if curl $\mathbf{F} = \nabla \times \mathbf{F} = \mathbf{0}$. Hence, a vector field $\mathbf{F}$ with curl $\mathbf{F} = \mathbf{0}$ is called *irrotational.* We have therefore proved that a vector field in $\mathbb{R}^3$ is irrotational if and only if it is a gradient field for some function, that is, if and only if $\mathbf{F} = \nabla f$. The function f is called a *potential* for $\mathbf{F}$.

EXAMPLE 1. Consider the vector field $\mathbf{F}$ on $\mathbb{R}^3$ defined by

$$\mathbf{F}(x, y, z) = y\mathbf{i} + (z \cos yz + x)\mathbf{j} + (y \cos yz)\mathbf{k}$$

Show that $\mathbf{F}$ is irrotational and find a scalar potential for $\mathbf{F}$.

We compute $\nabla \times \mathbf{F}$:

$$\nabla \times \mathbf{F} = \begin{vmatrix} \mathbf{i} & \mathbf{j} & \mathbf{k} \\ \dfrac{\partial}{\partial x} & \dfrac{\partial}{\partial y} & \dfrac{\partial}{\partial z} \\ y & x + z \cos yz & y \cos yz \end{vmatrix}$$

$$= (\cos yz - yz \sin yz - \cos yz + yz \sin yz)\mathbf{i}$$

$$+ (0 - 0)\mathbf{j} + (1 - 1)\mathbf{k}$$

$$= 0\mathbf{i} + 0\mathbf{j} + 0\mathbf{k} = \mathbf{0}$$

so $\mathbf{F}$ is irrotational. We can find a scalar potential in several ways.

Method 1. By the technique used to prove that (*ii*) implies (*iii*) in Theorem 7, we may set

$$f(x, y, z) = \int_0^x F_1(t, 0, 0)\, dt + \int_0^y F_2(x, t, 0)\, dt + \int_0^z F_3(x, y, t)\, dt$$

$$= \int_0^x 0\, dt + \int_0^y x\, dt + \int_0^z y \cos yt\, dt$$

$$= 0 + xy + \sin yz = xy + \sin yz$$

One easily verifies that $\nabla f = \mathbf{F}$ as required:

$$\nabla f = \frac{\partial f}{\partial x}\mathbf{i} + \frac{\partial f}{\partial y}\mathbf{j} + \frac{\partial f}{\partial z}\mathbf{k} = y\mathbf{i} + (x + z \cos yz)\mathbf{j} + (y \cos yz)\mathbf{k}$$

Method 2. Because we know that f exists, we know that it can solve the equations

$$\frac{\partial f}{\partial x} = y$$

$$\frac{\partial f}{\partial y} = x + z \cos yz$$

$$\frac{\partial f}{\partial z} = y \cos yz$$

for $f(x, y, z)$. These are equivalent to the simultaneous equations

(a) $f(x, y, z) = xy + h_1(y, z)$

(b) $f(x, y, z) = \sin yz + xy + h_2(x, z)$

(c) $f(x, y, z) = \sin yz + h_3(x, y)$

for functions h_1, h_2, h_3 independent of x, y, and z (respectively). When $h_1(y, z) = \sin yz$, $h_2(x, z) = 0$, and $h_3(x, y) = xy$ the three equations agree, and so yield a potential for $\mathbf{F}$. However, we have only guessed at the values of h_1, h_2, and h_3. To derive f more systematically we note that since $f(x, y, z) = xy + h_1(y, z)$ and $\partial f/\partial z = y \cos yz$, we find that

$$\frac{\partial h_1(y, z)}{\partial z} = y \cos yz$$

or

$$h_1(y, z) = \int y \cos yz \, dz + g(y)$$

$$= \sin yz + g(y)$$

Therefore, plugging this back into (a) we get

$$f(x, y, z) = xy + \sin yz + g(y)$$

but by (b)

$$g(y) = h_2(x, z)$$

Since the right side of this equation is a function of x and z, and the left side is a function of y alone, we conclude that they must equal some constant C. Thus

$$f(x, y, z) = xy + \sin yz + C$$

and we have determined f up to a constant.

EXAMPLE 2. A mass M at the origin in $\mathbb{R}^3$ exerts a force on a unit mass located at $\mathbf{r} = (x, y, z)$ with magnitude GM/r^2 and directed toward the origin. Here G is the gravitational constant, which depends on the units of measurement, and $r = \|\mathbf{r}\| = \sqrt{x^2 + y^2 + z^2}$. If we remember that $-\mathbf{r}/r$ is a unit vector directed towards the origin, then we can write the force field as

$$\mathbf{F}(x, y, z) = -\frac{GM\mathbf{r}}{r^3}$$

We shall show that $\mathbf{F}$ is irrotational and we shall find a scalar potential for $\mathbf{F}$. Notice that $\mathbf{F}$ is not defined at the origin, but Theorem 7 still applies since it allows an exceptional point.

First let us verify that $\nabla \times \mathbf{F} = \mathbf{0}$. Referring to formula 11 of the table in Section 3.5, we get

$$\nabla \times \mathbf{F} = -GM\left\{\nabla\left(\frac{1}{r^3}\right) \times \mathbf{r} + \frac{1}{r^3} \nabla \times \mathbf{r}\right\}$$

But $\nabla(1/r^3) = -3\mathbf{r}/r^5$ (see Exercise 8, Section 3.5) so the first term vanishes, since $\mathbf{r} \times \mathbf{r} = \mathbf{0}$. The second term vanishes because

$$\nabla \times \mathbf{r} = \begin{vmatrix} \mathbf{i} & \mathbf{j} & \mathbf{k} \\ \dfrac{\partial}{\partial x} & \dfrac{\partial}{\partial y} & \dfrac{\partial}{\partial z} \\ x & y & z \end{vmatrix} = \left(\frac{\partial z}{\partial y} - \frac{\partial y}{\partial z}\right)\mathbf{i} + \left(\frac{\partial x}{\partial z} - \frac{\partial z}{\partial x}\right)\mathbf{j} + \left(\frac{\partial y}{\partial x} - \frac{\partial x}{\partial y}\right)\mathbf{k} = \mathbf{0}$$

Hence $\nabla \times \mathbf{F} = \mathbf{0}$ (for $\mathbf{r} \neq \mathbf{0}$).

If we recall the formula $\nabla(r^n) = nr^{n-2}\mathbf{r}$ (Exercise 8, Section 3.5) then we can read off a scalar potential for $\mathbf{F}$ by inspection. We have $\mathbf{F} = -\nabla\phi$, where $\phi(x, y, z) = -GM/r$ is called the *gravitational potential energy*.

By Theorem 3 of Section 6.2, the work done by $\mathbf{F}$ in moving a unit mass particle from a point P_1 to P_2 is given by

$$\phi(P_1) - \phi(P_2) = GM\left(\frac{1}{r_2} - \frac{1}{r_1}\right)$$

where r_1 is the radial distance of P_1 from the origin, with r_2 similarly defined.

By the same proof, Theorem 7 is also true for smooth vector fields $\mathbf{F}$ on $\mathbb{R}^2$. However, in this case $\mathbf{F}$ cannot have any exceptional points; that is, $\mathbf{F}$ must be smooth everywhere (see Exercise 12).

If $\mathbf{F} = P\mathbf{i} + Q\mathbf{j}$ then

$$\nabla \times \mathbf{F} = \left(\frac{\partial Q}{\partial x} - \frac{\partial P}{\partial y}\right)\mathbf{k}$$

and so the condition $\nabla \times \mathbf{F} = \mathbf{0}$ reduces to

$$\frac{\partial P}{\partial y} = \frac{\partial Q}{\partial x}$$

Thus we have:

Corollary. *If $\mathbf{F}$ is a C^1 vector field on $\mathbb{R}^2$ of the form $P\mathbf{i} + Q\mathbf{j}$ with $\partial P/\partial y = \partial Q/\partial x$, then $\mathbf{F} = \nabla f$ for some f on $\mathbb{R}^2$.*

EXAMPLE 3. (a) Determine whether the vector field

$$\mathbf{F} = e^{xy}\mathbf{i} + e^{x+y}\mathbf{j}$$

is a gradient field.

Here $P(x, y) = e^{xy}$ and $Q(x, y) = e^{x+y}$, so we compute

$$\frac{\partial P}{\partial y} = xe^{xy}, \qquad \frac{\partial Q}{\partial x} = e^{x+y}$$

These are not equal so $\mathbf{F}$ cannot have a potential function.

(b) Repeat part (a) for

$$\mathbf{F} = (2x \cos y)\mathbf{i} - (x^2 \sin y)\mathbf{j}$$

In this case, we find

$$\frac{\partial P}{\partial y} = -2x \sin y = \frac{\partial Q}{\partial x}$$

and so $\mathbf{F}$ has a potential function f. To compute f we solve the equations

$$\frac{\partial f}{\partial x} = 2x \cos y, \qquad \frac{\partial f}{\partial y} = -x^2 \sin y$$

Thus

$$f(x, y) = x^2 \cos y + h_1(y)$$

and

$$f(x, y) = x^2 \cos y + h_2(x)$$

If $h_1 = h_2 = 0$ both equations are satisfied, and we find that $f(x, y) = x^2 \cos y$ is a potential for $\mathbf{F}$.

EXAMPLE 4. Let $\boldsymbol{\sigma}: [1, 2] \to \mathbb{R}^2$ be given by

$$x = e^{t-1}, \qquad y = \sin(\pi/t)$$

Compute the integral

$$\int_{\boldsymbol{\sigma}} \mathbf{F} = \int_{\boldsymbol{\sigma}} 2x \cos y \, dx - x^2 \sin y \, dy$$

where $\mathbf{F} = (2x \cos y)\mathbf{i} - (x^2 \sin y)\mathbf{j}$.

We have $\boldsymbol{\sigma}(1) = (1, 0)$ and $\boldsymbol{\sigma}(2) = (e, 1)$. Since $\partial(2x \cos y)/\partial y = \partial(-x^2 \sin y)/\partial x$, $\mathbf{F}$ is irrotational and hence a gradient vector field (as we saw in Example 3). Thus by Theorem 7 we can replace $\boldsymbol{\sigma}$ by any piecewise C^1 curve having the same endpoints, in particular by the polygonal path from $(1, 0)$ to $(e, 0)$ to $(e, 1)$. So the line integral must be equal to

$$\begin{aligned} \int_{\boldsymbol{\sigma}} \mathbf{F} &= \int_1^e 2t \cos 0 \, dt + \int_0^1 -e^2 \sin t \, dt \\ &= (e^2 - 1) + e^2(\cos 1 - 1) = e^2 \cos 1 - 1 \end{aligned}$$

Alternatively, using Theorem 3 of Section 6.2 we have

$$\int_{\boldsymbol{\sigma}} 2x \cos y \, dx - x^2 \sin y \, dy = \int_{\boldsymbol{\sigma}} \nabla f = f(\boldsymbol{\sigma}(2)) - f(\boldsymbol{\sigma}(1)) = e^2 \cos 1 - 1$$

since $f(x, y) = x^2 \cos y$ is a potential function for $\mathbf{F}$.

Evidently this technique is simpler than computing the integral directly.

We conclude this section with a theorem that is quite similar in spirit to Theorem 7. Theorem 7 was motivated partly as a converse to the result that curl $\nabla f = \mathbf{0}$ for any C^1 function $f: \mathbb{R}^3 \to \mathbb{R}$—or, if curl $\mathbf{F} = \mathbf{0}$ then $\mathbf{F} = \nabla f$. We also know (formula 10 in the table in Section 3.5) that div(curl $\mathbf{G}$) $= 0$ for any C^2 vector field $\mathbf{G}$. We may ask about the converse statement: if div $\mathbf{F} = 0$, is $\mathbf{F}$ the curl of a vector field $\mathbf{G}$? The following theorem answers in the affirmative.

Theorem 8. *If* $\mathbf{F}$ *is a* C^1 *vector field on* $\mathbb{R}^3$ *with* div $\mathbf{F} = 0$, *then there exists a* C^1 *vector field* $\mathbf{G}$ *with* $\mathbf{F} =$ curl $\mathbf{G}$.

The proof is outlined in Exercise 14. We should warn the reader at this point that, unlike the $\mathbf{F}$ in Theorem 7, the vector field $\mathbf{F}$ in Theorem 8 is not allowed to have an exceptional point. For example the gravitational force field $\mathbf{F} = -(GM\mathbf{r}/r^3)$ has the property that div $\mathbf{F} = 0$, and yet there is no $\mathbf{G}$ for which $\mathbf{F} =$ curl $\mathbf{G}$ (see Exercise 20). Theorem 8 does not apply because the gravitational force field $\mathbf{F}$ is not defined at $\mathbf{0} \in \mathbb{R}^3$.

EXERCISES

1. Show that any two potential functions for a vector field differ at most by a constant.
2. (a) Let $\mathbf{F}(x, y) = (xy, y^2)$ and let $\boldsymbol{\sigma}$ be the path $y = 2x^2$ joining $(0, 0)$ to $(1, 2)$ in $\mathbb{R}^2$. Evaluate $\int_{\boldsymbol{\sigma}} \mathbf{F}$.
   (b) Does the integral in (a) depend on the path joining $(0, 0)$ to $(1, 2)$?
3. Let $\mathbf{F}(x, y, z) = (2xyz + \sin x)\mathbf{i} + x^2z\mathbf{j} + x^2y\mathbf{k}$. Find a function f such that $\mathbf{F} = \nabla f$.
4. Evaluate $\int_{\boldsymbol{\sigma}} \mathbf{F}$, where $\boldsymbol{\sigma}(t) = (\cos^5 t, \sin^3 t, t^4)$, $0 \le t \le \pi$, and $\mathbf{F}$ is as in Exercise 3.
5. What is the work done by the force $\mathbf{F} = -\mathbf{r}/\|\mathbf{r}\|^3$ in moving a particle from a point $\mathbf{r}_0 \in \mathbb{R}^3$ "to ∞", where $\mathbf{r}(x, y, z) = (x, y, z)$?
6. In Exercise 5, show that $\mathbf{F} = \nabla(1/r)$, $r \neq 0$, $r = \|\mathbf{r}\|$. In what sense is the integral of $\mathbf{F}$ independent of path?
7. Let $\mathbf{F}(x, y, z) = xy\mathbf{i} + y\mathbf{j} + z\mathbf{k}$. Can there exist a function f such that $\mathbf{F} = \nabla f$?
8. Let $\mathbf{F} = F_1\mathbf{i} + F_2\mathbf{j} + F_3\mathbf{k}$ and suppose each F_1 satisfies the homogeneity condition

$$F_k(tx, ty, tz) = tF_k(x, y, z), \qquad k = 1, 2, 3$$

Suppose also $\nabla \times \mathbf{F} = \mathbf{0}$. Prove $\mathbf{F} = \nabla f$ where

$$2f(x, y, z) = xF_1(x, y, z) + yF_2(x, y, z) + zF_3(x, y, z)$$

9. Is each of the following vector fields the curl of some other vector field? If so, find the vector field.
   (a) $\mathbf{F} = x\mathbf{i} + y\mathbf{j} + z\mathbf{k}$
   (b) $\mathbf{F} = (x^2 + 1)\mathbf{i} + (z - 2xy)\mathbf{j} + y\mathbf{k}$
10. Let a fluid have the velocity field $\mathbf{F}(x, y, z) = xy\mathbf{i} + yz\mathbf{j} + xz\mathbf{k}$. What is the circulation around the unit circle? Interpret your answer.
11. Let $\mathbf{F}(x, y, z) = (e^x \sin y)\mathbf{i} + (e^x \cos y)\mathbf{j} + z^2\mathbf{k}$. Evaluate the integral $\int_{\boldsymbol{\sigma}} \mathbf{F}$, where $\boldsymbol{\sigma}(t) = (\sqrt{t}, t^3, \exp\sqrt{t})$, $0 \le t \le 1$.
12. (a) Show that $\int_C (x\,dy - y\,dx)/(x^2 + y^2) = 2\pi$, where C is the unit circle.
    (b) Conclude that the associated vector field $(x/(x^2 + y^2))\mathbf{i} - (y/(x^2 + y^2))\mathbf{j}$ is not a potential field.
    (c) Show, however, that $\partial P/\partial y = \partial Q/\partial x$. Does this contradict the corollary to Theorem 7? If not, why not?
13. The mass of the earth is approximately 6×10^{27} g and that of the sun is 330,000 times as much. The gravitational constant (in units of grams, seconds, and centimeters) is 6.7×10^{-8}. The distance of the earth from the sun is about 1.5×10^{12} cm. Compute, approximately, the work necessary to increase the distance of the earth from the sun by 1 cm.
14. Prove Theorem 8. (HINT: Define $\mathbf{G} = G_1\mathbf{i} + G_2\mathbf{j} + G_3\mathbf{k}$ by

$$G_1(x, y, z) = \int_0^z F_2(x, y, t)\,dt - \int_0^y F_3(x, t, 0)\,dt$$

$$G_2(x, y, z) = -\int_0^z F_1(x, y, t)\,dt$$

and $G_3(x, y, z) = 0$.)

15. Let $\mathbf{F} = xz\mathbf{i} - yz\mathbf{j} + y\mathbf{k}$. Verify that $\nabla \cdot \mathbf{F} = 0$. Find a $\mathbf{G}$ such that $\mathbf{F} = \nabla \times \mathbf{G}$.
16. Let $\mathbf{F} = (x \cos y)\mathbf{i} - (\sin y)\mathbf{j} + (\sin x)\mathbf{k}$. Find a $\mathbf{G}$ such that $\mathbf{F} = \nabla \times \mathbf{G}$.
17. Show that the function f defined in the proof of Theorem 7 for "(*ii*) implies (*iii*)" satisfies $\partial f/\partial x = F_1$, $\partial f/\partial y = F_2$.
18. Let $\mathbf{F}$ be a vector field on $\mathbb{R}^3$ given by $\mathbf{F} = -y\mathbf{i} + x\mathbf{j}$.
    (a) Show that $\mathbf{F}$ is rotational, i.e., $\mathbf{F}$ is not irrotational.
    (b) Suppose $\mathbf{F}$ represents the velocity vector field of a fluid. Show that if we place a cork in this fluid it will revolve in a plane parallel to the xy-plane, in a circular trajectory about the z-axis.
    (c) In what direction does the cork revolve?

*19. Let $\mathbf{G}$ be the vector field on $\mathbb{R}^3\backslash\{z\text{-axis}\}$ defined by

$$\mathbf{G} = \frac{-y}{x^2 + y^2}\mathbf{i} + \frac{x}{x^2 + y^2}\mathbf{j}$$

(a) Show that $\mathbf{G}$ is irrotational.
(b) Show that 18(b) holds for $\mathbf{G}$ also.

(c) How can we resolve the fact that the trajectories of **F** and **G** are both the same (circular about the z-axis) yet **F** is rotational and **G** is not? (HINT: The property of being rotational is a local condition, that is, a property of the fluid in the neighborhood of a point.)

*20. Let $\mathbf{F} = -(GM\mathbf{r}/r^3)$ be the gravitational force field on $\mathbb{R}^3$.
   (a) Show that div $\mathbf{F} = 0$.
   (b) Show that $\mathbf{F} \neq$ curl $\mathbf{G}$ for any C^1 vector field $\mathbf{G}$ on $\mathbb{R}^3\backslash\{\mathbf{0}\}$.

7.4 GAUSS' THEOREM

Gauss' Theorem states that the flux of a vector field out of a closed surface equals the integral of the divergence of that vector field over the volume enclosed by the surface. The result parallels Stokes' Theorem and Green's Theorem in that it relates an integral over a closed geometrical object (curve or surface) to an integral over a contained region (surface or volume).

We shall begin by asking the reader to review the various regions in space that were introduced when we considered the volume integral; these regions are illustrated in Figure 5.6.3. As that figure indicates, the boundary of a region of type 1, 2, or 3 in $\mathbb{R}^3$ is a surface made up of a finite number (at most six, at least two) of surfaces that can be described as graphs of functions from $\mathbb{R}^2$ to $\mathbb{R}$. This kind of surface is called a *closed surface*; it has no boundary. The surfaces $S_1, S_2, \ldots, S_N$ composing such a closed surface are called its *faces*.

EXAMPLE 1. The cube in Figure 7.4.1 is a region of type 4 (recall that this means it is simultaneously of types 1, 2, 3), with six squares composing its boundary. The sphere is the boundary of a solid ball, which is also a region of type 4.

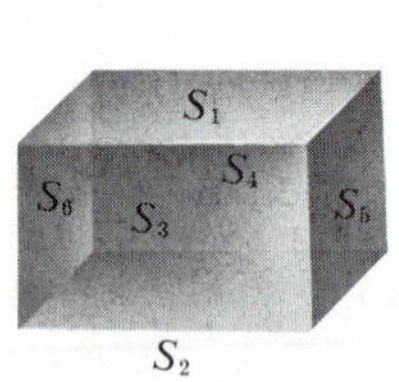

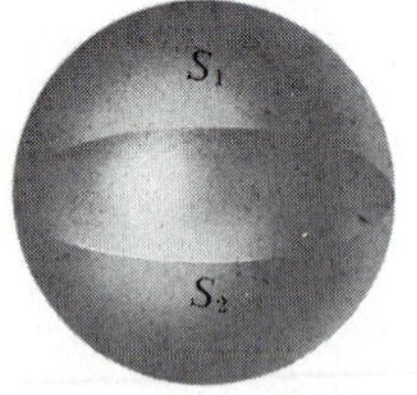

FIGURE 7.4.1
Regions of type 4 and the surfaces S_i composing their boundaries.

Closed surfaces can be oriented in two ways. The outward orientation makes the normal point outward into space, and the inward orientation makes the normal point into the bounded region (Figure 7.4.2).

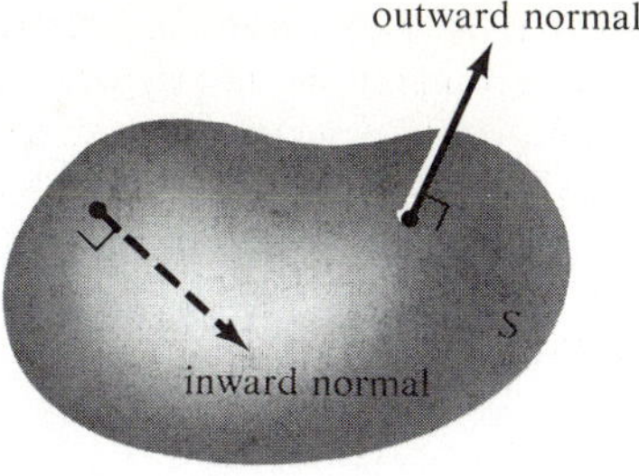

FIGURE 7.4.2
Two possible orientations for a closed surface.

Suppose S is a closed surface oriented in one of these two ways and $\mathbf{F}$ is a vector field on S. Then, as we defined above (p. 336),

$$\int_S \mathbf{F} = \sum_i \int_{S_i} \mathbf{F}$$

If S is given the outward orientation, the integral $\int_S \mathbf{F}$ measures the total flux of $\mathbf{F}$ outwards across S. That is, if we think of $\mathbf{F}$ as the velocity field of a fluid, $\int_S \mathbf{F}$ indicates the amount of fluid leaving the region bounded by S per unit time. If S is given the inward orientation, the integral $\int_S \mathbf{F}$ measures the total flux of $\mathbf{F}$ inwards across S.

There is another common way of writing these surface integrals, a way that explicitly specifies the orientation of S. Let the orientation of S be given by a unit normal vector $\mathbf{n}(x, y, z)$ at each point of S. Then we have the oriented integral

$$\int_S \mathbf{F} = \int_S (\mathbf{F} \cdot \mathbf{n})\, dS$$

that is, the integral of the normal component of $\mathbf{F}$ over S. In the remainder of this section, if S is a closed surface enclosing a region Ω, we adopt the convention that $S = \partial\Omega$ is given the outward orientation, with outward unit normal $\mathbf{n}(x, y, z)$ at each point $(x, y, z) \in S$. Furthermore, we denote the surface with the opposite (inward) orientation by $\partial\Omega_{\text{op}}$. Then the associated unit normal direction for this orientation is $-\mathbf{n}$. Thus

$$\int_{\partial\Omega} \mathbf{F} = \int_S (\mathbf{F} \cdot \mathbf{n})\, dS = -\int_S (\mathbf{F} \cdot (-\mathbf{n}))\, dS = -\int_{\partial\Omega_{\text{op}}} \mathbf{F}$$

EXAMPLE 2. The unit cube Ω given by

$$0 \le x \le 1, \qquad 0 \le y \le 1, \qquad 0 \le z \le 1$$

is a region in space of type 4 (see Figure 7.4.3). We write the faces as

$$
\begin{aligned}
&S_1: z = 1, && 0 \le x \le 1, && 0 \le y \le 1\\
&S_2: z = 0, && 0 \le x \le 1, && 0 \le y \le 1\\
&S_3: x = 1, && 0 \le y \le 1, && 0 \le z \le 1\\
&S_4: x = 0, && 0 \le y \le 1, && 0 \le z \le 1\\
&S_5: y = 1, && 0 \le x \le 1, && 0 \le z \le 1\\
&S_6: y = 0, && 0 \le x \le 1, && 0 \le z \le 1
\end{aligned}
$$

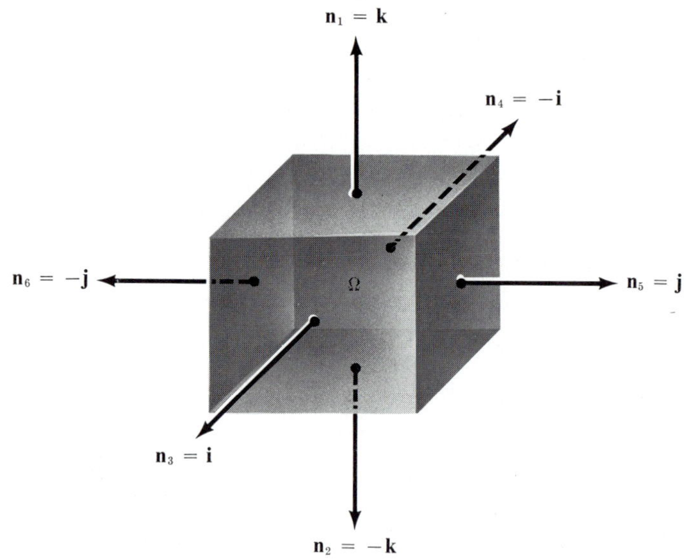

FIGURE 7.4.3
The outward orientation on the cube.

From Figure 7.4.3 we see that

$$
\begin{aligned}
\mathbf{n}_1 &= \mathbf{k} = -\mathbf{n}_2\\
\mathbf{n}_3 &= \mathbf{i} = -\mathbf{n}_4\\
\mathbf{n}_5 &= \mathbf{j} = -\mathbf{n}_6
\end{aligned}
$$

So for a continuous vector field $\mathbf{F} = F_1\mathbf{i} + F_2\mathbf{j} + F_3\mathbf{k}$

$$
\begin{aligned}
\int_{\partial\Omega} \mathbf{F} = \int_S \mathbf{F}\cdot\mathbf{n}\,dS = \int_{S_1} F_3\,dS - \int_{S_2} F_3\,dS + \int_{S_3} F_1\,dS\\
- \int_{S_4} F_1\,dS + \int_{S_5} F_2\,dS - \int_{S_6} F_2\,dS
\end{aligned}
$$

We have now come to the last of the three central theorems of this chapter. This theorem relates surface integrals to volume integrals; in words, the theorem states that if Ω is a region in $\mathbb{R}^3$, then the flux of a field **F** outward across the closed surface $\partial\Omega$ is equal to the integral of div **F** over Ω. (See p. 337 for the interpretation of surface integrals in terms of flux.)

***Theorem 9.* (*Gauss' Divergence Theorem*).** *Let Ω be a region in space of type 4. Denote by $\partial\Omega$ the oriented closed surface that bounds Ω. Let* **F** *be a smooth vector field defined on Ω. Then*

$$\int_{\Omega} \nabla \cdot \mathbf{F} = \int_{\partial\Omega} \mathbf{F}$$

or alternatively

$$\int_{\Omega} \operatorname{div} \mathbf{F} \, dV = \int_{\partial\Omega} (\mathbf{F} \cdot \mathbf{n}) \, dS$$

Proof. If $\mathbf{F} = P\mathbf{i} + Q\mathbf{j} + R\mathbf{k}$, then by definition, $\operatorname{div} \mathbf{F} = \partial P/\partial x + \partial Q/\partial y + \partial R/\partial z$, so we can write (using additivity of the volume integral)

$$\int_{\Omega} \operatorname{div} \mathbf{F} \, dV = \int_{\Omega} \frac{\partial P}{\partial x} \, dV + \int_{\Omega} \frac{\partial Q}{\partial y} \, dV + \int_{\Omega} \frac{\partial R}{\partial z} \, dV$$

On the other hand, the surface integral in question is

$$\int_{\partial\Omega} \mathbf{F} = \int_{\partial\Omega} \mathbf{F} \cdot \mathbf{n} \, dS = \int_{\partial\Omega} (P\mathbf{i} + Q\mathbf{j} + R\mathbf{k}) \cdot \mathbf{n} \, dS$$

$$= \int_{\partial\Omega} P\mathbf{i} \cdot \mathbf{n} \, dS + \int_{\partial\Omega} Q\mathbf{j} \cdot \mathbf{n} \, dS + \int_{\partial\Omega} R\mathbf{k} \cdot \mathbf{n} \, dS$$

The theorem will follow if we establish the three equalities

$$\int_{\partial\Omega} P\mathbf{i} \cdot \mathbf{n} \, dS = \int_{\Omega} \frac{\partial P}{\partial x} \, dV \tag{1}$$

$$\int_{\partial\Omega} Q\mathbf{j} \cdot \mathbf{n} \, dS = \int_{\Omega} \frac{\partial Q}{\partial y} \, dV \tag{2}$$

$$\int_{\partial\Omega} R\mathbf{k} \cdot \mathbf{n} \, dS = \int_{\Omega} \frac{\partial R}{\partial z} \, dV \tag{3}$$

We shall prove (3); the other two equalities can be proved in a perfectly analogous fashion.

Since Ω is a region of type 1 (as well as of types 2 and 3), there exists a pair of functions

$$z = f_1(x, y), \qquad z = f_2(x, y)$$

with common domain an elementary region D in the xy-plane, such that Ω is the set of all points (x, y, z) satisfying

$$f_2(x, y) \le z \le f_1(x, y), \qquad (x, y) \in D$$

By formula (4) of Section 5.6, we have

$$\int_{\Omega} \frac{\partial R}{\partial z} \, dV = \int_{D} \left(\int_{z=f_2(x,y)}^{z=f_1(x,y)} \frac{\partial R}{\partial z} \, dz \right) dx \, dy$$

and so

$$\int_\Omega \frac{\partial R}{\partial z}\,dV = \int_D [R(x, y, f_1(x, y)) - R(x, y, f_2(x, y))]\,dx\,dy \tag{4}$$

The boundary of Ω is a closed surface whose top S_1 is the graph of $z = f_1(x, y)$, $(x, y) \in D$, and whose bottom S_2 is the graph of $z = f_2(x, y)$, $(x, y) \in D$. The four other sides of $\partial\Omega$ consist of surfaces S_3, S_4, S_5, and S_6 whose normals are always perpendicular to the z-axis. (See, for example, Figure 7.4.4. Of course,

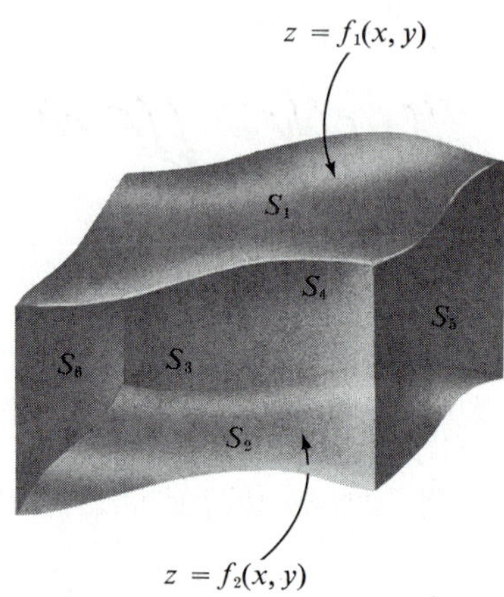

FIGURE 7.4.4
A region Ω of type 1 for which $\int_{\partial\Omega} R\mathbf{k}\cdot\mathbf{n}\,dS = \int_\Omega (\partial R/\partial z)\,dV$. The four sides of $\partial\Omega$ S_3, S_4, S_5, S_6 have normals perpendicular to the z-axis.

some of the other four sides might vanish—for instance, if Ω is a solid ball, and $\partial\Omega$ is a sphere—but this will not affect the argument.) By definition

$$\int_{\partial\Omega} R\mathbf{k}\cdot\mathbf{n}\,dS = \int_{S_1} R\mathbf{k}\cdot\mathbf{n}_1\,dS + \int_{S_2} R\mathbf{k}\cdot\mathbf{n}_2\,dS + \sum_{i=3}^{6}\int_{S_i} R\mathbf{k}\cdot\mathbf{n}_i\,dS$$

Since on each of S_3, S_4, S_5, and S_6 the normal $\mathbf{n}_i$ is perpendicular to $\mathbf{k}$, we have $\mathbf{k}\cdot\mathbf{n}_i = 0$ along these faces, and so the integral reduces to

$$\int_{\partial\Omega} R\mathbf{k}\cdot\mathbf{n}\,dS = \int_{S_1} R\mathbf{k}\cdot\mathbf{n}_1\,dS + \int_{S_2} R\mathbf{k}\cdot\mathbf{n}_2\,dS \tag{5}$$

The surface S_2 is defined by $z = f_2(x, y)$, so

$$\mathbf{n}_2 = \frac{\dfrac{\partial f_2}{\partial x}\mathbf{i} + \dfrac{\partial f_2}{\partial y}\mathbf{j} - \mathbf{k}}{\sqrt{\left(\dfrac{\partial f_2}{\partial x}\right)^2 + \left(\dfrac{\partial f_2}{\partial y}\right)^2 + 1}}$$

(since S_2 is the bottom portion of Ω, for $\mathbf{n}_2$ to point outward it must have a negative $\mathbf{k}$ component; see Example 2). Thus

$$\mathbf{n}_2\cdot\mathbf{k} = \frac{-1}{\sqrt{\left(\dfrac{\partial f_2}{\partial x}\right)^2 + \left(\dfrac{\partial f_2}{\partial y}\right)^2 + 1}}$$

and

$$\int_{S_2} R(\mathbf{k} \cdot \mathbf{n}_2)\, dS =$$

$$\int_D R(x, y, f_2(x, y)) \left(\frac{-1}{\sqrt{\left(\frac{\partial f_2}{\partial x}\right)^2 + \left(\frac{\partial f_2}{\partial y}\right)^2 + 1}} \right) \sqrt{\left(\frac{\partial f_2}{\partial x}\right)^2 + \left(\frac{\partial f_2}{\partial y}\right)^2 + 1}\, dA$$

$$= -\int_D R(x, y, f_2(x, y))\, dx\, dy \tag{6}$$

This equation also follows from formula (4), Section 6.5.

Similarly, on the top face S_1 we have

$$\mathbf{k} \cdot \mathbf{n}_1 = \frac{1}{\sqrt{\left(\frac{\partial f_1}{\partial x}\right)^2 + \left(\frac{\partial f_1}{\partial y}\right)^2 + 1}}$$

and so

$$\int_{S_1} R(\mathbf{k} \cdot \mathbf{n}_1)\, dS = \int_D R(x, y, f_1(x, y))\, dx\, dy \tag{7}$$

Substituting (6) and (7) into equation (5) and then comparing with (4) we obtain

$$\int_\Omega \frac{\partial R}{\partial z}\, dV = \int_{\partial\Omega} R(\mathbf{k} \cdot \mathbf{n})\, dS$$

The remaining equalities (1) and (2) can be established in exactly the same way to complete the proof. ▮

The reader should note that the proof is similar to that of Green's Theorem. By the procedure used in Exercise 8 of Section 7.1, we can extend Gauss' Theorem to any region that can be broken up into subregions of type 4. This includes all regions of interest to us here. As an example, consider the region between two closed surfaces, one inside the other. The surface of this region consists of two pieces oriented as shown in Figure 7.4.5. We shall apply the Divergence

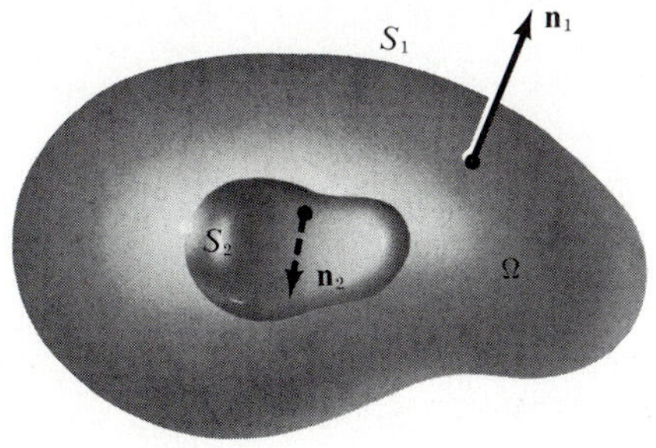

FIGURE 7.4.5
A more general region to which Gauss' Theorem applies.

Theorem to such a region when we prove Gauss' law in Example 6 below.

EXAMPLE 3. Consider $\mathbf{F} = 2x\mathbf{i} + y^2\mathbf{j} + z^2\mathbf{k}$. Let S be the unit sphere $x^2 + y^2 + z^2 = 1$. Evaluate $\int_S \mathbf{F} \cdot \mathbf{n}\, dS$.

By Gauss' Theorem,

$$\int_\Omega \operatorname{div} \mathbf{F} = \int_S \mathbf{F} \cdot \mathbf{n}\, dS$$

where Ω is the ball bounded by the sphere. The integral on the left is

$$2\int_\Omega (1 + y + z)\, dV = 2\int_\Omega dV + 2\int_\Omega y\, dV + 2\int_\Omega z\, dV$$

By symmetry we can argue that $\int_\Omega y\, dV = \int_\Omega z\, dV = 0$ (for example, see Exercise 11, Section 5.6). Thus

$$2\int_\Omega (1 + y + z)\, dV = 2\int_\Omega dV = \frac{8\pi}{3}$$

(since the unit ball has volume $4\pi/3$; see Example 1, Section 5.6). The reader can convince himself that direct computation of $\int_S \mathbf{F} \cdot \mathbf{n}\, dS$ proves unwieldy.

EXAMPLE 4. Use the Divergence Theorem to evaluate

$$\int_{\partial W} (x^2 + y + z)\, dS$$

where W is the solid ball $x^2 + y^2 + z^2 \leq 1$.

In order to apply Gauss' Divergence Theorem we must find some vector field

$$\mathbf{F} = F_1\mathbf{i} + F_2\,\mathbf{j} + F_3\,\mathbf{k}$$

on W with

$$\mathbf{F} \cdot \mathbf{n} = x^2 + y + z$$

At any point $(x, y, z) \in \partial W$ the outward unit normal $\mathbf{n}$ to ∂W is

$$\mathbf{n} = x\mathbf{i} + y\mathbf{j} + z\mathbf{k}$$

since on ∂W, $x^2 + y^2 + z^2 = 1$, and the radius vector $\mathbf{r} = x\mathbf{i} + y\mathbf{j} + z\mathbf{k}$ is normal to the sphere ∂W (Figure 7.4.6). Therefore, if $\mathbf{F}$ is the desired vector field, then

$$\mathbf{F} \cdot \mathbf{n} = F_1 x + F_2 y + F_3 z$$

We set

$$F_1 x = x^2, \qquad F_2 y = y, \qquad F_3 z = z$$

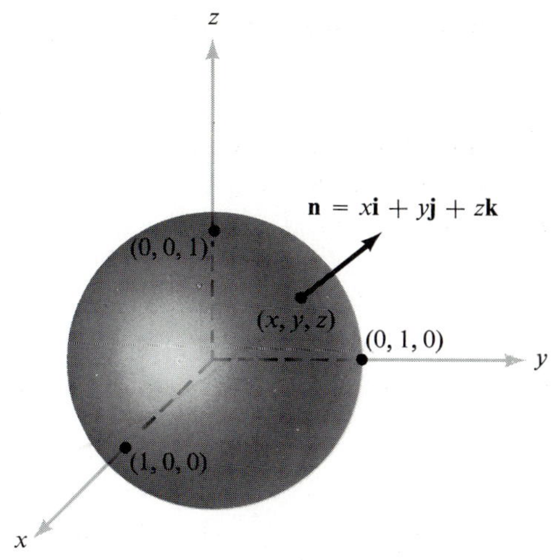

FIGURE 7.4.6
n *is the unit normal to* ∂W*, the boundary of the ball* W.

and solve for F_1, F_2, and F_3 to find that

$$\mathbf{F} = x\mathbf{i} + \mathbf{j} + \mathbf{k}$$

Computing div **F** we get

$$\text{div } \mathbf{F} = 1 + 0 + 0 = 1$$

Thus by Gauss' Divergence Theorem

$$\int_{\partial W} (x^2 + y + z)\, dS = \int_W dV = \text{volume } (W) = \tfrac{4}{3}\pi$$

The physical meaning of divergence is that at a point P, div $\mathbf{F}(P)$ is the rate of net outward flux at P per unit volume. This follows from Gauss' Theorem and the Mean Value Theorem for Integrals: If Ω_ρ is a ball in $\mathbb{R}^3$ of radius ρ centered at P, then there is a point $Q \in \Omega_\rho$ such that

$$\int_{\partial\Omega_\rho} \mathbf{F} \cdot \mathbf{n}\, dS = \int_{\Omega_\rho} \text{div } \mathbf{F}\, dV = \text{div } \mathbf{F}(Q) \cdot \text{volume } (\Omega_\rho)$$

and so

$$\text{div } \mathbf{F}(P) = \underset{\rho \to 0}{\text{limit}}\ \text{div } \mathbf{F}(Q) = \underset{\rho \to 0}{\text{limit}} \frac{1}{V(\Omega_\rho)} \int_{\partial\Omega_\rho} \mathbf{F} \cdot \mathbf{n}\, dS$$

This is analogous to the limit formulation of curl given at the end of Section 7.2. Thus, if div $\mathbf{F}(P) > 0$, we consider P to be a *source*, for there is a net outwards flow near P. If div $\mathbf{F}(P) < 0$, P is called a *sink* for $\mathbf{F}$.

A C^1 vector field $\mathbf{F}$ defined on $\mathbb{R}^3$ is called *divergence free* if div $\mathbf{F} = 0$. If $\mathbf{F}$ is divergence free, we have $\int_S \mathbf{F} = 0$ for all closed surfaces S. The converse can also be demonstrated readily using Gauss' Theorem: If $\int_S \mathbf{F} = 0$ for all closed surfaces S, then $\mathbf{F}$ is divergence free. If $\mathbf{F}$ is divergence free, we thus see that the flux of $\mathbf{F}$ across any closed surface S is 0, so if $\mathbf{F}$ is the velocity field of a fluid, the net amount of fluid to flow out of any region will be 0. Thus, exactly as much fluid must flow into the region as flows out (in unit time). A fluid with this property is therefore called *incompressible*.

EXAMPLE 5. Evaluate $\int_S \mathbf{F}$, where $\mathbf{F}(x, y, z) = xy^2\mathbf{i} + x^2y\mathbf{j} + y\mathbf{k}$ and S is the surface of the cylinder $x^2 + y^2 = 1$, $-1 < z < 1$, and $x^2 + y^2 \leq 1$ when $z = \pm 1$. Interpret physically.

One can compute this integral directly but, as in many other cases, it is easier to use the Divergence Theorem.

Now S is the boundary of the region Ω given by $x^2 + y^2 \leq 1$, $-1 \leq z \leq 1$. Thus $\int_S \mathbf{F} = \int_\Omega \text{div } \mathbf{F}$. Moreover,

$$\int_\Omega \text{div } \mathbf{F} = \int_\Omega (x^2 + y^2)\, dx\, dy\, dz = \int_{-1}^{1} \left(\int_{x^2+y^2 \leq 1} (x^2 + y^2)\, dx\, dy \right) dz$$

$$= 2 \int_{x^2+y^2 \leq 1} (x^2 + y^2)\, dx\, dy$$

Before evaluating the double integral, we note that $\int_{\partial\Omega} \mathbf{F} \cdot \mathbf{n}\, dS = 2 \int_{x^2+y^2 \leq 1} (x^2 + y^2)\, dx\, dy > 0$. This means that $\int_{\partial\Omega} \mathbf{F}$, the net flux of $\mathbf{F}$ out of the cylinder, is positive, which agrees with the fact that $0 \leq \text{div } \mathbf{F} = x^2 + y^2$ inside the cylinder.

We change variables to polar coordinates to evaluate the double integral:

$$x = r \cos \theta, \qquad y = r \sin \theta, \qquad 0 \leq r \leq 1, \qquad 0 \leq \theta \leq 2\pi$$

Hence, we have $\partial(x, y)/\partial(r, \theta) = r$ and $x^2 + y^2 = r^2$. Thus

$$\int_{x^2+y^2 \leq 1} (x^2 + y^2)\, dx\, dy = \int_0^{2\pi} \left(\int_0^1 r^3\, dr \right) d\theta = \tfrac{1}{2}\pi$$

Therefore, $\int_\Omega \text{div } \mathbf{F}\, dV = \pi$.

As we remarked above, Gauss' Divergence Theorem can be applied to regions in space more general than those of type 4. To conclude this section, we shall use this observation to prove an important result.

***Theorem 10.* (*Gauss' law*).** *Let M be a region in $\mathbb{R}^3$ of type 4. Then if $(0, 0, 0) \notin \partial M$ we have*

$$\int_{\partial M} \frac{\mathbf{r} \cdot \mathbf{n}}{r^3}\, dS = \begin{cases} 4\pi & \text{if } (0, 0, 0) \in M \\ 0 & \text{if } (0, 0, 0) \notin M \end{cases}$$

where

$$\mathbf{r}(x, y, z) = x\mathbf{i} + y\mathbf{j} + z\mathbf{k}$$

and

$$r(x, y, z) = \|\mathbf{r}(x, y, z)\| = \sqrt{x^2 + y^2 + z^2}$$

Proof. First suppose $(0, 0, 0) \notin M$. Then $\mathbf{r}/r^3$ is a C^1 vector field on M and ∂M, so by the Divergence Theorem

$$\int_{\partial M} \frac{\mathbf{r} \cdot \mathbf{n}}{r^3}\, dS = \int_M \nabla \cdot \left(\frac{\mathbf{r}}{r^3}\right) dV$$

But $\nabla \cdot (\mathbf{r}/r^3) = 0$ for $r \neq 0$, as the reader can easily verify (see Exercise 8, Section 3.5). Thus

$$\int_{\partial M} \frac{\mathbf{r} \cdot \mathbf{n}}{r^3}\, dS = 0$$

Now let us suppose $(0, 0, 0) \in M$. We can no longer use the above method, because $\mathbf{r}/r^3$ is not smooth on M, due to the singularity at $\mathbf{r} = (0, 0, 0)$. Since $(0, 0, 0) \in M$, there is an $\varepsilon > 0$ such that the ball N of radius ε centered at $(0, 0, 0)$ is contained completely inside M. Now let Ω be the region between M and N. Then Ω has boundary $\partial N \cup \partial M = S$. But the orientation on ∂N induced by the *outward* normal on Ω is the opposite of that obtained from N (see Figure 7.4.7). Now $\nabla \cdot (\mathbf{r}/r^3) = 0$ on Ω, so by the (generalized) Divergence Theorem,

$$\int_S \frac{\mathbf{r} \cdot \mathbf{n}}{r^3}\, dS = \int_\Omega \nabla \cdot \left(\frac{\mathbf{r}}{r^3}\right) = 0$$

Since

$$\int_S \frac{\mathbf{r} \cdot \mathbf{n}}{r^3}\, dS = \int_{\partial M} \frac{\mathbf{r} \cdot \mathbf{n}}{r^3}\, dS + \int_{\partial N} \frac{\mathbf{r} \cdot \mathbf{n}}{r^3}\, dS$$

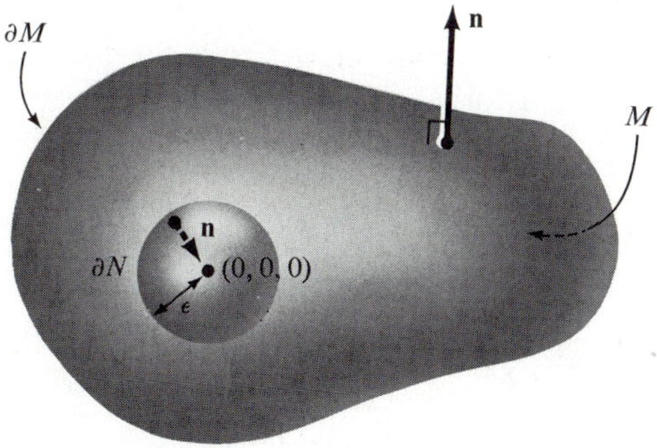

FIGURE 7.4.7
Induced outward orientation on S.

where **n** is the outward normal to S, we have

$$\int_{\partial M} \frac{\mathbf{r}\cdot\mathbf{n}}{r^3}\,dS = -\int_{\partial N} \frac{\mathbf{r}\cdot\mathbf{n}}{r^3}\,dS$$

(Note that these integrals of scalar functions are independent of orientation by Theorem 4, Section 6.6.)

Now on ∂N, $\mathbf{n} = -\mathbf{r}/r$ and $r = \varepsilon$, since ∂N is a sphere of radius ε, so

$$-\int_{\partial N} \frac{\mathbf{r}\cdot\mathbf{n}}{r^3}\,dS = \int_{\partial N} \frac{\varepsilon^2}{\varepsilon^4}\,dS = \frac{1}{\varepsilon^2}\int_{\partial N} dS$$

But $\int_{\partial N} dS = 4\pi\varepsilon^2$, the surface area of the sphere of radius ε. This proves the result. ∎

Gauss' law has the following physical interpretation. The potential due to a point charge Q at (0, 0, 0) is given by

$$\phi(x, y, z) = \frac{Q}{4\pi r} = \frac{Q}{4\pi\sqrt{x^2 + y^2 + z^2}}$$

and the corresponding electric field is

$$\mathbf{E} = -\nabla\phi = \frac{Q}{4\pi}\left(\frac{\mathbf{r}}{r^3}\right)$$

Thus Theorem 10 states that the total electric flux $\int_{\partial M} \mathbf{E}$ (that is, the flux of **E** out of a closed surface ∂M) equals Q if the charge lies inside M and zero otherwise. (A generalization is given in Exercise 10.) Note that even if $(0, 0, 0) \notin M$, **E** will still be nonzero on M.

For a continuous charge distribution described by a charge density ρ, the field **E** is related to the density ρ by

$$\operatorname{div} \mathbf{E} = \nabla \cdot \mathbf{E} = \rho$$

Thus by Gauss' Theorem,

$$\int_S \mathbf{E} = \int_\Omega \rho = Q$$

or the flux out of a surface is equal to the total charge inside.

EXERCISES

1. Let S be a closed surface. Then use Gauss' Theorem to show $\int_S \nabla \times \mathbf{F} = 0$. (Compare with Exercise 8 of Section 7.2.)
2. Let $\mathbf{F} = x^3\mathbf{i} + y^3\mathbf{j} + z^3\mathbf{k}$. Evaluate the surface integral of **F** over the unit sphere.
3. Evaluate $\int_{\partial\Omega} \mathbf{F}$, where $\mathbf{F} = x\mathbf{i} + y\mathbf{j} - z\mathbf{k}$ and Ω is the unit cube (in the first octant). Perform the calculation directly and check by using the Divergence Theorem.

4. Repeat Exercise 3 for (a) $\mathbf{F} = \mathbf{i} + \mathbf{j} + \mathbf{k}$; (b) $\mathbf{F} = x^2\mathbf{i} + y^2\mathbf{j} + z^2\mathbf{k}$.
5. Let $\mathbf{F} = y\mathbf{i} + z\mathbf{j} + xz\mathbf{k}$. Evaluate $\int_{\partial\Omega} \mathbf{F}$, where Ω is the set (x, y, z) with $x^2 + y^2 \leq z \leq 1$.
6. Repeat Exercise 5 for $\mathbf{F} = (x - y)\mathbf{i} + (y - z)\mathbf{j} + (z - x)\mathbf{k}$.
7. Let S be the surface of a region Ω. Show

$$\int_S \mathbf{r} \cdot \mathbf{n}\, dS = 3 \text{ volume}(\Omega)$$

Attempt to explain this geometrically. (HINT: Assume $(0, 0, 0) \in \Omega$ and consider the skew cone with its vertex at $(0, 0, 0)$ with base ΔS and altitude $\|\mathbf{r}\|$. Its volume is $\frac{1}{3}(\Delta S) \cdot (\mathbf{r} \cdot \mathbf{n})$.)

8. Evaluate $\int_S \mathbf{F} \cdot \mathbf{n}\, dS$, where $\mathbf{F} = 3xy^2\mathbf{i} + 3x^2y\mathbf{j} + z^3\mathbf{k}$ and S is the surface of the unit sphere.
9. Show $\int_\Omega 1/r^2\, dx\, dy\, dz = \int_{\partial\Omega} \mathbf{r} \cdot \mathbf{n}/r^2\, dS$ where $\mathbf{r} = x\mathbf{i} + y\mathbf{j} + z\mathbf{k}$.
10. Fix vectors $\mathbf{v}_1, \ldots, \mathbf{v}_k \in \mathbb{R}^3$ and numbers ("charges") $q_1, \ldots, q_k$. Set $\phi(x, y, z) = \sum_{i=1}^{k} q_i/(4\pi\|\mathbf{r} - \mathbf{v}_i\|)$, where $\mathbf{r} = (x, y, z)$. Show that for a closed surface S, and $\mathbf{E} = -\nabla\phi$,

$$\int_S \mathbf{E} = Q$$

where Q is the total charge inside S. (Assume Gauss' law from Theorem 10 and that none of the charges are on S.)

11. Prove *Green's identities*

$$\int_{\partial\Omega} f\nabla g \cdot \mathbf{n}\, dS = \int_\Omega [f\nabla^2 g + \nabla f \cdot \nabla g]\, dV$$

$$\int_{\partial\Omega} (f\nabla g - g\nabla f) \cdot \mathbf{n}\, dS = \int_\Omega (f\nabla^2 g - g\nabla^2 f)\, dV$$

12. Suppose $\mathbf{F}$ satisfies div $\mathbf{F} = 0$ and curl $\mathbf{F} = 0$. Show that we can write $\mathbf{F} = \nabla f$, where $\nabla^2 f = 0$.
13. Let ρ be a given (continuous) function on $\mathbb{R}^3$. Let p be a point in $\mathbb{R}^3$, and let $q \in \Omega$ be denoted by $(x, y, z) = q$. The *potential* of ρ is defined to be the function

$$\phi(p) = \int_\Omega \frac{\rho(q)}{4\pi\|p - q\|}\, dq$$

where $\|p - q\|$ is the distance between p and q. Argue that

$$\int_{\partial\Omega} \nabla\phi \cdot \mathbf{n}\, dS = -\int_\Omega \rho$$

for any region Ω, and hence ϕ satisfies *Poisson's equation*

$$\nabla^2\phi = -\rho$$

14. Suppose $\mathbf{F}$ is tangent to the closed surface S of a region Ω. Then prove

$$\int_\Omega \text{div}\, \mathbf{F} = 0$$

*15. Use Gauss' law and symmetry to prove that the electric field due to a charge Q evenly spread over the surface of a sphere is the same outside the surface as the field from a point charge Q located at the center of the sphere. What is the field inside the sphere?

*16. Reformulate Exercise 15 in terms of gravitational fields.

17. Show how Gauss' law can be used to solve part (b) of Exercise 20 in Section 7.3.

7.5 APPLICATIONS TO PHYSICS AND DIFFERENTIAL EQUATIONS*

We can apply the concepts developed in this chapter to the formulation of some physical theories. Let us first discuss an important equation that is referred to as a *conservation* equation. For fluids, it expresses the conservation of mass, and for electromagnetic theory, the conservation of charge. We shall apply the equation to heat conduction and to electromagnetism.

Let $\mathbf{V}(t, x, y, z)$ be a C^1 vector field on $\mathbb{R}^3$ for each t and let $\rho(t, x, y, z)$ be a C^1 real-valued function. By the *law of conservation of mass* for $\mathbf{V}$ and ρ, we shall mean that the condition

$$\frac{d}{dt}\int_\Omega \rho = -\int_{\partial\Omega} \mathbf{J}\cdot\mathbf{n}\, dS$$

holds for all regions Ω in $\mathbb{R}^3$, where $\mathbf{J} = \rho\mathbf{V}$.

If we think of ρ as a mass density (ρ could also be charge density), that is, the mass per unit volume, and $\mathbf{V}$ as the velocity field of a fluid, the condition just says that the rate of change of total mass in Ω equals the rate at which mass flows *into* Ω. Recall that $\int_{\partial\Omega} \mathbf{J}\cdot\mathbf{n}\, dS$ is called the flux of $\mathbf{J}$. We need the following result.

Theorem 11. *For $\mathbf{V}$ and ρ defined on $\mathbb{R}^3$, the law of conservation of mass for $\mathbf{V}$ and ρ is equivalent to the condition*

$$\operatorname{div}\mathbf{J} + \frac{\partial\rho}{\partial t} = 0 \tag{1}$$

that is

$$\rho \operatorname{div}\mathbf{V} + \mathbf{V}\cdot\nabla\rho + \frac{\partial\rho}{\partial t} = 0 \tag{1$'$}$$

NOTE. Here, div $\mathbf{J}$ means that we compute div $\mathbf{J}$ for t held fixed, and $\partial\rho/\partial t$ means we differentiate ρ with respect to t for x, y, z fixed.

Proof. First, observe that $(d/dt)\int \rho\, dx\, dy\, dz = \int (\partial\rho/\partial t)\, dx\, dy\, dz$, and

$$\int_{\partial\Omega} \mathbf{J}\cdot\mathbf{n}\, dS = \int_\Omega \operatorname{div}\mathbf{J}$$

* For additional examples the reader may profitably refer to H. M. Schey, *Div, Grad, Curl, and All That*, W. W. Norton, New York, 1973.

by the Divergence Theorem. Thus conservation of mass is equivalent to the condition

$$\int_{\Omega}\left(\operatorname{div}\mathbf{J}+\frac{\partial\rho}{\partial t}\right)dx\,dy\,dz=0$$

Since this is to hold for all regions Ω, this is equivalent to $\operatorname{div}\mathbf{J}+\partial\rho/\partial t=0$. ∎

The equation $\operatorname{div}\mathbf{J}+\partial\rho/\partial t=0$ is called the *equation of continuity*. This is not the only equation governing fluid motion and it does not determine the motion of the fluid, but is just one equation that must hold.

Notice that the fluids that this equation governs can be compressible. If $\operatorname{div}\mathbf{V}=0$ (incompressible case) and ρ is constant, equation (1′) follows automatically. But in general, even for incompressible fluids the equation is not automatic, because ρ can depend on (x, y, z) and t. Thus, while $\operatorname{div}\mathbf{V}=0$ may hold, $\operatorname{div}(\rho\mathbf{V})\neq 0$ may still be true.

We shall now turn our attention to the *heat equation*, one of the most important equations of applied mathematics. It has been, and remains, one of the prime motivations for the study of partial differential equations.

Let us argue intuitively. If $T(t, x, y, z)$ (a C^2 function) denotes the temperature in a body at time t, then ∇T represents the temperature gradient and heat "flows" with the vector field $-\nabla T=\mathbf{F}$. Note that ∇T points in the direction of increasing T (Chapter 3). Since heat flows from hot to cold, we have inserted a minus sign. The energy density, that is, the energy per unit volume, is $c\rho_0 T$, where c is a constant (specific heat) and ρ_0 is the mass density, assumed constant. (We accept these assertions from elementary physics.) The *energy flux vector* is $\mathbf{J}=\kappa\mathbf{F}$, where κ is a constant called the *conductivity*.

We propose that energy be conserved. Formally, this means that $\mathbf{J}$ and $\rho=c\rho_0 T$ should obey the law of conservation of mass with ρ playing the role of "mass"; that is,

$$\frac{d}{dt}\int_{\Omega}\rho=-\int_{\partial\Omega}\mathbf{J}\cdot\mathbf{n}\,dS$$

By Theorem 11 this assertion is equivalent to

$$\operatorname{div}\mathbf{J}+\frac{\partial\rho}{\partial t}=0$$

But $\operatorname{div}\mathbf{J}=\operatorname{div}(-\kappa\nabla T)=-\kappa\nabla^2 T$. (Recall that $\nabla^2 T=\partial^2 T/\partial x^2+\partial^2 T/\partial y^2+\partial^2 T/\partial z^2$ and ∇^2 is the Laplace operator.) Continuing, we have $\partial\rho/\partial t=\partial(c\rho_0 T)/\partial t=c\rho_0(\partial T/\partial t)$. Thus equation (1) becomes, in this case,

$$\frac{\partial T}{\partial t}=\frac{\kappa}{c\rho_0}\nabla^2 T=k\nabla^2 T \tag{2}$$

where $k=\kappa/c\rho_0$ is called the *diffusivity*. Equation (2) is the important heat equation.

This equation does completely govern the conduction of heat, unlike the flow of fluids, in the following sense. If $T(0, x, y, z)$ is a given initial temperature distribution, then a unique $T(t, x, y, z)$ is determined that satisfies equation (2). In other words the initial condition, at $t = 0$, gives us the result for $t > 0$. Notice that if T does not change with time (steady-state case) then we must have $\nabla^2 T = 0$ (Laplace's equation).

We shall now discuss *Maxwell's equations*, governing electromagnetic fields. The form of these equations depends on the physical units one is employing, and changing units introduces factors like 4π, the velocity of light, and so on. We assume in each case the system in which Maxwell's equations are simplest.

Let $\mathbf{E}(t, x, y, z)$ and $\mathbf{H}(t, x, y, z)$ be C^1 functions of (t, x, y, z) that are vector fields for each t. They satisfy (by definition) Maxwell's equations with charge density $\rho(t, x, y, z)$ and current density $\mathbf{J}(t, x, y, z)$ when the following hold:

$$\nabla \cdot \mathbf{E} = \rho \qquad \text{(Gauss' law)} \tag{3}$$

$$\nabla \cdot \mathbf{H} = 0 \qquad \text{(no magnetic sources)} \tag{4}$$

$$\nabla \times \mathbf{E} + \frac{\partial \mathbf{H}}{\partial t} = \mathbf{0} \qquad \text{(Faraday's law)} \tag{5}$$

and

$$\nabla \times \mathbf{H} - \frac{\partial \mathbf{E}}{\partial t} = \mathbf{J} \qquad \text{(Ampère's law)} \tag{6}$$

Of these laws, (3) and (5) were discussed earlier in Sections 7.4 and 7.2 in integral form; historically, they arose in these forms as physically observed laws. Ampère's law was mentioned for a special case in Example 12, Section 6.2.

Physically, one interprets $\mathbf{E}$ as the *electric field* and $\mathbf{H}$ as the *magnetic field*. As time t progresses, these fields interact with one another, charges and currents according to the above equations. For example, electromagnetic waves (light) propagate this way.

Since $\nabla \cdot \mathbf{H} = 0$ we can apply Theorem 8 of Section 7.3 to conclude that $\mathbf{H} = \nabla \times \mathbf{A}$ for some vector field $\mathbf{A}$. (We are assuming that $\mathbf{H}$ is defined on all of $\mathbb{R}^3$ for each time t.) This vector field $\mathbf{A}$ is not unique, and we can equally well use $\mathbf{A}' = \mathbf{A} + \nabla f$ for any function $f(t, x, y, z)$, since $\nabla \times \nabla f = \mathbf{0}$. For any such choice of $\mathbf{A}$, we have by (5)

$$\begin{aligned}\mathbf{0} = \nabla \times \mathbf{E} + \frac{\partial \mathbf{H}}{\partial t} &= \nabla \times \mathbf{E} + \frac{\partial}{\partial t} \nabla \times \mathbf{A} \\ &= \nabla \times \mathbf{E} + \nabla \times \frac{\partial \mathbf{A}}{\partial t} = \nabla \times \left(\mathbf{E} + \frac{\partial \mathbf{A}}{\partial t}\right)\end{aligned}$$

Hence applying Theorem 7, Section 7.3, there is a real-valued function ϕ on $\mathbb{R}^3$ such that

$$\mathbf{E} + \frac{\partial \mathbf{A}}{\partial t} = -\nabla \phi$$

Substituting this equation and $\mathbf{H} = \nabla \times \mathbf{A}$ into equation (6) and using the identity

$$\nabla \times (\nabla \times \mathbf{A}) = \nabla(\nabla \cdot \mathbf{A}) - \nabla^2 \mathbf{A}$$

we get

$$\mathbf{J} = \nabla \times \mathbf{H} - \frac{\partial \mathbf{E}}{\partial t} = \nabla \times (\nabla \times \mathbf{A}) - \frac{\partial}{\partial t}\left(-\frac{\partial \mathbf{A}}{\partial t} - \nabla\phi\right)$$

$$= \nabla(\nabla \cdot \mathbf{A}) - \nabla^2\mathbf{A} + \frac{\partial^2 \mathbf{A}}{\partial t^2} + \frac{\partial}{\partial t}(\nabla\phi)$$

Thus

$$\nabla^2\mathbf{A} - \frac{\partial^2 \mathbf{A}}{\partial t^2} = -\mathbf{J} + \nabla(\nabla \cdot \mathbf{A}) + \frac{\partial}{\partial t}(\nabla\phi)$$

that is

$$\nabla^2\mathbf{A} - \frac{\partial^2 \mathbf{A}}{\partial t^2} = -\mathbf{J} + \nabla\left(\nabla \cdot \mathbf{A} + \frac{\partial\phi}{\partial t}\right) \tag{7}$$

Again using the equation $\mathbf{E} + \partial\mathbf{A}/\partial t = -\nabla\phi$ and the equation $\nabla \cdot \mathbf{E} = \rho$, we obtain

$$\rho = \nabla \cdot \mathbf{E} = \nabla \cdot \left(-\nabla\phi - \frac{\partial \mathbf{A}}{\partial t}\right) = -\nabla^2\phi - \frac{\partial(\nabla \cdot \mathbf{A})}{\partial t}$$

that is

$$\nabla^2\phi = -\rho - \frac{\partial(\nabla \cdot \mathbf{A})}{\partial t} \tag{8}$$

Now let us exploit the freedom in our choice of **A**. We impose the "condition"

$$\nabla \cdot \mathbf{A} + \frac{\partial\phi}{\partial t} = 0 \tag{9}$$

We must be sure we can do this. Supposing we have a given $\mathbf{A}_0$ and a corresponding ϕ_0, can we choose a new $\mathbf{A} = \mathbf{A}_0 + \nabla f$ and then a new ϕ such that $\nabla \cdot \mathbf{A} + \partial\phi/\partial t = 0$? With this new **A**, the new ϕ is $\phi_0 - \partial f/\partial t$; we leave verification as an exercise for the reader. Condition (9) on f then becomes

$$0 = \nabla \cdot (\mathbf{A}_0 + \nabla f) + \frac{\partial(\phi_0 - \partial f/\partial t)}{\partial t} = \nabla \cdot \mathbf{A}_0 + \nabla^2 f + \frac{\partial\phi_0}{\partial t} - \frac{\partial^2 f}{\partial t^2}$$

or

$$\nabla^2 f - \frac{\partial^2 f}{\partial t^2} = -\left(\nabla \cdot \mathbf{A}_0 + \frac{\partial\phi_0}{\partial t}\right) \tag{10}$$

Thus to be able to choose **A** and ϕ satisfying $\nabla \cdot \mathbf{A} + \partial\phi/\partial t = 0$, we must be able to solve equation (10) for f. One can indeed do this under general conditions, although we do not prove it here. Equation (10) is called the *inhomogeneous wave equation.*

If we accept that **A** and ϕ can be chosen to satisfy $\nabla \cdot \mathbf{A} + \partial\phi/\partial t = 0$, then the equations (7) and (8) for **A** and ϕ become

$$\nabla^2\mathbf{A} - \frac{\partial^2 \mathbf{A}}{\partial t^2} = -\mathbf{J} \tag{7$'$}$$

$$\nabla^2\phi - \frac{\partial^2 \phi}{\partial t^2} = -\rho \tag{8$'$}$$

(8′) follows from (8) by substituting $-\partial\phi/\partial t$ for $\nabla \cdot \mathbf{A}$. Thus the wave equation appears again.

Conversely, if $\mathbf{A}$ and ϕ satisfy the equations $\nabla \cdot \mathbf{A} + \partial\phi/\partial t = 0$, $\nabla^2\phi - \partial^2\phi/\partial t^2 = -\rho$, and $\nabla^2\mathbf{A} - \partial^2\mathbf{A}/\partial t^2 = -\mathbf{J}$, then $\mathbf{E} = -\nabla\phi - \partial\mathbf{A}/\partial t$ and $\mathbf{H} = \nabla \times \mathbf{A}$ satisfy Maxwell's equations. *This procedure then reduces Maxwell's equations to a study of the wave equation.**

This is fortunate because the solutions to the wave equation have been well studied (one learns how to solve it in most courses in differential equations). To indicate the wavelike nature of the solutions, for example, observe that for any function f

$$\phi(t, x, y, z) = f(x - t)$$

solves the wave equation $\nabla^2\phi - (\partial^2\phi/\partial t^2) = 0$. This solution just propagates the graph of f like a wave; thus one might conjecture that solutions of Maxwell's equations are wavelike in nature. Historically, this was Maxwell's great achievement, and it soon led to Hertz's discovery of radio waves.

Next we shall show briefly how vector analysis can be used to solve differential equations by a method called "potential theory" or "the Green's-function method." The presentation will be quite informal; the reader may consult the aforementioned references (see preceding footnote) for further information.

Suppose we wish to solve Poisson's equation $\nabla^2 u = \rho$ for $u(x, y, z)$ where $\rho(x, y, z)$ is a given function (this equation arises from Gauss' law if $\mathbf{E} = \nabla u$).

A function $G(\mathbf{x}, \mathbf{y})$ that has the property

$$\nabla^2 G(\mathbf{x}, \mathbf{y}) = \delta(\mathbf{x} - \mathbf{y}) \tag{11}$$

(in this expression $\mathbf{y}$ is held fixed), that is, which solves the differential equation with ρ replaced by δ, is called the *Green's function* for this differential equation. Here $\delta(\mathbf{x} - \mathbf{y})$ represents the Dirac delta function, "defined" by†

(*i*) $\delta(\mathbf{x} - \mathbf{y}) = 0$, $\mathbf{x} \neq \mathbf{y}$

(*ii*) $\int_{\mathbb{R}^3} \delta(\mathbf{x} - \mathbf{y})\, d\mathbf{x} = 1$

It has the operational property: for any continuous function $f(x)$

$$\int_{\mathbb{R}^3} f(\mathbf{x})\delta(\mathbf{x} - \mathbf{y})\, d\mathbf{x} = f(\mathbf{y}) \tag{12}$$

This is sometimes called the *sifting property* of δ.

Theorem 12. *If $G(\mathbf{x}, \mathbf{y})$ satisfies the differential equation $\nabla^2 u = \rho$ with ρ replaced by $\delta(\mathbf{x} - \mathbf{y})$, then*

$$u(x) = \int G(\mathbf{x}, \mathbf{y})\rho(\mathbf{y})\, d\mathbf{y} \tag{13}$$

is a solution to $\nabla^2 u = \rho$.

* There are variations on this procedure. For further details see, for example, G. F. D. Duff and D. Naylor, *Differential Equations of Applied Mathematics*, Wiley & Sons, New York, 1966, or books on electromagnetic theory, such as J. D. Jackson, *Classical Electrodynamics*, Wiley & Sons, New York, 1962.

† This is not a precise definition; nevertheless, it is enough here to assume that δ is a symbolic expression with the operational property (12). See the references in the preceding footnote for a more careful definition of δ.

Proof. To see this, note that

$$\nabla^2 \int_{\mathbb{R}^3} G(\mathbf{x}, \mathbf{y})\rho(\mathbf{y})\, d\mathbf{y} = \int_{\mathbb{R}^3} (\nabla^2 G(\mathbf{x}, \mathbf{y}))\rho(\mathbf{y})\, d\mathbf{y}$$

$$= \int_{\mathbb{R}^3} \delta(\mathbf{x} - \mathbf{y})\rho(\mathbf{y})\, d\mathbf{y} \qquad \text{by (11)}$$

$$= \rho(\mathbf{x}) \qquad \text{by (12)} \quad \blacksquare$$

The "function" $\rho(\mathbf{x}) = \delta(\mathbf{x})$ represents a unit charge concentrated at a single point (see conditions (*i*) and (*ii*) above). Thus $G(\mathbf{x}, \mathbf{y})$ *represents the potential at* $\mathbf{x}$ *due to a charge placed at* $\mathbf{y}$. From Gauss' law in Theorem 10 we can infer that in all of space,

$$G(\mathbf{x}, \mathbf{y}) = \frac{-1}{4\pi\|\mathbf{x} - \mathbf{y}\|} \tag{14}$$

Thus, the solution of $\nabla^2 u = \rho$ is

$$u(\mathbf{x}) = \int_{\mathbb{R}^3} \frac{-\rho(\mathbf{y})}{4\pi\|\mathbf{x} - \mathbf{y}\|}\, d\mathbf{y} \tag{15}$$

by Theorem 12. (See also Exercise 13, p. 387, where $\phi = -u$.)

In two dimensions, one can similarly show that

$$G(\mathbf{x}, \mathbf{y}) = \frac{1}{2\pi} \log\|\mathbf{x} - \mathbf{y}\| \tag{16}$$

so the solution of $\nabla^2 u = \rho$ is

$$u(\mathbf{x}) = \frac{1}{2\pi} \int_{\mathbb{R}^2} \rho(\mathbf{y}) \log\|\mathbf{x} - \mathbf{y}\|\, d\mathbf{y} \tag{17}$$

We now turn to the problem of using Green's functions to solve Poisson's equation in a bounded region with given boundary conditions. To do this, we need Green's first and second identities, which can be obtained from the divergence theorem. We start with the identity

$$\int_V \nabla \cdot \mathbf{F}\, dV = \int_S \mathbf{F} \cdot \mathbf{n}\, dS$$

where V is a region in space, S is its boundary, and $\mathbf{n}$ is the outward unit normal vector at any point on S. Replacing $\mathbf{F}$ by $f\nabla g$, where f and g are scalar functions, we obtain

$$\int_V \nabla f \cdot \nabla g\, dV + \int_V f\nabla^2 g\, dV = \int_S f \frac{\partial g}{\partial n}\, dS \tag{18}$$

where $\partial g/\partial n = \nabla g \cdot \mathbf{n}$. This is *Green's first identity*. If we simply permute f and g, and subtract the result from the above equation, we obtain *Green's second identity*

$$\int_V (f\nabla^2 g - g\nabla^2 f)\, dV = \int_S \left(f \frac{\partial g}{\partial n} - g \frac{\partial f}{\partial n} \right) dS \tag{19}$$

It is this identity that we shall use.

Consider Poisson's equation

$$\nabla^2 u = \rho$$

in some region V, and the corresponding equation for the Green's function

$$\nabla^2 G(\mathbf{x}, \mathbf{y}) = \delta(\mathbf{x} - \mathbf{y})$$

Inserting u and G into (19) we obtain

$$\int_V (u\nabla^2 G - G\nabla^2 u)\, dV = \int_S \left(u\frac{\partial G}{\partial n} - G\frac{\partial u}{\partial n}\right) dS$$

which becomes

$$\int_V [u(\mathbf{x})\delta(\mathbf{x} - \mathbf{y}) - G(\mathbf{x}, \mathbf{y})\rho(\mathbf{x})]\, d\mathbf{x} = \int_S \left(u\frac{\partial G}{\partial n} - G\frac{\partial u}{\partial n}\right) dS$$

that is

$$u(\mathbf{y}) = \int_V G(\mathbf{x}, \mathbf{y})\rho(\mathbf{x})\, d\mathbf{x} + \int_S \left(u\frac{\partial G}{\partial n} - G\frac{\partial u}{\partial n}\right) dS \tag{20}$$

Note that for an unbounded region, this becomes identical to our previous result (15) for all of space. Equation (20) enables us to solve for u in a bounded region where $\rho = 0$ by incorporating the conditions that u must obey on S. If $\rho = 0$, this reduces to

$$u = \int_S \left(u\frac{\partial G}{\partial n} - G\frac{\partial u}{\partial n}\right) dS$$

or fully

$$u(\mathbf{y}) = \int_S \left[u(\mathbf{x})\frac{\partial G}{\partial n}(\mathbf{x}, \mathbf{y}) - G(\mathbf{x}, \mathbf{y})\frac{\partial u}{\partial n}(\mathbf{x})\right] dS(\mathbf{x}) \tag{21}$$

where u appears on both sides of the equation. The crucial point is that evaluation of the integral requires only that we know the behavior of u on S. Commonly either u is given on the boundary (*Dirichlet problem*) or $\partial u/\partial n$ is given on the boundary (*Neumann problem*). If we know u on the boundary, we want to make $G\, \partial u/\partial n$ vanish on the boundary so we can evaluate the integral. So if u is given on S we must find a G such that $G(\mathbf{x}, \mathbf{y})$ vanishes whenever $\mathbf{x}$ lies on S. This is called the *Dirichlet Green's function for the region V*. Conversely, if $\partial u/\partial n$ is given on S we must find a G such that $\partial G/\partial n$ vanishes on S. This is the *Neumann Green's function.*

Doing any particular Dirichlet or Neumann problem thus becomes the task of finding the appropriate Green's function. We shall do this by modifying the Green's function for Laplace's equations on all $\mathbb{R}^2$ or $\mathbb{R}^3$, namely (14) and (16).

As an example, we shall now use the two-dimensional Green's-function method to construct the Dirichlet Green's function for the disc of radius R (see Figure 7.5.1). This will enable us to solve $\nabla^2 u = 0$ (or $\nabla^2 u = \rho$) with u given on the boundary circle.

In Figure 7.5.1 we have drawn the point $\mathbf{x}$ on the circumference because that is where we want G to vanish. The Green's function $G(\mathbf{x}, \mathbf{y})$ that we shall find will, of course, be valid for all $\mathbf{x}$, $\mathbf{y}$ in the disc. The point $\mathbf{y}'$ represents the

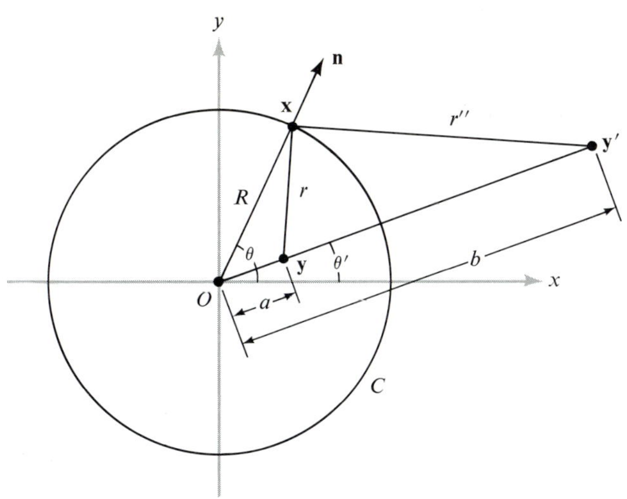

FIGURE 7.5.1
Geometry of the construction of the Green's function for a disc.

"reflection" of the point y into the region outside the circle, such that $ab = R^2$. Now when $\mathbf{x} \in C$, by the similarity of the triangles $\mathbf{x}O\mathbf{y}$ and $\mathbf{x}O\mathbf{y}'$

$$\frac{r}{R} = \frac{r''}{b}$$

or

$$r = \frac{r''R}{b} = \frac{r''a}{R}$$

So if we choose our Green's function as

$$G(\mathbf{x}, \mathbf{y}) = \frac{1}{2\pi}\left(\log r - \log \frac{r''a}{R}\right) \tag{22}$$

we see that G is zero if $\mathbf{x}$ is on C. If we can show that G satisfies $\nabla^2 G = \delta(\mathbf{x} - \mathbf{y})$ in the circle, then we will have proved that G is indeed the Dirichlet Green's function. From (16) we know that $\nabla^2(\log r/2\pi) = \delta(\mathbf{x} - \mathbf{y})$, so

$$\nabla^2 G(\mathbf{x}, \mathbf{y}) = \left[\delta(\mathbf{x} - \mathbf{y}) - \frac{a}{R}\,\delta(\mathbf{x} - \mathbf{y}')\right],$$

but $\mathbf{y}'$ is always outside the circle, so $\mathbf{x}$ can never be equal to $\mathbf{y}'$ and so $\delta(\mathbf{x} - \mathbf{y}')$ is always zero. Thus

$$\nabla^2 G(\mathbf{x}, \mathbf{y}) = \delta(\mathbf{x} - \mathbf{y})$$

and this G is the Dirichlet Green's function for the circle.

Now we shall consider the problem of solving

$$\nabla^2 u = 0$$

in this circle if $u(R, \theta) = f(\theta)$ is the boundary condition. By (21) we have a solution

$$u = \int_C \left(u \frac{\partial G}{\partial n} - G \frac{\partial u}{\partial n} \right) dS$$

but $G = 0$ on C, so we are left with the integral

$$u = \int_C u \frac{\partial G}{\partial n} \, dS$$

where we can replace u by $f(\theta)$ since the integral is around C. Thus the task of solving the Dirichlet problem in the circle is reduced to finding $\partial G/\partial n$. From (22) we can write

$$\frac{\partial G}{\partial n} = \frac{1}{2\pi} \left(\frac{1}{r} \frac{\partial r}{\partial n} - \frac{1}{r''} \frac{\partial r''}{\partial n} \right)$$

Now

$$\frac{\partial r}{\partial n} = \nabla r \cdot \mathbf{n}$$

and

$$\nabla r = \frac{\mathbf{r}}{r}$$

where $\mathbf{r} = \mathbf{x} - \mathbf{y}$, so

$$\frac{\partial r}{\partial n} = \frac{\mathbf{r} \cdot \mathbf{n}}{r} = \frac{nr \cos(nr)}{r} = \cos(nr)$$

where (nr) represents the angle between $\mathbf{n}$ and $\mathbf{r}$. Likewise

$$\frac{\partial r''}{\partial n} = \cos(nr'')$$

Now, in triangle $\mathbf{xy}O$, we have, by the cosine law

$$a^2 = r^2 + R^2 - 2rR \cos(nr)$$

and in triangle $\mathbf{xy}'O$, we get

$$b^2 = (r'')^2 + R^2 - 2r''R \cos(nr'')$$

and so

$$\frac{\partial r}{\partial n} = \cos(nr) = \frac{R^2 + r^2 - a^2}{2rR}$$

and

$$\frac{\partial r''}{\partial n} = \cos(nr'') = \frac{R^2 + (r'')^2 - b^2}{2r''R}$$

Hence

$$\frac{\partial G}{\partial n} = \frac{1}{2\pi} \left[\frac{R^2 + r^2 - a^2}{2r^2R} - \frac{R^2 + (r'')^2 - b^2}{2(r'')^2R} \right]$$

and by using the relationship between r and r'' when $\mathbf{x}$ is on C

$$\left.\frac{\partial G}{\partial n}\right|_{x \in C} = \frac{1}{2\pi}\left(\frac{R^2 - a^2}{Rr^2}\right)$$

So the solution can be written as

$$u = \frac{1}{2\pi}\int_C f(\theta)\frac{R^2 - a^2}{Rr^2}\,ds$$

Let us write this in a more explicit and tractable form. First, note that in triangle $\mathbf{xy}O$, we can write

$$r = [a^2 + R^2 - 2aR\cos(\theta - \theta')]^{1/2}$$

where θ and θ' are the polar angles in $\mathbf{x}$- and $\mathbf{y}$-space, respectively. Second, our solution must be valid for all $\mathbf{y}$ in the circle; hence the distance of $\mathbf{y}$ from the origin must now become a variable, which we shall call r'. Finally, note that $ds = R\,d\theta$ on C, so we can write the solution in polar coordinates as

$$u(r', \theta') = \frac{R^2 - (r')^2}{2\pi}\int_0^{2\pi}\frac{f(\theta)\,d\theta}{(r')^2 + R^2 - 2r'R\cos(\theta - \theta')}$$

This is known as *Poisson's formula in two dimensions.** As an exercise, the reader should write down the solution of $\nabla^2 u = \rho$ with u a given function $f(\theta)$ on the boundary.

The Dirichlet problem in the sphere of radius R has the Green's function

$$G(\mathbf{x}, \mathbf{y}) = \frac{1}{4\pi}\left(\frac{1}{r} - \frac{R}{ar''}\right)$$

where the same diagram applies. Poisson's formula in three dimensions is

$$u(r', \theta', \phi') = \frac{R(R^2 - (r')^2)}{4\pi}\int_0^{2\pi}\int_0^{\pi}\frac{f(\theta, \phi)\sin\theta\,d\theta\,d\phi}{(R^2 + (r')^2 - 2Rr'\cos\gamma)^{3/2}}$$

where $f(\theta, \phi)$ is the boundary condition and γ is the angle between r' and R. The ambitious reader can work out this case as an exercise.

7.6 DIFFERENTIAL FORMS

The theory of differential forms provides a convenient and elegant way of phrasing Green's, Stokes', and Gauss' theorems. In fact, the use of differential forms shows that these theorems are all manifestations of a single underlying mathematical theory, and provides the necessary language to generalize them to n dimensions. In this section we shall give a very elementary exposition of the theory of forms. Since our primary goal is to show that the theorems of Green, Stokes, and Gauss are all manifestations of the same phenomenon, we shall be satisfied with less than the strongest possible version of these theorems. Moreover, we shall introduce forms in a purely axiomatic and non-

* There are several ways of deriving this famous formula. For the method of complex variables, see J. Marsden, *Basic Complex Analysis*, W. H. Freeman and Company, San Francisco, 1973, p. 145. For the method of Fourier series, see J. Marsden, *Elementary Classical Analysis*, W. H. Freeman and Company, 1974, p. 466.

constructive manner, thereby avoiding the tremendous amount of formal algebraic preliminaries that is usually required for their construction. Thus to the purist our approach will be far from complete, but to the student it may be comprehensible. We hope that this will motivate some students to delve further into the theory of differential forms.

We shall begin by introducing the notion of a 0-form and a 1-form.

Definition. *Let K be an open set in* $\mathbb{R}^3$. *A **0-form** on K is a real-valued function* $f: K \to \mathbb{R}$. *When we differentiate f once, it is assumed to be* C^1, *and* C^2 *when we differentiate twice.*

Given two 0-forms f_1 and f_2 on K, we can add them in the usual way to get a new 0-form $f_1 + f_2$ or multiply them to get a 0-form $f_1 f_2$.

EXAMPLE 1. $f_1(x, y, z) = xy + yz$ and $f_2(x, y, z) = y \sin xz$ are 0-forms on $\mathbb{R}^3$.

$$(f_1 + f_2)(x, y, z) = xy + yz + y \sin xz$$

and

$$(f_1 f_2)(x, y, z) = y^2 x \sin xz + y^2 z \sin xz.$$

Definition. *The **basic** 1-forms are the expressions dx, dy, and dz. At present we consider these to be only formal symbols. A **1-form** ω on an open set K is a formal linear combination*

$$\omega = P(x, y, z)\, dx + Q(x, y, z)\, dy + R(x, y, z)\, dz$$

or simply

$$\omega = P\, dx + Q\, dy + R\, dz$$

where P, Q, and R are real-valued functions on K. By the expression P dx we mean the 1-form $P\,dx + 0 \cdot dy + 0 \cdot dz$, *and similarly for Q dy and R dz. Also the order of P dx, Q dy, and R dz is immaterial, so*

$$P\, dx + Q\, dy + R\, dz = R\, dz + P\, dx + Q\, dy \text{ etc.}$$

Given two 1-forms $\omega_1 = P_1\, dx + Q_1\, dy + R_1\, dz$ *and* $\omega_2 = P_2\, dx + Q_2\, dy + R_2\, dz$, *we can add them to get a new 1-form* $\omega_1 + \omega_2$ *defined by*

$$\omega_1 + \omega_2 = (P_1 + P_2)\, dx + (Q_1 + Q_2)\, dy + (R_1 + R_2)\, dz$$

and given a 0-form f, we can form the 1-form $f\omega_1$ *defined by*

$$f\omega_1 = (fP_1)\, dx + (fQ_1)\, dy + (fR_1)\, dz$$

EXAMPLE 2. Let $\omega_1 = (x + y^2)\, dx + (zy)\, dy + (e^{xyz})\, dz$ and $\omega_2 = \sin y\, dx + \sin x\, dy$ be 1-forms. Then

$$\omega_1 + \omega_2 = (x + y^2 + \sin y)\, dx + (zy + \sin x)\, dy + (e^{xyz})\, dz$$

If $f(x, y, z) = x$ then

$$f\omega_2 = x \sin y\, dx + x \sin x\, dy$$

Definition. *The **basic** 2-forms are the formal expressions dx dy, dy dz, and dz dx. These expressions should be thought of as products of dx and dy, dy and dz, and dz and dx.*

*A **2-form** η on K is a formal expression*

$$\eta = F\, dx\, dy + G\, dy\, dz + H\, dz\, dx$$

where F, G, and H are real-valued functions on K. The order of F dx dy, G dy dz, and H dz dx is immaterial; for example

$$F\, dx\, dy + G\, dy\, dz + H\, dz\, dx = H\, dz\, dx + F\, dx\, dy + G\, dy\, dz \text{ etc.}$$

At this point it is useful to note that in a 2-form the basic 1-forms dx, dy, and dz always appear in cyclic pairs (see Figure 7.6.1), that is, $dx\, dy$, $dy\, dz$, and $dz\, dx$.

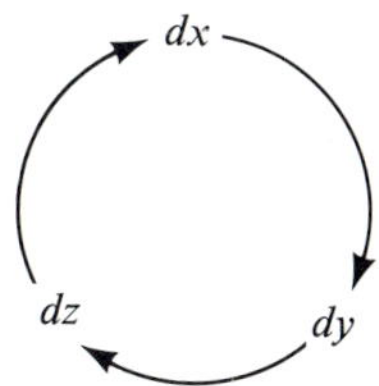

FIGURE 7.6.1
The cyclic order of dx, dy, and dz.

By analogy with 0-forms and 1-forms we can add two 2-forms

$$\eta_i = F_i\, dx\, dy + G_i\, dy\, dz + H_i\, dz\, dx,$$

$i = 1$ and 2, to obtain a new 2-form

$$\eta_1 + \eta_2 = (F_1 + F_2)\, dx\, dy + (G_1 + G_2)\, dy\, dz + (H_1 + H_2)\, dz\, dx$$

Similarly, if f is a 0-form and η is a 2-form we can take the product

$$f\eta = (fF)\, dx\, dy + (fG)\, dy\, dz + (fH)\, dz\, dx$$

Finally, by the expression $F\, dx\, dy$ we mean the 2-form $F\, dx\, dy + 0 \cdot dy\, dz + 0 \cdot dz\, dx$.

EXAMPLE 3. The expressions

$$\eta_1 = x^2\, dx\, dy + y^3x\, dy\, dz + \sin zy\, dz\, dx$$

and

$$\eta_2 = y\, dy\, dz$$

are 2-forms. Their sum is

$$\eta_1 + \eta_2 = x^2\, dx\, dy + (y^3x + y)\, dy\, dz + \sin zy\, dz\, dx$$

If $f(x, y, z) = xy$, then

$$f\eta_2 = xy^2\, dy\, dz$$

Definition. *A **basic** 3-form is a formal expression dx dy dz (on cyclic order, see Figure 7.6.1). A **3-form** ν on an open set $K \subset \mathbb{R}^3$ is an expression of the form $\nu = f(x, y, z)\, dx\, dy\, dz$, where f is a real-valued function on K.*

We can add two 3-forms and we can multiply them by 0-forms in the obvious way. There seems to be little difference between a 0-form and a 3-form, since both involve a single real-valued function. But we distinguish them for a purpose that will become clear when we multiply and differentiate forms.

EXAMPLE 4. Let $v_1 = y\,dx\,dy\,dz$, $v_2 = e^{x^2}\,dx\,dy\,dz$, and $f(x, y, z) = xyz$. Then $v_1 + v_2 = (y + e^{x^2})\,dx\,dy\,dz$ and $fv_1 = y^2xz\,dx\,dy\,dz$.

WARNING. Although we can add two 0-forms, two 1-forms, two 2-forms, or two 3-forms, we *never* add a k-form and a j-form if $k \neq j$. For example, we never write

$$f(x, y, z)\,dx\,dy + g(x, y, z)\,dz$$

Now that we have defined these formal objects (forms), one can legitimately ask what they are good for, how they are used, and, perhaps most important, what they mean. The answer to the first question will become clear as we proceed, but we can immediately describe how to use and interpret them.

A real-valued function on a domain K in $\mathbb{R}^3$ is a rule that assigns to each point in K a real number. Differential forms are, in some sense, generalizations of the real-valued functions we have studied in calculus. In fact, 0-forms on an open set K are just functions on K. Thus a 0-form f takes points in K to real numbers.

We should like to interpret differential k-forms (for $k \geq 1$) not as functions on points in K, but as functions on geometric objects such as curves and surfaces. Many of the early Greek geometers viewed lines and curves as being made up of infinitely many points, and planes and surfaces as being made up of infinitely many curves. Consequently there is at least some historical justification for applying this geometric hierarchy to the interpretation of differential forms.

Given an open subset $K \subset \mathbb{R}^3$, we shall distinguish four types of subsets of K (see Figure 7.6.2):.

(*i*) points in K,

(*ii*) oriented simple curves and oriented simple closed curves C in K,

(*iii*) oriented surfaces $S \subset K$,

(*iv*) elementary subregions (of types 1 to 4) $R \subset K$.

We shall begin with 1-forms. Let

$$\omega = P(x, y, z)\,dx + Q(x, y, z)\,dy + R(x, y, z)\,dz$$

be a 1-form on K and let C be an oriented simple curve as in Figure 7.6.2. The real number that ω assigns to C is given by the formula*

$$\int_C \omega = \int_C P(x, y, z)\,dx + Q(x, y, z)\,dy + R(x, y, z)\,dz \tag{1}$$

* Recall (see Section 6.2) that this integral is evaluated as follows. Let $\boldsymbol{\sigma}\colon [a, b] \to K$, $\boldsymbol{\sigma}(t) = (x(t), y(t), z(t))$, be an orientation-preserving parametrization of C. Then

$$\int_C \omega = \int_{\boldsymbol{\sigma}} \omega = \int_a^b \left[P(x(t), y(t), z(t)) \cdot \frac{dx}{dt} + Q(x(t), y(t), z(t)) \cdot \frac{dy}{dt} + R(x(t), y(t), z(t)) \cdot \frac{dz}{dt}\right] dt$$

Theorem 1 of Section 6.2 guarantees that $\int_C \omega$ does not depend on the choice of the parametrization $\boldsymbol{\sigma}$.

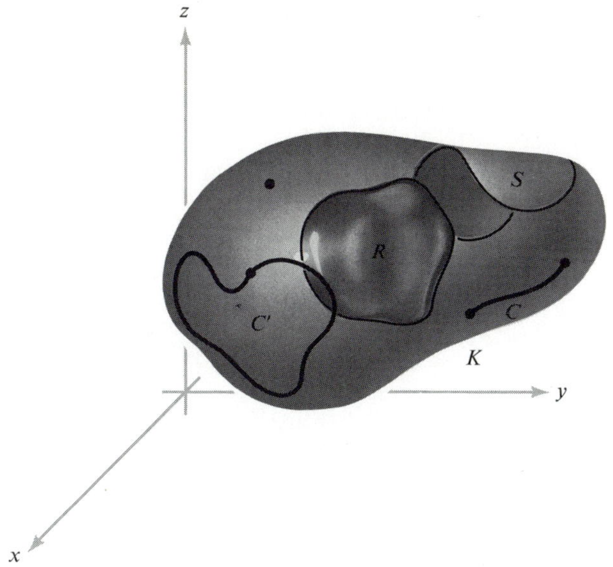

FIGURE 7.6.2
The four geometric types of subsets of an open set $K \subset \mathbb{R}^3$ to which our theory of forms applies.

We can thus interpret a 1-form ω on K as a rule assigning a real number to each curve $C \subset K$; a 2-form η will similarly be seen to be a rule assigning a real number to each surface $S \subset K$; and a 3-form ν as a rule assigning a real number to each elementary subregion of K. The rules for associating real numbers to curves, surfaces, and regions are completely contained in the formal expressions we have defined.

EXAMPLE 5. Let $\omega = xy\,dx + y^2\,dy + dz$ be a 1-form on $\mathbb{R}^3$ and let C be the oriented simple curve in $\mathbb{R}^3$ described by the parametrization $\boldsymbol{\sigma}(t) = (t^2, t^3, 1)$, $0 \le t \le 1$. C is oriented by choosing the positive direction of C to be the direction in which $\boldsymbol{\sigma}(t)$ traverses C as t goes from 0 to 1. Then by formula (1)

$$\int_C \omega = \int_0^1 [t^5(2t) + t^6(3t^2) + 0]\,dt = \int_0^1 [2t^6 + 3t^8]\,dt = \tfrac{13}{21}$$

Thus this 1-form ω assigns to each oriented simple curve and each oriented simple closed curve C in $\mathbb{R}^3$ the number $\int_C \omega$.

A 2-form η on an open set $K \subset \mathbb{R}^3$ may similarly be interpreted as a function that associates with each oriented surface $S \subset K$ a real number. This is accomplished by means of the notion of integration of 2-forms over surfaces. Let

$$\eta = F(x, y, z)\,dx\,dy + G(x, y, z)\,dy\,dz + H(x, y, z)\,dz\,dx$$

be a 2-form on K, and let $S \subset K$ be an oriented surface parametrized by a function $\boldsymbol{\Phi}\colon D \to \mathbb{R}^3$, $D \subset \mathbb{R}^2$, $\boldsymbol{\Phi}(u, v) = (x(u, v), y(u, v), z(u, v))$ (see Section 6.3).

Definition. *If S is such a surface and η is a 2-form on K we define $\int_S \eta$ by the formula*

$$
\begin{aligned}
\int_S \eta &= \int_S F\,dx\,dy + G\,dy\,dz + H\,dz\,dx \\
&= \int_D \left[F(x(u,v), y(u,v), z(u,v)) \cdot \frac{\partial(x,y)}{\partial(u,v)} \right. \\
&\quad + G(x(u,v), y(u,v), z(u,v)) \cdot \frac{\partial(y,z)}{\partial(u,v)} \\
&\quad \left. + H(x(u,v), y(u,v), z(u,v)) \cdot \frac{\partial(z,x)}{\partial(u,v)} \right] du\,dv
\end{aligned}
\tag{2}
$$

where

$$
\frac{\partial(x,y)}{\partial(u,v)} = \begin{vmatrix} \dfrac{\partial x}{\partial u} & \dfrac{\partial x}{\partial v} \\[2mm] \dfrac{\partial y}{\partial u} & \dfrac{\partial y}{\partial v} \end{vmatrix}, \quad
\frac{\partial(y,z)}{\partial(u,v)} = \begin{vmatrix} \dfrac{\partial y}{\partial u} & \dfrac{\partial y}{\partial v} \\[2mm] \dfrac{\partial z}{\partial u} & \dfrac{\partial z}{\partial v} \end{vmatrix}, \quad
\frac{\partial(z,x)}{\partial(u,v)} = \begin{vmatrix} \dfrac{\partial z}{\partial u} & \dfrac{\partial z}{\partial v} \\[2mm] \dfrac{\partial x}{\partial u} & \dfrac{\partial x}{\partial v} \end{vmatrix}
$$

If S is composed of several pieces S_i, $i = 1, \ldots, k$, as in Figure 7.4.4, each with its own parametrization $\mathbf{\Phi}_i$, we define

$$\int_S \eta = \sum_{i=1}^{k} \int_{S_i} \eta$$

One must verify that $\int_S \eta$ does not depend on the choice of parametrization $\mathbf{\Phi}$. This result is essentially (but not obviously) contained in Theorem 4, Section 6.6.

EXAMPLE 6. Let $\eta = z^2\,dx\,dy$ be a 2-form on $\mathbb{R}^3$, and let S be the upper unit hemisphere in $\mathbb{R}^3$. Find $\int_S \eta$.

Let us parametrize S by

$$\mathbf{\Phi}(u, v) = (\sin u \cos v, \sin u \sin v, \cos u),$$

where $(u, v) \in D = [0, \pi/2] \times [0, 2\pi]$. So by formula (2)

$$\int_S \eta = \int_D \cos^2 u \left[\frac{\partial(x,y)}{\partial(u,v)} \right] du\,dv$$

where

$$
\begin{aligned}
\frac{\partial(x,y)}{\partial(u,v)} &= \begin{vmatrix} \cos u \cos v & -\sin u \sin v \\ \cos u \sin v & \sin u \cos v \end{vmatrix} \\
&= \sin u \cos u \cos^2 v + \cos u \sin u \sin^2 v = \sin u \cos u
\end{aligned}
$$

Therefore

$$
\begin{aligned}
\int_S \eta &= \int_D \cos^2 u \cos u \sin u\,du\,dv \\
&= \int_0^{2\pi} \int_0^{\pi/2} \cos^3 u \sin u\,du\,dv = \int_0^{2\pi} \left[-\frac{\cos^4 u}{4} \right]_0^{\pi/2} dv = \frac{\pi}{2}
\end{aligned}
$$

EXAMPLE 7. Evaluate $\int_S x\,dy\,dz + y\,dx\,dy$, where S is the oriented surface described by the parametrization $x = u + v$, $y = u^2 - v^2$, $z = uv$, $(u, v) \in D = [0, 1] \times [0, 1]$.

By definition we have

$$\frac{\partial(y, z)}{\partial(u, v)} = \begin{vmatrix} 2u & -2v \\ v & u \end{vmatrix} = 2(u^2 + v^2)$$

$$\frac{\partial(x, y)}{\partial(u, v)} = \begin{vmatrix} 1 & 1 \\ 2u & -2v \end{vmatrix} = -2(u + v)$$

Consequently

$$\begin{aligned}\int_S x\,dy\,dz + y\,dx\,dy &= \int_D [(u + v)(2)(u^2 + v^2) + (u^2 - v^2)(-2)(u + v)]\,du\,dv \\ &= 4\int_D [v^3 + uv^2]\,du\,dv = 4\int_0^1\int_0^1 [v^3 + uv^2]\,du\,dv \\ &= 4\int_0^1 \left[uv^3 + \frac{u^2v^2}{2}\right]_0^1 dv = 4\int_0^1 \left[v^3 + \frac{v^2}{2}\right] dv \\ &= \left[v^4 + \frac{2v^3}{3}\right]_0^1 = 1 + \frac{2}{3} = \frac{5}{3}\end{aligned}$$

Finally, we must interpret 3-forms as functions on the elementary subregions (of types 1 to 4) of K. Let $\nu = f(x, y, z)\,dx\,dy\,dz$ be a 3-form and let $R \subset K$ be an elementary subregion of K. Then to each such $R \subset K$ we assign the number

$$\int_R \nu = \int_R f(x, y, z)\,dx\,dy\,dz \tag{3}$$

which is just the ordinary triple integral of f over R, as described in Section 5.6.

EXAMPLE 8. Suppose $\nu = (x + z)\,dx\,dy\,dz$ and $R = [0, 1] \times [0, 1] \times [0, 1]$. Then

$$\begin{aligned}\int_R \nu &= \int_R (x + z)\,dx\,dy\,dz = \int_0^1\int_0^1\int_0^1 (x + z)\,dx\,dy\,dz \\ &= \int_0^1\int_0^1 \left[\frac{x^2}{2} + zx\right]_0^1 dy\,dz = \int_0^1\int_0^1 \left[\frac{1}{2} + z\right] dy\,dz = \int_0^1 \left[\frac{1}{2} + z\right] dz \\ &= \left[\frac{z}{2} + \frac{z^2}{2}\right]_0^1 = 1\end{aligned}$$

We must now discuss the algebra (or rules of multiplication) of forms, which, together with differentiation of forms, will enable us to state Green's, Stokes', and Gauss' theorems in terms of differential forms.

If ω is a k-form and η is an l-form on K, $0 \le k + l \le 3$ there is a product called the *wedge product* $\omega \wedge \eta$ of ω and η, which is a $k + l$ form on K. The wedge product satisfies the following laws:

(*i*) For each k there is a zero k-form 0 with the property that $0 + \omega = \omega$ for all k-forms ω, and $0 \wedge \eta = 0$ for all l-forms η if $0 \le k + l \le 3$.

(*ii*) (*distributivity*) If f is a 0-form, then

$$(f\omega_1 + \omega_2) \wedge \eta = f(\omega_1 \wedge \eta) + (\omega_2 \wedge \eta)$$

(*iii*) (*anticommutativity*) $\omega \wedge \eta = (-1)^{kl}(\eta \wedge \omega)$.

(*iv*) (*associativity*) If ω_1, ω_2, ω_3 are k_1, k_2, k_3 forms, respectively, with $k_1 + k_2 + k_3 \leq 3$, then

$$\omega_1 \wedge (\omega_2 \wedge \omega_3) = (\omega_1 \wedge \omega_2) \wedge \omega_3$$

(*v*) (*homogeneity with respect to functions*) If f is a 0-form, then

$$\omega \wedge (f\eta) = (f\omega) \wedge \eta = f(\omega \wedge \eta)$$

Notice that (*ii*) and (*iii*) actually imply (*v*).

(*vi*) The following multiplication rules for 1-forms hold

$$dx \wedge dy = dx\, dy$$

$$dy \wedge dx = -dx\, dy = (-1)(dx \wedge dy)$$

$$dy \wedge dz = dy\, dz = (-1)(dz \wedge dy)$$

$$dz \wedge dx = dz\, dx = (-1)(dx \wedge dz)$$

$$dx \wedge dx = 0$$

$$dy \wedge dy = 0$$

$$dz \wedge dz = 0$$

$$dx \wedge (dy \wedge dz) = (dx \wedge dy) \wedge dz = dx\, dy\, dz$$

(*vii*) If f is a 0-form and ω is any k-form, then $f \wedge \omega = f\omega$.

Using laws (*i*) to (*vii*), we can now find a unique product of any l-form η and any k-form ω, if $0 \leq k + l \leq 3$.

EXAMPLE 9. Show that $dx \wedge dy\, dz = dx\, dy\, dz$.

Solution. By (*vi*), $dy\, dz = dy \wedge dz$. Therefore

$$dx \wedge dy\, dz = dx \wedge (dy \wedge dz) = dx\, dy\, dz$$

EXAMPLE 10. If $\omega = x\, dx + y\, dy$ and $\eta = zy\, dx + xz\, dy + xy\, dz$, find $\omega \wedge \eta$.

Solution.

$$\begin{aligned}
\omega \wedge \eta &= (x\, dx + y\, dy) \wedge (zy\, dx + xz\, dy + xy\, dz) \\
&= [(x\, dx + y\, dy) \wedge (zy\, dx)] + [(x\, dx + y\, dy) \wedge (xz\, dy)] \\
&\quad + [(x\, dx + y\, dy) \wedge (xy\, dz)] \\
&= xyz(dx \wedge dx) + zy^2(dy \wedge dx) + x^2z(dx \wedge dy) + xyz(dy \wedge dy) \\
&\quad + x^2y(dx \wedge dz) + xy^2(dy \wedge dz) \\
&= -zy^2\, dx\, dy + x^2z\, dx\, dy - x^2y\, dz\, dx + xy^2\, dy\, dz \\
&= (x^2z - y^2z)\, dx\, dy - x^2y\, dz\, dx + xy^2\, dy\, dz
\end{aligned}$$

EXAMPLE 11. If $\omega = x\,dx - y\,dy$ and $\eta = x\,dy\,dz + z\,dx\,dy$, find $\omega \wedge \eta$.

Solution.

$$
\begin{aligned}
\omega \wedge \eta &= (x\,dx - y\,dy) \wedge (x\,dy\,dz + z\,dx\,dy) \\
&= [(x\,dx - y\,dy) \wedge (x\,dy\,dz)] + [(x\,dx - y\,dy) \wedge (z\,dx\,dy)] \\
&= (x^2\,dx \wedge dy\,dz) - (xy\,dy \wedge dy\,dz) \\
&\quad + (xz\,dx \wedge dx\,dy) - (yz\,dy \wedge dx\,dy) \\
&= [x^2\,dx \wedge (dy \wedge dz)] - [xy\,dy \wedge (dy \wedge dz)] \\
&\quad + [xz\,dx \wedge (dx \wedge dy)] - [yz\,dy \wedge (dx \wedge dy)] \\
&= x^2\,dx\,dy\,dz - [xy(dy \wedge dy) \wedge dz] \\
&\quad + [xz(dx \wedge dx) \wedge dy] - [yz(dy \wedge dx) \wedge dy] \\
&= x^2\,dx\,dy\,dz - xy(0 \wedge dz) + xz(0 \wedge dy) + [yz(dy \wedge dy) \wedge dx] \\
&= x^2\,dx\,dy\,dz
\end{aligned}
$$

The last major step in the development of this theory is to show how to differentiate forms. The derivative of a k-form is a $(k + 1)$-form if $k < 3$, and the derivative of a 3-form is always zero. If ω is a k-form we shall denote the derivative of ω by $d\omega$. The operation d has the following properties:

(1) If $f\colon K \to \mathbb{R}$ is a 0-form, then

$$df = \frac{\partial f}{\partial x}dx + \frac{\partial f}{\partial y}dy + \frac{\partial f}{\partial z}dz$$

(2) (*linearity*) If ω_1 and ω_2 are k-forms, then

$$d(\omega_1 + \omega_2) = d\omega_1 + d\omega_2$$

(3) If ω is a k-form and η is an l-form

$$d(\omega \wedge \eta) = (d\omega \wedge \eta) + (-1)^l(\omega \wedge d\eta)$$

(4) $d(d\omega) = 0$ and $d(dx) = d(dy) = d(dz) = 0$ or, simply, $d^2 = 0$.

Properties (1) to (4) provide enough information to allow us to uniquely differentiate any form.

EXAMPLE 12. Let $\omega = P(x, y, z)\,dx + Q(x, y, z)\,dy$ be a 1-form on some open set $K \subset \mathbb{R}^3$. Find $d\omega$.

Solution.

$$
\begin{aligned}
&d[P(x, y, z)\,dx + Q(x, y, z)\,dy] \\
&\qquad = d[P(x, y, z) \wedge dx] + d[Q(x, y, z) \wedge dy] \\
&\quad \text{(using 2)} \\
&\qquad = [dP \wedge dx] + [(-1)P \wedge d(dx)] + [dQ \wedge dy] + [(-1)Q \wedge d(dy)]
\end{aligned}
$$

(using 3)

$$= (dP \wedge dx) + (dQ \wedge dy)$$

$$= \left[\frac{\partial P}{\partial x} dx + \frac{\partial P}{\partial y} dy + \frac{\partial P}{\partial z} dz\right] \wedge dx + \left[\frac{\partial Q}{\partial x} dx + \frac{\partial Q}{\partial y} dy + \frac{\partial Q}{\partial z} dz\right] \wedge dy$$

(using 1)

$$= \left(\frac{\partial P}{\partial x} dx \wedge dx\right) + \left(\frac{\partial P}{\partial y} dy \wedge dx\right) + \left(\frac{\partial P}{\partial z} dz \wedge dx\right)$$

$$+ \left(\frac{\partial Q}{\partial x} dx \wedge dy\right) + \left(\frac{\partial Q}{\partial y} dy \wedge dy\right) + \left(\frac{\partial Q}{\partial z} dz \wedge dy\right)$$

$$= -\frac{\partial P}{\partial y} dx\, dy + \frac{\partial P}{\partial z} dz\, dx + \frac{\partial Q}{\partial x} dx\, dy - \frac{\partial Q}{\partial z} dy\, dz$$

$$= \left(\frac{\partial Q}{\partial x} - \frac{\partial P}{\partial y}\right) dx\, dy + \frac{\partial P}{\partial z} dz\, dx - \frac{\partial Q}{\partial z} dy\, dz$$

EXAMPLE 13. Let f be a 0-form. Using only differentiation rules (1) to (3) and the fact that $d(dx) = d(dy) = d(dz) = 0$, show that $d(df) = 0$.

Solution. By (1)

$$df = \frac{\partial f}{\partial x} dx + \frac{\partial f}{\partial y} dy + \frac{\partial f}{\partial z} dz$$

$$d(df) = d\left(\frac{\partial f}{\partial x} dx\right) + d\left(\frac{\partial f}{\partial y} dy\right) + d\left(\frac{\partial f}{\partial z} dz\right)$$

Working only with the first term, using (3) we get

$$d\left(\frac{\partial f}{\partial x} dx\right) = d\left(\frac{\partial f}{\partial x} \wedge dx\right) = d\left(\frac{\partial f}{\partial x}\right) \wedge dx - \frac{\partial f}{\partial x} \wedge d(dx)$$

$$= \left(\frac{\partial^2 f}{\partial x^2} dx + \frac{\partial^2 f}{\partial y\, \partial x} dy + \frac{\partial^2 f}{\partial z\, \partial x} dz\right) \wedge dx + 0$$

$$= \frac{\partial^2 f}{\partial y\, \partial x} dy \wedge dx + \frac{\partial^2 f}{\partial z\, \partial x} dz \wedge dx$$

$$= -\frac{\partial^2 f}{\partial y\, \partial x} dx\, dy + \frac{\partial^2 f}{\partial z\, \partial x} dz\, dx$$

Similarly, we find that

$$d\left(\frac{\partial f}{\partial y} dy\right) = \frac{\partial^2 f}{\partial x\, \partial y} dx\, dy - \frac{\partial^2 f}{\partial z\, \partial y} dy\, dz$$

and

$$d\left(\frac{\partial f}{\partial z} dz\right) = -\frac{\partial^2 f}{\partial x\, \partial z} dz\, dx + \frac{\partial^2 f}{\partial y\, \partial z} dy\, dz$$

Adding these up, we get $d(df) = 0$.

EXAMPLE 14. Show that $d(dx\,dy)$, $d(dy\,dz)$, and $d(dz\,dx)$ are all zero.

Solution. To prove the first case, we use property (3):

$$d(dx\,dy) = d(dx \wedge dy) = [d(dx) \wedge dy - dx \wedge d(dy)] = 0$$

The other cases are similar.

EXAMPLE 15. If $\eta = F(x, y, z)\,dx\,dy + G(x, y, z)\,dy\,dz + H(x, y, z)\,dz\,dx$, find $d\eta$.

Solution. By property (2)

$$d\eta = d(F\,dx\,dy) + d(G\,dy\,dz) + d(H\,dz\,dx)$$

We shall compute $d(F\,dx\,dy)$. Using (3) again, we get

$$d(F\,dx\,dy) = d(F \wedge dx\,dy) = dF \wedge (dx\,dy) + (-1)^2 F \wedge d(dx\,dy)$$

By Example 14, $d(dx\,dy) = 0$, so we are left with

$$\begin{aligned} dF \wedge (dx\,dy) &= \left(\frac{\partial F}{\partial x}dx + \frac{\partial F}{\partial y}dy + \frac{\partial F}{\partial z}dz\right) \wedge (dx \wedge dy) \\ &= \left[\frac{\partial F}{\partial x}dx \wedge (dx \wedge dy)\right] + \left[\frac{\partial F}{\partial y}dy \wedge (dx \wedge dy)\right] \\ &\quad + \left[\frac{\partial F}{\partial z}dz \wedge (dx \wedge dy)\right] \end{aligned}$$

Now

$$\begin{aligned} dx \wedge (dx \wedge dy) &= (dx \wedge dx) \wedge dy = 0 \wedge dy = 0 \\ dy \wedge (dx \wedge dy) &= -dy \wedge (dy \wedge dx) \\ &= -(dy \wedge dy) \wedge dx = 0 \wedge dx = 0 \end{aligned}$$

and

$$dz \wedge (dx \wedge dy) = (-1)^2(dx \wedge dy) \wedge dz = dx\,dy\,dz$$

Consequently

$$d(F\,dx\,dy) = \frac{\partial F}{\partial z}dx\,dy\,dz$$

Analogously, we get that

$$d(G\,dy\,dz) = \frac{\partial G}{\partial x}dx\,dy\,dz$$

and

$$d(H\,dz\,dx) = \frac{\partial H}{\partial y}dx\,dy\,dz$$

Therefore

$$d\eta = \left(\frac{\partial F}{\partial z} + \frac{\partial G}{\partial x} + \frac{\partial H}{\partial y}\right) dx\,dy\,dz$$

We have now reviewed all the concepts needed to reformulate Green's, Stokes', and Gauss' theorem in the language of forms.

***Theorem 13* (*Green's Theorem*).** *Let D be an elementary region in the xy-plane, with ∂D given the counterclockwise orientation. Suppose $\omega = P(x, y)\,dx + Q(x, y)\,dy$ is a 1-form on some open set K in $\mathbb{R}^3$ that contains D. Then*

$$\int_{\partial D} \omega = \int_D d\omega$$

Here $d\omega$ is a 2-form on K and D is in fact a surface in $\mathbb{R}^3$ parametrized by $\mathbf{\Phi}: D \to \mathbb{R}^3, \mathbf{\Phi}(x, y) = (x, y, 0)$. Since P and Q are explicitly *not* functions of z, then $\partial P/\partial z$ and $\partial Q/\partial z = 0$, and by Example 12 $d\omega = (\partial Q/\partial x - \partial P/\partial y)\,dx\,dy$. Consequently, Theorem 13 means nothing more than

$$\int_{\partial D} P\,dx + Q\,dy = \int_D \left(\frac{\partial Q}{\partial x} - \frac{\partial P}{\partial y}\right) dx\,dy$$

which is precisely Green's Theorem of Section 7.1. Hence Theorem 13 holds.

***Theorem 14* (*Stokes' Theorem*).** *Let S be an oriented surface in $\mathbb{R}^3$ with a boundary consisting of a simple closed curve ∂S (Figure 7.6.3) oriented as the*

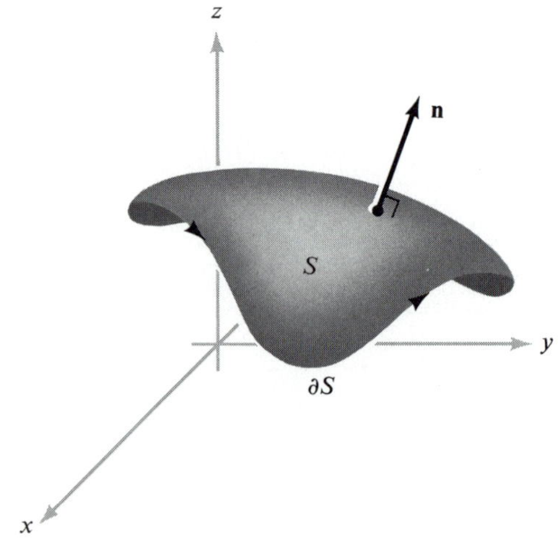

FIGURE 7.6.3
An oriented surface to which Stokes' Theorem applies.

boundary of S (see Figure 7.2.1). Suppose that ω is a 1-form on some open set K that contains S. Then

$$\int_{\partial S} \omega = \int_S d\omega$$

***Theorem 15* (*Gauss' Theorem*).** *Let $\Omega \subset \mathbb{R}^3$ be an elementary region with $\partial\Omega$*

given the outward orientation (see Section 7.4). If η is a 2-form on some region K containing Ω then

$$\int_{\partial\Omega} \eta = \int_{\Omega} d\eta$$

The reader has probably noticed the strong similarity in the statements of these theorems. In the vector field formulations, we used divergence for regions in $\mathbb{R}^3$ (Gauss' Theorem), and the curl for surfaces in $\mathbb{R}^3$ (Stokes' Theorem) and regions in $\mathbb{R}^2$ (Green's Theorem). Here we just use the unified notion of derivative of a differential form for all three theorems; and in fact, we can state all theorems as one by introducing a little more terminology.

By an *oriented 2-manifold with boundary* in $\mathbb{R}^3$ we mean a surface in $\mathbb{R}^3$ whose boundary is a simple closed curve with orientation as described in Section 7.2. By an *oriented 3-manifold* in $\mathbb{R}^3$ we mean an elementary region in $\mathbb{R}^3$ (we assume its boundary, which is a surface, is given the outward orientation discussed in Section 7.4). We call the following unified theorem "Stokes' Theorem," according to the current convention.

***Theorem 16* (*Stokes' Theorem*).** *Let M be an oriented k-manifold in $\mathbb{R}^3$ ($k = 2$ or 3) contained in some open set K. Suppose ω is a $(k-1)$-form on K. Then*

$$\int_{\partial M} \omega = \int_{M} d\omega$$

EXERCISES

1. Evaluate $\omega \wedge \eta$ if
   (a) $\omega = 2x\, dx + y\, dy$
   $\eta = x^3\, dx + y^2\, dy$
   (b) $\omega = x\, dx - y\, dy$
   $\eta = y\, dx + x\, dy$
   (c) $\omega = x\, dx + y\, dy + z\, dz$
   $\eta = z\, dx\, dy + x\, dy\, dz + y\, dz\, dx$
   (d) $\omega = xy\, dy\, dz + x^2\, dx\, dy$
   $\eta = dx + dz$
   (e) $\omega = e^{xyz}\, dx\, dy$
   $\eta = e^{-xyz}\, dz$
2. Prove that

$$(a_1\, dx + a_2\, dy + a_3\, dz) \wedge (b_1\, dy\, dz + b_2\, dz\, dx + b_3\, dx\, dy) = \left(\sum_{i=1}^{3} a_i b_i\right) dx\, dy\, dz$$

3. Find $d\omega$ in the following examples.
   (a) $\omega = x^2 y + y^3$
   (b) $\omega = y^2 \cos x\, dy + xy\, dx + dz$
   (c) $\omega = xy\, dy + (x + y)^2\, dx$
   (d) $\omega = x\, dx\, dy + z\, dy\, dz + y\, dz\, dx$

(e) $\omega = (x^2 + y^2)\, dy\, dz$
(f) $\omega = (x^2 + y^2 + z^2)\, dz$

(g) $\omega = \dfrac{-x}{x^2 + y^2}\, dx + \dfrac{y}{x^2 + y^2}\, dy$

(h) $\omega = x^2 y\, dy\, dz$

4. Let $\mathbf{V}$: $K \to \mathbb{R}^3$ be a vector field defined by $\mathbf{V}(x, y, z) = G(x, y, z)\mathbf{i} + H(x, y, z)\mathbf{j} + F(x, y, z)\mathbf{k}$, and let η be the 2-form on K given by

$$\eta = F\, dx\, dy + G\, dy\, dz + H\, dz\, dx$$

Show that $d\eta = (\text{div } \mathbf{V})\, dx\, dy\, dz$.

5. If $\mathbf{V} = A(x, y, z)\mathbf{i} + B(x, y, z)\mathbf{j} + C(x, y, z)\mathbf{k}$ is a vector field on $K \subset \mathbb{R}^3$, define the operation Form_2: Vector Fields $\to$ 2-forms by

$$\text{Form}_2(\mathbf{V}) = A\, dy\, dz + B\, dz\, dx + C\, dx\, dy$$

(a) Show that $\text{Form}_2(\alpha\mathbf{V}_1 + \mathbf{V}_2) = \alpha\, \text{Form}_2(\mathbf{V}_1) + \text{Form}_2(\mathbf{V}_2)$, where α is a real number.
(b) Show that $\text{Form}_2(\text{curl } \mathbf{V}) = d\omega$, where $\omega = A\, dx + B\, dy + C\, dz$.

6. Using the differential-form version of Stokes' Theorem, prove the vector-field version in Section 7.2. Repeat for Gauss' Theorem.

7. Interpret Theorem 16 in the case $k = 1$.

8. Let $\omega = (x + y)\, dz + (y + z)\, dx + (x + z)\, dy$, and let S be the upper part of the unit sphere; i.e., S is the set of (x, y, z) with $x^2 + y^2 + z^2 = 1$ and $z \geq 0$. ∂S is the unit circle in the xy-plane. Evaluate $\int_{\partial S} \omega$ both directly and by Stokes' Theorem.

*9. Let T be the triangular solid bounded by the xy-plane, the xz-plane, the yz-plane, and the plane $2x + 3y + 6z = 12$. Compute

$$\int_{\partial T} F_1\, dx\, dy + F_2\, dy\, dz + F_3\, dz\, dx$$

directly and by Gauss' Theorem, if
(a) $F_1 = 3y,\ F_2 = 18z,\ F_3 = -12$
(b) $F_1 = z,\ F_2 = x^2,\ F_3 = y$

10. Evaluate $\int_S \omega$ where $\omega = z\, dx\, dy + x\, dy\, dz + y\, dz\, dx$ and S is the unit sphere, directly and by Gauss' Theorem.

11. Let R be an elementary region in $\mathbb{R}^3$. Show that the volume of R is given by the formula

$$v(R) = \tfrac{1}{3} \int_{\partial R} x\, dy\, dz + y\, dz\, dx + z\, dx\, dy$$

*12. In Section 3.2 we saw that the length $l(\boldsymbol{\sigma})$ of a curve $\boldsymbol{\sigma}(t) = (x(t), y(t), z(t))$, $a \leq t \leq b$ was given by the formula

$$l(\boldsymbol{\sigma}) = \int ds = \int_a^b \left(\frac{ds}{dt}\right) dt$$

where, loosely speaking

$$(ds)^2 = (dx)^2 + (dy)^2 + (dz)^2$$

or

$$\frac{ds}{dt} = \sqrt{\left(\frac{dx}{dt}\right)^2 + \left(\frac{dy}{dt}\right)^2 + \left(\frac{dz}{dt}\right)^2}$$

Now suppose a surface S is given in parametrized form by $\mathbf{\Phi}(u, v) = (x(u, v), y(u, v), z(u, v))$, $(u, v) \in D$. Show that the area of S can be expressed as

$$A(S) = \int_D dS$$

where $(dS)^2 = (dx \wedge dy)^2 + (dy \wedge dz)^2 + (dz \wedge dx)^2$.
(HINT:

$$dx = \frac{\partial x}{\partial u}\, du + \frac{\partial x}{\partial v}\, dv$$

and similarly for dy and dz. Use the law of forms for the basic 1-forms du and dv. Then dS turns out to be a function times the basic 2-form $du\, dv$, which we can integrate over D.)

REVIEW EXERCISES FOR CHAPTER 7

1. Use Green's Theorem to find the area of the loop of the curve $x = a \sin\theta \cos\theta$, $y = a \sin^2\theta$, for $a > 0$ and $0 \le \theta \le \pi$.
2. Let $\mathbf{r}(x, y, z) = (x, y, z)$, $r = \|\mathbf{r}\|$. Show $\nabla^2(\log r) = 1/r^2$ and $\nabla^2(r^n) = n(n+1)r^{n-2}$.
3. If $\nabla \times \mathbf{F} = \mathbf{0}$ and $\nabla \times \mathbf{G} = \mathbf{0}$, prove $\nabla \cdot (\mathbf{F} \times \mathbf{G}) = 0$.
4. Let $\mathbf{m}$ be a constant vector field and let

$$\phi = \frac{\mathbf{m} \cdot \mathbf{r}}{r^3}, \qquad \mathbf{F} = \frac{\mathbf{m} \times \mathbf{r}}{r^3}$$

Compute $\nabla\phi$, $\nabla^2\phi$, $\nabla \cdot \mathbf{F}$, and $\nabla \times \mathbf{F}$.

5. Prove the identity

$$(\mathbf{F} \cdot \nabla)\mathbf{F} = \tfrac{1}{2}\nabla(\mathbf{F} \cdot \mathbf{F}) + (\nabla \times \mathbf{F}) \times \mathbf{F}$$

6. Prove that if $\nabla \times \mathbf{F} = \nabla \times \mathbf{G}$ then $\mathbf{F} = \mathbf{G} + \nabla f$ for some f.
7. Let $\mathbf{F} = 2yz\mathbf{i} + (-x + 3y + 2)\mathbf{j} + (x^2 + z)\mathbf{k}$. Evaluate $\int_S \nabla \times \mathbf{F}$, where S is the cylinder $x^2 + y^2 = a^2$, $0 \le z \le 1$ (without the top and bottom). What if the top and bottom are included?
8. Let $\mathbf{F} = x^2y\mathbf{i} + z^8\mathbf{j} - 2xyz\mathbf{k}$. Evaluate the integral of $\mathbf{F}$ over the surface of the unit cube.
9. Let Ω be a region in $\mathbb{R}^3$ with boundary $\partial\Omega$. Prove the identity

$$\int_{\partial\Omega} \mathbf{F} \times (\nabla \times \mathbf{G}) = \int_\Omega (\nabla \times \mathbf{F}) \cdot (\nabla \times \mathbf{G}) - \int_\Omega \mathbf{F} \cdot (\nabla \times \nabla \times \mathbf{G})$$

10. Verify Green's Theorem for the line integral

$$\int_C x^2 y\, dx + y\, dy$$

when C is the boundary of the region between the curves $y = x$ and $y = x^3$, $0 \le x \le 1$.

11. Can you derive Green's Theorem in the plane from Gauss' Theorem?

12. (a) Show that $\mathbf{F} = (x^3 - 2xy^3)\mathbf{i} - 3x^2y^2\mathbf{j}$ is a gradient vector field.
    (b) Evaluate the integral of this form along the path $x = \cos^3\theta$, $y = \sin^3\theta$, $0 \le \theta \le \pi/2$.

13. (a) Show that $\mathbf{F} = 6xy(\cos z)\mathbf{i} + 3x^2(\cos z)\mathbf{j} - 3x^2y(\sin z)\mathbf{k}$ is conservative (see Section 7.3).
    (b) Find f such that $\mathbf{F} = \nabla f$.
    (c) Evaluate the integral of $\mathbf{F}$ along the curve $x = \cos^3\theta$, $y = \sin^3\theta$, $z = 0$, $0 \le \theta \le \pi/2$.

14. Let the velocity of a fluid be described by $\mathbf{F} = 6xz\mathbf{i} + x^2y\mathbf{j} + yz\mathbf{k}$. Compute the rate at which fluid is leaving the unit cube.

15. Suppose $\nabla \cdot \mathbf{F}(x_0, y_0, z_0) > 0$. Show that for a sufficiently small sphere S centered at (x_0, y_0, z_0), the flux of $\mathbf{F}$ out of S is positive.

16. Let $\mathbf{F} = x^2\mathbf{i} + (x^2y - 2xy)\mathbf{j} - x^2z\mathbf{k}$. Does there exist a $\mathbf{G}$ such that $\mathbf{F} = \nabla \times \mathbf{G}$?

17. Let $\mathbf{a}$ be a constant vector and $\mathbf{F} = \mathbf{a} \times \mathbf{r}$ [as usual, $\mathbf{r}(x, y, z) = (x, y, z)$]. Is $\mathbf{F}$ conservative? If so, find a potential for it.

18. Consider the case of incompressible fluid flow with velocity field $\mathbf{F}$ and density ρ.
    (a) If ρ is constant for each fixed t then ρ is constant in t as well.
    (b) If ρ is constant in t then $\mathbf{F} \cdot \nabla\rho = 0$.

19. (a) Let $f(x, y, z) = 3xye^{z^2}$. Compute ∇f.
    (b) Let $\boldsymbol{\sigma}(t) = (3\cos^3 t, \sin^2 t, e^t)$, $0 \le t \le \pi$. Evaluate

$$\int_{\boldsymbol{\sigma}} \nabla f$$

    (c) Verify directly Stokes' Theorem for gradient vector fields $\mathbf{F} = \nabla f$.

20. Using Green's Theorem, or otherwise, evaluate $\int_C x^3\, dy - y^3\, dx$, where C is the unit circle $(x^2 + y^2 = 1)$.

21. Evaluate the integral $\iint_S \mathbf{F} \cdot d\mathbf{S}$ where $\mathbf{F} = x\mathbf{i} + y\mathbf{j} + 3\mathbf{k}$ and where S is the surface of the unit sphere $x^2 + y^2 + z^2 = 1$.

22. (a) State Stokes' Theorem for surfaces in $\mathbb{R}^3$.
    (b) Let $\mathbf{F}$ be a vector field on $\mathbb{R}^3$ satisfying $\nabla \times \mathbf{F} = \mathbf{0}$. Use Stokes' Theorem to show that $\int_C \mathbf{F} = 0$ where C is a closed curve.

23. True or false?
    (a) A continuous function is always differentiable.
    (b) If the partial derivatives of f exist and are continuous, then f is differentiable.
    (c) $\{(x, y) \in \mathbb{R}^2 \mid x^2 + y^2 > 0\}$ is an open set.
    (d) The magnitude of the curl of a vector field $\mathbf{F}$ represents the amount of circulation per unit area in a plane normal to $\nabla \times \mathbf{F}$.

(e) It is always true that $\mathbf{v} \cdot (\mathbf{v} \times \mathbf{w}) = 0$.

(f) It is always true that $\nabla \cdot (\nabla f) \geq 0$.

(g) $\int_0^1 \int_0^y f(x, y)\, dx\, dy = \int_0^1 \int_0^x f(x, y)\, dy\, dx$ holds for all continuous functions $f(x, y)$.

(h) The area of a region enclosed by a closed curve C is given by $A = \int_C x\, dy$.

(i) If S is a surface and $\mathbf{F}$ is orthogonal to the boundary of S, then it is true that $\int_S \nabla \times \mathbf{F} = 0$.

(j) If the height of a hill is described by a function $h(x, y)$, then a skier who wishes to descend the fastest should ski in a (compass) direction orthogonal to ∇h.

(k) $\nabla \times (\nabla \times \mathbf{F})$ is always equal to $\mathbf{0}$.

24. Show that in cylindrical coordinates

$$\nabla f = \frac{\partial f}{\partial r}\mathbf{e}_r + \frac{1}{r}\frac{\partial f}{\partial \theta}\mathbf{e}_\theta + \frac{\partial f}{\partial z}\mathbf{e}_z$$

$$\nabla \cdot \mathbf{F} = \frac{1}{r}\left[\frac{\partial}{\partial r}(rF_r) + \frac{\partial F_\theta}{\partial \theta} + \frac{\partial}{\partial z}(rF_z)\right]$$

and

$$\nabla \times \mathbf{F} = \frac{1}{r}\begin{vmatrix} \mathbf{e}_r & r\mathbf{e}_\theta & \mathbf{e}_z \\ \dfrac{\partial}{\partial r} & \dfrac{\partial}{\partial \theta} & \dfrac{\partial}{\partial z} \\ F_r & rF_\theta & F_z \end{vmatrix}$$

where $\mathbf{e}_r$, $\mathbf{e}_\theta$, $\mathbf{e}_z$ are the unit orthonormal vectors shown in Figure 7. Review. 1 and $\mathbf{F} = F_r\mathbf{e}_r + F_\theta\mathbf{e}_\theta + F_z\mathbf{e}_z$.

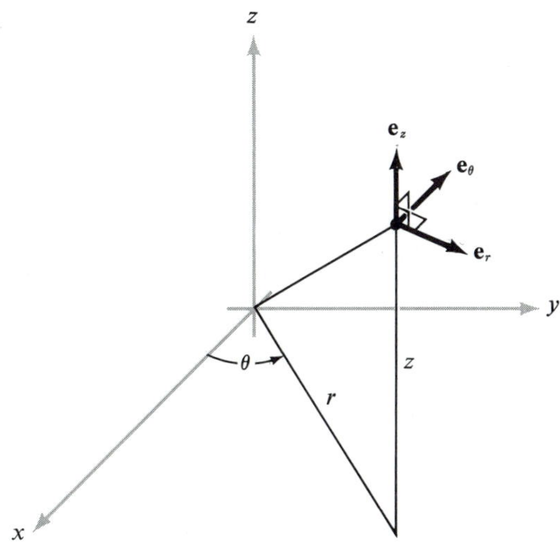

FIGURE 7. REVIEW. 1
Orthonormal vectors $\mathbf{e}_r$, $\mathbf{e}_\theta$, and $\mathbf{e}_z$ associated with cylindrical coordinates.

*25. Show that in spherical coordinates (see p. 274)

$$\nabla f = \frac{\partial f}{\partial \rho}\hat{\mathbf{e}}_\rho + \frac{1}{\rho}\frac{\partial f}{\partial \theta}\hat{\mathbf{e}}_\theta + \frac{1}{\rho \sin\theta}\frac{\partial f}{\partial \phi}\hat{\mathbf{e}}_\varphi$$

$$\nabla \cdot \mathbf{F} = \frac{1}{\rho^2}\frac{\partial}{\partial \rho}(\rho^2 F_\rho) + \frac{1}{\rho \sin\theta}\frac{\partial}{\partial \theta}(\sin\theta\, F_\theta) + \frac{1}{\rho\sin\theta}\frac{\partial F_\phi}{\partial \varphi}$$

$$\nabla \times \mathbf{F} = \left(\frac{1}{\rho\sin\theta}\frac{\partial}{\partial\theta}(\sin\theta\, F_\phi) - \frac{1}{\rho\sin\theta}\frac{\partial F_\theta}{\partial\varphi}\right)\hat{\mathbf{e}}_\rho$$

$$+ \left(\frac{1}{\rho\sin\theta}\frac{\partial F_\rho}{\partial\phi} - \frac{1}{\rho}\frac{\partial}{\partial\rho}(\rho F_\phi)\right)\hat{\mathbf{e}}_\theta + \left(\frac{1}{\rho}\frac{\partial}{\partial\rho}(\rho F_\theta) - \frac{1}{\rho}\frac{\partial F_\rho}{\partial\theta}\right)\hat{\mathbf{e}}_\varphi$$

where $\hat{\mathbf{e}}_\rho$, $\hat{\mathbf{e}}_\theta$, $\hat{\mathbf{e}}_\varphi$ are as shown in Figure 7. Review. 2.

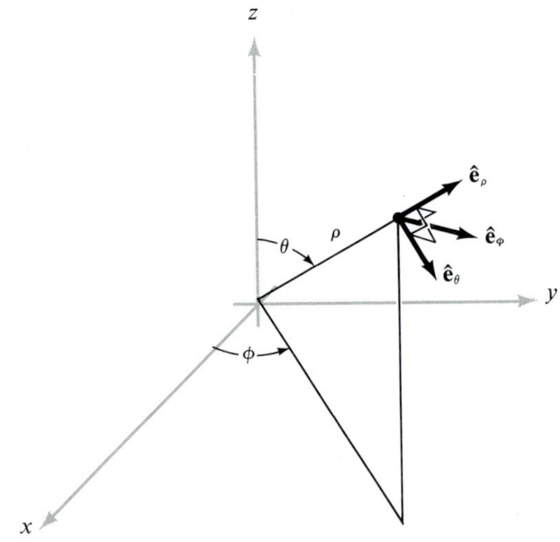

FIGURE 7. REVIEW. 2
Orthonormal vectors $\hat{\mathbf{e}}_\rho$, $\hat{\mathbf{e}}_\theta$, and $\hat{\mathbf{e}}_\phi$ associated with spherical coordinates.

APPENDIX A

TABLES

1. Trigonometric Functions

Angle					Angle				
Degree	*Radian*	*Sine*	*Cosine*	*Tangent*	*Degree*	*Radian*	*Sine*	*Cosine*	*Tangent*
0°	0.000	0.000	1.000	0.000					
1°	.017	.017	1.000	.017	46°	0.803	0.719	0.695	1.036
2°	.035	.035	0.999	.035	47°	.820	.731	.682	1.072
3°	.052	.052	.999	.052	48°	.838	.743	.669	1.111
4°	.070	.070	.998	.070	49°	.855	.755	.656	1.150
5°	.087	.087	.996	.087	50°	.873	.766	.643	1.192
6°	.105	.105	.995	.105	51°	.890	.777	.629	1.235
7°	.122	.122	.993	.123	52°	.908	.788	.616	1.280
8°	.140	.139	.990	.141	53°	.925	.799	.602	1.327
9°	.157	.156	.988	.158	54°	.942	.809	.588	1.376
10°	.175	.174	.985	.176	55°	.960	.819	.574	1.428
11°	.192	.191	.982	.194	56°	.977	.829	.559	1.483
12°	.209	.208	.978	.213	57°	.995	.839	.545	1.540
13°	.227	.225	.974	.231	58°	1.012	.848	.530	1.600
14°	.244	.242	.970	.249	59°	1.030	.857	.515	1.664
15°	.262	.259	.966	.268	60°	1.047	.866	.500	1.732
16°	.279	.276	.961	.287	61°	1.065	.875	.485	1.804
17°	.297	.292	.956	.306	62°	1.082	.883	.470	1.881
18°	.314	.309	.951	.325	63°	1.100	.891	.454	1.963
19°	.332	.326	.946	.344	64°	1.117	.899	.438	2.050
20°	.349	.342	.940	.364	65°	1.134	.906	.423	2.145
21°	.367	.358	.934	.384	66°	1.152	.914	.407	2.246
22°	.384	.375	.927	.404	67°	1.169	.921	.391	2.356
23°	.401	.391	.921	.425	68°	1.187	.927	.375	2.475
24°	.419	.407	.914	.445	69°	1.204	.934	.358	2.605
25°	.436	.423	.906	.466	70°	1.222	.940	.342	2.747
26°	.454	.438	.899	.488	71°	1.239	.946	.326	2.904
27°	.471	.454	.891	.510	72°	1.257	.951	.309	3.078
28°	.489	.470	.883	.532	73°	1.274	.956	.292	3.271
29°	.506	.485	.875	.554	74°	1.292	.961	.276	3.487
30°	.524	.500	.866	.577	75°	1.309	.966	.259	3.732
31°	.541	.515	.857	.601	76°	1.326	.970	.242	4.011
32°	.559	.530	.848	.625	77°	1.344	.974	.225	4.331
33°	.576	.545	.839	.649	78°	1.361	.978	.208	4.705
34°	.593	.559	.829	.675	79°	1.379	.982	.191	5.145
35°	.611	.574	.819	.700	80°	1.396	.985	.174	5.671
36°	.628	.588	.809	.727	81°	1.414	.988	.156	6.314
37°	.646	.602	.799	.754	82°	1.431	.990	.139	7.115
38°	.663	.616	.788	.781	83°	1.449	.993	.122	8.144
39°	.681	.629	.777	.810	84°	1.466	.995	.105	9.514
40°	.698	.643	.766	.839	85°	1.484	.996	.087	11.43
41°	.716	.656	.755	.869	86°	1.501	.998	.070	14.30
42°	.733	.669	.743	.900	87°	1.518	.999	.052	19.08
43°	.750	.682	.731	.933	88°	1.536	.999	.035	28.64
44°	.768	.695	.719	.966	89°	1.553	1.000	.017	57.29
45°	.785	.707	.707	1.000	90°	1.571	1.000	.000	∞

2. Derivatives

1. $\frac{dau}{dx} = a\frac{du}{dx}$

2. $\frac{d(u+v-w)}{dx} = \frac{du}{dx} + \frac{dv}{dx} - \frac{dw}{dx}$

3. $\frac{d(uv)}{dx} = u\frac{dv}{dx} + v\frac{du}{dx}$

4. $\frac{d(u/v)}{dx} = \frac{v(du/dx) - u(dv/dx)}{v^2}$

5. $\frac{d(u^n)}{dx} = nu^{n-1}\frac{du}{dx}$

6. $\frac{d(u^v)}{dx} = vu^{v-1}\frac{du}{dx} + u^v(\log u)\frac{dv}{dx}$

7. $\frac{d(e^u)}{dx} = e^u\frac{du}{dx}$

8. $\frac{d(e^{au})}{dx} = ae^{au}\frac{du}{dx}$

9. $\frac{da^u}{dx} = a^u(\log a)\frac{du}{dx}$

10. $\frac{d(\log u)}{dx} = \frac{1}{u}\frac{du}{dx}$

11. $\frac{d(\log_a u)}{dx} = \frac{1}{u(\log a)}\frac{du}{dx}$

12. $\frac{d \sin u}{dx} = \cos u\frac{du}{dx}$

13. $\frac{d \cos u}{dx} = -\sin u\frac{du}{dx}$

14. $\frac{d \tan u}{dx} = \sec^2 u\frac{du}{dx}$

15. $\frac{d \cot u}{dx} = -\csc^2 u\frac{du}{dx}$

16. $\frac{d \sec u}{dx} = \tan u \sec u\frac{du}{dx}$

17. $\frac{d \csc u}{dx} = -(\cot u)(\csc u)\frac{du}{dx}$

18. $\frac{d \text{ arc sin } u}{dx} = \frac{1}{\sqrt{1-u^2}}\frac{du}{dx}$

19. $\frac{d \text{ arc cos } u}{dx} = \frac{-1}{\sqrt{1-u^2}}\frac{du}{dx}$

20. $\frac{d \text{ arc tan } u}{dx} = \frac{1}{1+u^2}\frac{du}{dx}$

21. $\frac{d \text{ arc cot } u}{dx} = \frac{-1}{1+u^2}\frac{du}{dx}$

22. $\frac{d \text{ arc sec } u}{dx} = \frac{+1}{u\sqrt{u^2-1}}\frac{du}{dx}$

23. $\frac{d \text{ arc csc } u}{dx} = \frac{-1}{u\sqrt{u^2-1}}\frac{du}{dx}$

24. $\frac{d \sinh u}{dx} = \cosh u\frac{du}{dx}$

25. $\frac{d \cosh u}{dx} = \sinh u\frac{du}{dx}$

26. $\frac{d \tanh u}{dx} = \text{sech}^2 u\frac{du}{dx}$

27. $\frac{d \coth u}{dx} = -(\text{csch}^2 u)\frac{du}{dx}$

28. $\frac{d \text{ sech } u}{dx} = -(\text{sech } u)(\tanh u)\frac{du}{dx}$

29. $\frac{d \text{ csch } u}{dx} = -(\text{csch } u)(\coth u)\frac{du}{dx}$

30. $\frac{d \sinh^{-1} u}{dx} = \frac{1}{\sqrt{1+u^2}}\frac{du}{dx}$

31. $\frac{d \cosh^{-1} u}{dx} = \frac{1}{\sqrt{u^2-1}}\frac{du}{dx}$

32. $\frac{d \tanh^{-1} u}{dx} = \frac{1}{1-u^2}\frac{du}{dx}$

33. $\frac{d \coth^{-1} u}{dx} = \frac{1}{1-u^2}\frac{du}{dx}$

34. $\frac{d \text{ sech}^{-1} u}{dx} = \frac{-1}{u\sqrt{1-u^2}}\frac{du}{dx}$

35. $\frac{d \text{ csch}^{-1} u}{dx} = \frac{-1}{|u|\sqrt{1+u^2}}\frac{du}{dx}$

3. Integrals

(*An arbitrary constant may be added to each integral*).

1. $\int x^n\,dx = \dfrac{1}{n+1}x^{n+1} \quad (n \neq -1)$

2. $\int \dfrac{1}{x}\,dx = \log|x|$

3. $\int e^x\,dx = e^x$

4. $\int a^x\,dx = \dfrac{a^x}{\log a}$

5. $\int \sin x\,dx = -\cos x$

6. $\int \cos x\,dx = \sin x$

7. $\int \tan x\,dx = -\log|\cos x|$

8. $\int \cot x\,dx = \log|\sin x|$

9. $\int \sec x\,dx = \log|\sec x + \tan x| = \log|\tan(\tfrac{1}{2}x + \tfrac{1}{4}\pi)|$

10. $\int \csc x\,dx = \log|\csc x - \cot x| = \log|\tan \tfrac{1}{2}x|$

11. $\int \arcsin\dfrac{x}{a}\,dx = x \arcsin\dfrac{x}{a} + \sqrt{a^2 - x^2} \quad (a > 0)$

12. $\int \arccos\dfrac{x}{a}\,dx = x \arccos\dfrac{x}{a} - \sqrt{a^2 - x^2} \quad (a > 0)$

13. $\int \arctan\dfrac{x}{a}\,dx = x \arctan\dfrac{x}{a} - \dfrac{a}{2}\log(a^2 + x^2) \quad (a > 0)$

14. $\int \sin^2 mx\,dx = \dfrac{1}{2m}(mx - \sin mx \cos mx)$

15. $\int \cos^2 mx\,dx = \dfrac{1}{2m}(mx + \sin mx \cos mx)$

16. $\int \sec^2 x\,dx = \tan x$

17. $\int \csc^2 x\,dx = -\cot x$

18. $\int \sin^n x\,dx = -\dfrac{\sin^{n-1} x \cos x}{n} + \dfrac{n-1}{n}\int \sin^{n-2} x\,dx$

19. $\int \cos^n x\,dx = \dfrac{\cos^{n-1} x \sin x}{n} + \dfrac{n-1}{n}\int \cos^{n-2} x\,dx$

20. $\int \tan^n x\,dx = \dfrac{\tan^{n-1} x}{n-1} - \int \tan^{n-2} x\,dx \quad (n \neq 1)$

21. $\int \cot^n x\,dx = -\dfrac{\cot^{n-1} x}{n-1} - \int \cot^{n-2} x\,dx \quad (n \neq 1)$

22. $\int \sec^n x\,dx = \dfrac{\tan x \sec^{n-2} x}{n-1} + \dfrac{n-2}{n-1}\int \sec^{n-2} x\,dx \quad (n \neq 1)$

23. $\int \csc^n x\, dx = -\dfrac{\cot x \csc^{n-2} x}{n-1} + \dfrac{n-2}{n-1}\int \csc^{n-2} x\, dx \quad (n \neq 1)$

24. $\int \sinh x\, dx = \cosh x$

25. $\int \cosh x\, dx = \sinh x$

26. $\int \tanh x\, dx = \log|\cosh x|$

27. $\int \coth x\, dx = \log|\sinh x|$

28. $\int \operatorname{sech} x\, dx = \arctan(\sinh x)$

29. $\int \operatorname{csch} x\, dx = \log\left|\tanh \dfrac{x}{2}\right| = -\dfrac{1}{2}\log \dfrac{\cosh x + 1}{\cosh x - 1}$

30. $\int \sinh^2 x\, dx = \frac{1}{4}\sinh 2x - \frac{1}{2}x$

31. $\int \cosh^2 x\, dx = \frac{1}{4}\sinh 2x + \frac{1}{2}x$

32. $\int \operatorname{sech}^2 x\, dx = \tanh x$

33. $\int \sinh^{-1}\dfrac{x}{a}\, dx = x \sinh^{-1}\dfrac{x}{a} - \sqrt{x^2 + a^2} \quad (a > 0)$

34. $\int \cosh^{-1}\dfrac{x}{a}\, dx = \begin{cases} x\cosh^{-1}\dfrac{x}{a} - \sqrt{x^2 - a^2} & \left[\cosh^{-1}\left(\dfrac{x}{a}\right) > 0,\ a > 0\right] \\ x\cosh^{-1}\dfrac{x}{a} + \sqrt{x^2 - a^2} & \left[\cosh^{-1}\left(\dfrac{x}{a}\right) < 0,\ a > 0\right] \end{cases}$

35. $\int \tanh^{-1}\dfrac{x}{a}\, dx = x \tanh^{-1}\dfrac{x}{a} + \dfrac{a}{2}\log|a^2 - x^2|$

36. $\int \dfrac{1}{\sqrt{a^2 + x^2}}\, dx = (x + \sqrt{a^2 + x^2}) = \sinh^{-1}\dfrac{x}{a} \quad (a > 0)$

37. $\int \dfrac{1}{a^2 + x^2}\, dx = \dfrac{1}{a}\arctan \dfrac{x}{a} \quad (a > 0)$

38. $\int \sqrt{a^2 - x^2}\, dx = \dfrac{x}{2}\sqrt{a^2 - x^2} + \dfrac{a^2}{2}\arcsin \dfrac{x}{a} \quad (a > 0)$

39. $\int (a^2 - x^2)^{3/2}\, dx = \dfrac{x}{8}(5a^2 - 2x^2)\sqrt{a^2 - x^2} + \dfrac{3a^4}{8}\arcsin \dfrac{x}{a} \quad (a > 0)$

40. $\int \dfrac{1}{\sqrt{a^2 - x^2}}\, dx = \arcsin \dfrac{x}{a} \quad (a > 0)$

41. $\int \dfrac{1}{a^2 - x^2}\, dx = \dfrac{1}{2a}\log\left|\dfrac{a + x}{a - x}\right|$

42. $\int \dfrac{1}{(a^2 - x^2)^{3/2}}\, dx = \dfrac{x}{a^2\sqrt{a^2 - x^2}}$

43. $\int \sqrt{x^2 \pm a^2}\, dx = \dfrac{x}{2}\sqrt{x^2 \pm a^2} \pm \dfrac{a^2}{2}\log|x + \sqrt{x^2 \pm a^2}|$

44. $\int \dfrac{1}{\sqrt{x^2 - a^2}}\, dx = \log|x + \sqrt{x^2 - a^2}| = \cosh^{-1}\dfrac{x}{a} \quad (a > 0)$

45. $$\int \frac{1}{x(a+bx)}\,dx = \frac{1}{a}\log\left|\frac{x}{a+bx}\right|$$

46. $$\int x\sqrt{a+bx}\,dx = \frac{2(3bx-2a)(a+bx)^{3/2}}{15b^2}$$

47. $$\int \frac{\sqrt{a+bx}}{x}\,dx = 2\sqrt{a+bx} + a\int \frac{1}{x\sqrt{a+bx}}\,dx$$

48. $$\int \frac{x}{\sqrt{a+bx}}\,dx = \frac{2(bx-2a)\sqrt{a+bx}}{3b^2}$$

49. $$\int \frac{1}{x\sqrt{a+bx}}\,dx = \frac{1}{\sqrt{a}}\log\left|\frac{\sqrt{a+bx}-\sqrt{a}}{\sqrt{a+bx}+\sqrt{a}}\right| \quad (a>0)$$
$$= \frac{2}{\sqrt{-a}}\arctan\sqrt{\frac{a+bx}{-a}} \quad (a<0)$$

50. $$\int \frac{\sqrt{a^2-x^2}}{x}\,dx = \sqrt{a^2-x^2} - a\log\left|\frac{a+\sqrt{a^2-x^2}}{x}\right|$$

51. $$\int x\sqrt{a^2-x^2}\,dx = -\tfrac{1}{3}(a^2-x^2)^{3/2}$$

52. $$\int x^2\sqrt{a^2-x^2}\,dx = \frac{x}{8}(2x^2-a^2)\sqrt{a^2-x^2} + \frac{a^4}{8}\arcsin\frac{x}{a} \quad (a>0)$$

53. $$\int \frac{1}{x\sqrt{a^2-x^2}}\,dx = -\frac{1}{a}\log\left|\frac{a+\sqrt{a^2-x^2}}{x}\right|$$

54. $$\int \frac{x}{\sqrt{a^2-x^2}}\,dx = -\sqrt{a^2-x^2}$$

55. $$\int \frac{x^2}{\sqrt{a^2-x^2}}\,dx = -\frac{x}{2}\sqrt{a^2-x^2} + \frac{a^2}{2}\arcsin\frac{x}{a} \quad (a>0)$$

56. $$\int \frac{\sqrt{x^2+a^2}}{x}\,dx = \sqrt{x^2+a^2} - a\log\left|\frac{a+\sqrt{x^2+a^2}}{x}\right|$$

57. $$\int \frac{\sqrt{x^2-a^2}}{x}\,dx = \sqrt{x^2-a^2} - a\arccos\frac{a}{|x|}$$
$$= \sqrt{x^2-a^2} - a\,\mathrm{arc\,sec}\left(\frac{x}{a}\right) \quad (a>0)$$

58. $$\int x\sqrt{x^2\pm a^2}\,dx = \tfrac{1}{3}(x^2\pm a^2)^{3/2}$$

59. $$\int \frac{1}{x\sqrt{x^2+a^2}}\,dx = \frac{1}{a}\log\left|\frac{x}{a+\sqrt{x^2+a^2}}\right|$$

60. $$\int \frac{1}{x\sqrt{x^2-a^2}}\,dx = \frac{1}{a}\arccos\frac{a}{|x|} \quad (a>0)$$

61. $$\int \frac{1}{x^2\sqrt{x^2\pm a^2}}\,dx = \mp\frac{\sqrt{x^2\pm a^2}}{a^2x}$$

62. $$\int \frac{x}{\sqrt{x^2\pm a^2}}\,dx = \sqrt{x^2\pm a^2}$$

63. $$\int \frac{1}{ax^2+bx+c}\,dx = \frac{1}{\sqrt{b^2-4ac}}\log\left|\frac{2ax+b-\sqrt{b^2-4ac}}{2ax+b+\sqrt{b^2-4ac}}\right| \quad (b^2>4ac)$$
$$= \frac{2}{\sqrt{4ac-b^2}}\arctan\frac{2ax+b}{\sqrt{4ac-b^2}} \quad (b^2<4ac)$$

64. $\displaystyle\int \frac{x}{ax^2+bx+c}\,dx = \frac{1}{2a}\log|ax^2+bx+c| - \frac{b}{2a}\int \frac{1}{ax^2+bx+c}\,dx$

65. $\displaystyle\int \frac{1}{\sqrt{ax^2+bx+c}}\,dx$

$$= \frac{1}{\sqrt{a}}\log|2ax+b+2\sqrt{a}\sqrt{ax^2+bx+c}| \quad (a>0)$$

$$= \frac{1}{\sqrt{-a}}\arcsin\frac{-2ax-b}{\sqrt{b^2-4ac}} \quad (a<0)$$

66. $\displaystyle\int \sqrt{ax^2+bx+c}\,dx$

$$= \frac{2ax+b}{4a}\sqrt{ax^2+bx+c} + \frac{4ac-b^2}{8a}\int \frac{1}{\sqrt{ax^2+b+c}}\,dx$$

67. $\displaystyle\int \frac{x}{\sqrt{ax^2+bx+c}}\,dx = \frac{\sqrt{ax^2+bx+c}}{a} - \frac{b}{2a}\int \frac{1}{\sqrt{ax^2+bx+c}}\,dx$

68. $\displaystyle\int \frac{1}{x\sqrt{ax^2+bx+c}}\,dx$

$$= \frac{-1}{\sqrt{c}}\log\left|\frac{2\sqrt{c}\sqrt{ax^2+bx+c}+bx+2c}{x}\right| \quad (c>0)$$

$$= \frac{1}{\sqrt{-c}}\arcsin\frac{bx+2c}{|x|\sqrt{b^2-4ac}} \quad (c<0)$$

69. $\displaystyle\int x^3\sqrt{x^2+a^2}\,dx = (\tfrac{1}{5}x^2 - \tfrac{2}{15}a^2)\sqrt{(a^2+x^2)^3}$

70. $\displaystyle\int \frac{\sqrt{x^2\pm a^2}}{x^4}\,dx = \frac{\mp\sqrt{(x^2\pm a^2)^3}}{3a^2x^3}$

71. $\displaystyle\int \sin ax\sin bx\,dx = \frac{\sin(a-b)x}{2(a-b)} - \frac{\sin(a+b)x}{2(a+b)} \quad (a^2\neq b^2)$

72. $\displaystyle\int \sin ax\cos bx\,dx = -\frac{\cos(a-b)x}{2(a-b)} - \frac{\cos(a+b)x}{2(a+b)} \quad (a^2\neq b^2)$

73. $\displaystyle\int \cos ax\cos bx\,dx = \frac{\sin(a-b)x}{2(a-b)} + \frac{\sin(a+b)x}{2(a+b)} \quad (a^2\neq b^2)$

74. $\displaystyle\int \sec x\tan x\,dx = \sec x$

75. $\displaystyle\int \csc x\cot x\,dx = -\csc x$

76. $\displaystyle\int \cos^m x\sin^n x\,dx = \frac{\cos^{m-1}x\sin^{n+1}x}{m+n} + \frac{m-1}{m+n}\int \cos^{m-2}x\sin^n x\,dx$

$$= -\frac{\sin^{n-1}x\cos^{m+1}x}{m+n} + \frac{n-1}{m+n}\int \cos^m x\sin^{n-2}x\,dx$$

77. $\displaystyle\int x^n\sin ax\,dx = -\frac{1}{a}x^n\cos ax + \frac{n}{a}\int x^{n-1}\cos ax\,dx$

78. $\displaystyle\int x^n\cos ax\,dx = \frac{1}{a}x^n\sin ax - \frac{n}{a}\int x^{n-1}\sin ax\,dx$

79. $\int x^n e^{ax}\,dx = \frac{x^n e^{ax}}{a} - \frac{n}{a}\int x^{n-1}e^{ax}\,dx$

80. $\int x^n \log ax\,dx = x^{n+1}\left[\frac{\log ax}{n+1} - \frac{1}{(n+1)^2}\right]$

81. $\int x^n(\log ax)^m\,dx = \frac{x^{n+1}}{n+1}(\log ax)^m - \frac{m}{n+1}\int x^n(\log ax)^{m-1}\,dx$

82. $\int e^{ax}\sin bx\,dx = \frac{e^{ax}(a\sin bx - b\cos bx)}{a^2+b^2}$

83. $\int e^{ax}\cos bx\,dx = \frac{e^{ax}(b\sin bx + a\cos bx)}{a^2+b^2}$

84. $\int \operatorname{sech} x \tanh x\,dx = -\operatorname{sech} x$

85. $\int \operatorname{csch} x \coth x\,dx = -\operatorname{csch} x$

APPENDIX B

ANSWERS TO SELECTED EXERCISES

1.1

1. $x = 0, z = 0, y \in \mathbb{R}; x = 0, y = 0, z \in \mathbb{R}; y = 0, x, z \in \mathbb{R}; x = 0, y, z \in \mathbb{R}$
4. $(9, 6, -1)$
5. $4; 17$
6. $(0, 0, 0)$
7. $(-104 + 16a, -24 - 4b, -22 + 26c)$
8. $(-2, 6, 5)$
9. $24\mathbf{i} + 0\mathbf{j} + 0\mathbf{k}$
10. $P = t(\mathbf{i} + 3\mathbf{k}) + s(-2\mathbf{j}), 0 \le t \le 1, 0 \le s \le 1$
13. $\mathbf{l}(t) = (2t - 1)\mathbf{i} - \mathbf{j} + (3t - 1)\mathbf{k}$
14. $P = \alpha\mathbf{a} + \beta\mathbf{b} + \delta\mathbf{c}, 0 \le \alpha \le 1, 0 \le \beta \le 1, 0 \le \delta \le 1$
16. $P = (x_0, y_0, z_0) + t(x_1 - x_0, y_1 - y_0, z_1 - z_0)$
    $+ s(x_2 - x_0, y_2 - y_0, z_2 - z_0), t, s \in \mathbb{R}$

1.2

2. -54
3. 99°
5. 75.7
7. $\|\mathbf{u}\| = \sqrt{5}, \|\mathbf{v}\| = \sqrt{2}, \mathbf{u} \cdot \mathbf{v} = -3$
9. $\|\mathbf{u}\| = \sqrt{11}, \|\mathbf{v}\| = \sqrt{62}, \mathbf{u} \cdot \mathbf{v} = -14$
11. $\|\mathbf{u}\| = \sqrt{14}, \|\mathbf{v}\| = \sqrt{26}, \mathbf{u} \cdot \mathbf{v} = -17$
12. (7) $\mathbf{u}/\|\mathbf{u}\| = (-\mathbf{i} + 2\mathbf{j})/\sqrt{5}, \mathbf{v}/\|\mathbf{v}\| = (+\mathbf{i} - \mathbf{j})/\sqrt{2}$

1.3

2. (a) 4
2. (c) -8
5. $\sqrt{35}$
6. 0
9. $\pm(113\mathbf{i} + 17\mathbf{j} - 103\mathbf{k})/\sqrt{23667}$
11. $2/\sqrt{338}$
13. $\mathbf{u} + \mathbf{v} = 3\mathbf{i} - 3\mathbf{j} + 3\mathbf{k}$; $\mathbf{u} \cdot \mathbf{v} = 6$; $\|\mathbf{u}\| = \sqrt{6}$; $\|\mathbf{v}\| = 3$; $\mathbf{u} \times \mathbf{v} = -3\mathbf{i} + 3\mathbf{k}$
15. (a) $x + y + z - 1 = 0$
16. $2x + 3y + 4z = 0.$
17. HINT: Use the triple-vector-product identities $(\mathbf{A} \times \mathbf{B}) \times \mathbf{C} = (\mathbf{A} \cdot \mathbf{C})\mathbf{B} - (\mathbf{B} \cdot \mathbf{C})\mathbf{A}$ and $\mathbf{A} \times (\mathbf{B} \times \mathbf{C}) = (\mathbf{A} \cdot \mathbf{C})\mathbf{B} - (\mathbf{A} \cdot \mathbf{B})\mathbf{C}$

1.4

2. (b) HINT: $(\|\mathbf{x} + \mathbf{y}\| \, \|\mathbf{x} - \mathbf{y}\|)^2 = \|\mathbf{x} + \mathbf{y}\|^2 \|\mathbf{x} - \mathbf{y}\|^2 = [(\mathbf{x} + \mathbf{y}) \cdot (\mathbf{x} + \mathbf{y})][(\mathbf{x} - \mathbf{y}) \cdot (\mathbf{x} - \mathbf{y})]$
6. $$AB = \begin{bmatrix} 3 & 1 & -3 \\ 5 & -1 & 1 \\ 1 & 1 & -1 \end{bmatrix}$$
   $\det A = 4$; $\det B = -3$; $\det(AB) = -12$
9. HINT: For $k = 2$ use the triangle inequality to show that $\|\mathbf{x}_1 + \mathbf{x}_2\| \leq \|\mathbf{x}_1\| + \|\mathbf{x}_2\|$; then for $k = i + 1$ note that $\|\mathbf{x}_1 + \mathbf{x}_2 + \cdots + \mathbf{x}_{i+1}\| \leq \|\mathbf{x}_1 + \mathbf{x}_2 + \cdots + \mathbf{x}_i\| + \|\mathbf{x}_{i+1}\|$

Review Exercises for Chapter 1

1. $\mathbf{v} + \mathbf{w} = 4\mathbf{i} + 3\mathbf{j} + 6\mathbf{k}$; $3\mathbf{v} = 9\mathbf{i} + 12\mathbf{j} + 15\mathbf{k}$; $6\mathbf{v} + 8\mathbf{w} = 26\mathbf{i} + 16\mathbf{j} + 38\mathbf{k}$; $-2\mathbf{v} = -6\mathbf{i} - 8\mathbf{j} - 10\mathbf{k}$; $\mathbf{v} \cdot \mathbf{w} = 4$; $\mathbf{v} \times \mathbf{w} = 9\mathbf{i} + 2\mathbf{j} - 7\mathbf{k}$
2. (b) $\mathbf{l}(t) = (0, 1, t)$ or $\mathbf{l}(t) = (0, 1, 1 - t)$
5. Let $\mathbf{v} = (a_1, a_2, a_3)$, $\mathbf{w} = (b_1, b_2, b_3)$, and apply the CBS inequality.
7. The area is the absolute value of
$$\begin{vmatrix} a_1 & a_2 \\ b_1 & b_2 \end{vmatrix} = \begin{vmatrix} a_1 & a_2 \\ b_1 + \lambda a_1 & b_2 + \lambda a_2 \end{vmatrix}$$
   (A multiple of one row of a determinant may be added to another row without changing its value.)
8. 1
13. (b) $\sqrt{2}$
15. (a) Note that
$$\frac{1}{2}\begin{vmatrix} 1 & 1 & 1 \\ x_1 & x_2 & x_3 \\ y_1 & y_2 & y_3 \end{vmatrix} = \frac{1}{2}\begin{vmatrix} 1 & 0 & 0 \\ x_1 & x_2 - x_1 & x_3 - x_1 \\ y_1 & y_2 - y_1 & y_3 - y_1 \end{vmatrix} = \frac{1}{2}\begin{vmatrix} x_2 - x_1 & x_3 - x_1 \\ y_2 - y_1 & y_3 - y_1 \end{vmatrix}$$
17. $$AB = \begin{bmatrix} 3 & 0 & 4 \\ 2 & 0 & 3 \\ 1 & 0 & 2 \end{bmatrix} \text{ and } BA = \begin{bmatrix} 4 & 0 & 2 \\ 6 & 0 & 3 \\ 1 & 0 & 1 \end{bmatrix}, \text{ so } AB \neq BA.$$
19. yes

1. The level curves and graphs are sketched below.
(a)

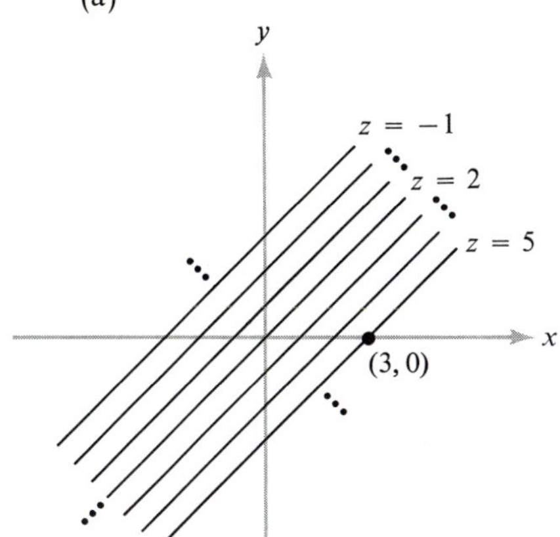

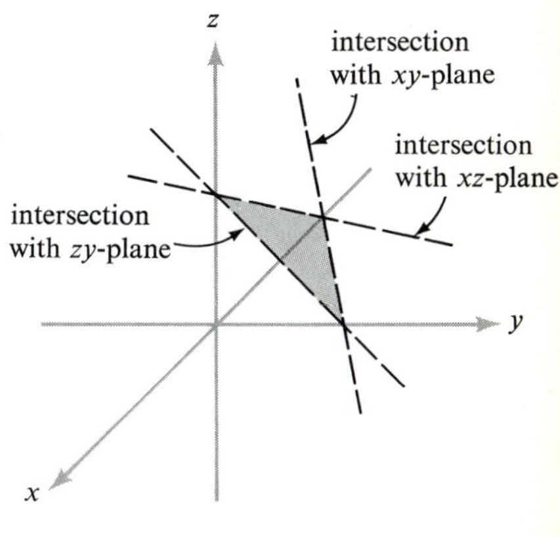

(b)

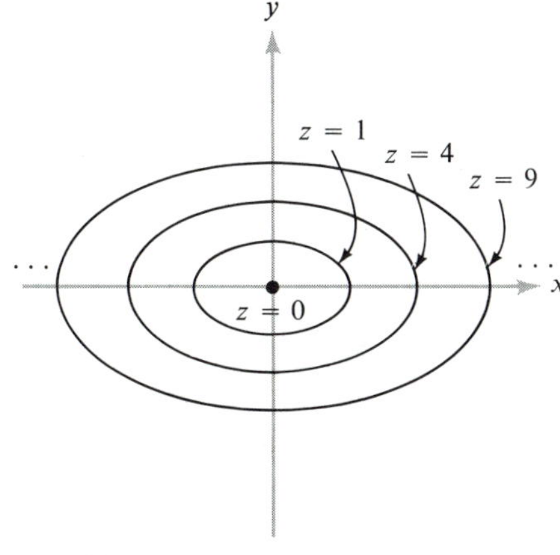

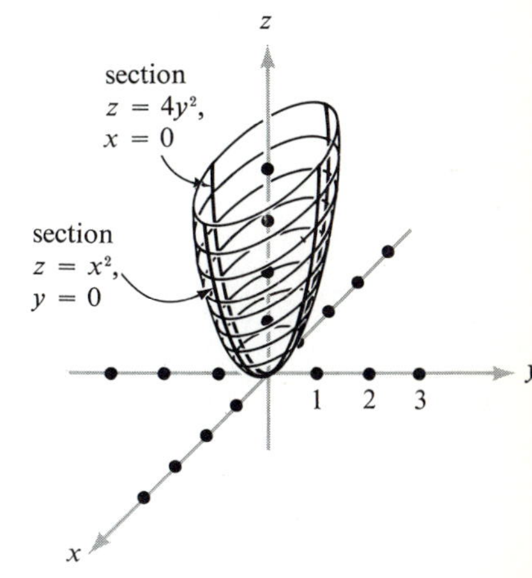

(c)

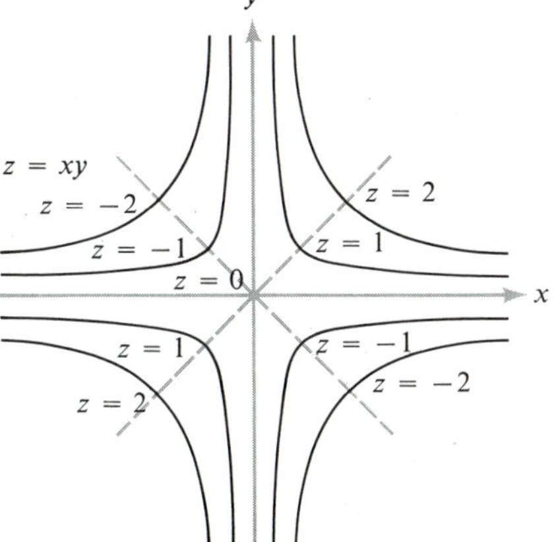

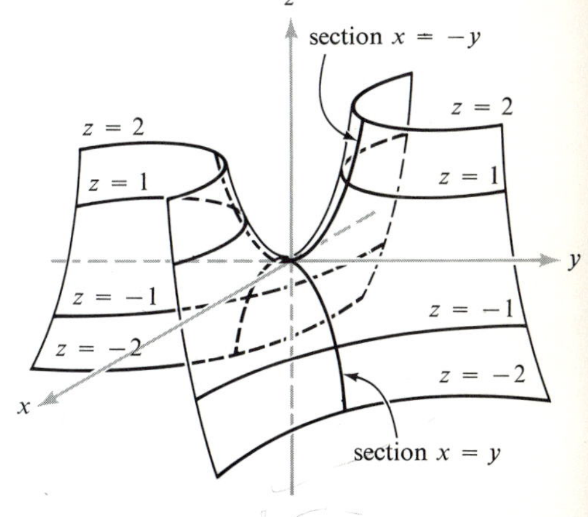

4. The level curves are circles centered at the origin. The graph is a cone.

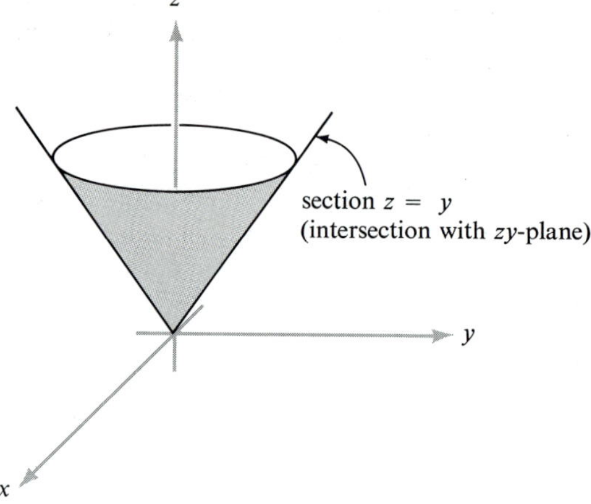

10. For $c > 0$ the level surfaces are ellipsoids. They are empty for $c < 0$, and a point if $c = 0$.
$S_{y=0} \cap$ graph is $t = 4x^2 + 9z^2$, an elliptic paraboloid in x, z, t space.
$S_{x=0} \cap$ graph and $S_{z=0} \cap$ graph are elliptic paraboloids.
$S_{t=0} \cap$ graph is a point.
15. The graph is sketched here.

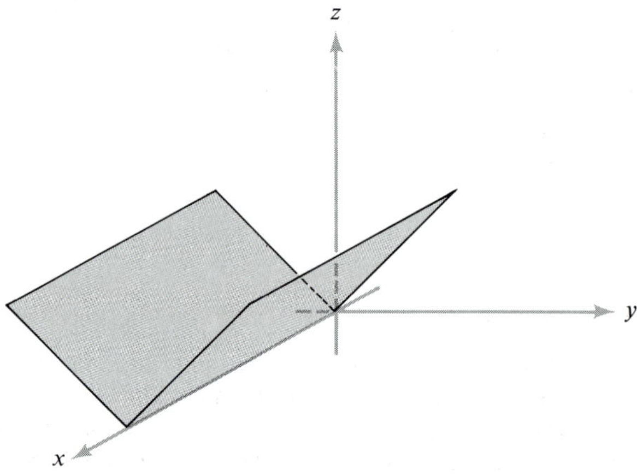

17. The level curves are rays $x = \lambda y$.
18. The level curves are sketched in this figure.

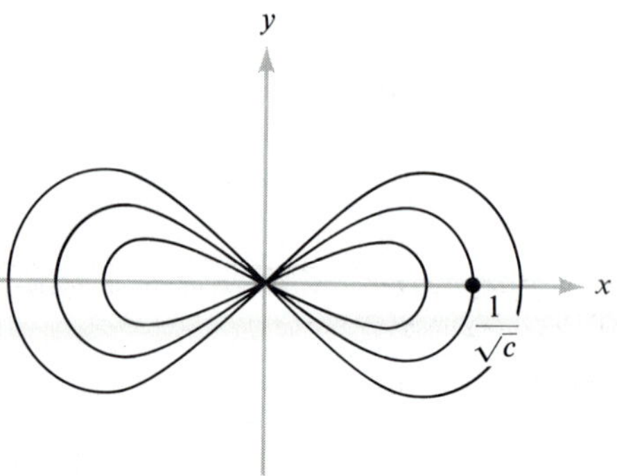

2.2

1. (a) If $(x_0, y_0) \in A$ then $|x_0| < 1$ and $|y_0| < 1$. Let $r < 1 - |x_0|$ and $r < 1 - |y_0|$. Then prove that $D_r(x_0, y_0) \subset A$ either analytically or by drawing a figure.
   (b) If $(x_0, y_0) \in B$ and $0 < r < x_0$ (e.g., $r = x_0/2$), then $D_r(x_0, y_0) \subset B$ (prove analytically or by drawing a figure).
4. Use the assertion in italics on p. 71. Note that any interval about a or b overlaps $]a, b[$, but if c is a point not in $[a, b]$, say $c < a$, then the interval $]-\infty, a[$ contains c but doesn't overlap $]a, b[$.
7. (a) 1; (d) limit doesn't exist (look at the limits for $x = 0$ and $y = 0$ separately)
9. 0

10. (b) no

11. (b) $\lim_{x \to 0-} (1/x) = -\infty$, $\lim_{t \to -\infty} e^t = 0$, so $\lim_{x \to 0-} e^{1/x} = 0$. Hence $\lim_{x \to 0-} 1/(1 + e^{1/x}) = 1$. The other limit is 0.
    (c)

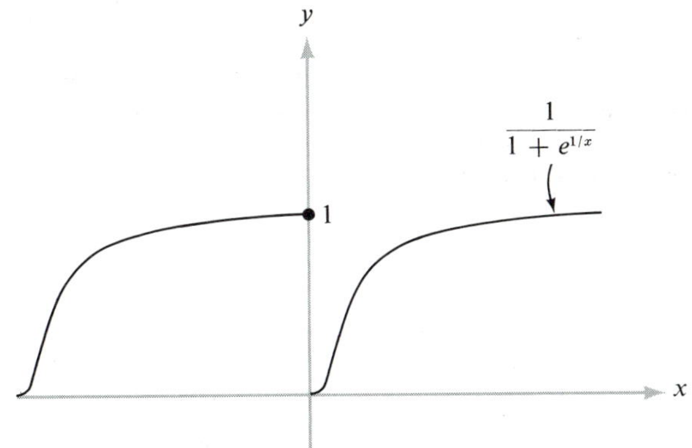

12. (a) The ε-δ definition of limits with $\varepsilon = 1/100$ and $\lim_{a \to 0} a^3 + 3a^2 + a = 0$ guarantees the existence of a δ. (One can actually find δ—any number less than 1/500 is such a δ.)
13. (d) 1; (e) $-1/2$

14. (a) 3; (b) 0; (c) 1

18. Use the fact that polynomials, the cosine function, and sums and compositions of continuous functions are continuous to conclude that f is continuous.

2.3

1. (a) $\partial f/\partial x = y$
   (b) $\partial f/\partial y = xe^{xy}$
   (d) $\partial f/\partial x = 2x(1 + \log(x^2 + y^2))$, $(x, y) \neq (0, 0)$
2. (a) $\partial z/\partial x = 0$ at $(0, 0)$, $-\sqrt{2}/2$ at $(a/2, a/2)$
   (b) $\frac{\partial z}{\partial x}(1, 2) = \frac{1}{3}$, $\frac{\partial z}{\partial y}(1, 2) = \frac{1}{6}$

(c) $\frac{\partial z}{\partial x}\left(\frac{2\pi}{b}, 0\right) = ae^{2\pi a/b}, \qquad \frac{\partial z}{\partial y}\left(\frac{2\pi}{b}, 0\right) = 0$

3. (b) $\frac{\partial w}{\partial x} = -\frac{4xy^2}{(x^2 - y^2)^2}$

(e) $\frac{\partial w}{\partial x} = -y^2 e^{xy} \sin ye^{xy} \sin x + \cos ye^{xy} \cos x$

$\frac{\partial w}{\partial y} = (xye^{xy} + e^{xy})(-\sin ye^{xy} \sin x)$

4. (a) f is C^1 on its domain, namely $\mathbb{R}^2\backslash\{(0, 0)\}$.

6. (a) $z = 0$

(c) $z = -x$

(d) $z = 2y - 2$

7. (a) $\begin{bmatrix} 1 & 0 \\ 0 & 1 \end{bmatrix}$ (c) $\begin{bmatrix} 1 & 1 & e^z \\ 2xy & x^2 & 0 \end{bmatrix}$

10. 1.08

11. (a) $\nabla f = (e^{-x^2-y^2-z^2}(-2x^2 + 1), -2xye^{-x^2-y^2-z^2}, -2xze^{-x^2-y^2-z^2})$

2.4

1. Let $g(\mathbf{x}) = f(\mathbf{x})^2 + 2f(\mathbf{x})$. Then $Dg(\mathbf{x}) = 2f(\mathbf{x})Df(\mathbf{x}) + 2Df(\mathbf{x})$.

2. (a) $Df(x, y) = [0 \;\; 0]$

(d) $Df(x) = [2x \;\; 2y]$

3. (a) $h(x, y) = f(x, u(x, y)) = f(p(x), u(x, y))$

We introduce p here solely as notation: $p(x) = x$

Written out: $\frac{\partial h}{\partial x} = \frac{\partial f}{\partial p}\frac{dp}{dx} + \frac{\partial f}{\partial u}\frac{\partial u}{\partial x} = \frac{\partial f}{\partial p} + \frac{\partial f}{\partial u}\frac{\partial u}{\partial x}$

since $\frac{dp}{dx} = \frac{dx}{dx} = 1$

Justification: Call (p, u) the variables of f. In order to use the Chain Rule we must express h as a composition of functions; i.e., first find g such that $h(x, y) = f(g(x, y))$. Let $g(x, y) = (p(x), u(x, y))$. Therefore $Dh = Df \cdot Dg$. Then

$$\begin{bmatrix} \frac{\partial h}{\partial x} & \frac{\partial h}{\partial y} \end{bmatrix} = \begin{bmatrix} \frac{\partial f}{\partial p} & \frac{\partial f}{\partial u} \end{bmatrix} \begin{bmatrix} \frac{\partial g_1}{\partial x} & \frac{\partial g_1}{\partial y} \\ \frac{\partial g_2}{\partial x} & \frac{\partial g_2}{\partial y} \end{bmatrix} = \begin{bmatrix} \frac{\partial f}{\partial p} & \frac{\partial f}{\partial u} \end{bmatrix} \begin{bmatrix} 1 & 0 \\ \frac{\partial u}{\partial x} & \frac{\partial u}{\partial y} \end{bmatrix}$$

$$= \begin{bmatrix} \frac{\partial f}{\partial p} + \frac{\partial f}{\partial u}\frac{\partial u}{\partial x} & \frac{\partial f}{\partial u}\frac{\partial u}{\partial y} \end{bmatrix}$$

So $\frac{\partial h}{\partial x} = \frac{\partial f}{\partial p} + \frac{\partial f}{\partial u}\frac{\partial u}{\partial x}$

You may see $\frac{\partial h}{\partial x} = \frac{\partial f}{\partial x} + \frac{\partial f}{\partial u}\frac{\partial u}{\partial x}$ as an answer.

This requires careful interpretation because of possible ambiguity about the meaning of $\partial f/\partial x$, which is why the name p was used.

3. (c) $\dfrac{\partial h}{\partial x} = \dfrac{\partial f}{\partial u}\dfrac{\partial u}{\partial x} + \dfrac{\partial f}{\partial v}\dfrac{\partial v}{\partial x} + \dfrac{\partial f}{\partial w}\dfrac{dw}{dx}$

6. (b) $\mathbf{c}'(t) = (6t, 3t^2)$
7. Use Theorem 10 (*iii*) and replace matrices by vectors.

11. (a) $G(x, y(x)) = 0$ so $\dfrac{\partial G}{\partial x} + \dfrac{\partial G}{\partial y}\dfrac{dy}{dx} = 0$

(b) $\begin{bmatrix} \dfrac{dy_1}{dx} \\ \dfrac{dy_2}{dx} \end{bmatrix} = -\begin{bmatrix} \dfrac{\partial G_1}{\partial y_1} & \dfrac{\partial G_1}{\partial y_2} \\ \dfrac{\partial G_2}{\partial y_1} & \dfrac{\partial G_2}{\partial y_2} \end{bmatrix}^{-1} \begin{bmatrix} \dfrac{\partial G_1}{\partial x} \\ \dfrac{\partial G_2}{\partial x} \end{bmatrix}$ where $^{-1}$ means the inverse matrix,

or e.g.

$$\frac{dy_1}{dx} = \frac{-\dfrac{\partial G_1}{\partial x}\dfrac{\partial G_2}{\partial y_2} + \dfrac{\partial G_2}{\partial x}\dfrac{\partial G_1}{\partial y_2}}{\dfrac{\partial G_1}{\partial y_1}\dfrac{\partial G_2}{\partial y_2} - \dfrac{\partial G_2}{\partial y_1}\dfrac{\partial G_1}{\partial y_2}}$$

(c) $\dfrac{dy}{dx} = \dfrac{-2x}{3y^2 + e^y}$

13. Define $R_1(\mathbf{h}) = f(\mathbf{x}_0 + \mathbf{h}) - f(\mathbf{x}_0) - Df(\mathbf{x}_0) \cdot \mathbf{h}$.
14. The derivative exists if $p = 2$, exists and is continuous if $p > 2$. The derivative doesn't exist at $x = 0$ if $p < 2$.

2.5

2. (a) -5
   (b) $2/\sqrt{5}$
   (c) $1/\sqrt{5}$

3. (c) $x + y + z = 3$
   (f) $z = 1$

6. $\nabla f(0, 0, 1) = (0, 1, -1) = \mathbf{j} - \mathbf{k}$

8. (a) $\left(\dfrac{x}{(x^2 + y^2 + z^2)^{1/2}}, \dfrac{y}{(x^2 + y^2 + z^2)^{1/2}}, \dfrac{z}{(x^2 + y^2 + z^2)^{1/2}}\right)$

(b) $(y + z, x + z, x + y)$

(c) $\left(\dfrac{-2x}{(x^2 + y^2 + z^2)^2}, \dfrac{-2y}{(x^2 + y^2 + z^2)^2}, \dfrac{-2z}{(x^2 + y^2 + z^2)^2}\right)$

9. (b) $\nabla f = (yze^{xyz}, xze^{xyz}, xye^{xyz})$, $g'(t) = [6\ 6t\ 3t^2]$, $(f \circ g)'(1) = 108e^{18}$
12. $(e^{-1}, 2e^{-2}, 3e^{-3})$

2.6

1. (a) $\dfrac{\partial^2 f}{\partial x^2} = 24\dfrac{x^3y - xy^3}{(x^2 + y^2)^4}$, $\dfrac{\partial^2 f}{\partial y^2} = 24\dfrac{-x^3y + xy^3}{(x^2 + y^2)^4}$,

$$\frac{\partial^2 f}{\partial x\,\partial y} = \frac{\partial^2 f}{\partial y\,\partial x} = \frac{-6x^4 + 36x^2y^2 - 6y^4}{(x^2 + y^2)^4}$$

(b) $\dfrac{\partial^2 f}{\partial x^2} = \dfrac{2}{x^3}, \dfrac{\partial^2 f}{\partial y^2} = xe^{-y}, \dfrac{\partial^2 f}{\partial x\, \partial y} = \dfrac{\partial^2 f}{\partial y\, \partial x} = -e^{-y}$

3. $\dfrac{\partial^2 f}{\partial x^2}\left(\dfrac{dx}{dt}\right)^2 + 2\dfrac{\partial^2 f}{\partial x\, \partial y}\dfrac{dx}{dt}\dfrac{dy}{dt} + \dfrac{\partial^2 f}{\partial y^2}\left(\dfrac{dy}{dt}\right)^2 + \dfrac{\partial f}{\partial x}\dfrac{d^2 x}{dt^2} + \dfrac{\partial f}{\partial y}\dfrac{d^2 y}{dt^2},$

where $\mathbf{c}(t) = (x(t), y(t))$

4. $\dfrac{\partial^2 u}{\partial x^2} = e^x \sin y = -\dfrac{\partial^2 u}{\partial y^2}$

5. $2x + 6y - z = 5$

8. (b) $-\sin \sqrt{2}$

2.7

1. $Df(x, y, z) = \begin{bmatrix} e^x & 0 & 0 \\ 0 & -\sin y & 0 \\ 0 & 0 & \cos z \end{bmatrix}$

Df is diagonal if f_n is a function of x_n alone.

2. (a) For each i, $|y_i| \le \left(\sum_{j=1}^{n} A_{ij}^2\right)^{1/2}\left(\sum_{j=1}^{n} x_j^2\right)^{1/2}$

by the CBS inequality. Hence

$$y_i^2 \le \left(\sum_{j=1}^{n} A_{ij}^2\right)\left(\sum_{j=1}^{n} x_j^2\right);$$

so adding,

$$\|A\mathbf{x}\| \le \left\{\left(\sum_{i=1}^{n}\sum_{j=1}^{n} A_{ij}^2\right)\left(\sum_{j=1}^{n} x_j^2\right)\right\}^{1/2} = M\|\mathbf{x}\|.$$

(b) Given $\varepsilon > 0$, let $\delta = \varepsilon/M$.

(c) Let $\mathbf{y}_0 = A\mathbf{x}_0/\|\mathbf{x}_0\|$. Note that $\mathbf{y}_0 = A(\lambda\mathbf{x}_0)/\|\lambda\mathbf{x}_0\|$ if $\lambda > 0$, and $\lambda\mathbf{x}_0 \to \mathbf{0}$ as $\lambda \to 0$. Therefore, $\mathbf{y}_0 \to \mathbf{0}$ as $\lambda \to 0$ since $\lim_{\mathbf{x}\to\mathbf{0}} A\mathbf{x}/\|\mathbf{x}\| = \mathbf{0}$. Hence $\mathbf{y}_0 = \mathbf{0}$, so $A = 0$. (Part (b) is not needed for (c).)

3. (a) Let $A = B = C = \mathbb{R}$ with $f(x) = 0$ and $g(x) = 0$ if $x \ne 0$ and $g(0) = 1$. Then $w = 0$ and $g(f(x)) = 1$ for all x.

4. (a) See 2(b).

6. $\dfrac{\partial f}{\partial x} = \dfrac{-2x^2y^2 + 2y^6}{(x^2 + y^4)^2}$ if $(x, y) \ne (0, 0)$

$\dfrac{\partial f}{\partial x} = 0$ if $(x, y) = (0, 0)$

Show f is not continuous by approaching $(0, 0)$ along $x = y^2$ (to get a limit 1) and along $y = 0$ (to get a limit 0). Since differentiable functions are continuous, f is not differentiable.

7. Show that $T = Df(x_0, y_0) = \begin{bmatrix} \dfrac{\partial g}{\partial x}(x_0) & \dfrac{\partial h}{\partial y}(y_0) \end{bmatrix}$

satisfies the definition of differentiability (that is, formula (1)).

8. Use $|\langle \mathbf{v}, \mathbf{x}\rangle - \langle \mathbf{v}, \mathbf{x}_0\rangle| \le \|\mathbf{v}\|\,\|\mathbf{x} - \mathbf{x}_0\|$.

9. Use the limit theorems and the fact that the function $g(x) = \sqrt{|x|}$ is continuous. (Prove the last statement.)
10. This requires an examination of the proof.
11. For continuity at (0, 0), use the fact that

$$\left|\frac{xy}{(x^2+y^2)^{1/2}}\right| \le \frac{|xy|}{(x^2)^{1/2}} = |y|$$

or $|xy| \le (x^2+y^2)/2$. See also Exercise 13, Section 2.7.

Review Exercises for Chapter 2

1. (a) elliptic paraboloid
   (b) Let $y' = y + 3$ and write $z = xy'$. This is a (shifted) hyperbolic paraboloid.
2. (a) The level surfaces are ellipsoids of revolution. Plane sections through the origin are elliptic paraboloids, (or points).
3. (b) $Df(x) = \begin{bmatrix} 1 \\ 1 \end{bmatrix}$

   (c) $Df(x, y, z) = [e^x \quad e^y \quad e^z]$
5. The tangent plane to a sphere at (x_0, y_0, z_0) is normal to the line from the center to (x_0, y_0, z_0).
6. (b) $z = 4x - 8y - 8$
   (d) $4z = -6 + \pi + 10x + 6y$
   (e) $2z = \sqrt{2}\,x + \sqrt{2}\,y$
8. (b) $x - 2y + z = 0$
   (c) $x = \pi/2$
11. (a) 0
    (b) limit doesn't exist
12. (b) $\dfrac{\partial f}{\partial y} = 10(x + y + z)^9$
17. $(\mathbf{i} + \mathbf{j})/\sqrt{2}$
18. (a) 2/3 (NOTE: $\mathbf{v}$ must be normalized.)
20. $(-1, -1)$
27. (b) Neither function is continuous at (0, 0).
28. (b) $$\nabla f(x, y) = \left(y \sin\left(\frac{1}{x^2+y^2}\right) - \frac{2x^2y}{(x^2+y^2)^2}\cos\left(\frac{1}{x^2+y^2}\right),\right.$$
$$\left. x \sin\left(\frac{1}{x^2+y^2}\right) - \frac{2y^2x}{(x^2+y^2)^2}\cos\left(\frac{1}{x^2+y^2}\right)\right),$$

    $(x, y) \neq (0, 0)$
    $\nabla f(0, 0) = (0, 0)$
31. $(-4e^{-1}, 0)$

3.1

1. (a) $\boldsymbol{\sigma}'(t) = (2\pi \cos 2\pi t, \ -2\pi \sin 2\pi t, \ 2 - 2t)$, $\boldsymbol{\sigma}'(0) = (2\pi, 0, 2)$
   (c) $\boldsymbol{\sigma}'(t) = (2t, 3t^2 - 4, 0)$, $\boldsymbol{\sigma}'(0) = (0, -4, 0)$
2. (a) $\mathbf{r}'(0) = 6\mathbf{i}$, $\mathbf{r}''(0) = 6\mathbf{j}$, tangent line is $\mathbf{l}(t) = 6t\mathbf{i}$
   (d) $\boldsymbol{\sigma}'(1) = (0, 0, 1)$, $\boldsymbol{\sigma}''(1) = (0, 0, 0)$, tangent line is $\mathbf{l}(t) = (0, 0, 1 + t)$
3. $6m\mathbf{j}$, where m is the particle's mass
5. $\boldsymbol{\sigma}(t) = \left(\dfrac{t^2}{2}, e^t - 6, \dfrac{t^3}{3} + 1\right)$

6. (b) $\boldsymbol{\sigma}(t) = (\frac{1}{2}\cos t, \sin t)$, an ellipse
7. $T = 5662$ seconds $= 1.57$ hours

3.2

1. (a) 7; (b) $4\sqrt{2} - 2$; (d) $\sqrt{21} + \frac{5}{4}[\log(4 + \sqrt{21}) - \log\sqrt{5}]$
   (f) Use the substitution $u = e^t$ to show that the integral is $2(e - e^{-1})$.
2. The arc-length function for β is $s(t) = \sqrt{2}\,t$, $t > 0$. For α see the hint in 1(f).
4. (8, 8, 0)

3.3

1. (a)
$$\frac{dE}{dt} = \frac{1}{2}m \cdot 2\langle \mathbf{r}'(t), \mathbf{r}''(t)\rangle + \langle \text{grad } V(\mathbf{r}(t)), \mathbf{r}'(t)\rangle$$
$$= \langle \mathbf{r}'(t), -\text{grad } V(\mathbf{r}(t))\rangle + \langle \text{grad } V(\mathbf{r}(t)), \mathbf{r}'(t)\rangle$$
$$= 0.$$
   (b) Use (a)
2. (a) Flow lines are circles flowing clockwise around the origin.
   (c) Flow lines are parabolas $y = ax^2 + b$.
3.
$$\frac{d}{dt}V(\mathbf{c}(t)) = \langle \nabla V(\mathbf{c}(t)), \mathbf{c}'(t)\rangle$$
$$= -\langle \nabla V(\mathbf{c}(t)), \nabla V(\mathbf{c}(t))\rangle \leq 0$$
4. The equipotential surfaccs are circles centered at $(1/2c, 1/2c)$ passing through the origin, and $-\nabla V$ is perpendicular to these.
5. Use the fact that $-\nabla T$ is perpendicular to the surface $T =$ constant.

3.4

1. (a) $\nabla f = \mathbf{r}/\|\mathbf{r}\|$, $\mathbf{r} = (x, y, z)$
   (b) $\nabla f = (y + z, x + z, y + x)$
2. (c) $6x + 8y + 10z$
3. (b) $\mathbf{0}$
6. (b) $x^3y + y^4/4 + c$
9. Show $\nabla \times \mathbf{F} \neq \mathbf{0}$. ($\nabla \times \mathbf{F}(0, 0) = -\mathbf{k} \neq \mathbf{0}$).

3.5

1. only (a)
6. (b) $3y^2zx\mathbf{i} + (-y^3z + 4xz)\mathbf{j}$
   (e) $-y^3zx^3\mathbf{i} + 2x^2y^4z\mathbf{j} + (2x^3z^2 - 2xy)\mathbf{k}$
7. No, consider $\mathbf{F} = x\mathbf{i} + xy\mathbf{j} + \mathbf{k}$; $\nabla \times \mathbf{F} = y\mathbf{k}$, which is not perpendicular to $\mathbf{F}$.

Review Exercises for Chapter 3

1. (a) 2; (c) 14
2. (b) $-\mathbf{i} - \mathbf{j} - \mathbf{k}$
3. (c) $(2xyz^3 - 3z^2xy^2)\mathbf{i} - (y^2z^3 - 2x^2y^2z)\mathbf{j} + (y^2z^3 - 2x^2yz^2)\mathbf{k}$
4. (b) $\nabla \cdot \mathbf{F} = 2x + 2y + 2z$, $\nabla \times \mathbf{F} = \mathbf{0}$
5. (c) $z = 10x + 10y - 25$
6. 35,830 kilometers
8. (a) Use the fact that $t'(s) = 1/s'(t) = 1/\|\boldsymbol{\alpha}'(t)\|$
   (b) $\boldsymbol{\alpha}(t) = (a\cos(t/\sqrt{a^2 + b^2}), a\sin(t/\sqrt{a^2 + b^2}), bt/\sqrt{a^2 + b^2})$

9. $\mathbf{F} = (0, 0, -m)$
10. (a) HINT: Differentiate $\langle \mathbf{c}(t), \mathbf{c}(t)\rangle = 1$ with respect to t.
11. $(2\pi, 3\pi^2, -2\pi)$
13. (a) $6/\sqrt{14}$
14. (c) $5x + y + z - 6 = 0$; (d) $15°48' = \arccos(5\sqrt{3}/9)$
16. (b) $3\sqrt{3}$
17. (a) no
18. (a) $(3, 0, 0)$
19. (b) $8z + 6\sqrt{2}\,y = 20$
20. (b) $f(x, y, z) = x^2ye^z + z^3/3$

4.1

1. $f(h_1, h_2) = h_1^2 + 2h_1h_2 + h_2^2$ ($R_2(\mathbf{h}, \mathbf{0}) = 0$ in this case)
2. $f(h_1, h_2) = 1 + h_1 + h_2 + \dfrac{h_1^2}{2} + h_1h_2 + \dfrac{h_2^2}{2} + R_2(\mathbf{h}, \mathbf{0})$
5. $f(h_1, h_2) = 1 + h_1h_2 + R_2(\mathbf{h}, \mathbf{0})$
7. (a) Show that $|R_k(x, a)| \le AB^{k+1}/(k+1)!$ for constants A, B and x in a fixed interval $[a, b]$. Prove $R_k \to 0$ as $k \to \infty$. (Use convergence of the series $\sum c^k/k! = e^c$ and use Taylor's Theorem.)

4.2

1. $(0, 0)$, saddle point
3. The critical points are on the line $y = -x$; they are local minima because $f(x, y) = (x + y)^2 \ge 0$, equaling zero only when $x = -y$.
5. $(0, 0)$, saddle point
7. $(-1/4, -1/4)$, local minimum
9. $(0, 0)$, local maximum (The tests fail, but use the fact that $\cos(z) \le 1$.)
   $(\sqrt{\pi/2}, \sqrt{\pi/2})$, local minimum
   $(0, \sqrt{\pi})$, local minimum
11. (b) Show that $f(g(t)) = 0$ at $t = 0$, and $f(g(t)) \ge 0$ if $|t| < |b|/3a^2$.
    (c) f is negative on the parabola $y = 2x^2$.
13. The critical points are on the line $y = x$ and they are local minima (see Exercise 3).
15. The only critical point is $(0, 0, 0)$. It is a minimum since

$$f(x, y, z) \ge \frac{x^2 + y^2}{2} + z^2 + xy = \frac{1}{2}(x + y)^2 + z^2 \ge 0$$

4.3

1. maximum at $\sqrt{\tfrac{2}{3}}(1, -1, 1)$, minimum at $\sqrt{\tfrac{2}{3}}(-1, 1, -1)$
2. no critical points
4. maximum at $\left(\dfrac{9}{\sqrt{70}}, \dfrac{4}{\sqrt{70}}\right)$, minimum at $\left(-\dfrac{9}{\sqrt{70}}, -\dfrac{4}{\sqrt{70}}\right)$
6. minimum at $(0, 2)$
7. Use a geometrical picture rather than Lagrange multipliers.
8. $(0, 1)$ is a minimum of f; it is the only critical point.
9. $(0, 0, 2)$ is a minimum of f.
11. $r = \sqrt[3]{1/2\pi}$, $h = 2\sqrt[3]{1/2\pi}$

4.4

1. $(-1/4, -1/4)$

2. (0, 0) is not stable. (V increases along the ray $y = x$ for $x > 0$ and decreases for $x < 0$.)
3. stable equilibrium point $(2 + m^2g^2)^{-1/2}(-1, -1, -mg)$
5. no critical points

Review Exercises for Chapter 4

2. (a) saddle point
   (b) if $|C| < 2$, strict minimum; if $|C| > 2$, saddle point; if $C = \pm 2$, minimum
3. (a) $z = 3x - 1$
   (b) $\sqrt{3}z = 3x + 1$
4. (b) (1, 1) local minimum
   (c) (0, 0) is a minimum (The only critical point is (0, 0); to show $x^2 + xy^2 + y^4 \geq 0$, suppose $x^2 + xy^2 + y^4 = -c$ for $c > 0$ and solve for x.)
5. (a) 1
   (b) $\sqrt{83}/6$
7. (a) $(yz^2, xz^2, 2xyz)$
   (c) $\nabla \cdot \mathbf{F} = 9$
   (d) $\nabla \times \mathbf{F} = \mathbf{0}$
   (e) $\nabla \cdot \mathbf{F} = 3$
8. $1 - \dfrac{h_1^2}{2} + h_1 h_2 + R_2(\mathbf{h}, \mathbf{0})$
9. $z = 1/4$
10. $(0, 0, \pm 1)$
13. If $b \geq 2$, the minimum distance is $2\sqrt{b-1}$; if $b \leq 2$, the minimum distance is $|b|$.
14. (a) $|b_1 a_1 + b_2 a_2 + b_3 a_3 + b_0| / \sqrt{b_1^2 + b_2^2 + b_3^2}$
15. not stable

5.1

1. (a) 13/15; (b) $\pi + 1/2$; (c) 1; (d) $\log 2 - 1/2$
3. In order to show that the volume of the two cylinders are equal, show that their area functions are equal.
5. (a) 26/9; (b) $2/\pi$; (c) $(2/\pi)(e^2 + 1)$
6. 11/2
7. 196/15

5.2

3. (a) 7/12; (b) $e - 2$; (c) $\frac{1}{9}\sin 1$
   (d) $1/(m + 1)(n + 1)$; (e) $\frac{1}{2}(a + b) + c$;
   (f) $2 \sin 1 - \sin 2$
4. 8/15
5. Use Fubini's Theorem to represent

$$\int_R [f(x)g(y)]\, dx\, dy \quad \text{by} \quad \int_c^d g(y)\left[\int_a^b f(x)\, dx\right] dy$$

and evaluate, using the Fundamental Theorem, to get $[G(d) - G(c)] \times [F(b) - F(a)] = [\int_c^d g(y)\, dy][\int_a^b f(x)\, dx]$, where $dF(x)/dx = f(x)$ and $dG(y)/dy = g(y)$.

6. 1
7. 11/6

10. Use the fact that the volume of a solid with base area A and vertical walls of height h is Ah
12. The discontinuities of f do not form a set of area zero. This is because the discontinuity set is composed of an infinite number of graphs $y = \phi(x) = 1/m$ and $x = \psi(y) = 1/n$, where m and $n \geq 1$.

5.3

1. (a) 1/3, both; (b) 5/2, type 1; (c) $(e^2 - 1)/4$, both; (d) 1/35, both
2. (a) 7895/84, type 2; (b) $\frac{1}{3}(e^{-3} - 1) + \frac{1}{2}(e^2 - 1) + e^{-1}$, type 1; (c) $\pi/4$, type 3; (d) 1/6, type 3; (e) $1/(2n + 3)(n + 2) + 1/(m + 2)(m + 3)$, type 3; (f) $-2/3$, type 3
5. 28,000 ft^3
6. 15625/1296
8. 33/140; D is the region $0 \leq x \leq 1, 0 \leq y \leq x^2$
9. $\pi/2$
10. 50π
14. no

5.4

1. (a) 1/8; (b) $\pi/4$; (c) $-17/12$
   (d) $G(b) - G(a)$, where $\partial G/\partial y = F(y, y) - F(a, y)$ and $\partial F/\partial x = f(x, y)$
2. (a) 2/3
4. Note that the maximum value of f on D is e and the minimum value of f on D is $1/e$. Use the ideas in the proof of Theorem 4 to show that

$$\frac{1}{e} \leq \frac{1}{4\pi^2} \int f(x, y)\, dA \leq e$$

6. $\frac{4}{3}\pi abc$

5.5

1. (a) 4; (b) 8/3; (c) 3/16
2. (b) $\frac{1}{4}(1 - 1/e)$
4. $\pi/4$
7. Use the fact that

$$\frac{\sin^2(x - y)}{\sqrt{1 - x^2 - y^2}} \leq \frac{1}{\sqrt{1 - x^2 - y^2}}$$

9. Use the fact that $e^{x^2 + y^2}/(x - y) \geq 1/(x - y)$ on the given region.

5.6

1. 1/3
2. $1/e$
3. 0
5. $-1/6$
8. $(4\pi/3)(1 - \sqrt{2}/2)$

5.7

1. $S =$ the unit disc minus its center
3. $D = [0, 5] \times [0, 1]$
5. D is the set of (x, y, z) with $x^2 + y^2 + z^2 \leq 1$ (the unit ball). T is not one-to-one, but is one-to-one on $]0, 1] \times]0, \pi[\times]0, 2\pi]$.

5.8

1. $\pi(e-1)$
2. $1/2$
3. D is the region $0 \le x \le 4$, $\frac{1}{2}x + 3 \le y \le \frac{1}{2}x + 6$; (a) 140; (b) -42
4. D is the region $0 \le x \le 1$, $x + 1 \le y \le 2x + 2$; (a) $17/8$; (b) $-8/3$
5. D^* is the region $0 \le u \le 1$, $0 \le v \le 2$; $\frac{2}{3}(9 - 2\sqrt{2} - 3\sqrt{3})$
6. D is the region $-1 \le x \le 1$, $0 \le y \le \sqrt{1-x^2}$; $\pi/2$
7. π
13. $2a^2$
14. π

5.9

1. (a)

Cylindrical			Rectangular			Spherical		
r	θ	z	x	y	z	ρ	θ	ϕ
1	45°	1	$\sqrt{2}/2$	$\sqrt{2}/2$	1	$\sqrt{2}$	45°	45°
2	$\pi/2$	-4	0	2	-4	$2\sqrt{5}$	$\pi/2$	$\pi - \arccos\sqrt{5}/5$
0	45°	10	0	0	10	10	45°	0
3	$\pi/6$	4	$3\sqrt{3}/2$	$3/2$	4	5	$\pi/6$	$\pi - \arccos 2/\sqrt{5}$

(b)

Rectangular			Spherical		
x	y	z	ρ	θ	ϕ
2	1	-2	3	$\arctan 1/2$	$\pi/2 + \arccos\sqrt{5}/3$
0	3	4	5	$\pi/2$	$\arcsin 3/5$
$\sqrt{2}$	1	1	2	$\arcsin\sqrt{3}/3$	$\arccos 1/2$
$-2\sqrt{3}$	-2	3	5	$\pi + \arctan\sqrt{3}/3$	$\arccos 3/5$

Cylindrical		
r	θ	z
$\sqrt{5}$	$\arctan 1/2$	-2
3	$\pi/2$	4
$\sqrt{3}$	$\arcsin\sqrt{3}/3$	1
4	$\pi + \arctan\sqrt{3}/3$	3

2. (a) reflection in the xy-plane
   (b) rotation by π around the z-axis followed by reflection in the xy-plane
3. (a) rotation by π about the z-axis
4. (a) a cylinder centered on the z-axis
   a plane perpendicular to the xy-plane, through the z-axis
   a plane parallel to the xy-plane

(b) a sphere centered at the origin
a plane perpendicular to the xy-plane, containing the z-axis
a cone centered on the z-axis

5. no
6. The "onto" argument is like that for spherical coordinates except u is the radius, v is the angle from the y-axis in the xy-plane, and w is the angle from the xy-plane. Periodicity of sin and cos destroys one-to-oneness.
7. (a) spherical:
$b^2c^2 \cos^2\theta \sin^2\phi + a^2c^2 \sin^2\theta \sin^2\phi + a^2b^2 \cos^2\phi = a^2b^2c^2/\rho^2$
cylindrical: $b^2c^2r^2 \cos^2\theta + a^2c^2r^2 \sin^2\theta + a^2b^2z^2 = a^2b^2c^2$
(b) spherical: $\cos^2\phi = \sin^2\phi$ or $\phi = \pi/4$ or $3\pi/4$
cylindrical: $z^2 = r^2$
(c) spherical: $\theta = \pi/4$, $\cot\phi = \sqrt{2}/2$
cylindrical: $\theta = \pi/4$, $z = \sqrt{2}\,r/2$
(d) spherical: $\rho\cos\phi = \theta$, $\rho^2\sin^2\phi = 1$
cylindrical: $r = 1$, $z = \theta$
9. This is a "spiral cylinder," $\sqrt{x^2 + y^2 + z^2} = \arctan y/x$.
10. $x^2 + y^2 = xz$
11. $4\pi(\sqrt{3}/2 - \log(1 + \sqrt{3}) + \log\sqrt{2})$
12. $\frac{2}{9}\pi[(1 + a^3)^{3/2} - a^{9/2} - 1]$ 13. 4π

Review Exercises for Chapter 5

1. (c) $\frac{1}{4}e^2 - e + \frac{9}{4}$ 4. $\frac{1}{3}\pi(4\sqrt{2} - 7/2)$
6. 2π 12. no
18. (a) $6xy$; $(xe^{xy} - x)\mathbf{i} - (ye^{xy})\mathbf{j} + (z - 3x^2)\mathbf{k}$
19. $16\pi/3$ 20. $13/6$
21. $\frac{1}{2}\pi(1 - 1/e)$; volume under graph of $z = e^{-x^2-y^2}$ over that part of the unit disc with $y \le 0$
22. $9/40$

6.1

1. $\int_{\sigma} f(x, y, z)\, ds = \int_I f(x(t), y(t), z(t)) \cdot |\sigma'(t)|\, dt$
$= \int_0^1 y(t) \cdot 1\, dt = \int_0^1 0 \cdot 1\, dt = 0.$
3. (a) 2 4. $-\frac{1}{3}(1 + 1/e^2)^{3/2} + \frac{1}{3}(2^{3/2})$
(b) $52\sqrt{14}$
(c) $16/3 - 2\sqrt{3}$
7. $(5\sqrt{5} - 1)/(6\sqrt{5} + 3\log(2 + \sqrt{5}))$

6.2

1. (a) $3/2$; (b) 0 2. (a) 2π; (b) 0
3. 9 6. $34/15$
7. $\frac{3}{4} - (n - 1)/(n + 1)$ 8. 0
11. the length of $\boldsymbol{\sigma}$ 14. 0
15. 7

6.3

1. $z = 2(y - 1) + 1$
2. $4x + z - 1 = 0$
3. $18(z - 1) - 4(y + 2) - (x - 13) = 0$
6. $(\cos u \cos v)\mathbf{i} + (\sin u \cos v)\mathbf{j} - (\sin v)\mathbf{k}$
8. (a) $z - x - y = 0$
   (b) $z - 3 = 6(x - 1) + 8y$
   (c) $4x + 3y - 5 = 0$

6.4

1. 4π
2. When $-\pi/2 \leq \phi \leq \pi/2$, we are computing twice the area of the upper hemisphere. When $0 \leq \phi \leq 2\pi$, we are computing twice the area of the sphere.
4. $\frac{3}{2}\pi(\sqrt{2} + \log|1 + \sqrt{2}|)$
5. $\frac{1}{3}\pi(6\sqrt{6} - 8)$
6. $2\pi(1 - \sqrt{2}/2)$
7. $(2\pi + 4)/(2\pi - 4)$
9. $A(E) = \int_0^{2\pi} \int_0^{\pi} \sqrt{a^2b^2 \sin^2 \phi \cos^2 \phi + b^2c^2 \sin^4 \phi \cos^2 \theta + a^2c^2 \sin^4 \phi \sin^2 \theta}\, d\phi\, d\theta$
12. $\pi(5\sqrt{5} - 1)/6$
14. $\pi\sqrt{6}/2$
15. $4\sqrt{5}$

6.5

1. $\dfrac{5\sqrt{2} + 3}{24}$
2. (b) $\int_D \sqrt{EG}\, du\, dv$
   (c) $4\pi a^2$
3. πa^3
4. 0
5. $\sqrt{6}/30$
7. $\pi(10\sqrt{5}/3 + 2/15)/8$
8. $40/3$
9. $16\pi R^3/3$
10. $(\frac{1}{2}R, \frac{1}{2}R, \frac{1}{2}R)$
11. $\frac{4}{3}\pi R^4$
13. $R/2$

6.6

2. 0, there is no net flux
5. 2π (or -2π, if you choose a different orientation)
6. 16π
8. For a cylinder of radius $R = 1$ and normal component F_r

$$\int_S \mathbf{E} = \int_a^b \int_0^{2\pi} F_r\, d\theta\, dz$$

10. $2\pi/3$

Review Exercises for Chapter 6

1. (a) $3\sqrt{2}(1 - e^{6\pi})/13$
   (b) $-\pi\sqrt{2}/2$
   (c) $(236158\sqrt{26} - 8)/35 \cdot (25)^3$
   (d) $8\sqrt{2}/189$
2. $\sqrt{3}$
3. $\frac{1}{2}\int_0^{2\pi} \sqrt{3\cos^2\theta + 5}\, d\theta$
4. $11\sqrt{3}/6$
5. $\sqrt{2}/3$
6. 0
7. $5\sqrt{5}/6$

8. (a) $e^y(\cos \pi z)\mathbf{i} + xe^y(\cos \pi z)\mathbf{j} - \pi xe^y(\sin \pi z)\mathbf{k}$
   (b) 0
9. 2π
10. $\frac{1}{2}(e^2 + 1)$
12. $2z - x - 1 = 0$
13. 0
14. (a) $3x^2 + 3y^2 + 3z^2$

7.1

1. πR^2
2. -8
4. (a) 0
   (b) $-\pi R^2$
   (c) 0
5. $3\pi a^2$
7. $3\pi/2$
9. $3\pi(b^2 - a^2)/2$
11. (a) 2π
    (b) 0
13. 0
14. πab
16. $-5/6$
19. $9\pi/8$

7.2

3. 0
5. 0
6. 0
9. The orientations of $\partial S_1 = \partial S_2$ must agree.

7.3

2. (a) 19/6
   (b) yes
3. $x^2yz - \cos x$
4. 0
5. $-\dfrac{1}{\|\mathbf{r}_0\|}$
7. no
9. (a) no
   (b) $(\frac{1}{2}z^2, xy - z, x^2y)$ or $(\frac{1}{2}z^2 - 2xyz - \frac{1}{2}y^2, -x^2z - z, 0)$
11. $e \sin 1 + \frac{1}{3}e^3 - \frac{1}{3}$
13. 3.54×10^{29} ergs
15. $-\frac{1}{2}(yz^2 + y^2, xz^2, 0)$
18. (b) Let $\boldsymbol{\sigma}(t)$ be the path of an object in the fluid. Then $\mathbf{F}(t) = \boldsymbol{\sigma}'(t)$. Let $\boldsymbol{\sigma}(t) = (x(t), y(t))$. Then $x' = -y$ and $y' = x$, so $x'' + x = 0$, $y'' + y = 0$. Thus $x = A \cos t + B \sin t$ and $y = C \cos t + D \sin t$. Substituting these values in $x' = -y$, $y' = x$, we get $C = -B$, $D = A$, so $x^2 + y^2 = A^2 + B^2$ and we have a circle.
    (c) counterclockwise

7.4

2. $12\pi/5$
3. 1
4. (a) 0
   (b) 3
5. 0
6. $3\pi/2$
8. $12\pi/5$
15. The field inside is $\mathbf{0}$.

7.6

1. (a) $(2xy^2 - yx^3)\,dx\,dy$
   (b) $(x^2 + y^2)\,dx\,dy$
   (c) $(x^2 + y^2 + z^2)\,dx\,dy\,dz$
   (d) $(xy + x^2)\,dx\,dy\,dz$
   (e) $dx\,dy\,dz$

3. (a) $2xy\,dx + (x^2 + 3y^2)\,dy$
   (b) $-(x + y^2 \sin x)\,dx\,dy$
   (c) $-(2x + y)\,dx\,dy$
   (d) $dx\,dy\,dz$
   (e) $2x\,dx\,dy\,dz$
   (f) $2y\,dy\,dz - 2x\,dz\,dx$
   (g) $-\dfrac{4xy}{(x^2 + y^2)^2}\,dx\,dy$
   (h) $2xy\,dx\,dy\,dz$

8. 0

9. (a) 0
   (b) 40

10. 4π

Review Exercises for Chapter 7

1. $\pi a^2/4$

4. (a) $\nabla\phi = \dfrac{\mathbf{m}}{r^3} - \dfrac{3\mathbf{m}\cdot\mathbf{r}}{r^5}\mathbf{r}$
   (b) $\nabla^2\phi = 0$
   (c) $\nabla \cdot \mathbf{F} = 0$
   (d) $\nabla \times \mathbf{F} = \dfrac{2}{r^3}\mathbf{m} - \dfrac{3}{r^5}\mathbf{r} \times (\mathbf{m} \times \mathbf{r})$

7. (a) $2\pi a^2$
   (b) 0

8. 0

10. $-1/12$

12. (a) $f = x^4/4 - x^2y^3$
    (b) $-1/4$

13. (b) $f = 3x^2y \cos z$
    (c) 0

14. 23/6

16. yes

17. no

19. (a) $\nabla f = 3ye^{z^2}\mathbf{i} + 3xe^{z^2}\mathbf{j} + 6xyze^{z^2}\mathbf{k}$
    (b) 0
    (c) both sides are 0

20. $3\pi/2$

21. $8\pi/3$

23. (a) False
    (b) True
    (c) True
    (d) True
    (e) True
    (f) False
    (g) False
    (h) True
    (i) True
    (j) False
    (k) False

SYMBOLS

(in order of appearance)

INDEX

Ex 12 Sec 6.2